Cortical Spreading Depression of Leao

Avraham Mayevsky • Judith Sonn

Cortical Spreading Depression of Leao

From Mitochondrial Function to Brain Metabolic Score (BMS)

Springer

Avraham Mayevsky
Faculty of Life Sciences
Bar-Ilan University
Ramat-Gan, Israel

Judith Sonn
Faculty of Life Sciences
Bar-Ilan University
Ramat-Gan, Israel

ISBN 978-3-031-08067-8 ISBN 978-3-031-08068-5 (eBook)
https://doi.org/10.1007/978-3-031-08068-5

This Springer imprint is published by the registered company Springer Nature Switzerland AG
The registered company address is: Gewerbestrasse 11, 6330 Cham, Switzerland

Preface 1

Between the middle of 1913 and mid-1914, three very important people that affected dramatically my life were born around the world. My parents Shabtai and Rachel Mayevsky were born in Poland; Britton Chance was born in Pennsylvania, USA; and Aristides Azevedo Pacheco Leão was born in Rio de Janeiro, Brazil.

At the end of 1940, I was born in Tel Aviv, Israel, about the same time that Britton Chance earned his PhD and started his interest and research on mitochondrial biochemistry, and Aristides Leão was accepted at Harvard University in Cambridge, MA, as a graduate student in physiology.

After earning my BSc (1964) and MSc (1966) at the Hebrew University in Rehovot Israel, I studied for the PhD degree at the Weizmann Institute of Science, Rehovot, Israel. The title of my doctoral thesis was "The Use of Isotopes in Studying the Effect of Various Behavioral Situations on Oxygen and Phosphorous Metabolism in the Rat Brain" under the supervisor Professor David Samuel.

My first meeting with Prof. Britton Chance was in Israel during his attendance of the Biophysical Society meeting in 1971. He came to visit my advisor for my PhD thesis, Prof. David Samuel at the Isotope Department of the Weizmann Institute of Science, Israel. He saw the brain in vivo monitoring system for the evaluation in radioactive phosphate and immediately offered me to come to the Johnson Research Foundation in Philadelphia and join his group as a postdoctoral fellow. After my graduation, from the Weizmann Institute (Rehovot, Israel) in October 1972, my family (wife and three children) moved to Philadelphia and stayed there for 2 years. I started my post-doctorate training with Prof. Britton Chance at the University of Pennsylvania, Philadelphia, studying brain mitochondrial function in vivo using NADH fluorescence in real time. A few months later I studied the mitochondrial NADH in vivo and tried to test the effect of stimulation of brain energy metabolism on the redox state of the mitochondria. We tested the effect of CSD in the rat model using the fiber optic–based fluorometer/reflectometer in the un-anesthetized animal.

In 1973, 30 years after the discovery of the cortical spreading depression (CSD) event by Aristides Leão,[1] I published my first paper, with Britton Chance, describing the effect of Leão CSD on mitochondrial function monitored by the measurement of NADH fluorescence in real time. Since then, I published more than 70 papers related to CSD in animal models as well as in patients.

Ramat Gan, Israel Avraham Mayevsky

Preface 2

I was born in Romania in November 1953 to Holocaust survivor parents, Ides and Wolf Leib Ghenuth. In May 1964, we immigrated to Israel.

I studied for my academic degrees in the Department of Life Sciences at Bar Ilan University, Ramat Gan, Israel. I graduated with a BSc degree in 1974 and an MSc in Cardiac Physiology in 1979. The thesis of the MSc was "Oxygen Balance in the Myocardium" under the supervision of Prof. Joseph Kedem.

The MSc thesis dealt with the development and implementation of a new approach for quantitative evaluation of myocardial oxygen balance, measuring oxygen supply, cardiac contractility, and mitochondrial NADH levels in a local region of the myocardial surface in open-chest in vivo. This research approach helped us to evaluate and understand the myocardial oxygen balance under different physiological and pathophysiological conditions which were described in about 15 articles.

In the early 1990s I joined Prof. Avraham Mayevsky's research team and later as a PhD student under his supervision. The title of my doctoral dissertation (1996) was "Effect of Spreading Depression on the Interrelation Between Brain Energy Metabolism and Ion Homeostasis In Vivo". The research was based on experimental studies that were unique because a large number of parameters that were suspected to change during the cortical spreading depression (CSD) were measured simultaneously and continuously. Therefore, the aim of the study was to investigate the interaction between physiological and biochemical mechanisms that are activated during a single wave or multiple waves of CSDs, under various levels of oxygenation in the rat cerebral cortex in vivo. I was part of Prof. Mayevsky's research team for more than two decades.

Ganei Tikwa, Israel

Judith Sonn

Aim of the Book

This book aims to review our studies on Cortical Spreading Depression (CSD) that follows the original discovery and approach of Aristides Azevedo Pacheco Leão as published in the 1940s (1943; Leão, 1944a; 1944b). His discovery was the basis for his PhD thesis submitted to Harvard University in 1943.

Figure 1A show the title front page of the PhD thesis, and the approval of the examination committee members is shown in Part B.

Figure 2 is the acknowledgments page written by A.A.P. Leão.

In Fig. 3, Leão, in the introduction page, described the original subject proposed to him by Dr. Hallowell Davis and the change of the subject as seen in the underlined lines in red.

The first description of the new response named "Cortical Spreading Depression" is presented in Fig. 4A and B.

Figure 5A shows the legend to Fig. $5B_1$ and $5B_2$ presenting the effect of CSD on the diameter of the surface pial blood vessel and the correlated spontaneous electrical activity recorded before and during the wave of CSD.

Figure 6 demonstrates in a schematic way the main elements that build a typical functional cerebral tissue. The cellular compartment contains typical neurons and glial cells as well as the synapses. The small blood vessels – the capillaries – provide the oxygen and other nutrients. As seen in Fig. 6, oxygen is supplied by the blood flow (1) containing the oxygenated hemoglobin (2) that releases oxygen to the extracellular (3) and intracellular (3a) compartments reaching finally and consumed by the mitochondria (4) and water is formed. The ATP formed is used for different physiological and biochemical activities. The majority of the ATP is consumed by the ion pumps such as Na^+K^+ ATPase (5) that keeps the ionic homeostasis (6) and low intracellular calcium levels (7). The recharging of the ATP pool is done by oxidative phosphorylation in the mitochondria (8). There are four pathological events that are developed in the brain of patients as seen in Fig. 6. The major initial event is (9) cerebral ischemia (core area) due to the decrease in CBF and oxygen supply to the neuronal tissue. The second event is cerebral hypoxia developed under systemic decrease in blood oxygenation (10). The third event is epileptic activity (11) that may have developed under various pathologies. Cortical spreading depression (12) was developed initially by Leão in 1944 (1944b) by direct electrical stimulation and later was found in the penumbra zone around the ischemic core area or other pathological states.

The four neuropathological events shown in Fig. 6 are presented in Fig. 7 and are divided into two groups according to their relationship to various physiological and biochemical parameters monitored in vivo. The upper two factors (A and B), namely, systemic hypoxia and cerebral ischemia, affect the availability of oxygen so that energy metabolism is inhibited (blue arrows). The other two perturbations shown are epileptic activity (C) and cortical spreading depression (D) that induced cortical depolarization (purple arrows). All four perturbations are affecting the same type of processes in the brain such as CBF, mitochondrial NADH, electrical activities, and ionic homeostasis but in the opposite direction as seen in the scheme. Under hypoxia and ischemia, the production of ATP by the mitochondria is decreased and becomes a limiting factor in the system. Due to the inhibition of the ionic pumps, the ionic homeostasis is disrupted and hypoxic or ischemic cortical depolarization is developed. The end points of this process are changes in the electrical activities recorded as a depression of the ECoG and a negative shift in the DC steady potential. In addition, due to the depolarization event, the extracellular potassium is elevated. If this situation persists for more than a few minutes, irreversible damage to the cortex may develop.

When the brain is exposed to epileptic activity or CSD, the events will propagate toward the activation of the brain and energy consumption is increased (purple arrows). The coupling between the two pathological events is clear and manifested by the electrical activity (ECoG) as well as extracellular K^+ levels and NADH redox state. During the epileptic stage, extracellular potassium increased and started to recover to the baseline, but then a larger elevation was recorded. There is a clear correlation between the various parameters during epilepsy and CSD. The decrease in NADH was smaller during the first stage, followed by an oxidation cycle typical for CSD.

We decided to describe and discuss the Leão CSD according to his original discovery, namely, that the CSD was developed in a well-oxygenated rabbit or rat brain that was exposed to electrical stimulation expected to develop epileptic activity. Also, we will describe the effects of the three pathological events A, B, and D shown in Fig. 7 on the development and responses of the cortical spreading depression event (Part C).

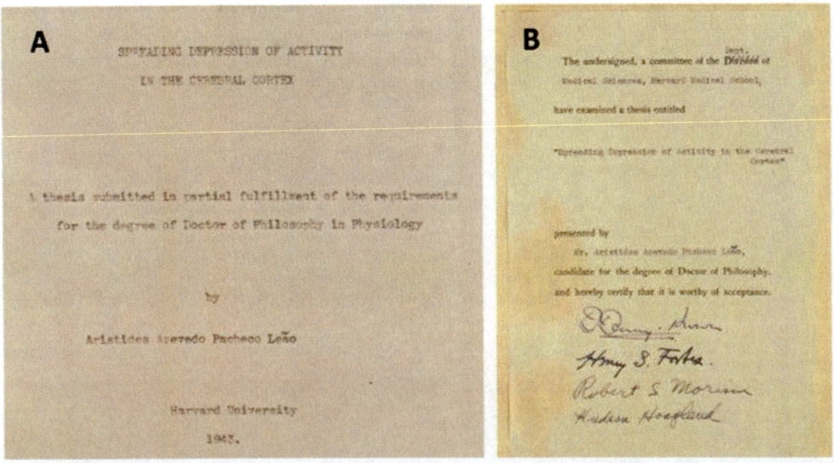

A

SPREADING DEPRESSION OF ACTIVITY

IN THE CEREBRAL CORTEX

A thesis submitted in partial fulfillment of the requirements

for the degree of Doctor of Philosophy in Physiology

by

Aristides Azevedo Pacheco Leão

Harvard University

1943.

B

The undersigned, a committee of the Dept. of Medical Sciences, Harvard Medical School,

have examined a thesis entitled

"Spreading Depression of Activity in the Cerebral Cortex"

presented by

Mr. Aristides Azevedo Pacheco Leão,

candidate for the degree of Doctor of Philosophy,

and hereby certify that it is worthy of acceptance.

Henry S. Forbes

Robert S. Morison

Hudson Hoagland

Fig. 1 (**A**) The front page of A.A.P Leao Ph.D. thesis submitted to Harvard University in 1943. (**B**) The signatures of the Ph.D. examination committee, approving the acceptance of the Ph.D. thesis submitted by Dr. A.A.P. Leao. (Reprinted with kind permission of "The Brazilian Academy of Sciences Archive" (Leão, 1943))

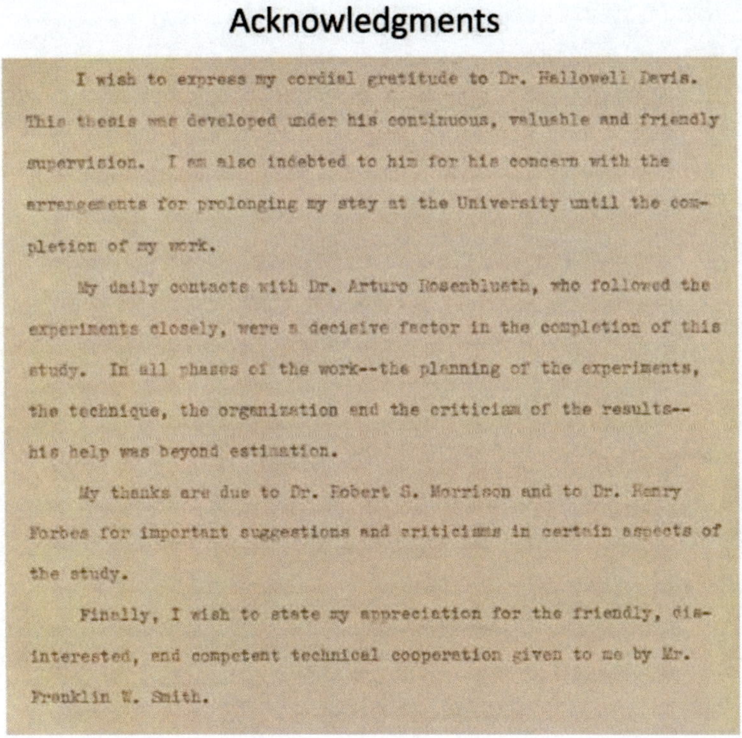

Acknowledgments

I wish to express my cordial gratitude to Dr. Hallowell Davis. This thesis was developed under his continuous, valuable and friendly supervision. I am also indebted to him for his concern with the arrangements for prolonging my stay at the University until the completion of my work.

My daily contacts with Dr. Arturo Rosenblueth, who followed the experiments closely, were a decisive factor in the completion of this study. In all phases of the work--the planning of the experiments, the technique, the organization and the criticism of the results--his help was beyond estimation.

My thanks are due to Dr. Robert S. Morrison and to Dr. Henry Forbes for important suggestions and criticisms in certain aspects of the study.

Finally, I wish to state my appreciation for the friendly, disinterested, and competent technical cooperation given to me by Mr. Franklin W. Smith.

Fig. 2 The acknowledgments page of the Thesis submitted. (Reprinted with kind permission of "The Brazilian Academy of Sciences Archive" (Leão, 1943))

Introduction

This study originated in an attempt to secure more data for the understanding of the cortical electrogram which occurs in "experimental epilepsy", and of the conditions in which it is brought forth by electrical stimulation. A study of the subject had just been carried out in this laboratory by Rosenblueth and Cannon (1942), who experimented mainly upon monkeys. A comparative survey of the responses in different orders of mammals was the initial aim of the present study. Experiments on birds and reptiles were also planned. Early in the development of the study, however, an interesting response, elicited by electrical stimulation, was noticed in the cortex of rabbits. The distinctive feature of this response was a marked enduring, reduction of the "spontaneous" electrical activity of the cortex. The elucidation of some of the characteristics of this response required a series of experiments. These results constitute the subject presented and discussed in the following pages.

Fig. 3 The introduction page of the Thesis submitted. (Reprinted with kind permission of "The Brazilian Academy of Sciences Archive" (Leão, 1943))

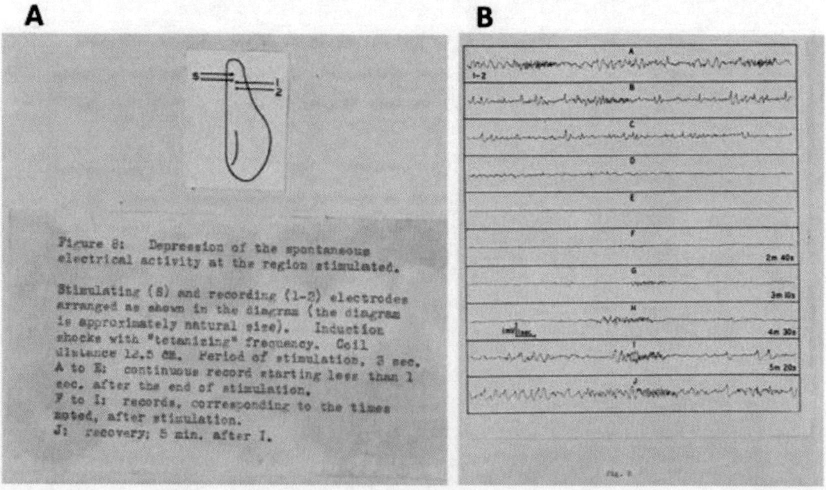

Fig. 4 The first experimental demonstration of the new dramatic discovery of the CSD by A.A.P. Leao (1943). (Reprinted with kind permission of "The Brazilian Academy of Sciences Archive")

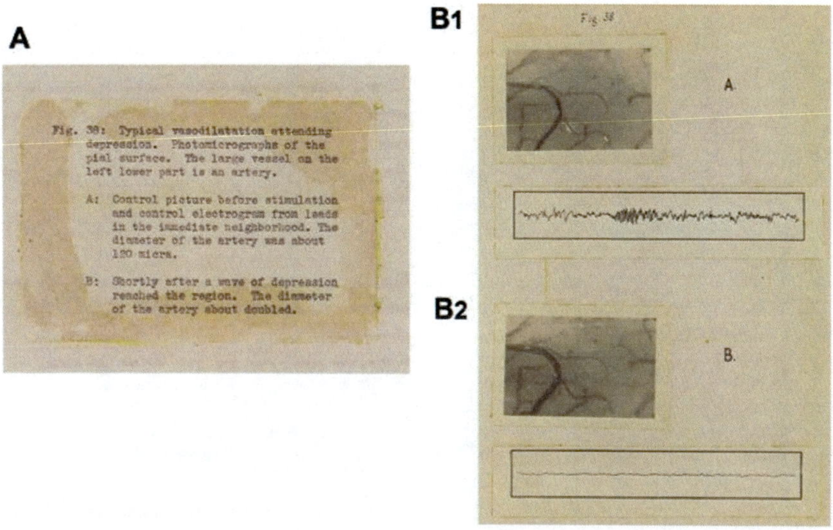

Fig. 5 The second very significant discovery related to CSD discovered by A.A.P. Leao. (Reprinted with kind permission of "The Brazilian Academy of Sciences Archive" (Leão, 1943))

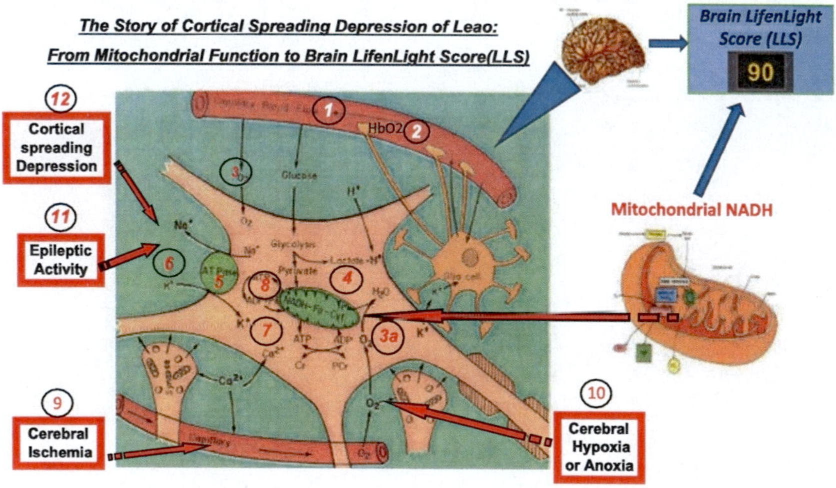

Fig. 6 Schematic presentation of the "basic building stones" of a typical cerebral cortex tissue. The four pathological states (9–12) are related directly or indirectly to the CSD. (Mayevsky, 1984)

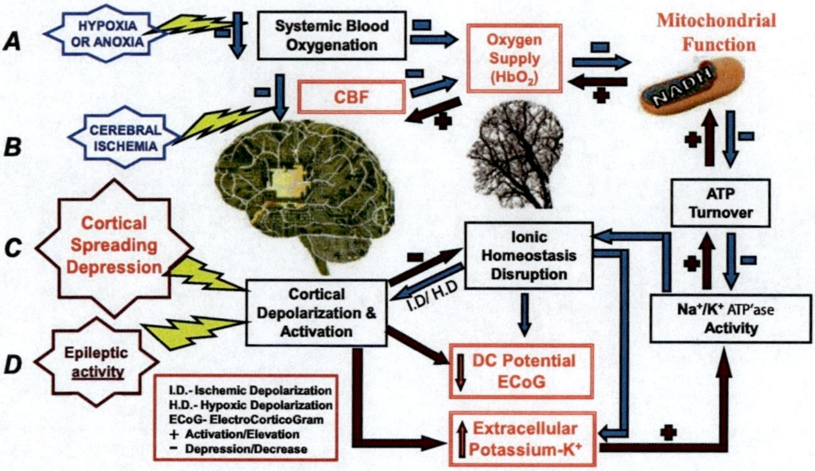

Fig. 7 A schematic representation of the effects of various pathological conditions on the energetic-mitochondrial function, ionic and hemodynamic state of the cerebral cortex (details are in the text). (Mayevsky and Barbiro-Michaely, 2013a)

Acknowledgments

I

This book is dedicated to my dearest wife, Zipora, for her vision, patience, and support for more than six decades. Without her encouragement, my academic achievements and writing of this book were impossible to be accomplished.

I dedicate this book to my four children, Amotz, Sarit, Hagit, and Shalev, that understood my devotion to the unusual career and enabled me to fulfill all of my obligations and academic activities.

My deep appreciation is given to my parents, Shabtai and Rachel Mayevsky, for their devotion and support throughout my life span.

My appreciation is given to, the late, Prof. Britton Chance, an outstanding scientist, teacher, and colleague for almost four decades of fruitful collaboration.

A special appreciation is given to Judith Sonn, PhD, that joined my laboratory as a PhD student and recently joined me as a coauthor in writing this book.

We both (AM and JS) are thanking Mrs. Avrille Goldreich for her dedication to help and bring this book to press in its excellent form.

During almost 60 years of research activities at Bar Ilan University, I trained many MSc and PhD students. I want to thank all of them for their confidence and the efforts that they did during the experimental work as well as in preparation and publication of the manuscripts.

Spccial appreciation is given to Nili Zarchin, MSc, and Efrat Barbiro-Michaeli, PhD, that joined my research team after their graduation. Their assistance in training new undergraduate and graduate students is appreciated very much.

Many thanks are given to Prof. Gennady Rogatsky who immigrated to Israel in the early 1990s and was active in my laboratory until his retirement.

II

I dedicate this book to my dear parents, Holocaust survivors, my late father Wolf Leib Ghenuth and my mother Ides Ghenuth who should be spared for a long life. I thank them for their love, help, support, and total devotion and dedication throughout my life.

My deep love and appreciation is given to my husband Moshe, my daughters Anat and Liat, and their families.

I thank Prof. Mayevsky for being a wonderful and purposeful teacher, for the privilege of learning from an excellent researcher and for his cooperation for more than two decades. I am honored to be a part in writing this book which deals with the complicated and interesting Leão's Cortical Spreading Depression event, that I have researched during the PhD study period, as well as additional 10 years of activity.

Contents

Chapter 1
Introduction and Historical Background

Abstract In this chapter, we will introduce the four main subjects that will be covered in the book and will described and discuss the cortical spreading depression (CSD) since its discovery by Aristides Leão in 1942–1944.

1. Leão's CSD.
2. Brain Oxygen Homeostasis.
3. Monitoring of Brain Mitochondrial Function In Vivo.
4. Brain Energy Metabolism and Mitochondrial Function During CSD.

Keywords Cortical Spreading Depression-CSD · Electroencephalography · Brain Activation · Propagation of depression · Brain DC steady potential

1.1 Leão' Cortical Spreading Depression (CSD)

Aristides Leão was born on August 3, 1914, to his father Manoel Pacheco Leão and mother Francisca Azevedo Leão in Rio de Janeiro, Brazil (Maranhao-Filho & Vincent, 2009). In 1914, two Brazilian scientists, Eliasz Engelhardt and Marleide da Mota Gomes, wrote a historical note ("Aristides Leão: A Birth Centennial Homage with Comments on His Spreading Depression") and published it in the journal "*Arduinos de Neuro-Psiquiatria*, 73, 544–546" (Engelhardt & Gomes Mda, 2015). With the permission of the two authors, we are citing in the following three sections, a few paragraphs that were published in their paper.

A. Mayevsky, J. Sonn, *Cortical Spreading Depression of Leao*,
https://doi.org/10.1007/978-3-031-08068-5_1

1.1.1 Aristides Leão: Life History

The year of 2014 is the time to commemorate the birth centennial of Aristides Azevedo Pacheco Leão (August third, 1914–December 14th, 1993 – born in Rio de Janeiro, deceased in São Paulo, and buried in Rio de Janeiro, according information of the widow, Mrs. Elisabeth Raja Gabaglia Leão) (Maranhao-Filho & Vincent, 2009). This paper intends to present some facets of Aristides Leão's early academic formation, his scientific achievements, and some of their unfolding. He ingresses in the School of Medicine of Sao Paulo (1932). However, while in the second year, he interrupted the course due to a physical ailment. On recovering he decided to be a researcher. He left to the USA (1941), being admitted in the Medical Science Department (Physiology) of Harvard University. There, he achieved the degrees of MA (1942) and PhD in 1943 (Maranhao-Filho & Vincent, 2009; Teive et al., 2005). In the same year, he became a Research Fellow of the Department of Anatomy at the Harvard Medical School. Despite the opportunity to remain in the USA and develop further his studies, he opted to return to Brazil (Teive et al., 2005). Back to Brazil (1944), Aristides Leão was invited by Carlos Chagas Filho to compose the basic staff of the Institute of Biophysics he was structuring. He was immediately joined by Gustavo de Oliveira Castro and Romualdo José do Carmo, and soon also by Hiss Martins-Ferreira (Teive et al., 2005). Aristides Leão, besides his career as neurophysiologist, reached top administrative positions: Director (1966–1970) and Chief emeritus of the Department of Neurobiology (IBCCF) (1984–1993), and President (1966–1981), and President Emeritus (posthumous) (1994) of the Brazilian Academy of Science, taking also part in governmental and educational committees. He became one of the most distinguished Brazilian scientist, with incontestable international recognition (Maranhao-Filho & Vincent, 2009). He was an extraordinary person, with a very rich personal life, amid his family and friends, besides manifold no-scientific and scientific extra-neurophysiological interests. Maranhão Filho and Vincent's paper on his personal life is worth reading (Maranhao-Filho & Vincent, 2009).

1.1.2 The Discovery of the "Spreading Depression"

Once in Harvard (1941), Aristides Leão's was involved in a study on experimental epilepsy, conducted by Arturo Rosenblueth and Walter Cannon. They were analyzing the responses to electrical stimulation of the cortex of anesthetized rabbits, with electrodes placed on the cortical surface, and the changes were recorded by electrocorticography (ECoG). They agreed that some of the electrical epileptiform after discharges should be further studied by the newcomer. The analysis of the cycle of excitability of cortical neurons after the convulsive state, under supervision of Arturo Rosenblueth and Hallowell Davis, permitted him to identify a decrease of the excitability spreading in increasing circles around the initial focus, a phenomenon he named "spreading depression" (CSD), the theme of his PhD thesis, the "Spreading depression of electrical activity in the cerebral cortex" (Leão, 1943). The following year the study was published (1944), and soon the phenomenon was named after him ("Leão's wave" or "spreading depression of Leão") (Gattass & Leão, 2013; Leão, 1944b; Maranhao-Filho & Vincent, 2009; Teive et al., 2005). In sequence, he published another paper, the same year, focusing on the vasodilatation accompanying CSD (Leão, 1944a). Before leaving back to Brazil, he submitted, together with Robert S. Morison, the "Propagation of spreading cortical depression," published in 1945 (Leão & Morison, 1945). Later, his first international paper developed in his Brazilian laboratory (1947) appeared (Leão, 1947). These papers may be envisaged as seminal. He kept producing numerous papers, mainly on his theme, the last one in 1992. His studies were very valued, and in the 1970s, his publications were responsible for the largest number of the citations received by the academics of the Brazilian Academy of Science (Gattass & Leão, 2013), its importance remaining in the medical literature until the present days. In conclusion, following Leão's description of CSD, in the 40s, numerous studies followed in the next decades for a better understanding of this and related phenomena, seen in normal and pathological conditions. The concept of spreading depolarization, a generic term for the whole spectrum of waves in the CNS, came in vogue. It was characterized by abrupt, near-complete sustained depolarization of neurons, related to massive ion translocation between intra- and extracellular space, neuronal swelling, distortion of dendritic spines, and glial depolarization. Such mechanism, preventing neurons to fire action potentials, causes as a consequence brain electrical silencing – the CSD of Leão (1944), an epiphenomenon of spreading depolarization. Here, the studies advanced in relation to Leão's findings, by identifying the main electrophysiological occurrence, and its underpinnings, where CSD continues as an important electrophysiological feature. Since Leão's discovery, CSD has been implicated by him, and soon by other authors, in migraine (Leão, 1944b, 1947). Although it has been known that ischemia could also trigger CSD, as shown by Leão (1947; Leão & Morison, 1945), it was not until the 1980s that recurrent CSD-like depolarization waves were demonstrated in focal brain ischemia, which occurred as peri-infarct depolarizations (injury depolarizations, spreading depolarization), at the edge between the ischemic core and the penumbra rim. Such changes propagate often into tissue not attained by ischemia, where injurious metabolic and hemodynamic effects might be exerted, worsening the outcome. Thus, strong experimental and clinical evidence emerged to suggest that these cortical electrophysiological phenomena are involved in the mechanisms of migraine aura, acute cerebrovascular diseases (including spreading ischemia), traumatic brain injury, transient global amnesia, epileptic seizures (including spreading convulsion), and their pathophysiological characteristics offering novel therapeutic targets (Dreier, 2011; Lauritzen et al., 2011).

Figure 1.1 shows the number of publications that appeared in the PubMed system since the discovery of the cortical spreading depression (CSD) by Aristides

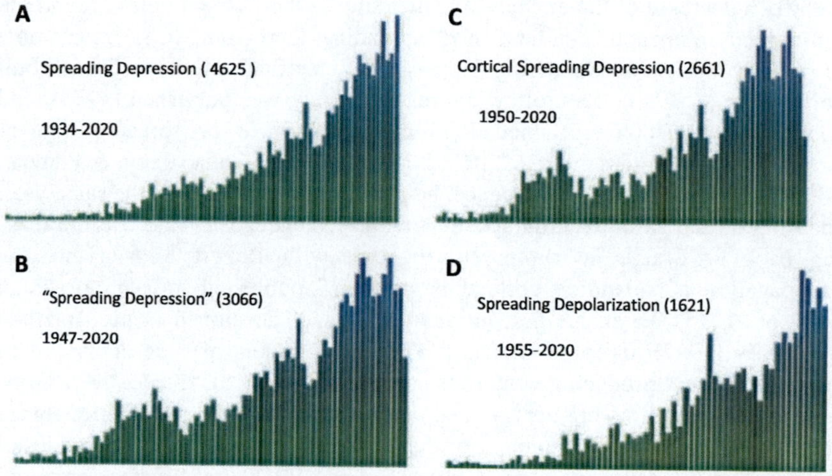

Fig. 1.1 The number of appearances of 4 names of the event called by Leão "Cortical Spreading Depression" in the papers cited in the PubMed system

Leão in the early 1940s. The number of publications varied between the selected word combinations used in the search process ranging from 4625 references in A to 1621 references in D.

1.1.3 Back to Brazil

Back to Brazil (1944), Aristides Leão was invited by Carlos Chagas Filho to compose the basic staff of the Institute of Biophysics he was structuring. He was immediately joined by Gustavo de Oliveira Castro and Romualdo José do Carmo, and soon also by Hiss Martins-Ferreira (Dreier, 2011; Gattass & Leão, 2013; Teive et al., 2005) (Fig. 1.2).

Aristides Leão, besides his career as neurophysiologist, reached top administrative positions: Director (1966–1970) and Chief emeritus of the Department of Neurobiology (IBCCF) (1984–1993), and President (1966–1981) and President emeritus (posthumous) (1994) of the Brazilian Academy of Science, taking also part in governmental and educational committees. He became one of the most distinguished Brazilian scientist, with incontestable international recognition (Chagas, 2000; Gattass & Leão, 2013; Maranhao-Filho & Vincent, 2009).

Aristides Leão was an extraordinary person, with a very rich personal life, amid his family and friends, besides manifold no-scientific and scientific extra-neurophysiological interests. Maranhão Filho and Vincent's paper on his personal life is worth reading (Maranhao-Filho & Vincent, 2009).

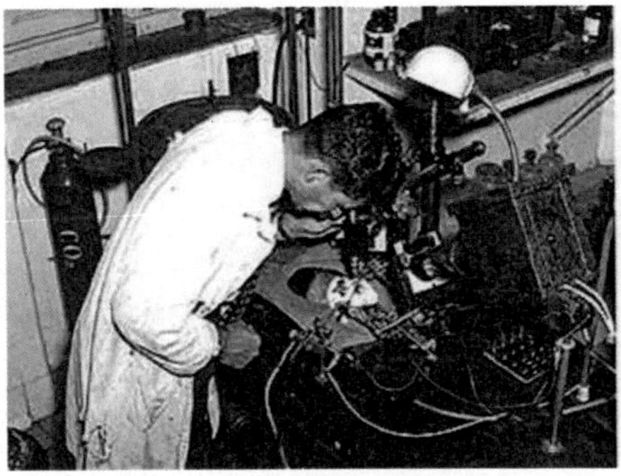

Leao'sat his laboratory

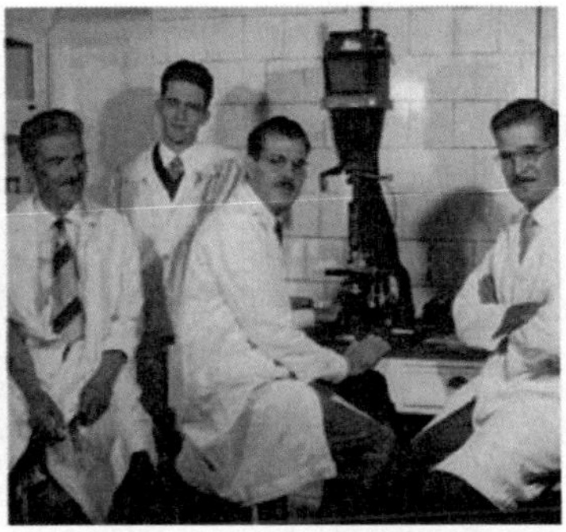

Laboratory in the IBCCF in the Faculdade Nacional de Medicina-Universidade do Brasil (FNM-UB).From left to right: Gustavo de Oliveira Castro, Gustavo Castro de Oliveira Filho, Antonio Moreira Couceiro, and Aristides Azevedo Pacheco Leão, after his return to Brazil.

Fig. 1.2 Aristides Azevedo Pacheco Leão in his laboratory. (Reprinted with kind permission of "The Brazilian Academy of Sciences Archive")

1.1.4 Conclusion

The CSD and the accompanying vascular changes, revealed by Aristides Leão, triggered the study of important related phenomena. This knowledge allowed to better understanding of the pathophysiology of several acute brain disorders, offering opportunities for early therapeutic interventions to change their outcome. This discovery granted him international scientific notoriety, and his seminal papers became classics in the field. Aristides departed at the age of 79 years, after a very productive life, scientific and affective, leaving numerous friends, disciples, and followers, in Brazil and abroad. One of them, Jan Bures, in a posthumous homage of the Academy of Science of the Czech Republic, wrote about the master in 1994, "I feel to be an old alumnus of Leão's invisible college, of the informal scientific community the members of which are scattered all over the world, but are nevertheless unified by the mysterious spell of a challenging problem, by the strong bonds formed in the joint pursuit of knowledge."

1.1.5 Aristides Leão: The Four Seminal Publications (1944–1947)

At the age of 18 (in 1932), Aristides Leão was admitted to the medical school in Sao Paulo University but he became ill forcing him to delay his studies for 2 years. After recovering, he went to Harvard University and graduated from the Department of Physiology with Master of Arts and Doctor of Philosophy degrees (1942 and 1943).

His first published paper was not part of his PhD thesis and was in collaboration with E.C. Del Pozo (Del Pozo & Leão, 1943).

The *Journal of Neurophysiology* published in 1994 a commentary (by RJ do Carmo and GG Somjen) celebrating 50 years since the seminal discovery of the "Spreading Depression" by Aristides Leão (do Carmo & Somjen, 1994). In 2005, G.G. Somjen (2005) described this very important discovery of Aristides Leão as follows:

> The discovery of spreading depression (CSD) illustrates how an unexpected chance observation, if made by an alert experimenter, can open a new major area of investigation. Aristides Leão (Fig. 1.3) was working on his PhD thesis at Harvard Medical School when his experiment turned out quite differently from what was expected. He was working under the supervision of Hallowell Davis, and he also had advice from Arturo Rosenblueth and later collaborated with R. S. Morison. In other words, he was the right person at the right time in the right environment to work on the electrophysiology of the brain, a field at the time as yet in its infancy. Originally, Leão intended to study 'experimental epilepsy', specifically the propagation of electrically provoked seizure discharges in the cerebral cortex (Bures et al., 1974).

The unexpected discovery of cortical spreading depression (CSD) by Aristides Leão, was published in four seminal papers in 1944–1947 (Leão, 1944a, b, 1947; Leão & Morison, 1945). Engelhardt and Gomes Mda (2015) reviewed the four papers as seen in Fig. 1.4. The original typical results are presented in Figs. 1.5, 1.6, 1.7, 1.8, and 1.9.

Fig. 1.3 Aristides Azevedo Pacheco Leão in his laboratory, office and his home in Rio de Janeiro. (Reprinted with kind permission of "The Brazilian Academy of Sciences Archive")

These studies, published in the Journal of Neurophysiology, were performed mainly in rabbits, under dial anesthesia, and pairs of electrodes were placed on their exposed cortical surface, one pair for stimulation, and the others, at increasing distance, for detetecting the electrographic changes recorded by ECoG.

1st paper (Leao, 1944) ["Spreading depression of activity in the cerebral cortex"]
Following the stimulation he observed often, instead seizure-like discharges, a flattening of the spontaneous ECoG waves. The activity was depressed first in the electrodes nearest to the stimulated area, and then slowly spreaded in all directions, affecting adjacent areas, eventually
covering almost all the dorsolateral neocortex – the "spreading depression", a slow response that starting near the frontal pole might take more
than 5 minutes to reach near the occipital pole. His experiments showed that the phenomenon was a cortico-cortical process.

2nd paper (Leao, 1944) ["Pial circulation and spreading depression of activity in the cerebral cortex"]
Besides the SD, he described an accompanying vasodilatation – a vasodilatation and increased blood flow, in arteries and veins, correlated with local activity of the cortical neurons. Two mechanisms were suggested for such occurrence – release of a chemical vasodilator agent, or activation of local vasomotor nerves. Whatever the mechanism of the vasodilation, he commented, the marked increase in blood flow probably influences in turn the activity of the cortical neurons.

3rd paper (Leao and Morison, 1945) ["Propagation of spreading cortical depression"]
Further investigation of the SD showed the attainment of the dorsolateral and medial-basal cortex, independent of subcortical connections and of the long association cortico-cortical tracts, the integrity of the corpus callosum for inter-hemispheric spread was needed. The destruction of the upper two layers of the cortex interrupted the spread. Changes in the permeability of the blood brain barrier to injected vital dyes, as ferrocyanide, trypan blue, and fuchsin, were not detected. According the investigators, the experiments leaved 'completely dark' the nature underlying the process of the cortical cells depression, but they were satisfied that "several obvious possibilities have been ruled out and attention directed to other paths of inquiry".

4th paper (Leao, 1947) ["Further observations on the spreading depression of activity in the cerebral cortex"]
The development of a specific electrical activity during the depression, which, when intense, closely resembled the "seizure pattern" of experimental epilepsy, incited to analyze the occurrence of a slow voltage variation – initially described as accompanying the SD, then considered as due to cortical anemia of short and prolonged duration, stating: "The results seem to indicate that in the SD of activity, a change of the same nature as one resulting from prolonged interruption of the circulation, occurs in the cerebral cortex,... suggesting that the negative voltage shifts of SD and of cortical ischemia are the result of 'some change of the same nature'.".

Fig. 1.4 Summary of the 4 initial papers published by Leão during the 1940's

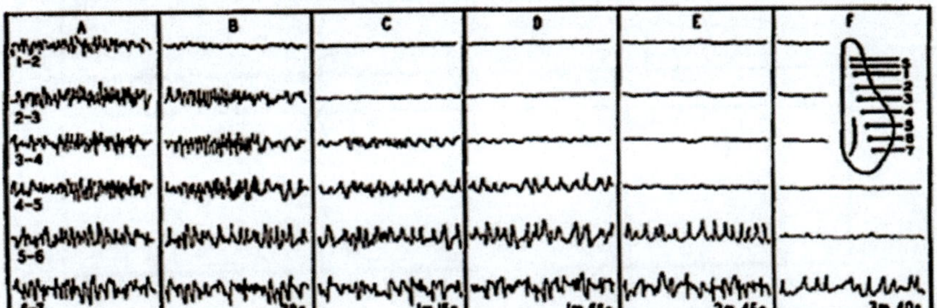

Fig. 1.5 Gradual spread of depression. Electrodes arranged as shown in the inset. A– Before stimulation. L– 7 min after K. In this and the following figures, when a strip is timed, that time indicates the interval from the end of the stimulation period. Upward excursions in the records denote negativity of the more frontal electrode (lower number) with respect to the more occipital electrode (higher number). Unless otherwise mentioned, the stimuli were induction shocks at "tetanizing" frequency, applied for 3 to 5 sec. Through electrodes S. (Leão, 1944b)

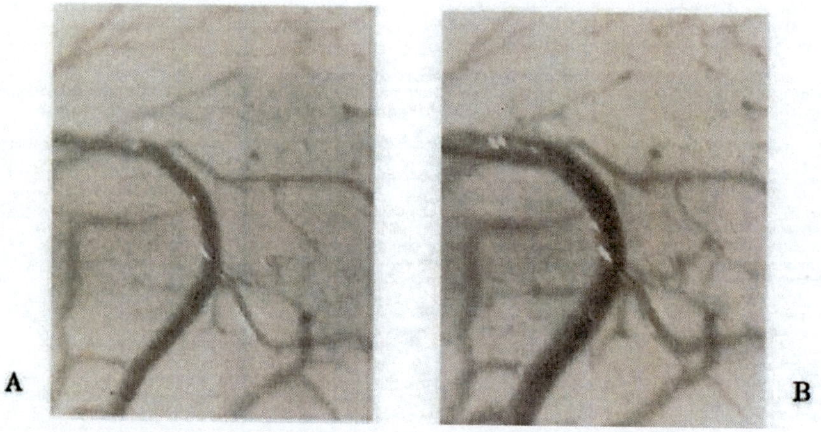

Fig. 1.6 **(A)** Control before stimulation. The large vessel is an artery, of about 120 μ diameter. **(B)** The wave of depression has reached the region (see Fig. 1.7B). The diameter of the artery is approximately doubled during the wave of CSD. (Leão, 1944a)

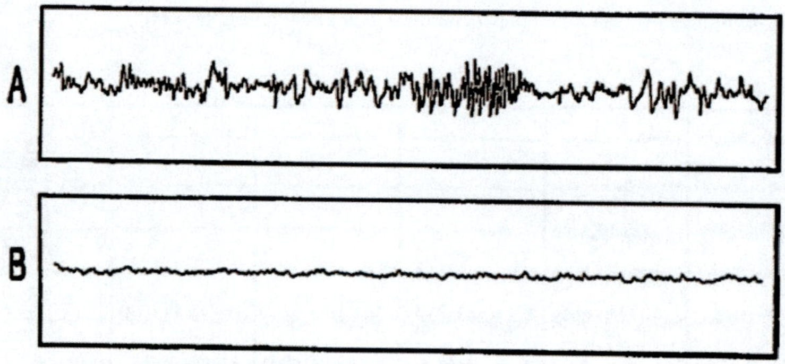

Fig. 1.7 The electrocorticogram measured from electrodes located in the immediate area of the region shown in Fig. 1.6. **(A)** Control before stimulation. **(B)** The depression of the electrical activity taken during the exposure seen in Fig. 1.6B. (Leão, 1944a)

In the first paper seen in Fig. 1.5, Leão described the depression of the spontaneous electrical activity (ECoG or EEG) propagated from the electrical or mechanical stimulated area to the rest of the stimulated hemisphere.

In the second paper (Figs. 1.6, 1.7, and 1.8), Leão described the vasodilatation of blood vessels that accompanied the CSD wave. He used microscopy of pial blood vessels that was interpreted as an increase in cerebral blood flow. Leão concluded that "the vascular response is apparently secondary to a local change in the activity of nervous elements."

In the third paper (Fig. 1.4), the nature of the propagation of the CSD wave was described by Leão and Morison (1945).

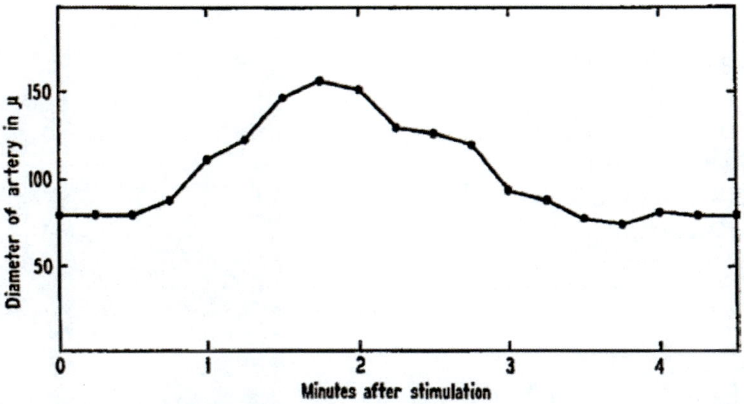

Fig. 1.8 Graph showing the time course and magnitude of arterial vasodilatation attending a depression. (American Physiological Society, reprinted with permission (Leão, 1944a))

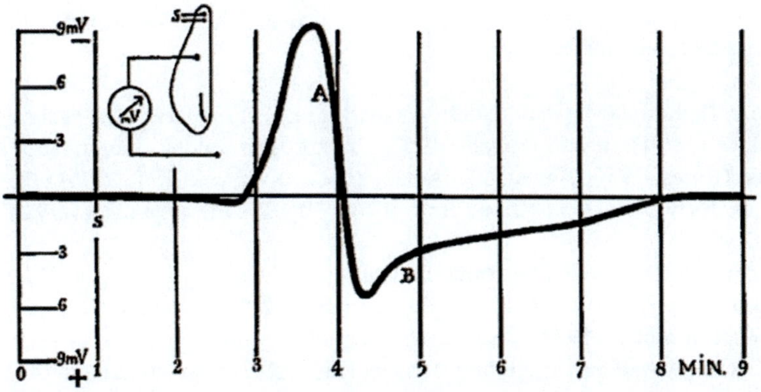

Fig. 1.9 A representative experiment on the slow voltage variation accompanying the spreading depression of activity. The curve was drawn from voltage readings taken with 5 second intervals. In this and the other two figures, an upward deflection denotes negativity of the cortex, with respect to the extra cortical reference electrode. Electrodes arranged as shown in the inset (s: stimulating electrodes). Stimulation, 5 seconds of "tetanizing" current from an induction coil, delivered at the time marked S. In this representative curve is indicated the time of occurrence (from **A** to **B**) of the specific electrical activity which often develops during the depression of the "spontaneous" activity. (American Physiological Society, reprinted with permission (Leão, 1947))

In the fourth paper (Leão, 1947) published in 1947, Leão described a slow voltage change being part of the response to CSD and recorded by a Direct Current (DC) amplifier (Fig. 1.9). This paper was published after Leão returned to Brazil. After Leão's return to Brazil, a group of scientists joined him and continued the studies of CSD.

Three more centers around the world started to study CSD and kept close relations and information exchanges with the group in Brazil. The leading groups were

Fig. 1.10 The front page of the book (left side **A**) on CSD published in 1974 by Bures J, Buresova O (right side **B**) and Krivanek J. (Bures et al., 1974)

active in Bethesda (Marshall, 1959; Marshall et al., 1951; Marshall & Essig, 1951), Pasadena (Van Harreveld & Ochs, 1957a, b; Van Harreveld & Stamm, 1952), and Prague (Bures, 1956; Bures & Krivanek, 1960a; Krivanek, 1958, 1961) that also published a book in 1974 (Bures et al., 1974) 30 years after the discovery of CSD by Leão.

Figures 1.10B and 1.11B show the leaders of the group in Prague (Bures and Buresova). Together with Jiri Krivanek, they published the book on CSD (Fig. 1.10A) in 1974 (Bures et al., 1974). They showed very clearly that typical responses to CSD were similar when they used the rat model (Fig. 1.11A) as compared to the rabbit used by Leão. In 1960, Bures and Krivanek (1960b) developed a washing cannula (Fig. 1.12A), attached to the cortical surface, in order to measure ion movement in the brain during the waves of CSD. Krivanek and Bures described the kinetics of potassium release from the cerebral cortex (Fig. 1.12B) during waves of CSD (Krivanek & Bures, 1960). In the late 1940s to early 1950s, Wade W. Marshall (Fig. 1.13A) published a few papers related to CSD. He established a group that was active in the NIH in Bethesda (Fig. 1.13B). In a paper published in 1951 (Marshall et al., 1951), they showed that "cooling the surface of the rabbit or monkey cortex decisively favors the spreading cortical depression of Leão," as seen in Fig. 1.14.

Figure 1.15A shows the leading scientist in Pasadena (Van Harreveld) in his laboratory. In part B, Van Harreveld is standing with the Caltech Biology Faculty in 1947. He published a few important papers on brain hemodynamics and CSD (Van Harreveld & Ochs, 1957a; Van Harreveld & Stamm, 1951, 1952). Figure 1.16A shows the wave of vasoconstriction moving in the cortex during the CSD event (Van Harreveld & Stamm, 1952) . The amount of blood in the cortical vessels during the wave of CSD is presented in Fig. 1.16B (Van Harreveld & Ochs, 1957a).

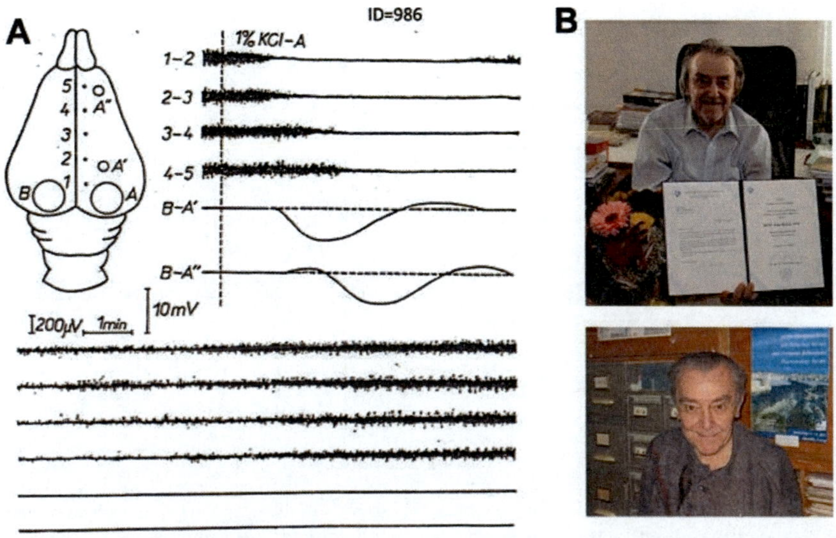

Fig. 1.11 (**A**) Typical responses of the EEG signal and DC steady potential recorded in a rat model during a wave of CSD and published by Bures, Buresova and Krivanek (Fig. 1.10B) (Bures et al., 1974). (**A**) Position of electrodes indicated on the brain diagram (1, 2,3, 4,5). Steel needle electrodes for EEG recording (*A'*, *A"*). Trephine openings for application of wick calomel cell electrodes for slow potential recording. (A) Large trephine opening for application of KCl solution (1–2, 2–3, 3–4, 4–5). Bipoloar EEG recordings with the respective electrode pairs (*B-A'*, *B-A"*). Slow potential recording from electrodes (*A'*) and *A"*) against the common reference electrode situated in the contralateral hemisphere (*B*). The upper and lower recordings are continuous. Note that the onset of the negative SPC coincides with the onset of the EEG depression in the respective areas (**B**) Prof J. Bures in his office

It's important to note that Leão in his second seminal study on CSD (Leão, 1944a) decided to investigate the connection between the changes in neuronal electrical activity and the pial microcirculation as an indicator of tissue energy metabolism. Therefore, we would like to discuss in the next section the history of the research dealing with the homeostasis of oxygen in the brain in general and especially during CSD.

1.2 Oxygen Homeostasis in the Brain

There are various scientific opinions regarding the evolution of oxygen levels in the atmosphere on earth (Lindahl, 2008; Lyons, 2007). In Trayhurn (2019) cited Lyons et al. (2014) that "The level of O_2 in the atmosphere is believed to have hovered at around the current 21% since about 600 million years ago." Actually, Trayhurn published two papers and claimed that oxygen could be considered as a nutrient (Trayhurn, 2017, 2019). This level of oxygen in the atmosphere (21%) must be kept in a narrow range in order to avoid the development of pathological events in the human body, especially the brain (hypoxia or hyperoxia).

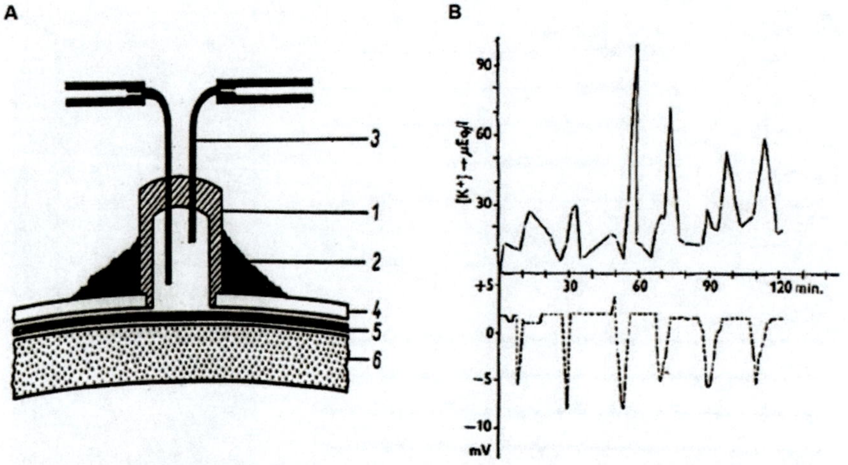

Fig. 1.12 (**A**) Diagram of the epidural cannula. (1) the body of the cannula; (2) phosphate cement or acrylic resin; (3) tubes through which the washing fluid flows; (4) cranial bone; (5) dura; (6) cortex. Reprinted with permission from the Czech Academy of Sciences (Bures & Krivanek, 1960b). (**B**) Time relation between the course of SPCs (below, ordinate in mV) and K^+ release into 0.9% NaCl (above, ordinate in μM × 10^2. (Reprinted with permission from the Czech Academy of Sciences (Krivanek & Bures, 1960))

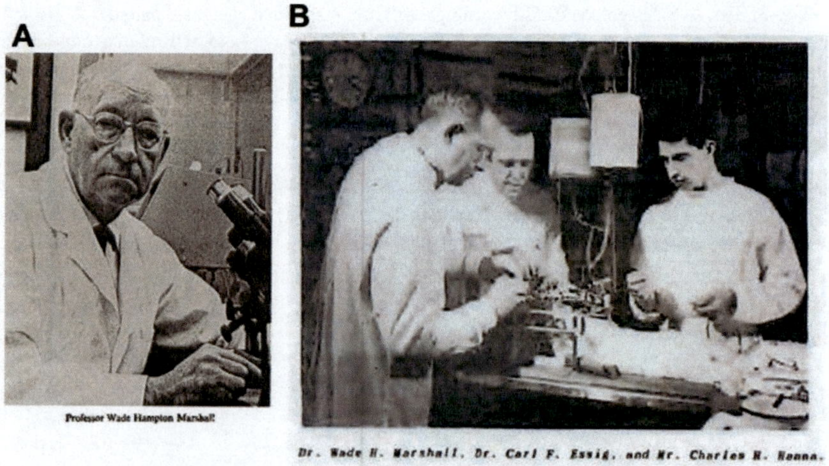

Fig. 1.13 (**A**) Dr. Wade H. Marshall, physiologist of the National Institute of Mental Health who established our first laboratory of neurophysiology, has been mapping the different areas of the cerebral cortex of animals such as the monkey and the cat. (**B**) Dr. Marshall during an in vivo experiment in his laboratory

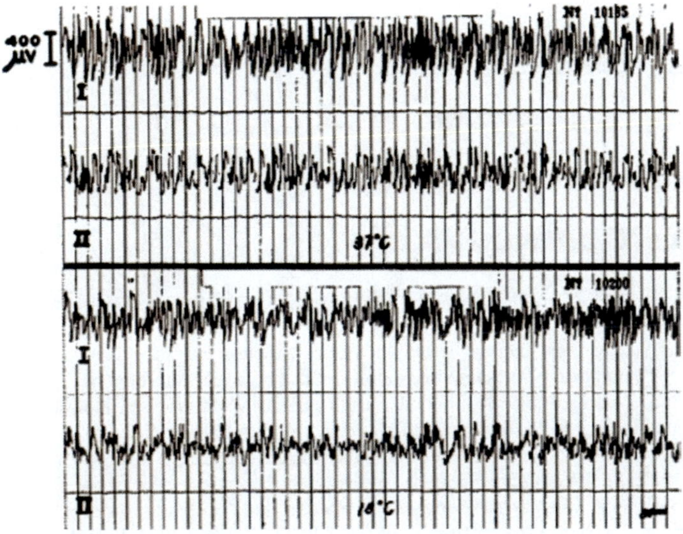

Fig. 1.14 Monkey, light nembutal anesthesia, "monopolar" records from monkey at points indicated. Upper pair of records at 37 °C, lower pair exit oil temperature of 18 °C (an arachnoid surface temperature in neighborhood of 25 °C or a little higher). Sample record in cold 5 minutes after cooling to stated temperature. Each record strip is 30 sec. Positive is upward deflection in all records. (Marshall et al., 1951)

Fig. 1.15 (**A**) Anthonie Van Harreveld, Professor of physiology, was interested in mammalian physiology, particularly the structural and physiological interactions of neurons in the central nervous system. (**B**) Caltech Biology Faculty, 1947. Standing: Keighley, Sturtevant, ENT, Haagen-Smit, Wildman, Beadle, Lewis, Wiersma, Mitchell, Van Harreveld, Alles, Anderson; seated: (top step) Borsook, Emerson; (bottom row) Dubnoff, Bonner, Tyler, Horowitz

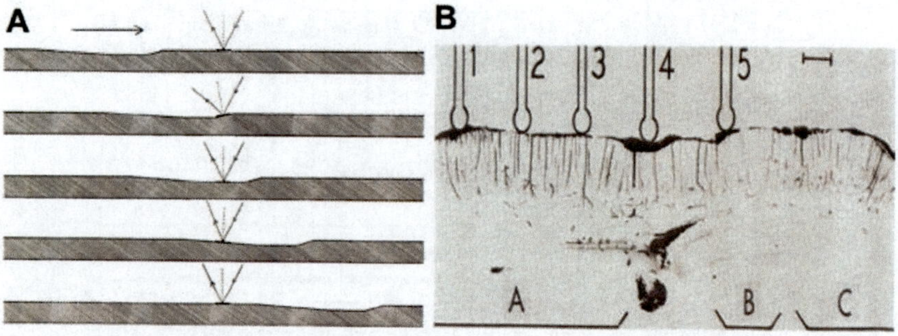

Fig. 1.16 (**A**) Conceptual diagram of wave of vasoconstriction travelling through cortex in direction of arrow. Small heavy line shows mirror on cortex which reflects a light beam. Deflections of light beam as waves of vasoconstriction travels underneath mirror are indicated. (American Physiological Society, reprinted with permission (Van Harreveld & Stamm, 1952)). (**B**) Amount of blood in the cortical vessels, while a spreading depression is in progress. Electrode *1* is the stimulating electrode. The electrocorticogram led off between electrodes 2 and 3 showed that this region was fully depressed, whereas that between electrodes 4 and 5 did not show signs of depression. The slow potential change led off from electrode 3 had gone through the negative phase and had almost reached its maximum positivity. At the moment the cortex was frozen the slow potential change was just developing under electrode 5. The vessels in the cortical region indicated by *C* are considered to be filled with the normal amount of blood. In the region (**B**) they contain considerably less blood than normal (vasoconstriction); in (**A**) the vessels were wider than normal (vasodilatation). The horizontal line indicates 1 mm. (American Physiological Society, reprinted with permission (Connor et al., 1976))

Two historical discoveries made in the eighteenth and nineteenth centuries led to our current understanding of the processes involved in the intracellular respiration providing the majority of ATP needed for cellular functions. The first event occurred about 250 years ago (1771–1775) when *Oxygen* was discovered by three scientists namely, Carl Wilhelm Scheele, Joseph Priestley, and Antoine de Lavoisier (Wilkinson, 2004). The second event was around 110–120 years later by Richards Altmann (1890) and Carl Benda (1898) that discovered the *Mitochondria* (Altmann, 1890; Benda, 1898).

1.2.1 The Discovery of Oxygen

Despite the critical need of oxygen and its significant use in body energy metabolism, the final steps of oxygen discovery in the atmosphere happened about 250 years ago (1771–1775).

There are two aspects that are related to the discovery of oxygen. Until about 25 years ago, the story of oxygen discovery was described and discussed in terms of oxygen as a chemical element and most important discovery of chemistry and science (Waltemath 1970; Cassebaum & Schufle, 1975; Marshall & Marshall, 2005; Palmer, 2006; Smeaton, 1992; West, 2014).

In many published papers and internet sites, the credit of oxygen discovery was given either to the American Joseph Priestley (Miller, 1987; Partington, 1962; Williams, 2003) or to the German-Swedish Carl Wilhelm Scheele (Partington, 1962; Smeaton, 1963; Taylor, 2019). The available information indicates that Scheele was the first one to discover oxygen in 1771–1772, but he published it officially a few years later in 1777. He claimed that in 1774 he sent a letter to the famous French chemist Antoine de Lavoisier, regarding his discovery, but Lavoisier claimed that he never received the letter. Priestley discovered oxygen, and in 1774, he notified his findings to Antoine de Lavoisier who studied oxygen chemistry for a number of years. Priestley actually published his discovery in 1775. In 2004, Wilkinson published a paper (Wilkinson, 2004) describing the significant contribution of Lavoisier, Scheele, and Priestley, seen in Fig. 1.17A, to the understanding of respiratory physiology in the eighteenth century. Here are his conclusions:

> There appear to be a great desire in history to define who was the first to do something. In reality it makes no great difference to today's work who was first to define oxygen in the eighteenth century. What can be seen from the above is the ability of these three great men to set new paradigms and work to prove their theories. Some were more accurate than others but it is the ability to take that lateral step that sets these characters apart from their peers. It could be considered that if Scheele first isolated oxygen then Priestley defined its properties and Lavoisier showed it was an element but in fact they all added a significant set of pieces to the physiological jigsaw puzzle of which oxygen was just a part.

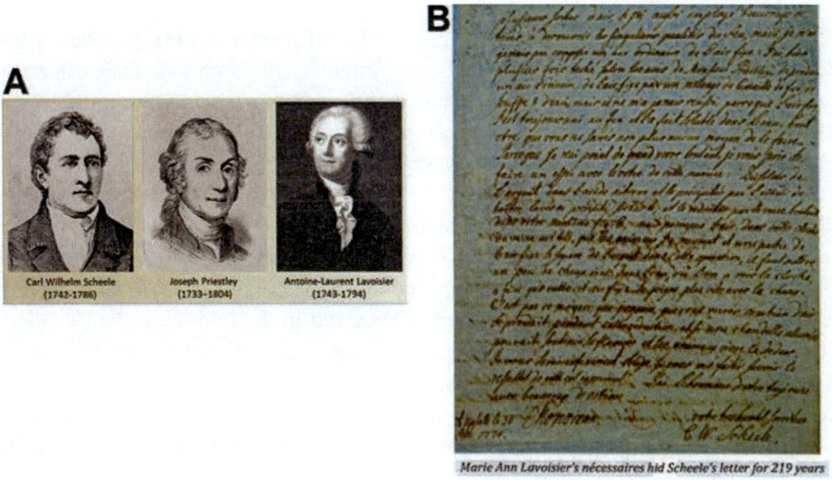

Fig. 1.17 (**A**) The three Chemists that were involved in the discovery of oxygen during the eighteenth century. (**B**) The document that Marie Ann Lavoisier's necessaires hid Scheele's letter for 219 years

In 1996, John Severinghaus became interested in the story of the discovery of oxygen and published his first paper on this subject (Severinghaus, 2002). In the introduction to the cited paper, Severinghaus wrote the next paragraph:

> AMERICANS are taught that the English Unitarian minister Joseph Priestley discovered oxygen in 1774. Scandinavians are taught that the Swedish apothecary Carl Wilhelm Scheele generated oxygen in Uppsala in 1771–2, several years before Priestley. Scheele claimed that he wrote Lavoisier, describing the experiments in September, 1774. However, Lavoisier denied seeing or receiving his letter. Among Nordic historians, Lavoisier has never been forgiven for this rebuff. Some historians have doubted whether such a letter was ever sent or received, because Scheele first published in 1777. Priestley brought the news of his discovery to Lavoisier in October 1774. It was undoubtedly Priestley who triggered Lavoisier to realize over the next few years that air contained a new element, oxygen, which combined with combustible materials in fire. Lavoisier spent most of a decade in experiments to clarify the chemistry of oxygen and in publications overturning the conventional Phlogiston Theory, causing a revolution in chemistry.

One year later, Severinghaus published an extended paper on this topic (Severinghaus, 2003) and came to the conclusion that Mary Ann Lavoisier received the original letter of Scheele and hid it until 219 years later the letter was found (Fig. 1.17B). Later, Severinghaus combined the discovery of Oxygen as an element, in the eighteenth century to the description of the pulmonary circulation in the thirteenth century (Severinghaus, 2016a, b). This subject, seen in Fig. 1.18, was discussed in other papers describing the Islamic Golden Age in the thirteenth century (Haddad & Khairallah, 1936; West, 2008).

1.2.2 *The Discovery of the Mitochondrial Function*

As mentioned, Richards Altmann (1890) and Carl Benda (1898) discovered the *Mitochondria.* It took another 60–70 years of intensive research of the mitochondria to conclude that 90–95% of the consumed molecular oxygen by the organism is utilized by the cytochrome oxidase acting at the final step in the oxidative phosphorylation process performed by the respiratory chain located in the mitochondria (Jobsis, 1964; Mason, 1957; Pereira et al., 2001; Waltemath, 1970; Wikstrom et al., 2018). The rest of the consumed O_2 is utilized by other oxidases such as the tyrosine hydroxylase converting the tyrosine to DOPA in the synthesis of dopamine in the brain (Mayevsky et al., 1973). Although the understanding of the exact mechanism of oxygen utilization by the mitochondria occurred in the middle of the twentieth century, the use of oxygen in respiratory therapy was initiated by the physician, Thomas Beddoes and the inventor, James Watt toward the end of the eighteenth century (1794–1798). The technology to generate oxygen and other gases enabled to open the first oxygen therapy site, the "Bristol Pneumatic Institute" in 1798 (Stansfield & Stansfield, 1986).

Now we would like to discuss the processes of oxygen uptake, transport by the blood, and utilization by the Mitochondria in healthy and sick patients. Those processes are the building stones of the basic concept named tissue oxygen balance that

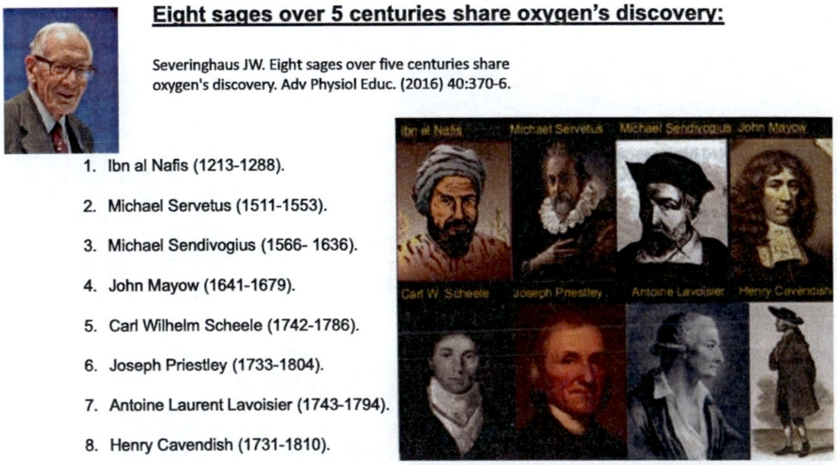

Fig. 1.18 The 8 scientists that were involved in the discovery of oxygen according to Severinghaus (2016a)

keeps the homeostasis of oxygen in the body. The concept of "Physiological Homeostasis" that was coined by Cannon in 1929 (Cannon, 1929) was based on the general idea of Claude Bernard in 1878 that the organism is keeping the internal environment stable (Bernard, 1878). One of the critical elements in the body is the homeostasis of oxygen that should be kept stable. It means that tissue oxygen balance that represents the relationship between oxygen supply and oxygen demand will vary in a narrow range. Figure 1.19 demonstrates the homeostasis of oxygen in the brain. Oxygen demand (right side) is determined by the specific physiological processes that consume oxygen in the various physiological and biochemical processes in the brain. Actually, in all organs of the body, the main energy source is ATP synthetized in the mitochondria. The basic processes involved in oxygen supply and its regulation (left side) are very similar in all organs of the body and the details will be discussed later. This concept oxygen homeostasis in the brain was investigated at the end of the nineteenth century and was published by Roy and Sherrington in 1890 (Roy & Sherrington, 1890). They studied the effect of various factors on the regulation of blood flow to the brain and came to the following conclusion:

> We conclude then, that the chemical products of cerebral metabolism contained in the lymph which bathes the walls of the arterioles of the brain can cause variations of the caliber of the cerebral vessels: that in this reaction the brain possesses an intrinsic mechanism by which its vascular supply can be varied locally in correspondence with local variations of functional activity.

The interpretation of their conclusion is that, in the brain, changes in oxygen consumption due to changes in functional activity are due to variations in blood vessels diameter. The next step in the development of the concept Oxygen Balance Homeostasis was made by the leading English physiologist Joseph Barcroft that published the book *The Respiratory Function of Blood* in 1914 (Barcroft, 1914). In

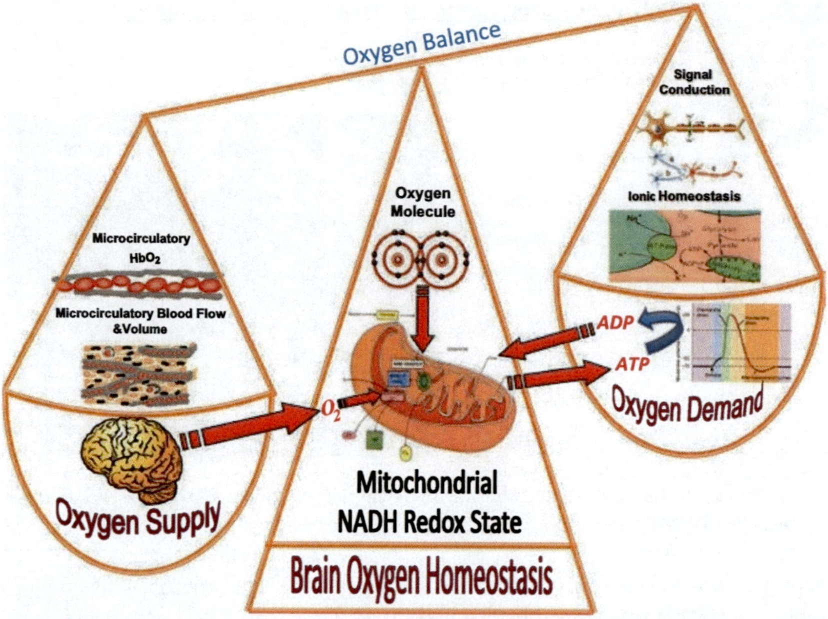

Fig. 1.19 Schematic presentation of brain oxygen balance evaluated by oxygen supply and demand. Oxygen supply could be evaluated by monitoring brain blood flow (CBF), brain blood volume (CBV), hemoglobin oxygenation (HbO_2) and brain tissue oxygen partial pressure. Oxygen demand includes Ionic Homeostasis, Signal Conduction and Biochemical synthesis. Mitochondrial NADH redox state served as an indicator for brain oxygen balance or brain oxygen homeostasis. (Mayevsky et al., 2020b)

Chap. XI of the book, he discussed the subject of "The call for Oxygen by the tissues." He proved that "There is no instance in which it can be proved that an organ increases its activity, under physiological conditions, without also increasing its demand for oxygen." Barcroft came to the second conclusion that "in no organ excited by any form of stimulus can it be shown that positive work is done without the blood supply having to respond to a call for oxygen." Those conclusions are presented also in Fig. 1.20. In 2016, Longo published a review entitled "Sir Joseph Barcroft: One Victorian physiologist's Contributions to a Half Century of Discovery." "Barcroft's scientific career may be divided into four major phases of inquiry: that of oxygen (O_2) consumption in tissues; the definition of the oxyhaemoglobin saturation [HbO_2] curve; establishing the limits of human tolerance to high altitude; and pioneering the exploration of various aspects of oxygenation of the fetus in utero."

It is important to remember that oxygen is a drug that may need to be optimized in the level given to patients, thus avoiding under as well as overdosing (Branson & Robinson, 2011; Martin & Grocott, 2013).

As seen in Fig. 1.19, the mitochondrial NADH redox state represents the tissue oxygen balance. The discovery of the intracellular organelle-mitochondrial happened toward the end of the nineteenth century by Altmann in 1890 (Altmann,

Sir Joseph Barcroft 1872-1947.

Barcroft J (1914) The Respiratory Function of Blood. Cambridge University Press, Cambridge.

Chapter XI: THE CALL FOR OXYGEN BY THE TISSUES

Page 73: The classical work of Pflüger (1) on the combustion of living material settled for all time, it seems to me, the logical order in which the constituent processes of respiration should be treated.
The issue before Pflüger may be stated in a few words. Is the quantity of oxygen taken up by the cell conditioned primarily by the needs of the cell, or by the supply of oxygen ? The answer was clear, the cell takes what it needs and leaves the rest.
The present chapter will deal with the following theme:
"There is no instance in which it can be proved that an organ increases its activity, under physiological conditions, without also increasing its demand for oxygen."

Page 105: In our discussion of the call for oxygen we have reviewed the activity of many organs of the body, muscle, heart, kidney, secreting glands and absorbing epithelium; these organs are excited by the most diverse forms of stimulus, electrical stimuli, hormones, drugs, etc. and evidence their activity by doing work of the most diverse kinds; in one respect only do they resemble one another, namely that in no organ excited by any form of stimulus can it be shown that positive work is done without the blood supply having to respond to a call for oxygen.

(1) Pflüger, "Ueber die physiologische Verbrennung in den lebendigen Organismen," Pflüger's Arch, x, p. 350, 1875.

Fig. 1.20 Citations from the book published in 1914 by Barcroft regarding the call for oxygen by tissues at rest or during activation. (Barcroft, 1914)

1890) and Benda in 1898 (Benda, 1898). Its involvement in representing the tissue oxygen balance was formulated only close to 100 years later (Mayevsky, 1992; Rampil et al., 1992). At this stage, we were able to monitor simultaneously, microcirculatory blood flow, blood volume, hemoglobin oxygenation, and mitochondrial NADH redox state. As we know today, the best monitored parameter that represents mitochondrial function in vivo is the redox state of NADH; therefore, I will describe in this section the historical background of its monitoring under in vivo conditions.

Chance et al. (1973) reviewed the historical studies that made the connection between the activity of mitochondria and oxygen utilization as follows: "The accumulation of evidence since the pioneer work of Otto Warburg an 'Atmungsferment' (Warburg, 1949) and David Keilin on cytochromes (Keilin, 1925), as the keystones of cellular oxygen utilization led us to the study of the redox states of electron carriers in isolated mitochondria as a function of oxygen concentration and to develop techniques to measure the states of anoxia and normoxia in living tissues."

During the early 1950s, the study of mitochondrial activity in vitro and after 10 years in vivo was accelerated by the development of new optical monitoring devices. The UV-A light was applied to measure NADH absorption (Chance, 1952, 1953; Theorell & Bonnichsen, 1951; Theorell & Chance, 1951), and later on, the measurement of NADH fluorescence was started (Chance & Baltscheffsky, 1958; Duysens & Amesz, 1957). During those years, a large number of papers on the monitoring of NADH were published from the laboratory of Prof. B. Chance, and he became the leader of in vitro and in vivo monitoring of NADH fluorescence.

The detailed descriptions of the respiratory chain and oxidative phosphorylation in the mitochondria – published in 1955 by Chance and Williams – established our basic knowledge of the mitochondrial function (Chance et al., 1955; Chance & Williams, 1955b, c, d, e). In those five papers, Chance and Williams defined, for the first time, the metabolic states of isolated mitochondria in vitro, depending on the substrate, oxygen, and ADP levels. In addition, they correlated those metabolic

A

States of Respiratory Pigments in Mitochondria

	State 1	State 2	State 3	State 4	State 5
Characteristics	Aerobic	Aerobic	Aerobic	Aerobic	Anaerobic
ADP level	Low	High	High	Low	High
Substrate level	Low endogenous	Approaching 0	High	High	High
Respiration rate	Slow	Slow	Fast	Slow	0
Rate-limiting component	Phosphate acceptor	Substrate	Respiratory chain	Phosphate acceptor	Oxygen

B

Otto Heinrich Warburg 8.10.1883-1.8.1970

David Keilin 21.3.1887-27.2.1963

Dietrich W. Lubbers 1917-2006

Britton Chance 24.7.1913-16.11.2010

Frans F. Jobsis 1.4.1929-29.8.2006

ID=1228

Fig. 1.21 (**A**) Mitochondrial metabolic state, defined in vitro, by Chance and Williams, and opened up a new era in measurements of respiratory chain enzyme's redox state in vitro as well as in vivo (Chance & Williams, 1955d). (**B**) The five scientists significantly affected our knowledge on mitochondrial function and NADH monitoring in vitro and in vivo. (Mayevsky, 2015)

states to the oxidation-reduction levels of the respiratory enzymes as seen in Figure 1.21A.The physiological significance of those metabolic states was discussed in 1955–1956 by Chance and Williams (1955d, 1956).

The "resting" state of the mitochondria in vitro was defined as "State 4," where NADH was 99% in the reduced form, and ADP was the rate limiting substance. If ADP is added to a suspension of mitochondria, ATP synthesis will be stimulated, oxygen consumption will increase, and the rate limit will be determined by the activity of the respiratory chain. During this State 3, or the "active state," the NADH redox state will decrease or become more oxidized (about 50%). When the "resting" mitochondria are deprived of oxygen, the activity of the mitochondria will stop and NADH will reach its maximum redox state (State 5).

A definitive description of the mitochondrial metabolic state has never been given for in vivo conditions. Therefore, we described the in vivo mitochondria conditions as recorded by NADH fluorescence in a representative tissue or organ – e.g., the brain. While the range between minimal NADH (~0) and its maximal level was determined in vitro, it is almost impossible to determine in the intact brain or other organs in vivo. For example, State 2, with a substrate free medium, could not be achieved in vivo since the tissue would die. On the other hand, the maximal level of NADH (State 5) could be monitored in vivo under complete deprivation of oxygen by anoxia or complete ischemia.

An intensive use of the real time in vivo NADH monitoring approach started in 1962. The "classical" paper on in vivo monitoring of NADH was published in 1962 by Chance et al. (1962c). They were able to simultaneously monitor the brain and kidney of anesthetized rats using two microfluorometers. Chance and collaborators elaborated on this kind of in vivo monitoring and used it in other rat organs (Chance et al., 1962a, b; Chance & Schoener, 1962).

Figure 1.21B presents the pictures of the five cited scientists that affected significantly the development of the theoretical and experimental technology for the monitoring of mitochondrial NADH function in vitro and in vivo.

1.3 Monitoring of Brain Mitochondrial Function in Vivo

The discovery of the optical properties of reduced nicotineamide adenine dinucleotide – NADH (earlier names: DPNH – diphosphopyridine nucleotide, or PN – pyridine nucleotide) – has led to a very intensive research since the early 1950s. The reduced form of this molecule, NADH (Fig. 1.22A), absorbs light at 320–380 nm (Fig. 1.22B) and emits fluorescent light at 420–480 nm range (Fig. 1.22C).

As the oxidized form NAD^+ (Fig. 1.22B) does not absorb light in this range, it was possible to evaluate the redox state of the mitochondria by monitoring the UV absorbance or blue fluorescence of NADH. Figure 1.22D presents the fluorescence spectra of the NADH measured from the exposed cortex of the rat brain. Under depletion of oxygen (anoxia), there is a significant increase in fluorescence intensity.

Table 1.1 summarizes the historical milestones in the development of NADH monitoring after its discovery in 1905–1906 by Harden and Young (1906, 1905).

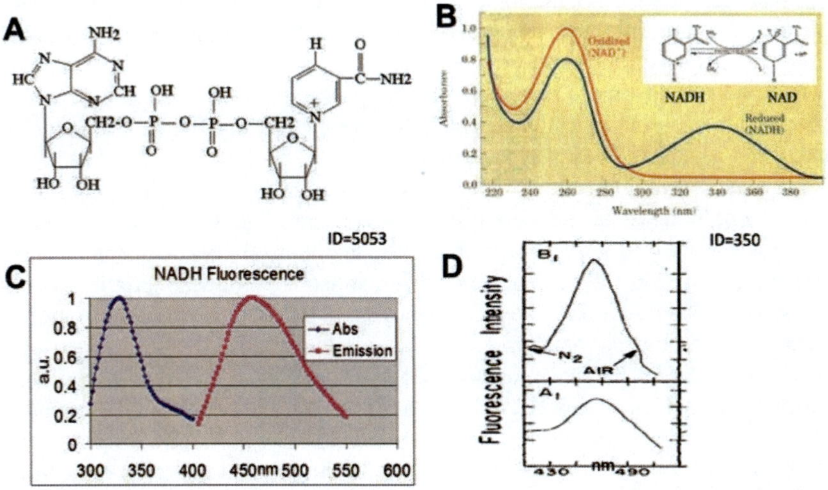

Fig. 1.22 (**A**) Structures of NAD^+. The nicotineamide group (broken ring) is the functional part of both molecules, i.e. the portion of molecules where oxidation and reduction take place (see arrow). (**B**) Absorption spectra of NAD^+ and NADH in solution. (**C**) Absorption and emission spectra of NADH illuminated by UV light (Mayevsky & Barbiro-Michaely, 2013a; Mayevsky & Meirovithz, 2020). (**D**) The emission spectra of rat cerebral cortex under aerobic conditions (lower trace) and under anoxic conditions (upper trace). (Mayevsky, 1984; Mayevsky & Meirovithz, 2020)

Table 1.1 Historical Milestones in Early period of NADH Monitoring

Event	Discovery	Author(s)
1	Involvement of adenine containing nucleotide in fermentation by yeast	(Harden & Young, 1905; Harden & Young, 1906)
2	Description of full structure of "hydrogen transferring coenzyme in erythrocyte. Naming the two cofactors DPN and TPN	(Warburg et al., 1935)
3	A shift in the absorption spectrum of DPNH with alcohol dehydrogenase	(Theorell & Bonnichsen, 1951)
4	Development of a rapid and sensitive spectrophotometer	(Chance & Legallias, 1951)
5	Monitoring of pyridine nucleotide enzymes	(Chance & Neilands, 1952)
6	The development of double beam spectrophotometer	(Chance, 1954)
7	The first detailed study on NADH using fluorescence Spectrophotometry	(Duysens & Amesz, 1957)
8	Measurement of NADH fluorescence in isolated Mitochondria	(Chance & Baltscheffsky, 1958)
9	Measurements of muscle NADH fluorescence in vitro	(Chance & Jobsis, 1959)
10	In vivo monitoring of NADH fluorescence from the brain And kidney	(Chance et al., 1962c)
11	Comparison between NADH fluorescence in vivo and enzymatic analysis of tissue NADH	(Chance et al., 1966)

The possibility of monitoring mitochondrial function in vivo, in experimental animals and patients, is of great importance and could contribute significantly to the understanding of various pathological processes. Most of the information on mitochondrial function has been accumulated from in vitro studies. A relatively small portion of published papers dealt with the monitoring of mitochondrial function in vivo and in real time. During the past 50 years, we have published over 150 papers in this very significant area, including the largest number of studies using NADH redox state monitoring in patients as well.

The need for an intracellular pO_2 indicator, as a physiological and biochemical parameter of living tissue, has emerged close to 60 years ago (Chance et al., 1962d). Mitochondria are the intracellular organelles that consume most of the oxygen. Therefore, the redox state of electron carriers in isolated mitochondria in vitro as well as in vivo as a function of oxygen concentration has been extensively studied. Chance et al. concluded that "for a system at equilibrium, NADH is at the extreme low potential end of the chain, and this may be the oxygen indicator of choice in mitochondria and tissue as well" (Chance et al., 1973). Lubbers in 1995 concluded that "the most important intrinsic luminescence indicator is NADH, an enzyme of which the reaction is connected with tissue respiration and energy metabolism" (Lubbers, 1995).

One of the main questions is, why monitoring of NADH fluorescence is the leading technology in monitoring mitochondrial function in vivo and in real time. It is clear that other electron carriers in the respiratory chain may have an advantage since they are localized only in the mitochondria. The NADH appears in addition to the mitochondria also in the cytosol. We have tried many options that will enable us to monitor flavoproteins (FP) fluorescence together with NADH. It was found that

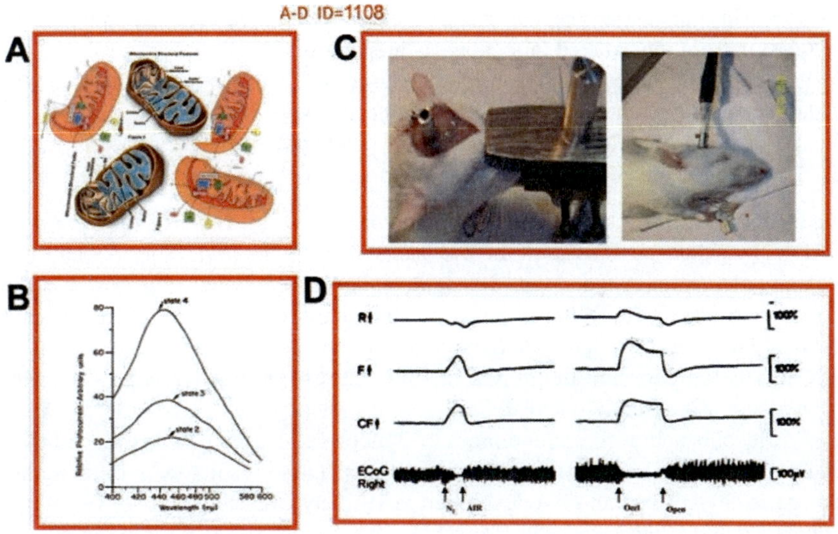

Fig. 1.23 The comparison between in vitro and in vivo monitoring of NADH fluorescence. The spectra in (**B**) was measured from isolated mitochondria (**A**) and the response of the intact brain in vivo (**C**) to anoxia (**D**). (Mayevsky et al., 2020a)

FP measurement is dramatically affected and sensitive to hemodynamic artifacts and mainly to changes in hemoglobin oxygenation. Cytochrome oxidase monitoring suffers from the same problems due to overlapping with hemoglobin absorption spectrum. So in the near future, the NADH is the best candidate to be monitored for mitochondrial function in patients as well as in experimental animals.

The pioneer work of Chance, Williams, Connelly, and other collaborators, in the early 1950's, opened up new possibilities enabling the study of mitochondrial function in vitro, and later on in vivo monitoring became a reality (Chance & Williams, 1954, 1955a, d, 1956; Connelly & Chance, 1954). The shift from studying mitochondrial NADH redox state in isolated mitochondria to higher cellular and tissue organization levels started in the late 1950s (Chance & Jobsis, 1959). The main breakthrough occurred in 1962, a year that at least six papers described the in vivo monitoring of various organs in the anesthetized rat (Chance et al., 1962a, b, c, d; Chance & Schoener, 1962, 1963). The crucial issue is: Can we extrapolate from the results obtained in monitoring of mitochondrial NADH in vitro to in vivo conditions? Mitochondrial function and its metabolic states under in vivo condition are completely different from the definition made for isolated mitochondria by Chance and Williams in 1955 and Chance & Williams, 1956. The main differences between the two situations presented in Figs. 1.23A and C are as follows:

1. The isolated mitochondria are not inter-connected to other intracellular organelles and components.
2. The amount of oxygen available in the mitochondrial medium is not regulated by changes in blood flow or hemoglobin oxygenation under in vitro conditions.

3. One of the largest oxygen or energy consumers, i.e., various ions pumps, are missing from the isolated mitochondria preparation.
4. When ADP is added to State 4 mitochondria, ATP will be synthesized but will not be consumed in parallel as occurred under in vivo conditions.
5. The isolated mitochondria preparation is not exposed to systemic changes in hemodynamic and other physiological parameters occurring in the organism.

Figure 1.23 illustrates the data interpretation dilemma. Part B shows the spectra of NADH fluorescence measured in the isolated mitochondria (Part A) and the responses to shifts in the metabolic state of the mitochondria. Part D shows the response to anoxia (left) and ischemia (right) measured from the intact rat brain in vivo (part C) under "State 4 to State 5" transition.

Can we compare and define the resting state (4) and active state (3) under in vivo situations to in vitro definitions? As seen in Fig. 1.24, the maximal NADH level is achieved under complete O_2 deprivation that can be induced both under in vitro (A1) and in vivo (B1) conditions. These results were obtained by using the devices (A_2 and B_2) developed in 1951 by Chance (Chance, 1951) and Mayevsky and Chance in 1973 (Mayevsky & Chance, 1973). This signifies that this definitive point can be used to determine State 5 in vivo as well. As can be seen in the table (Fig. 1.24 A_1 and B_1), the steady-state NADH level at State 4 is 99%, namely that most of the NAD is in the reduced form NADH. Another possible explanation is that the NADH almost reaches maximum allowed in the system, but this option of explanation is not in agreement with the technology used. When the mitochondria are shifted to State 3, by adding ADP, the NADH becomes more oxidized and the NADH level is decreased to 53% of the reduced form. Moving from State 4 to State 5 and eliminating the oxygen from the media

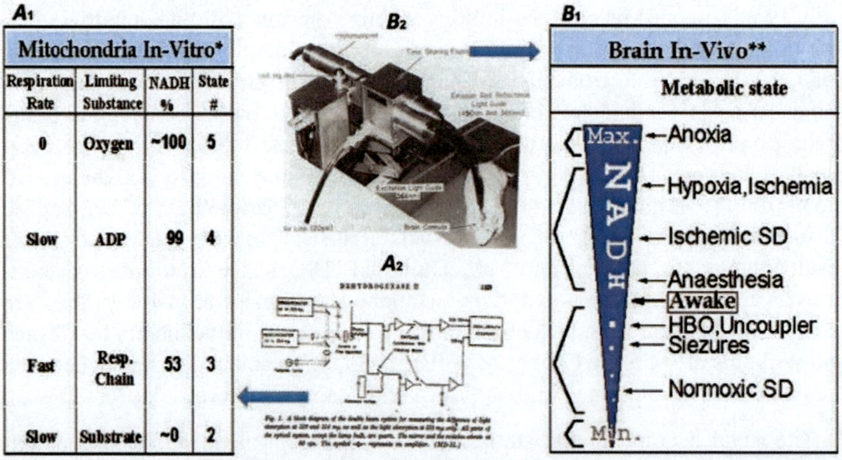

Fig. 1.24 Comparison between mitochondrial metabolic state (in vitro), defined by Chance & Williams in 1955 (**A**$_1$ and **A**$_2$) and responses of the in vivo brain to changes in O_2 supply and demand induced by brain activation described by Mayevsky (**B**$_1$ and **B**$_2$). (Mayevsky et al., 2020a)

will have a very small effect on the NADH level namely, changing from 99% to 100%. It is important to note that other members of the respiratory chain behave somewhat similarly to the NADH. When the mitochondria are shifted to State 2, NADH level is 0 since the NAD is the dominant state in the absence of a substrate.

The problem is to determine the "resting" metabolic state of a tissue in an in vivo situation. If we adopt the in vitro value of a resting state (State 4), this would signify that the increase in NADH during State 5, induced by anoxia (0% O_2), would be only 1% namely from 99% to 100%. According to all in vivo studies, this is not the case, and during anoxia, the increase in NADH is much larger than the decrease under State 4 to 3 transition.

When we started our monitoring of NADH redox state using fiber optic–based fluorometry/reflectometry in 1972, we used the brain in vivo exposed to various conditions in real time (Mayevsky & Chance, 1973). We found that the NADH level could either increase or decrease from the baseline level calibrated in the normoxic animal (Mayevsky, 1984). The initial results were plotted against the perturbation used as shown in the right side of Fig. 1.24B_1. The relative changes in brain NADH under various perturbations as compared to the Chance and Williams definitions (Fig. 1.24A_1) are presented. Although the maximal level of NADH could be achieved in vivo by exposing the animal to anoxia (100% nitrogen), the minimal level that could be monitored in vivo is very hard to achieve. We tried to build the scale between maximal and minimal levels of NADH by exposing the brain to many types of perturbations. In the table, the horizontal lines represent the relative change in NADH recorded during the perturbation. The values are not accurate but rather trying to map the scale of NADH changes. The question is how to locate the in vivo scale relative to the in vitro definition shown on the left side. According to the changes recorded, it seems that the so-called "resting brain" is not at State 4 (99% NADH) but rather around state 3.5–3.6. As seen, NADH will increase by various factors such as anesthesia, hypoxia, and partial ischemia. Under complete ischemia or anoxia, the NADH will reach its maximal level – State 5. Exposing the animal to hyperoxia or uncoupler led to a decrease in NADH toward State 3. Activation of the brain by inducing seizures or cortical spreading depression led to a very large decrease in NADH. The main issue in the scaling or the responses is how to "push" the brain in vivo to minimal NADH levels. As of now, it seems that this question remains open and unsolved.

Activation of the brain (such as CSD) under conditions of restricted amounts of oxygen, such as partial ischemia, will lead to an increase in NADH (reduction) instead of a decrease (oxidation) as found in the normoxic brain. Those results found in various experimental conditions led to the dilemma. According to Chance and Williams, ATP production related to State 4 to State 3 transition is always correlated to a decrease or oxidation of NADH from 99% to 53%. The question is how ATP turnover could increase under NADH increase (reduction) and not decrease (oxidation). We want to make sure that the experimental results do not represent a technical artifact related to the monitoring itself. We assumed

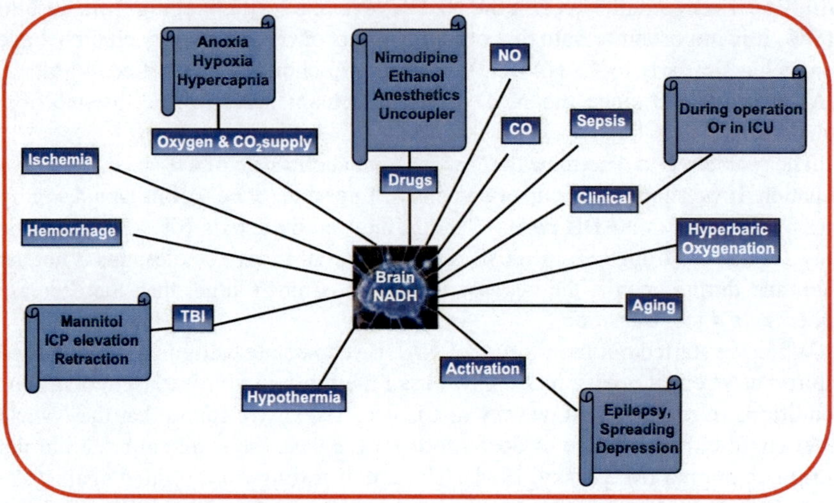

Fig. 1.25 Presentation of NADH redox state monitored in various organs and perturbations used by Mayevsky et al. in experimental animals as well as in patients. (Mayevsky et al., 2020a)

that by using appropriate modeling methods the dilemma could be clarified or resolved (Vatov et al., 2006).

During the years 1972–2014, we were able to monitor brain NADH redox state in many pathophysiological conditions in experimental animals as well as in neurosurgical patients in the OR and in the ICU (Mayevsky & Barbiro-Michaely, 2013a, b). As seen in Fig. 1.25, the experimental conditions included changes in oxygen supply (hypoxia, ischemia, hemorrhage and hyperoxia). Brain activation effects were tested in models of epileptic activity as well as CSD. The effects of various pharmacological agents were tested as well (Mayevsky & Rogatsky, 2007).

1.4 Brain Energy Metabolism and Mitochondrial Function During CSD

As mentioned before, Leão in one of his seminal studies published in 1944 found that blood flow in the pial vessels was "elevated significantly probably due to a local change in the activity of nervous elements" (Leão, 1944a). Between the period 1944–1973, other investigators studied the effects of CSD on various physiological and biochemical parameters related to brain energy metabolism. All details of those studies (Bures, 1956; Krivanek, 1961; Lukyanov & Bures, 1967; Van Harreveld & Ochs, 1957a; Van Harreveld & Stamm, 1952) were described in the book published by Bures et al. in (1974).

As shown in Fig. 1.19, the parameter that represents the balance between oxygen supply and oxygen demand is the redox state of mitochondrial NADH. Nevertheless, the first study on the effect of CSD on mitochondrial function, evaluated by NADH fluorescence, was published in 1973 by Rosenthal and Somjen (Rosenthal & Somjen, 1973) after preliminary presentation in a meeting in 1971. The second paper on this subject was published by Mayevsky and Chance et al. (1973).

Figure 1.26 shows the effect of KCl solution placed on the cerebral cortex of cats on the NADH redox state (Rosenthal & Somjen, 1973). Solutions containing an excess of KCl flowing over the observed cortical area also provoked spreading depression. Figure 1.26 shows that within 1 min after Ringer solution containing 12 mEq KCl was made to flow over the cortical surface, there was a rise in the cortical potential (recorded at 0.6 mm below the surface) which coincided with a drop of corrected and uncorrected fluorescence signals. The latter continued to decrease as the potential trace reversed and moved toward a negative trough. During this negative phase, the reflectance began to rise. While the potential reversed again and returned to base line, the fluorometric signals continued to decrease. Only after the steady potential reached its baseline level did the metabolic trace begin to recover. In the case of spreading depression provoked from the surface, microelectrodes placed deeply (about 2.0 mm depth) did not, as a rule, register large negative SP shifts, but sometimes smaller positive-going deflections (see also ref. Marshall, 1959).

In 1973, we performed preliminary experiments in order to correlate the redox state of NADH with the level of electrical activity of the brain of rats. In these experiments, we perfused a small area of the dura (2 mm^2) with 0.4 M KCl solution. The distance from the perfused area to the NADH fluorescence measurement was 2 mm. A typical response to spreading depression is shown in Fig. 1.27. The animal was breathing pure oxygen or normal air during this part of the experiment, so the

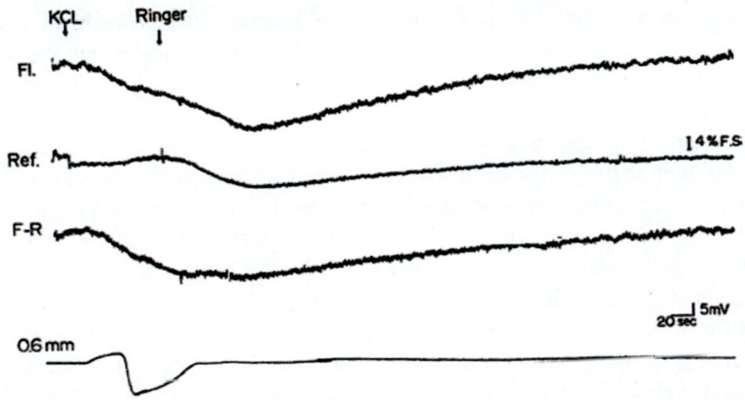

Fig. 1.26 Spreading depression of the suprasylvian gyrus provoked by topical application of K$^+$ ions. At arrow marked KCl, the fluid irrigating the cortical surface was changed from normal Ringer solution to one containing a high (12 mm) concentration of K$^+$. At second arrow the superfusion was returned to normal Ringer solution. Breaks in lines caused by mechanical disturbances. American Physiological Society, reprinted with permission (Rosenthal & Somjen, 1973). L, Ref. and F-R represent the NADH fluorescence, Reflectance and corrected NADH signals

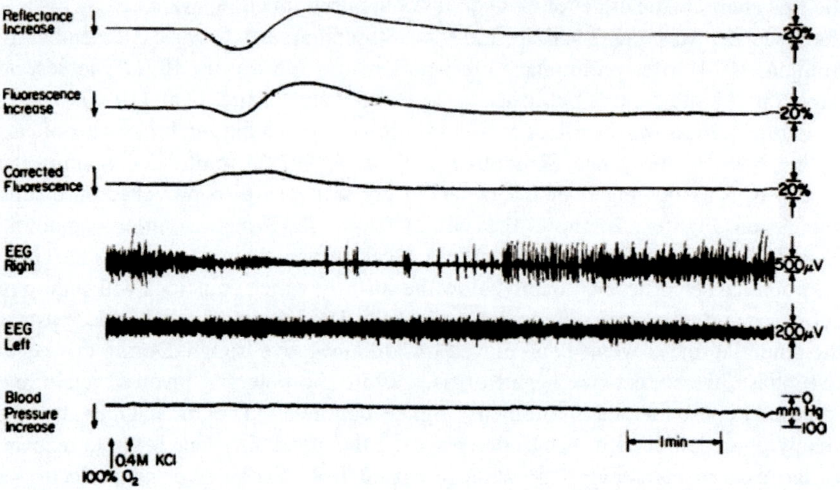

Fig. 1.27 The response of the brain to CSD induced by application of KCl (0.4 M) on the dural surface. The direction of the arrows indicates an increase of the signals. (Mayevsky & Chance, 1973)

oxygen supply was in the normoxic range. The change in the reflectance and the NADH fluorescence appears 30–40 seconds after the application of the KCl solution. The changes in the right hemisphere EEG (upper) begin even earlier since the recording was made from a point near the KCl application site. From the changes in the reflected light, we assume that the response to CSD is biphasic; in the beginning, there is a short increase in reflectance, followed by a longer decrease. The large reflectance changes introduce an artifact into the fluorescence measurement, and the corrected trace shows a small NADH oxidation. In some animals where the reflectance changes were smaller, we saw an NADH oxidation even in the uncorrected fluorescence trace.

The recovery of the corrected fluorescence trace from spreading depression is much earlier than that of the EEG. The recovery of the reflectance and of the uncorrected fluorescence trace is very close to that of the EEG. It seems that the response to CSD begins in a short vasoconstriction response, followed by a longer vasodilation, as suggested also by van Harreveld and Ochs (Tschirgi et al., 1957). It is also apparent that EEG activity in the contralateral side of the brain (left side) was unaffected by the CSD on the right side. The blood pressure was stable during this interval. We assume that the response to CSD contains changes in blood flow or volume as well as changes in oxidation reduction state of NADH in the brain. That the oxidation of NADH is due to an increase in oxygen consumption is in agreement with other findings in relation to the energy change during CSD. Bures (Bures, 1956; Lukyanov & Bures, 1967) found an increase in oxygen consumption during CSD.

References

Altmann, R. (1890). *Die Elementarorganismen Und Ihre Beziehungen Zu Den Zellen*. Veit & comp.

Barcroft, J. (1914). *The Respiratory Function of Blood*. Cambridge University Press.

Benda, C. (1898). Ueber die Spermatogenese der Vertebraten und höherer Evertebraten, II. Theil: Die Histiogenese der Spermien. *Archives of Anatomy and Physiology, 73*, 393–398.

Bernard, C., 1878. Leçons sur les phénomènes de la vie communs aux animaux et aux végétaux.

Branson, R. D., & Robinson, B. R. (2011). Oxygen: when is more the enemy of good? *Intensive Care Medicine, 37*, 1–3.

Bures, J. (1956). Some metabolic aspects of Leão's spreading depression. *Journal of Neurochemistry, 1*, 153–158.

Bures, J., & Krivanek, J. (1960a). Ionic movements in the brain as studied with the aid of washing the cortical surface with an epidural cannula. *Physiologia Bohemoslovaca, 9*, 488–493.

Bures, J., & Krivanek, J. (1960b). Ionic movements in the brain as studied with the aid of washing the cortical surface with an epidural cannula. *Physiologia Bohemoslovenica, 9*, 488–493.

Bures, J., Buresova, O., & Krivanek, J. (1974). *The Mechanism and Applications of Leão's Spreading Depression of Electroencephalographic Activity*. Academic.

Cannon, W. B. (1929). Organization for physiological homeostasis. *Physiological Reviews*, 399–430.

Carl Wilhelm Scheele. (1970). (1742–1786) Swedish apothecary. *JAMA, 212*, 2258–2259.

Cassebaum, H., & Schufle, J. A. (1975). Scheele's priority for the discovery of oxygen. *Journal of Chemical Education, 52*, 442.

Chagas, F. C. (2000). *Um aprendiz de ciência* (pp. 93–137). Instituto de Biofísica Carlos Chagas Filho.

Chance, B. (1951). Rapid and sensetive spectrophotometry. I. The accelerated and stopped-flow methods for the measurement of the reaction kinetics and spectra of unstable compounds in the visible region of the spectrum. *Review of Scientific Instruments*, 22, 619–626.

Chance, B. (1952). Spectra and reaction kinetics of respiratory pigments of homogenized and intact cells. *Nature, 169*, 215–221.

Chance, B. (1953). Dynamics of respiratory pigments of ascites tumor cells. Trans. N.Y. *Academy of Sciences, 16*, 74–75.

Chance, B. (1954). Spectrophotometry of intracellular respiratory pigments. *Science, 120*, 767–775.

Chance, B., & Baltscheffsky, H. (1958). Respiratory enzymes in oxidative phosphorylation (VII - Binding of intramitochondrial reduced pyridine nucleotide). *The Journal of Biological Chemistry, 233*, 736–739.

Chance, B., & Jobsis, F. (1959). Changes in fluorescence in a frog sartorius muscle following a twitch. *Nature, 184*, 195–196.

Chance, B., & Legallias, V. (1951). Rapid and sensetive spectrophotometry. II. A stopped-flow attachment for a stabilized quartz spectrophotometer. *Review of Scientific Instruments, 22*, 627–638.

Chance, B., & Neilands, J. B. (1952). Studies on lactic dehydrogenase of heart. II. A compound of lactic dehydrogenase and reduced pyridine nucleotide. *The Journal of Biological Chemistry, 199*, 383–387.

Chance, B., & Schoener, B. (1962). Correlation of oxidation-reduction changes of intracellular reduced pyridine nucleotide and changes in electro-encephalogram of the rat in anoxia. *Nature, 195*, 956–958.

Chance, B., & Schoener, B. (1963). Control of oxidation-reduction state of NADH in the liver of anesthetized rats. Symptoms of Regul. Enzyme Act. Synth. Norm. Neoplast. Tissues Proc. Advances in Enzyme Regulation, 1, 169–185.

Chance, B., & Williams, G. R. (1954). Steady-state of reduced pyridine nucleotides in phosphorylating rat liver mitochondria. *American Society for Biochemistry and Molecular Biology, 13*, Abstract 633.

Chance, B., & Williams, G. R. (1955a). A method for the localization of sites for oxidative phosphorylation. *Nature, 176*, 250–254.
Chance, B., & Williams, G. R. (1955b). Respiratory enzymes in oxidative phosphorylation (I-Kinetics of oxygen utilization). *The Journal of Biological Chemistry, 217*, 383–393.
Chance, B., & Williams, G. R. (1955c). Respiratory enzymes in oxidative phosphorylation (II-Difference spectra). *The Journal of Biological Chemistry, 217*, 395–407.
Chance, B., & Williams, G. R. (1955d). Respiratory enzymes in oxidative phosphorylation (III-The steady state). *The Journal of Biological Chemistry, 217*, 409–427.
Chance, B., & Williams, G. R. (1955e). Respiratory enzymes in oxidative phosphorylation (IV-The respiratory chain). *The Journal of Biological Chemistry, 217*, 429–438.
Chance, B., & Williams, G. R. (1956). The respiratory chain and oxidative phosphorylation. In F. F. Nord (Ed.), *Advances in Enzymology* (XVII ed., pp. 65–134). Interscience Publisher, Inc..
Chance, B., Williams, G. R., Holmes, W. F., & Higgins, J. (1955). Respiratory enzymes in oxidative phosphorylation (V- A mechanism for oxidative phosphorylation). *The Journal of Biological Chemistry, 217*, 439–451.
Chance, B., Legallias, V., & Schoener, B. (1962a). Metabolically linked changes in fluorescence emission spectra of cortex of rat brain, kidney and adrenal gland. *Nature, 195*, 1073–1075.
Chance, B., Schoener, B., & Fergusson, J. J. (1962b). In vivo induced oxidation by adrenocorticotrophic hormone of reduced pyridine nucleotide in the adrenal cortex of hypophysectomized rats. *Nature, 195*, 776–778.
Chance, B., Cohen, P., Jobsis, F., & Schoener, B. (1962c). Intracellular oxidation-reduction states in vivo. *Science, 137*, 499–508.
Chance, B., Cohen, P., Jobsis, F., & Schoener, B. (1962d). Localized fluorometry of oxidation-reduction states of intracellular pyridine nucleotide in brain and kidney cortex of the anesthetized rat. *Science, 136*, 325.
Chance, B., Jamieson, D., & Williamson, J. R. (1966). *Control of the oxidation-reduction state of reduced pyridine nucleotides in vivo and in vitro by hyperbaric oxygen* (pp. 15–41). National Academy of Sciences.
Chance, B., Oshino, N., Sugano, T., & Mayevsky, A. (1973). Basic principles of tissue oxygen determination from mitochondrial signals. Oxygen Transport to Tissue. *Advances in Experimental Medicine and Biology, 37A*, 277–292.
Connelly, C. M., & Chance, B. (1954). Kinetics of reduced pyridine nucleotides in stimulated frog muscle and nerve. *American Physiological Society, 13*, 29.
Connor, J. A., Kreulen, D. L., & Prosser, C. L. (1976). Relation between oxidative metabolism and slow rhythmic potentials in mammalian intestinal muscle. *Proceedings of the National Academy of Sciences of the United States of America, 73*, 4239–4243.
Del Pozo, E. C., & Leão, A. A. P. (1943). The effects of hyperventilation and of blood-pressure changes on the self-sustained responses of the cerebral cortex. *American Journal of Physiology-Legacy Content, 139*, 335–342.
do Carmo, R. J., & Somjen, G. G. (1994). Spreading depression of Leão: 50 years since a seminal discovery. *Journal of Neurophysiology, 72*, 1–2.
Dreier, J. P. (2011). The role of spreading depression, spreading depolarization and spreading ischemia in neurological disease. *Nature Medicine, 17*, 439–447.
Duysens, L. N. M., & Amesz, J. (1957). Fluorescence spectrophotometry of reduced phosphopyridine nucleotide in intact cells in the near-ultraviolet and visible region. *Biochimica et Biophysica Acta, 24*, 19–26.
Engelhardt, E., & Gomes Mda, M. (2015). Aristides Leão: a birth centennial homage with comments on his spreading depression. *Arquivos de Neuro-Psiquiatria, 73*, 544–546.
Gattass, R., & Leão, A. A. P. (2013). In D. F. Almeida & W. Souza (Eds.), *Construtores do Instituto de Biofsica Carlos Chagas ilho* (pp. 31–40). Corba.
Haddad, S. I., & Khairallah, A. A. (1936). A Forgotten Chapter in the History of the Circulation of the Blood. *Annals of Surgery, 104*, 1–8.

Harden, A., & Young, W. J. (1905). The influence of phosphates on the fermentation of glucose by yeast-juice: preliminary communication. *Proceedings of the Chemical Society (London), 21*, 189–191.

Harden, A., & Young, W. (1906). Alcoholic ferment of yeast-juice Part II Co-ferment of yeast-juice. *Proceedings of the Royal Society A, B78*, 369–375.

Jobsis, F. F. (1964). Basic processes in cellular respiration. In R. Fenn (Ed.), *Handbook of physiology - Respiration* (1st ed., pp. 63–124).

Keilin, D. (1925). On cytochrome, a respiratory pigment, common to animals, yeast, and higher plants. *Proceedings of the Royal Society of London - Series B: Biological Sciences, 98*, 312–339.

Krivanek, J. (1958). Changes of brain glycogen in the spreading EEG-depression of Leão. *Journal of Neurochemistry, 2*, 337–343.

Krivanek, J. (1961). Some metabolic changes accompanying Leão's spreading cortical depression in the rat. *Journal of Neurochemistry, 6*, 183–189.

Krivanek, J., & Bures, J. (1960). Ion shifts during Leão spreading cortical depression. *Physiologia Bohemoslovenica, 9*, 494–503.

Lauritzen, M., Dreier, J. P., Fabricius, M., Hartings, J. A., Graf, R., & Strong, A. J. (2011). Clinical relevance of cortical spreading depression in neurological disorders: migraine, malignant stroke, subarachnoid and intracranial hemorrhage, and traumatic brain injury. *Journal of Cerebral Blood Flow and Metabolism, 31*, 17–35.

Leão, A. A. P. (1943). *Spreading depression of activity in the cerebral cortex*. Harvard University.

Leão, A. A. P. (1944a). Pial circulation and spreading depression of activity in the cerebral cortex. *Journal of Neurophysiology, 7*, 391–396.

Leão, A. A. P. (1944b). Spreading depression of activity in cerebral cortex. *Journal of Neurophysiology, 7*, 359–390.

Leão, A. A. P. (1947). Further observations on the spreading depression of activity in the cerebral cortex. *Journal of Neurophysiology, 10*, 409–414.

Leão, A. A. P., & Morison, R. (1945). Propagation of spreading cortical depression. *Journal of Neurophysiology, 8*, 33–45.

Lindahl, S. G. (2008). Oxygen and life on earth: an anesthesiologist's views on oxygen evolution, discovery, sensing, and utilization. *Anesthesiology, 109*, 7–13.

Lubbers, D. W. (1995). Optical sensors for clinical monitoring. *Acta Anaesthesiologica Scandinavica Supplementum, 39*, 37–54.

Lukyanov, L. D., & Bures, J. (1967). Changes in pO_2 due to spreading depression in cortex and nucleus caudatus of rat. *Physiologia Bohemoslovaca, 16*, 449–455.

Lyons, T. W. (2007). Palaeoclimate: oxygen's rise reduced. *Nature, 448*, 1005–1006.

Lyons, T. W., Reinhard, C. T., & Planavsky, N. J. (2014). The rise of oxygen in Earth's early ocean and atmosphere. *Nature, 506*, 307–315.

Maranhao-Filho, P., & Vincent, M. (2009). Professor Aristides Leão. Much more than spreading depression. *Headache, 49*, 110–116.

Marshall, W. H. (1959). Spreading cortical depression of Leão. *Physiological Reviews, 39*, 239–279.

Marshall, W. H., & Essig, C. F. (1951). Relation of air exposure of cortex to spreading depression of Leão. *Journal of Neurophysiology, 14*, 265–273.

Marshall, J. L., & Marshall, V. R. (2005). *Rediscovery of the elements: Carl Wilhelm Scheele. University of North Texas Libraries, UNT digital library*. Crediting UNT College of Arts and Sciences.

Marshall, W. H., Essig, C. F., & Dubroff, S. J. (1951). Relation of temperature of cerebral cortex to spreading depression of Leão. *Journal of Neurophysiology, 14*, 153–166.

Martin, D. S., & Grocott, M. P. (2013). III. Oxygen therapy in anaesthesia: the yin and yang of O_2. *British Journal of Anaesthesia, 111*, 867–871.

Mason, H. S. (1957). Mechanisms of oxygen metabolism. *Advances in Enzymology and Related Subjects of Biochemistry, 19*, 79–233.

Mayevsky, A. (1984). Brain NADH redox state monitored in vivo by fiber optic surface fluorometry. *Brain Research Reviews, 7*, 49–68.

Mayevsky, A. (1992). Cerebral blood flow and brain mitochondrial redox state responses to various perturbations in gerbils. *Adv. Exp. Med. Biol., 317*, 707–716.

Mayevsky, A. (2015). *Mitochondrial Function In Vivo Evaluated by NADH Fluorescence*. Springer International Switzerland.

Mayevsky, A., & Barbiro-Michaely, E. (2013a). Shedding light on mitochondrial function by real time monitoring of NADH fluorescence: I. Basic methodology and animal studies. *Journal of Clinical Monitoring and Computing, 27*, 1–34.

Mayevsky, A., & Barbiro-Michaely, E. (2013b). Shedding light on mitochondrial function by real time monitoring of NADH fluorescence: II: Human studies. *Journal of Clinical Monitoring and Computing, 27*, 125–145.

Mayevsky, A., & Chance, B. (1973). In H. I. Bicher & D. F. Bruley (Eds.), *A new long-term method for the measurement of NADH fluorescence in intact rat brain with implanted cannula* (pp. 239–244). Plenum Press.

Mayevsky, A., & Meirovithz, E. (2020). Oxygen Homeostasis and Mitochondrial Function in COVID-19 and ARDS Patients: The Ultimate Vital Signs. *Journal of Bioscience and Bioengineering, 1*, 1–24.

Mayevsky, A., & Rogatsky, G. (2007). Mitochondrial function in vivo evaluated by NADH fluorescence: From animal models to human studies. *American Journal of Physiology. Cell Physiology, 292*, C615–C640.

Mayevsky, A., Sjoquist, B., Fri, C. G., Samuel, D., & Sedvall, G. (1973). Incorporation of ^{18}O into homovanillic acid of rat brain during exposure to oxygen 18 containing atmospheres. *Biochemical and Biophysical Research Communications, 51*, 746–755.

Mayevsky, A., Kutai-Asis, H., & Tolmasov, M. (2020a). Mitochondrial function and brain Metabolic Score (BMS) in ischemic Stroke: Evaluation of "neuroprotectants" safety and efficacy. *Mitochondrion, 50*, 170–194.

Mayevsky, A., Meirovithz, E., Livnat, A., Tolmasov, M., Kutai-Asis, H., & Mandelbaum, M. (2020b). Urethral Tissue Metabolic Score (TMS) as a Surrogate Marker of Brain Oxygen Balance in Stroke. *ARDS and Critical Care. EC Emergency Medicine and Critical care, SI.01*, 05–40.

Miller, F. A. (1987). *Joseph Priestley, preeminent amateur chemist* (p. 64). J. Chem.

Palmer, W. (2006). *Carl Wilhelm Scheele (1742–1786): The life of a great chemist, IPS-USA-2006 Internet conference on advances in the Internet*. Processing.

Partington, J. R. (1962). The discovery of oxygen. *Journal of Chemical Education, 39*, 123–125.

Pereira, M. M., Santana, M., & Teixeira, M. (2001). A novel scenario for the evolution of haem-copper oxygen reductases. *Biochimica et Biophysica Acta, 1505*, 185–208.

Rampil, I. J., Litt, L., & Mayevsky, A. (1992). Correlated, simultaneous, multiple-wavelength optical monitoring in vivo of localized cerebrocortical NADH and brain microvessel hemoglobin oxygen saturation. *Journal of Clinical Monitoring, 8*, 216–225.

Rosenthal, M., & Somjen, G. (1973). Spreading depression, sustained potential shifts, and metabolic activity of cerebral cortex of cats. *Journal of Neurophysiology, 36*, 739–749.

Roy, C. S., & Sherrington, C. S. (1890). On the regulation of the blood-supply of the brain. *The Journal of Physiology, 11*(85–158), 117.

Severinghaus, J. W. (2002). Priestley, the furious free thinker of the enlightenment, and Scheele, the taciturn apothecary of Uppsala. *Acta Anaesthesiologica Scandinavica, 46*, 2–9.

Severinghaus, J. W. (2003). Fire-air and dephlogistication. Revisionisms of oxygen's discovery. *Adv. Exp. Med. Biol., 543*, 7–19.

Severinghaus, J. W. (2016a). Eight sages over five centuries share oxygen's discovery. *Advances in Physiology Education, 40*, 370–376.

Severinghaus, J. W. (2016b). The most important discovery of science. In C. E. Elwell, T. S. Leung, & D. K. Harrison (Eds.), *Oxygen transport to tissue XXXVII. Advances in experimental medicine and biology* (pp. 1–16).

Smeaton, W. A. (1963). New Light on Lavoisier: The Research of the Last Ten Years. *History of Science, 2*, 51–69.
Smeaton, W. A. (1992). Carl Wilhelm Scheele (1742–1786): provincial Swedish pharmacist and world-famous chemist. *Endeavour, 16*, 128–131.
Somjen, G. G. (2005). Aristides Leão's discovery of cortical spreading depression. *Journal of Neurophysiology, 94*, 2–4.
Stansfield, D. A., & Stansfield, R. G. (1986). Dr. Thomas Beddoes and James Watt: preparatory work 1794–96 for the Bristol Pneumatic Institute. *Medical History, 30*, 276–302.
Taylor, G. (2019). The Chemical Works of Carl Wilhelm Scheele. *Ambix, 66*, 90–92.
Teive, H. A., Kowacs, P. A., Maranhao Filho, P., Piovesan, E. J., & Werneck, L. C. (2005). Leão's cortical spreading depression: from experimental "artifact" to physiological principle. *Neurology, 65*, 1455–1459.
Theorell, H., & Bonnichsen, R. (1951). Studies on liver alcohol dehydrogenase I. Equilibria and initial reaction velocities. *Acta Chemica Scandinavica, 5*, 1105–1126.
Theorell, H., & Chance, B. (1951). Studies on liver alcohol dehydrogenase II. The kinetics of the compound of horse liver alcohol dehydrogenase and reduced diphosphopyridine nucleotide. *Acta Chemica Scandinavica, 5*, 1127–1144.
Trayhurn, P. (2017). Oxygen - the forgotten nutrient. *Journal of Nutritional Science, 6*, e47.
Trayhurn, P. (2019). Oxygen-A Critical, but Overlooked. *Nutri-Score Front-of-Pack Nutrition, 6*, 10.
Tschirgi, R. D., Inanaga, K., Taylor, J. L., Walker, R. M., & Sonnenschein, R. R. (1957). Changes in cortical pH and blood flow accompanying spreading cortical depression and convulsion. *The American Journal of Physiology, 190*, 557–562.
Van Harreveld, A., & Ochs, S. (1957a). Electrical and vascular concomitants of spreading depression. *The American Journal of Physiology, 189*, 159–166.
Van Harreveld, A., & Ochs, S. (1957b). Electrical and vascular concomitants of spreading depression. *The American Journal of Physiology, 189*, 180–192.
Van Harreveld, A., & Stamm, J. S. (1951). On the conditions for the recording of Leão's spreading depression. *Electroencephalography and Clinical Neurophysiology, 3*, 323–328.
Van Harreveld, A., & Stamm, J. S. (1952). Vascular concomitants of spreading cortical depression. *Journal of Neurophysiology, 15*, 487–496.
Vatov, L., Kizner, Z., Ruppin, E., Meilin, S., Manor, T., & Mayevsky, A. (2006). Modeling brain energy metabolism and function: A multiparametric monitoring approach. *Bulletin of Mathematical Biology, 68*, 275–291.
Waltemath, C. L. (1970). Oxygen, uptake, transport, and tissue utilization. *Anesthesia and Analgesia, 49*, 184–203.
Warburg, O. (1949). Heavy Metal Prosthetic Groups and Enzyme Action. In T. A. Lawson (Ed.), *Clarendon Press*. Oxford.
Warburg, O., Christian, W., & Griese, A. (1935). Wasserstoff-Ubertragendes Co-ferment, seine Zusamensetzung und Wirkungsweise. *Biochemische Zeitschrift, 282*, 157.
West, J. B. (2008). Ibn al-Nafis, the pulmonary circulation, and the Islamic Golden Age. *Journal of Applied Physiology, 105*, 1877–1880.
West, J. B. (2014). Carl Wilhelm Scheele, the discoverer of oxygen, and a very productive chemist. *American Journal of Physiology. Lung Cellular and Molecular Physiology, 307*, L811–L816.
Wikstrom, M., Krab, K., & Sharma, V. (2018). Oxygen activation and energy conservation by Cytochrome c oxidase. *Chemical Reviews, 118*, 2469–2490.
Wilkinson, D. J. (2004). Resuscitation Great. The contributions of Lavoisier, Scheele and Priestley to the early understanding of respiratory physiology in the Eighteenth Century. *Resuscitation, 61*, 249–255.
Williams, K. R. (2003). The Discovery of Oxygen and Other Priestley Matters. *Journal of Chemical Education, 80*, 1129–1131.

Chapter 2
Basic Principles and Technology of NADH and Multiparametric Monitoring Used in Studying Cortical Spreading Depression (CSD)

Abstract The basic principles and technology of NADH monitoring were discussed in more detail in the book published in 2015 (Mayevsky A (2015) Mitochondrial function in vivo evaluated by NADH fluorescence. Springer). Nevertheless, in this chapter the more specific details of NADH monitoring during cortical spreading depression (CSD) are presented. In addition to the monitoring of mitochondrial NADH, this chapter will present the development of the multiparametric monitoring system that provides real-time information on the cerebral hemodynamic, ionic, and electrical activities during CSD. The various combinations of multiparametric assemblies will be discussed in detail.

Keywords NADH fluorometry · Ionic homeostasis monitoring · Brain oxygen level monitoring · Hyperbaric brain oxygenation · Oxygen toxicity · Multisite monitoring of NADH

2.1 Introduction

As described in the front matter, Fig. 1.22, NADH could be measured by utilizing its absorption spectrum in the UV range, or by the blue fluorescence spectrum emitted under UV illumination. In the early stages, NADH monitoring was based on the difference in the absorption of NADH and NAD^+. At the range of 320 nm to 380 nm, only the reduced form—NADH absorbs light, while NAD^+ does not (Fig. 1.22B). Therefore, when a mixture of NADH and NAD^+ is illuminated in a cuvette by 320–380 nm, only NADH will affect the absorption spectrum peak at 340 nm. The absorption approach is not practical for measuring NADH in a thick tissue; hence, another property of NADH was used. Since the early 1950s, fluorescence spectrophotometry of NADH has been employed in various in vitro and in vivo models. The emission of NADH fluorescence, under illumination at 320–380 nm, has a very wide spectrum (420–480) with a peak at 450–460 nm (Fig. 1.22C). NADH

A. Mayevsky, J. Sonn, *Cortical Spreading Depression of Leao*,
https://doi.org/10.1007/978-3-031-08068-5_2

fluorescence has been identified by Chance and his collaborators as a good indicator of the intramitochondrial oxidation–reduction state (Fig. 1.22D) (Chance et al., 1973).

Different groups have developed a number of fluorometers adapted to their specific experimental needs and protocols. Most of the groups using NADH fluorometers constructed the devices in house, since no suitable commercial products are available.

The basic features of NADH fluorometers used consist of the following components:

1. A light source unit (including appropriate filters).
2. An optical path to the preparation and back to the detection unit.
3. Detection and signal processing units.
4. Signal recording and storage units.

In our earlier review published in 1984 and in 2007 and the book published in 2015, we extensively specified the light-guide-based fluorometry used in our studies (Mayevsky, 1984).

In the present chapter, we will provide the reader with extensive reference information, gathered during the years, on the construction of fluorometers. Priority will be given to articles mainly dealing with NADH monitoring in vivo, though in vitro fluorometer types will be also considered. Duysens and Amesz (1957) schematized the first fluorescence spectrophotometer used for intact cells. They utilized the "classical" light source—the mercury arc—providing a very strong band at 366 nm, even though not at the maximal NADH absorption peak of NADH (340 nm).

In 1959, Chance and Legallias (1959) described a differential fluorometer that heralded a new era in monitoring NADH fluorescence in vivo as an indicator of mitochondrial function. They used a microscope, serving as the fluorometer basis, with two light sources: tungsten and mercury lamps with appropriate filters. In 1959, Chance and Jobsis (1959) and Chance (1959) proved that mechanical muscle activity is associated with NADH oxidation measured in excised muscle.

The first in vivo NADH monitoring device was presented in the late 1950s and early 1960s. At that stage, the effects of scattered light and tissue absorption due to blood were not taken into consideration when monitoring NADH fluorescence. The first detailed results of in vivo NADH fluorescence measurements were published in 1962 (Chance et al., 1962a).

These "classical" papers described two microfluorometers that were modifications of previous designs (Chance & Thorell, 1959; Duysens & Amesz, 1957). The same instrumentation was used in other in vivo studies, including those of Chance's group (Chance, 1966; Chance et al., 1962b, 1963, 1965, 1970, 1971; Chance & Schoener, 1965), Kobayashi et al. (1971), Dora and Kovach's group (Dora, 1984; Dora & Kovach, 1982), Rievich's group (Ginsberg et al., 1976; Harbig et al., 1976), Jobsis and collaborators (Jobsis, 1964; Jobsis & Duffield, 1967; Jobsis et al., 1971a, b; Rosenthal et al., 1976; Rosenthal & Jobsis, 1971), Schuette et al. (1976), Gosalvez et al. (1972a, b), Anderson (1975, 1978) and Sundt & Anderson (1975). A new model of time-sharing spectrofluorometer and fluorometer was developed at the Johenson Research Foundation in 1971 (Mayer et al., 1971).

The effect of blood on NADH fluorescence was discussed early by Chance et al. The monitoring of the second parameter in tissue fluorometry in vivo was reported by Chance et al. (1963). They showed that "changes due to the deoxygenation of oxyhemoglobin do not interfere with measurement of the time course of fluorescence changes in the tissue studies."

The addition of a second monitoring signal, namely, tissue reflectance at the excitation wavelength was reported in 1968 by Jobsis and Stainsby (1968). It was based on a previous model described by Jobsis et al. (1966). In another two papers by Jobsis and collaborators (Jobsis et al., 1971a, b), the measurement of 366 nm reflectance was used for the correction of the NADH fluorescence signal from the brain (Gyulai et al., 1982).

2.2 From NADH Monitoring to Multiparametric Monitoring Systems

During the years 1973–2003, various NADH monitoring systems that were used in studying CSD were developed in our laboratory. During those years, we developed and used 10 different monitoring systems (Table 2.1) that were used in studying the effects of CSD on the metabolic, hemodynamic, ionic, and electrical activities in the brain in vivo, in real time.

In this Chapter, we will describe, in detail, the technical aspects of each monitoring system including a typical recording representing the effect of CSD on the monitored parameters.

2.2.1 Fiber-Optic-Based Fluorometer/Reflectometer

In order to enable the monitoring of brain NADH fluorescence in anesthetized or unanesthetized animals, a flexible means was needed to connect the fluorometer with the tested brain. This was achieved in 1972, when UV-transmitting quartz fibers became available (Schott Jena Glass, Germany). We have used the light-guide-based fluorometer for in vivo monitoring of the brain (Chance et al., 1973; Mayevsky & Chance, 1973b) subjected to anoxia or CSD. The historical development of light-guide-based fluorometry reflectometry is shown in Fig. 2.1. The original device functioned on the time-sharing principle (Fig. 2.1A), where 4 filters were placed in front of a 2 arms light guide. Filters 1 and 3 enabled the measurement of NADH fluorescence, while filters 2 and 4 were used to measure tissue reflectance at the excitation wavelength. The reflectance trace was used to correct the NADH signal for hemodynamic artifacts and to indicate relative changes in the blood

Table 2.1 Milestones in the development of brain multiparametric monitoring of NADH fluorescence and other physiological parameters in vivo during CSD by *Mayevsky* et al

Protocol	Year	Discovery/ Activity	Reference
1	1973	The first fiber-optic-based time-sharing fluorometer–reflectometer used in the brain of an anesthetized animal for the monitoring of NADH redox state	(Mayevsky & Chance, 1973a)
2	1974	Simultaneous monitoring of NADH by DC fluorometer reflectometer together with monitoring of systemic blood pressure and electrocortical activity (ECoG)	(Chance et al., 1973; Mayevsky & Chance, 1973b)(157)
3	1974	Simultaneous monitoring of NADH in vivo together with extracellular K^+, DC steady potential, and ECoG	(Mayevsky et al., 1974b)
4	1980	Monitoring of brain NADH together with tissue pO_2 and ECoG	(Mayevsky et al., 1980b)
5	1982	The first multiparametric assembly for NADH, extracellular K^+, H^+, DC steady potential, and ECoG	(Friedli et al., 1982)
6	1983	Monitoring of NADH, pO2, extracellular K+, DC, and ECoG, outside and inside hyperbaric chamber	(Mayevsky, 1983)
7	1990	Simultaneous real-time monitoring of brain NADH, HbO_2, ECoG, DC potential, extracellular K^+ and Ca^{+2}	(Mayevsky et al., 1990b, 1992)
8	1992	Simultaneous real-time monitoring of brain NADH, CBF, ECoG, DC potential, extracellular K^+ and Ca^{+2}	(Mayevsky, 1992)
9	1995	Simultaneous monitoring of brain NADH, CBF, ECoG, DC potential, extracellular K^+, Ca^{+2}, H^+	(Mayevsky et al., 1995)
10	1996	Multiparametric monitoring of neurosurgical patients	(Mayevsky et al., 1996)
11	2000	Monitoring of the mechanism of CSD propagation	(Sonn & Mayevsky, 2000)
12	2003	Multisite monitoring of NADH and DC potential in rats under various conditions	(Rogatsky et al., 2003)
13	2008	Multisite multiparametric monitoring of NADH, CBF, and DC potential in rats under experimental conditions	(Manor et al., 2008)

volume of the sampled tissue. In this original system, only one photomultiplier tube was used for the detection of the two signals. Figure 2.1B presents one of the first in vivo brain monitoring time-sharing setups, connected to the brain of an anesthetized rat (Mayevsky & Chance, 1976) using a special light guide holder developed. The plexiglass cannula which fits the common part of the light pipe is shown in Fig. 2.1A_1. The thread in the bottom of the cannula enables the cannula to be screwed into the skull and to make a better connection with the acrylic cement. A 0.2-mm step holds the light pipe in at a constant distance from the brain.

The transition from normoxia to anoxia, seen in Fig. 2.1C, is slow because the oxygen tension is reduced from the normoxic value to a very low value by a combination of washout of the oxygen and tissue oxygen consumption. The return from anoxia to normoxia is very rapid, since the "front" of oxygen concentration in the freshly oxygenated hemoglobin is high, and the very small span of mitochondrial oxidation–reduction states. The conditions of this experiment are very similar to

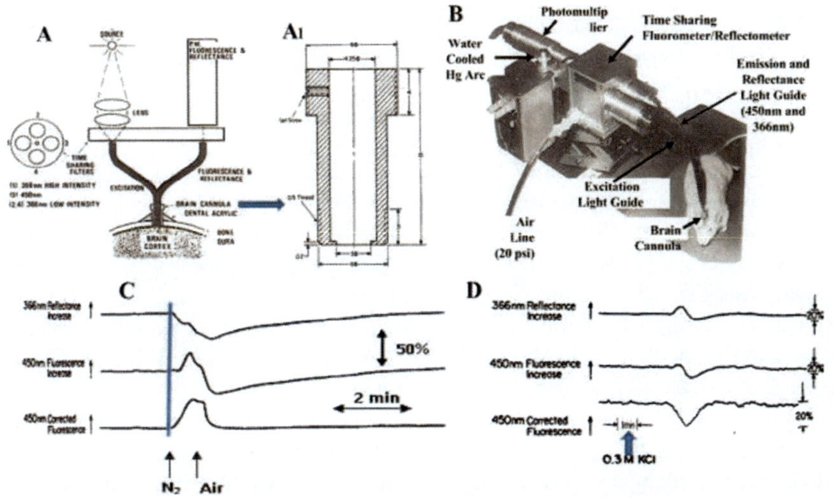

Fig. 2.1 The structure of the first type of fiber-optic-based fluorometers developed. In part (**A**) (Mayevsky, 1984), the time-sharing device is presented, and in part (**B**) (Mayevsky & Chance, 1976), the picture of the device is presented. Part A_1 shows the first cannula constructed to hold the fiber-optic probe and is cemented to the skull (Mayevsky & Chance, 1973b). In the picture seen in part B, the fiber-optic probe is connected to the skull of a small animal. (**C**)The effect of anoxia is shown in the 3 traces using a 2-mm-diameter probe (Mayevsky, 1984). (**D**) Typical responses to CSD are presented. (Mayevsky & Chance, 1974)

those employed by Chance and Schoener (1962) whose uncorrected fluorescence traces were also taken in later phases of the anoxic response where reflectance artifacts are negligible.

Preliminary experiments have been done in order to correlate the redox state of NADH with the level of electrical activity of the brain. In these experiments, we perfused a small area of the dura (2 mm diameter) with 0.4 KCl solution. The distance from the perfused area to the NADH fluorescence measurement was 2 mm. A typical response to CSD is shown in Fig. 2.1D. The animal was breathing normal air during this part of the experiment, so the oxygen supply was in the normoxic range. The change in the reflectance and the NADH fluorescence appears 30 to 40 seconds after the application of the KCl solution. From the changes in the reflected light, we assume that the response to CSD is biphasic; in the beginning, there is a short increase in reflectance, followed by a longer decrease. The large reflectance changes introducc an artifact into the fluorescence measurement, and the corrected trace shows a small NADH oxidation. In some animals where the reflectance changes were smaller, we saw an NADH oxidation even in the uncorrected fluorescence trace.

2.2.2 *Simultaneous Monitoring of NADH by DC Fluorometer Reflectometer Together with Monitoring of Systemic Blood Pressure and Electrocortical Activity (ECoG)*

In order to simplify the monitoring system, the time-sharing approach (AC mode) was replaced by splitting the light emitted from the tissue into two unequal fractions for the measurement of fluorescence and reflectance signals. This model, named the DC type fluorometer, had originally a 3-way light guide which was later replaced by a two arms light guide probe (Fig. 2.2A). In the 2 configurations, the reflectance signal was used for the correction of the fluorescence signal. The model shown in Fig. 2.2A was used to study the brain and other organs as well (Chance & Schoener, 1962).

Our group developed and used the model shown in Fig. 2.2A in the late 1970s. This model is still being used in our laboratory to monitor the brain, heart (Acad et al., 1986; Kedem et al., 1981; Mayevsky & Chance, 1976), liver, kidney, as also in multisite or multiorgan monitoring.

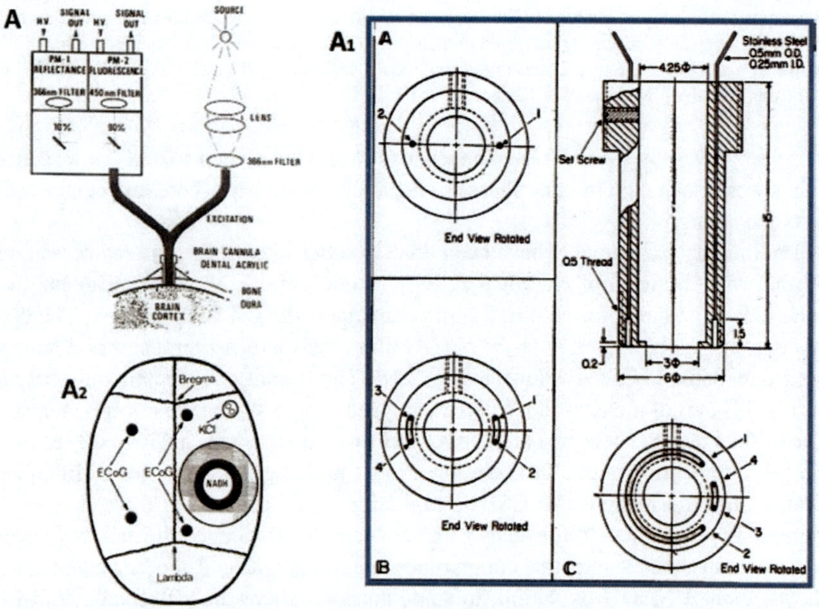

Fig. 2.2 (**A**) standard DC fluorometer/reflectometer developed after using the time-sharing model for few years. (*Ex* Excitation, *Em* emission optical fibers for the monitoring of NADH redox state. *H.V.* high voltage, *PM* photomultiplier) (Mayevsky & Chance, 1975). (**A_1**) The construction of the light guide holders (cannulas) for measurement of NADH fluorescence from the surface of the brain of an awake rat in experiments in which topical application of drugs is needed (Mayevsky & Chance, 1975). (**A_2**) Schematic representation of location of cannulas and ECoG electrodes on rat's skull. (Mayevsky & Shaya, 1980)

In order to monitor electrocorticogram—ECoG, we used two approaches as seen in Fig. 2.2A_1 and A_2 dependent on the aims and the protocols used.

Light Guide Holder

Three types of light pipe holder cannula were developed. Figure 2.2A_1 shows the construction of the cannulas. A step of 0.2 mm in the bottom of the cannula held the light pipe in a constant distance from the brain. The thread outside the bottom of this cannula enables screwing it into the skull and also gave a better connection between the cannula and the cement. Type A_1A was used in experiments in which the animals were exposed to nitrogen oxygen-breathing cycles (also ref. Mayevsky & Chance, 1973b) or to hyperbaric pressure of oxygen (Mayevsky et al., 1974a). The EEG was measured between the two electrodes, 1–2. Cannula type A_1B was used in all experiments in which CSD was elicited by the application of KCl solution above the dura (see also ref. Mayevsky & Chance, 1974). The KCI was washed through the tubes 1–2 or 3–4. The EEG was measured between the electrodes 1,2-3,4. The third type, A_1C, has two compartments in the bottom, the small one for KCl application 3–4 and the large compartment, 1–2, where chemicals such as Metrazol were applied and were affecting larger area. A fifth electrode was located 180° to electrodes 3–4, so the EEG was measured at the same time.

In many protocols, the light guide holder seen in Fig. 2.1A_1 was used. The ECoG was measured using 2 stainless screws (about 1 mm in diameter) connected to a short electrical wire that were placed in each hemisphere. Figure 2.2A_2 shows the location of the cannula and the screws for the monitoring of NADH and the ECoG. Blood pressure was measured from the femoral artery using a standard blood pressure transducer connected to a 6-channel Offner "Dynograph" (Backman).

Effects of Anoxia

In these experiments, the rats were slightly anesthetized by Equithesin. The nitrogen was applied via a nasal mask. Figure 2.3A shows the effect of N_2 on the NADH fluorescence, reflectance, EEG, and blood pressure. The top trace shows the reflectance which in all animals decreases during the N_2 cycle. This decrease of reflectance was in two phases. The first decrease was small (in comparison to the second) and occurred while the animal was breathing spontaneously. A second decrease occurred after the animal stopped breathing (SB). The recovery of the reflectance to the baseline occurred about 10 min after the rat started breathing again. The second trace from the top, the fluorescence, shows a large increase in NADH fluorescence during the first phase of the N_2 cycle. In order to correct for hemodynamic artifacts induced by anoxia, we used a correction technique suggested by Jobsis et al. (1971a) and Harbig et al. (1976). The reflectance signal at 366 nm was subtracted from the fluorescence signal at 450 nm at a 1:1 ratio. The difference between the fluorescence and reflectance signals is shown in the third trace, the "corrected" fluorescence. After the cessation of respiration, a large decrease in reflectance occurs; an apparent decrease in fluorescence (oxidation) is observed which is almost undetectable in the corrected trace. The small decrease shown in the corrected trace is due to imperfection of the correction factor in this special animal. After the N_2 administration had been discontinued (SN), artificial respiration (AR) was applied to induce

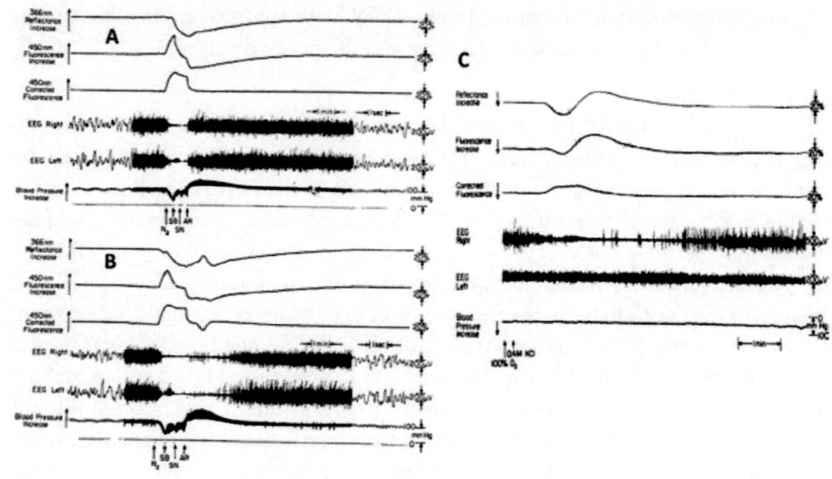

Fig. 2.3 (**A + B**) The effects of anoxia on NADH fluorescence, 366 mm reflectance, EEG, and blood pressure. In part (**A**), the duration of the anoxia was 70 seconds and in part (**B**) 100 sec. (*SB* the animal stopped breathing, *SN* stop nitrogen, *AR* short artificial respiration) (Mayevsky & Chance, 1975). (**C**) The response of the brain to the application of KCl (0.4 M) on the dura surface. The arrow indicates the direction of increase of the signal. (Mayevsky & Chance, 1973b)

spontaneous breathing. After the animal started breathing, a fast decrease of NADH is observed in the uncorrected fluorescence as well as in the corrected. The recovery of the NADH level to the baseline is very fast in comparison with the recovery of the reflectance. The EEG of both hemispheres reaches a very low amplitude when the NADH level reaches 80–90% of the maximum increase during the N_2 cycle. The response of the two hemispheres was identical. The recovery of the EEG follows the NADH recovery to the normoxic level. Figure 2.3B shows the response of the same animal to a longer N_2 cycle. The animal was exposed to N_2 for 100 sec. The response of the systemic blood pressure indicates that due to the lack of oxygen during the anoxia, the pressure decreased until the end of the anoxia. During the recovery stage, the pressure shows a short hypertension episode followed by recovery to normal pressure. The main differences between the two cycles are that after the recovery of the NADH to the normoxic level, a further decrease in NADH occurred (third trace), and at this time, the EEG was depressed and recovered to normal only later. This cycle of oxidation of NADH following the N_2 cycle (very similar to CSD response) was observed in many animals after exposure to a long N_2 cycle. The pattern of changes in reflectance, fluorescence, and the corrected traces was similar to those observed in the CSD elicited by KCl as shown in Fig. 2.3C and also in Fig. 1.27. CSD was induced by exposing the tissue to high KCl concentration (0.4 M). In this study, we used a recorder with reverse polarity and the increase in signals is downwards and not upwards. The NADH levels were decreased in parallel to the depression of the EEG in the right hemisphere. The EEG in the left hemisphere remains stable as seen also in the blood pressure trace. Those results suggest that the CSD event remains only in the stimulated hemisphere.

2.2.3 Simultaneous Monitoring of NADH Together with Extracellular K^+, DC Steady Potential, and ECoG

According to Leão (1944a) and our early studies on CSD (Mayevsky & Chance, 1973b, 1974), it was clear that energy metabolism is stimulated during the passage of a CSD wave in the cerebral cortex. As shown in Fig. 1.16A and B, Bures and Krivanek (1960) showed that during a passage of a CSD wave in the cerebral cortex, a leak of potassium was recorded. In 1972, Vyskocil et al. used a potassium selective microelectrode and showed the leakage of K^+ from the cells during the wave of CSD (Vyskocil et al., 1972). In order to correlate the responses of NADH and ECoG to changes in extracellular K^+, we adopted two approaches. Initially, we used a double-barreled microelectrode (Zeuthen, 1971; Zeuthen et al., 1974), seen in Fig. 2.4A, inserted 1 mm into the cerebral cortex of a rat (Mayevsky, et al., 1974b). This technology was based on the development of potassium-sensitive ion-exchanger (Walker, 1971) filled in one barrel, while the other barrel was filled with 2 M KCl for the measurement of DC steady potential. Typical recording of the response to CSD presented in Fig. 2.4B was made with a double-barreled microelectrode in the cortical tissue. The waves recorded with the microelectrode were more sharply defined, with fast rise times, fast signal-decay times, and a short duration.

The measurement of extracellular K^+ was combined with that of the NADH and ECoG and the responses are seen in Fig. 2.4C (Mayevsky, et al., 1974b). A CSD wave was initiated by passing a solution 0.6 M KCI epidurally through the cannula. We repeated the same experiment in 5 different animals and the results were almost the same. Three CSD cycles were developed and then stopped by washing the dura with 0.85% NaCl. The first cycle was recorded 2 min after the application of KCI. The DC potential showed a negative shift followed by a plateau and a slower recovery phase which had a small positive wave. The extracellular potassium concentration increased very quickly, and after a plateau phase, it decreased slowly to the original level. The ECoG showed depression responses and recovered to the normal pattern 10–15 min after washing the dura with NaCI. The 366 nm reflectance, NADH fluorescence, and the corrected fluorescence show the response to CSD as described previously (Mayevsky & Chance, 1974). The corrected fluorescence trace shows a decrease of 25% in NADH in comparison with the normoxic level. This decrease in NADH which represents an oxidation (25%) is identical in the three cycles. The first cycle was longer but the level of oxidation was the same. In our present experiments, a larger increase in reflectance was observed in most animals tested but the subsequent decrease in the reflectance response described previously was either smaller or obscure. Since the size of the light guide (0.3 mm) and the microelectrode (~1 μm) are not the same and the fields they sample are not identical, it is difficult to correlate the onset of changes in the NADH and K^+. However, the duration of the K^+ cycle and the NADH oxidation cycle is the same. In each set of cycles elicited by KCl, the first cycle was longer than the following cycles which were identical one to the other. The main differences between the

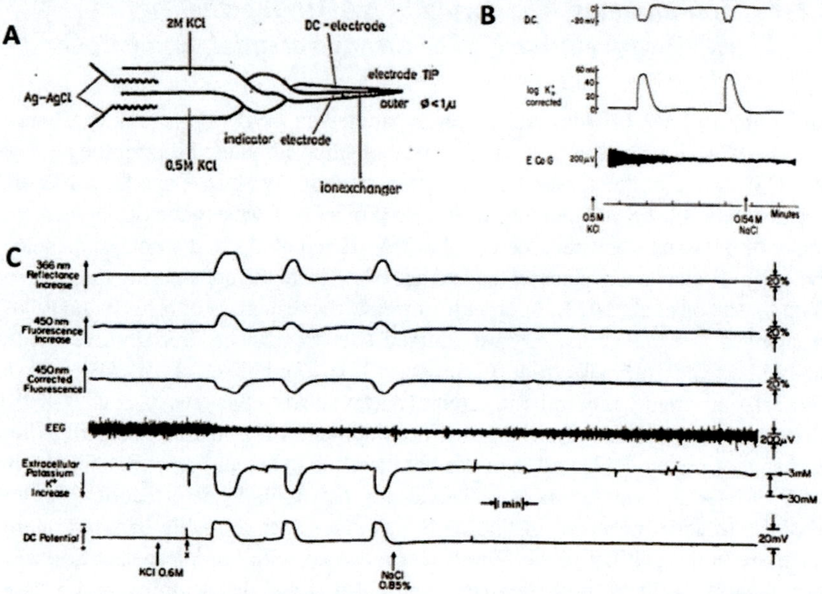

Fig. 2.4 **(A)** Schematic presentation of the microelectrodes used in order to measure brain extracellular K^+ and DC steady potential in a rat model. Reprinted with permission by the first author T. Zeuthen et al. (1979). **(B)** Microelectrode recording from brain in CSD using a two-barrel microelectrode and separate ECoG electrodes (Crowe et al., 1977). **(C)**The effect of CSD (elicited by 0.6 M KCI) on the extracellular potassium, DC potential, ECoG, and NADH fluorescence of the exposed rat brain cortex. (Mayevsky et al., 1974b)

results obtained with the two light guides are the size of the reflectance response and the duration of the CSD cycle which is twice as long as using the larger light guide. In one animal, we cemented the microelectrode to the skull and the animal was released from the head holder and allowed to become awake. The duration of the K^+ cycle in the awake animal was shorter than when the same animal was anesthetized.

In order to avoid damage to the brain tissue by inserting the microelectrode to the cortex, we decided to develop a K^+-sensitive surface electrode based on impregnated polymer membrane. All technical details were given in our published paper (Crowe et al., 1977). In the experiments described here, a Plexiglas cannula containing the combination electrode and having a pair of ECoG leads incorporated in the walls (see Fig. 2.5A and B) was fitted in place to make a tight fit with the skull and communicate with the cortex through a layer of cerebrospinal fluid. The cannula was sealed in place using dental acrylic cement. A second skull cannula (2 mm OD), this containing two stainless-steel tubes, was implanted in the skull above the dura mater at a distance 3 mm from the electrode (Fig. 2.5A_1 and A_2). This cannula was connected through one line of polyethylene tubing to a reservoir filled with 0.3 M KCl solution and through the other line to a syringe or a Harvard infusion/

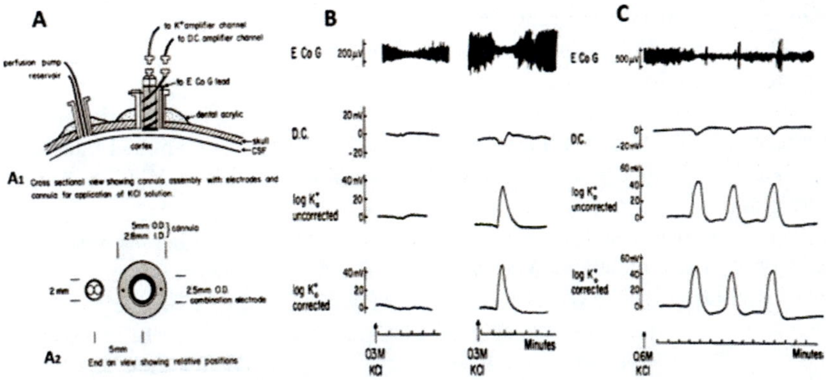

Fig. 2.5 **(A)** Representation of combination electrode ECoG leads and cannulas for four-channel recording from surface of brain. **(B)** Recording from brain using surface combination electrode before (Left side) and after (right side) removal of dura mater. **(C)** Recording made with a combination surface electrode during CSD. (Crowe et al., 1977)

withdrawal pump, allowing KCl to be drawn over the intact dura mater above the cortex to initiate CSD in the cortex (Bures et al., 1974). In some cases, a higher concentration of KCl solution was needed to elicit CSD.

In three recordings, measurements were attempted with the dura mater intact to determine whether K^+ freely diffuses through this membrane. There would be a marked improvement in the animal preparation if the dura could be left in place. However, as Fig. 2.5B (left side) indicates that the K^+ signal is severely attenuated by the dura, Fig. 2.5B (right side) shows the results immediately after removing the dura and measuring the same area of the brain. Typical four-channel recording is shown in Fig. 2.5C. The top line shows the ECoG as recorded from the two stainless steel leads in the wall of the support cannula. Depression of this trace is one of the indications that CSD has been elicited. The ECoG leads are positioned to be on a line roughly parallel to the advancing wave front of the CSD wave. The second and third traces give the outputs from the combination electrode. The DC electrode shows the characteristic surface negativity as the CSD wave passes.

A combination surface electrode was used to provide continuous measurements of potassium activity and slow wave surface potential. The active elements of the electrode were a valinomycin impregnated polymer membrane for measuring K^+ and a Ag/AgCl wire with salt bridge for measuring slow wave potential. An electronic subtraction was done to provide a measure of K^+ activity corrected for slow wave electrical changes.

Potassium electrodes were calibrated in isotonic solutions containing varying proportion of NaCl and KCl so that K^+_e values could be interpreted directly in units of concentration. Crowe et al. (1977) have discussed in detail the preparation and use of these electrodes.

Electrode signals were measured on voltage amplifiers (gain set at 1) with input impedance greater than 10^{13} ohms. Recordings were made on a Beckman Offner

Dynagraph 8 channel recorder. ECoG, recorded differentially from bipolar leads (stainless steel, 0.5 mm diameter), was amplified with a Beckman EEG preamplifier module and amplifier module.

The energy state of the brain was monitored by measuring the oxidation–reduction state of NADH by surface fluorometry. The excitation and emitted light from the fluorometer to the brain and back was transmitted via a Y-shape flexible light guide.

The fluorometer used in our experiments was a "DC type"; this instrument has been described in detail by Mayevsky (1978c). A schematic presentation of the special cannulae used in our experiments is shown in Fig. 2.6A. The localization of the various cannulae implanted above the brain is shown in Fig. 2.6B. In most gerbils used in this study, one combination electrode for K^+/DC and one light guide were implanted on the right hemisphere.

The anesthetic used was Equithesin (Jensen-Salsberg Laboratories, USA). Each animal was given 0.3 ml/100gr by intraperitoneal injections to provide sufficient anesthesia for the implantation of the cannulae. Holes were drilled in the calvarium to accommodate the cannulae, and the dura mater was removed at the site of K^+ measurement. The combination electrode/light pipe was located in place with micromanipulator and cemented to the calvarium together with the other cannula with dental acrylic cement. During the experiment, each gerbil was kept minimally anesthetized by injection of small amounts (0.03–0.05 ml) of Equithesin every 30 min.

Figure 2.6C shows a typical response of a normal brain to CSD elicited by 0.5 M KCl solution applied epidurally with a push-pull cannula. The response of the DC potential was very small so that K^+_e-uncorrected and K^+_e-corrected traces were quite similar. K^+_e increased quickly, reaching its maximum level within 10–20 sec. The recovery of the K^+_e took about 2 min and simultaneously the NADH trace (CFlu) showed a typical oxidation cycle. In this animal, the reflected light trace showed a decrease that indicates an increase of blood volume in the tissue. The ECoG trace showed a typical depression during CSD.

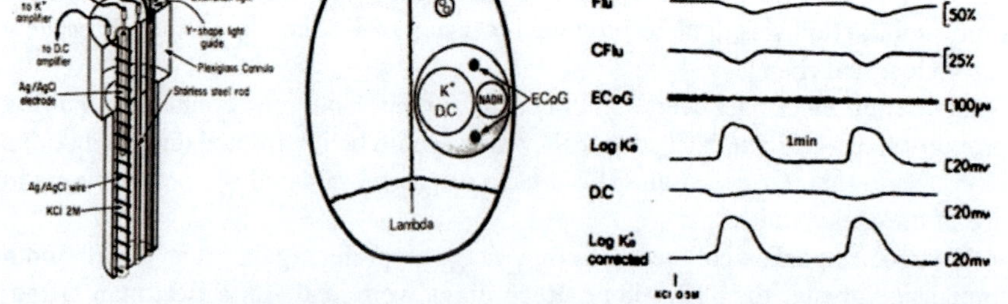

Fig. 2.6 (A) Schematic presentation of the longitudinal section of the combined cannula and (B) a topical view of the skull of a gerbil implanted with cannulae and electrodes. (C) The effects CSD on the metabolic, electrical, and extracellular K^+e activity in the gerbil. (Mayevsky et al., 1980a)

2.2.4 Monitoring of Brain NADH Together with Tissue pO_2 and ECoG

When O_2 consumption was evaluated, Leão and Morison (1945) found a decreased O_2 concentration in blood samples taken from the sagittal sinus during CSD. Van Harreveld and Stamm (1952) measured a decreased pO_2 in the depressed region. The same fall in pO_2, during CSD, was also described by Marshall (1959) and Lukyanov and Bures (1967). The described hemodynamic and pO_2 responses to CSD are indicators of increased metabolic activity during the recovery phase of CSD.

In the present study (Lukyanov & Bures, 1967), we correlated for the first time the changes in NADH fluorescence, changes in the reflected light, and the changes in pO_2 measured with a surface oxygen electrode. The interrelation between the three parameters measured was tested under various conditions, such as anoxia, hypoxia, CSD, anesthesia, and ischemia.

Male gerbils (60–70 gm) were used in our study. The animal was anesthetized by intraperitoneal injection of Equithesin (a mixture of pentobarbital, chloral hydrate, magnesium sulfate) 0.3 ml/100 gm. The skull was exposed by a midline incision and a 5–6-mm-diameter hole was drilled in the parietal bone to fit a combined NADH-pO_2 cannula (Fig. 2.7A). A 2-mm-diameter hole was drilled in the frontal bone to fit a push-pull cannula for epidural application of KCl in order to elicit cortical CSD. Two stainless-steel screws were implanted on the left parietal bone for holding the cement to the skull. The cannula and the screws were cemented to the skull by dental acrylic cement. Oxidation–reduction state of NADH was monitored using a standard DC fluorometer/reflectometer with a 1.5-mm-diameter, Y-shape light guide. The pO_2 was measured by an open-type oxygen electrode 2.5 mm in diameter. It consisted of a platinum wire 25 microns in diameter and a silver wire 250 microns in diameter embedded in epoxy resin and ground so that their surfaces were flush with the glass capillary tube. Polarization potential was 0.80 V. The electrode was dipped in 5% cellulose diacetate to decrease sensitivity to stirring under calibration conditions. ECoG was measured by two stainless-steel rods (0.5 mm in diameter) inserted through the wall of the combined cannula. Recordings were made on a Beckmann multichannel dynograph.

In Part B of Fig. 2.7, two responses to CSD are shown after elicitation by topical application of KCl solution. The reflectance trace (R) shows a typical biphasic response and the corrected fluorescence (CF) shows the oxidation–reduction cycle. The pO_2 trace presents a decreased level of oxygen during the CSD cycles. The ECoG was depressed during the cycles, although the recovery to the original amplitude was very slow and not shown in the figure.

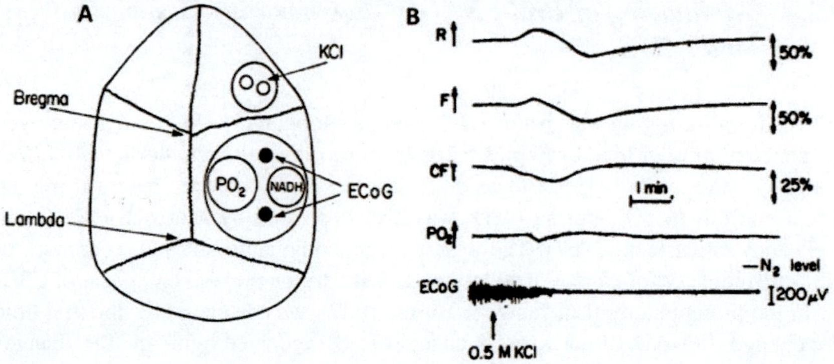

Fig. 2.7 **(A)**—Schematic presentation of the various probe location above the brain of the gerbil. The large cannula contained the light guide for NADH measurement, the PO_2, and ECoG electrodes. The cannula for KCI application was located 2–3 mm anterior to the large cannula. John Wiley and sons, reprinted with permission (Mayevsky et al., 1980b). **(B)**—The effects of CSD on NADH, PO_2, and ECoG in the awake gerbil brain. John Wiley and sons, reprinted with permission. (Mayevsky et al., 1980b)

2.2.5 *Multiparametric Assembly for Recording of NADH, Extracellular K^+, H^+, DC Steady Potential, and ECoG*

Surface probes became available for the monitoring of parameters reflecting normal metabolic and electrical activity of living tissues as well as pathological disturbances in their homeostatic mechanisms. Such probes include ion-selective electrodes for extracellular K^+ activity (Crowe et al., 1977), flat glass pH electrodes, surface pO_2 electrodes, and light guides for monitoring changes in the oxidation–reduction (redox) state of fluorescent components of the respiratory chain. Our aim was to construct a new holding system allowing an easy implantation of surface probes in various combinations on the brain of small animal model.

Ideally, the multiprobe assembly designed for our project had to be rugged enough to withstand routine use by non-specialists and yet be miniaturized enough to fit the limited surface available for implantation on rat and gerbil skulls (about 6–7 mm in diameter on each hemisphere). For ischemic studies, interest was focused in monitoring mean values over large areas, suggesting the use of larger electrodes. In contrast, following propagated events such as CSD would have required the tip of multiprobe assembly to be concentrated in a tiny space seeing the same phase of the wave unless each an auxiliary signal. The second option was retained by recording the wave of the DC potential concentrically to each sensor. In addition, the complete assembly had to provide adequate protection and shielding of the high-impedance ion-sensitive electrodes (K^+ and pH) and a stable enough reference junction for use in unanesthetized animals. With K^+ surface electrodes, concentric DC potential has been proposed as the best approximation for the DC bipotential component to be subtracted from the sensor signal (Crowe et al., 1977). In this study, we

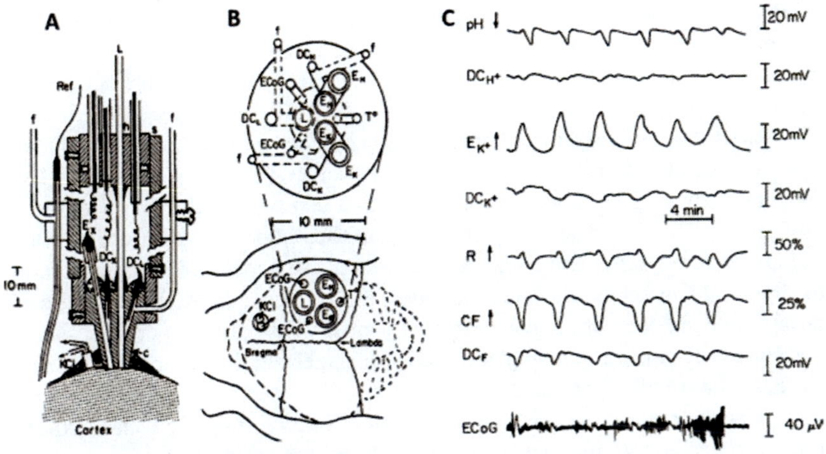

Fig. 2.8 Schematic presentation of multiprobe assembly used in our study. ***Left*** **(A)**: longitudinal section of cannula and assembly located above cortex. A small push-pull cannula was placed near assembly to initiate CSD. ***Lower right*** **(B)**: Topical view of gerbil skull together with location of various electrodes. ***Top right*** **(B)**: Cross section of wide end of cannula. **(C)** Repetitive responses of awake brain to CSD initiated by washing dura with KCl solution (not shown). For abbreviations, see text. (Friedli et al., 1982)

also attempted to check the accuracy of the method by recording DC potential differentially between a central barrel and the peripheral slit and to reevaluate its usefulness in correcting surface K^+ and pH measurements under various conditions.

2.2.5.1 Electrode Assembly

The solution adopted for the electrode holder is basically a modification of the Lucite cannula described by Mayevsky et al. (1980a) for the light guide and the potassium-sensitive electrodes. To offer space for more probes, the new cannula was shaped as a truncated cone instead of a cylinder (Fig. 2.8A). The holes accommodating short electrode probes (K^+, pH) are made convergent toward the lower surface to occupy less space on the brain and divergent at the top to occupy less space on the brain and divergent at the top to facilitate handling and sealing of the probes. An additional holc is drilled obliquely from the upper surface to merge with each sensor channel at about mid-distance of the lower surface. This hole accommodates an AgCl wire used to record the local DC potential concentrically to the sensor.

The long and rigid steel stem of the light guide (L) used in this study occupies a straight vertical hole in the cannula and serves as an axis to hold the cannula (c) and the cable holder (h) at a convenient distance of each other (4 cm). Steel rods preferentially threaded) can be used as additional or replacement pillars to fix the cannula to the cable holder. The arrangement leaves optimal access to the electrodes and electrical connections for assembling and replacement. The complete assembly is protected and shielded by a sliver-pointed Lucite sleeve sliding over the cable holder. If a stronger construction is desired, the sleeve can also be permanently screwed into the cannula and cable holder, with a half-cylinder piece cut out as a removable cover.

When the holder is assembled, the fixed steel pillars are screwed and/or glued in first. To protect the solder point, chloridized silver wires are glued with two-component epoxy glue in short pieces of Teflon that fit over the sensor electrode shaft and into the DC channel in the cannula. The Teflon-sleeved probes (Yellow Springs Instrument temperature probe 511) with paraffin melted by a thermocautery tip. To avoid tension on the electrode wires, they are connected to the input cable through a flexible coil of 36-guage isolated copper wire.

Once the electrodes were in place, the DC channel and the lower part of the sensor channel were filled with fluid from a lateral hold connected to a reservoir syringe. Thus, it can be ensured that each sensor and DC electrode is tightly sealed to the top of the cannula and will not let saline reflow back into the cable housing compartment. The system also allows the extracellular fluid and blood to be cleaned from the slit around each probe after the cannula is removed from an implantation site. In the final assembly (Fig. 2.8A and B), a protective sleeve is used as an anchoring point for the refill tubes to the DC and sensor channels (f) and for the reference liquid junction (Ref in Fig. 2.8A, B) (Crowe et al., 1977).

2.2.5.2 Correction for Local DC Biopotential

In the measurement of electrode transmembrane potential, an additional electromotive force may appear in series in the circuit: the tissue biopotential existing between the ion-selective membrane and the reference junction. With microelectrodes, this potential can be picked up within a few microns of the sensor tip by an adjacent DC glass microelectrode and subtracted from the potential recorded through the sensor. Applying the same correction principle with surface electrodes may introduce an error as the ion-sensing and DC-sensing area are not superimposed. In 1977, Crowe et al. (1977) proposed using the DC potential recorded from a concentric slit around K^+-sensitive electrodes as the best approximation from the DC component to subtract from the sensor potential. Although the peripheral slit sees CSD waves somewhat sooner than later than the central electrode, Crowe et al. (1977) considered that the correction will not be seriously compromised because the width of the CSD wave is as large or larger than the width of the electrode assembly. To reevaluate this assumption, we measured DC potential changes simultaneously from central electrodes and from their concentric slits, replacing the central ion-sensitive electrodes by a glass or polyethylene barrel of the same diameter filled with agar-saline.

2.2.5.3 NADH Fluorescence Measurement

In the present study, we used a one-channel DC fluorometer reflectometer (Chance et al., 1975; Mayevsky, 1978c). The source for the 366 nm excitation light was a 100 W water-cooled mercury arc passing via the excitation fiber-optic bundle to the brain. The emitted light from the brain transmitted through another fiber-optic bundle was split into 90:10 ratios in order to measure the fluorescence light (F; 90%) and the reflected light (R; 10%). The calibration of the signals as well as other details was published previously (Mayevsky, 1978a; Mayevsky et al., 1978, 1980a).

The potential use of the multiprobe assembly is shown in Fig. 2.8C in which the repetitive effect of CSD was tested. The brain was washed with 0.5 M KCl, which remained on the brain. As a result, six CSD cycles were recorded in all parameters measured. The ECoG was depressed during the entire period of the CSD, and toward the end, the beginning of epileptic activity was recorded. The three DC channels also showed cycles of negative shift. The main change in the pH was an alkalinization response. The extracellular K^+ showed the typical increase-decrease cycle, and as a result, the NADH showed the oxidation cycles.

In conclusion, the new multiprobe approach described in the present study opens new possibilities for studying mechanisms of brain pathology in a very detailed manner. We now use this approach in our laboratory to study hyperbaric oxygen toxicity, long-term ischemia, and recovery.

2.2.6 *Monitoring of NADH, pO_2, Extracellular K^+, DC, and ECoG, Outside and Inside Hyperbaric Chamber*

The normal functioning brain requires a continuous supply of blood in order to obtain glucose as well as oxygen. Any change in the oxygenation of the tissue will result in an impairment of the normal function. Most of the oxygen taken up by the brain is consumed by the mitochondria in order to supply the large amount of ATP needed. A major part of the ATP is used by the ATPase system to keep the neurons in the polarized state. Figure 1 in the current book ("Aim of the book") shows in a schematic way the interrelation between the energy production system and the pumping activity going on in the membranes. In order to understand the connection between various events occurring in the brain in vivo, one has to measure as many parameters as possible from the same site.

Using surface probes, we monitored the metabolic, ionic, and electrical activities from a small area of a rat or gerbil brain (6 mm in diameter). The metabolic state and activity were evaluated by monitoring partial pressure of oxygen (pO_2) as well as the intramitochondrial NADH redox state by surface fluorometry. Extracellular K^+ activity was measured by a surface valinomycin electrode. For electrical activities, we measure DC potential and ECoG. Using the multiprobe approach, we

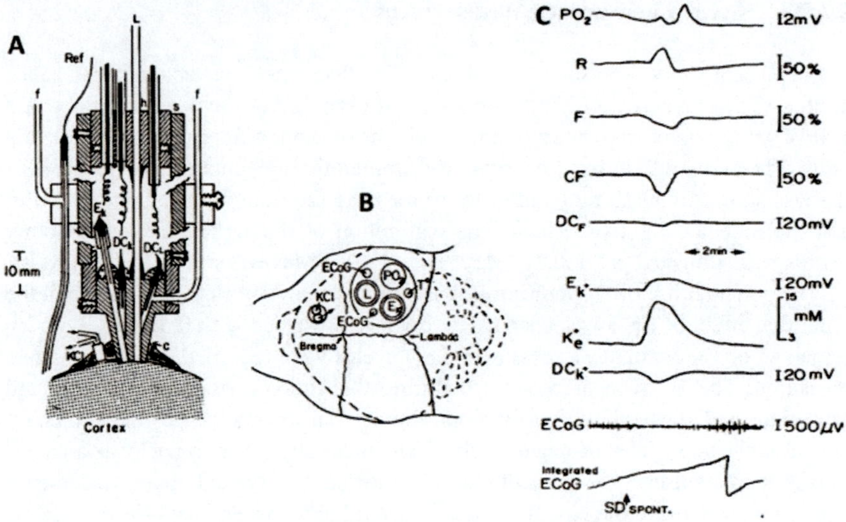

Fig. 2.9 Scheme of the multiprobe assembly (**A**) and its location on the gerbil brain (**B**). All details are in the text (Mayevsky & Sclarsky, 1983). (**C**) The metabolic, ionic, and electrical responses to CSD developed spontaneously. (Mayevsky & Sclarsky, 1983)

exposed the animal to anoxia, ischemia, and CSD and recorded the various responses from the brain.

All these electrodes were held in a multiprobe assembly described in detail elsewhere (Friedli et al., 1982; Mayevsky, 1983; Mayevsky et al., 1986). A general description follows.

The surface probes were mounted in a Lucite holder which can be cemented directly to the skull after removing a bone disc. Figure 2.9A shows the holder or "cannula" (c) shaped as a truncated cone (Crowe et al., 1977). The holes accommodating the specific sensing electrodes converge at the implantation site to occupy less space on the brain, and diverge at the top of the cannula to leave maximum room for mounting and replacing the electrodes. Another hole was drilled obliquely from the top to intersect the first chamber approximately half-way down. This hole houses an Ag/AgCl wire to record DC potential concentrically to the sensing electrodes.

One more hole was drilled horizontally into the chamber of the DC wire for filling (f) this chamber with saline and for cleaning ECF and blood from around the electrodes after removal from the animal. The steel light guide stem (L) and 2 threaded steel rods (not shown) were cemented with epoxy to the cable holder (h) and cannula to keep them 4 cm apart, leaving space for the electrode connections. A silver-painted Lucite sleeve (S) slides over the entire assembly and was held in place with a set of screws. This sleeve holds the assembly together as one unit, protects the electrodes and wiring, and also provides shielding made necessary by the K^+ electrode's high impedance which makes it very sensitive to electrical pick up.

The sensing electrodes, wires for DC measurement, stainless-steel wires for ECoG measurement and a polyethylene sleeved temperature probe (YSI 511) are held in, and sealed to the cannula with 5-minute epoxy, or by melting wax around them. All the sensing wires were attached to amplifier cables through 36-gauge insulated copper wire. Another block of Lucite was attached to the other shield (S) which holds the reference assembly and the tubes for filling the DC chambers (f). The recovery of the brain after an episode of CSD is dependent upon energy availability. In the rat shown in Fig. 2.9C, waves of CSD developed spontaneously at a certain stage of the experiment (after a few hypoxic episodes) and the results of the figure show the regular response to CSD. The efflux of K^+ reached a level of 14–15 mM and was then pumped back into the cells and the NADH (CF) shows a clear oxidation cycle.

2.2.7 *Simultaneous Real-Time Monitoring of Brain NADH, HbO_2, ECoG, DC Potential, and Extracellular K^+ and Ca^{2+}*

The online evaluation of brain physiological and biochemical functions in vivo is expected to contribute to a better understanding of the development of a pathophysiological state. The development and usage of a multiprobe assembly (MPA) for the monitoring of the metabolic, ionic, and electrical activities from the cerebral cortex of experimental animals were the subject of our previous publications (Friedli et al., 1982; Mayevsky et al., 1980b, 1985; Mayevsky & Zarchin, 1987). In the present study, we further extended the MPA to include a second fiber-optic bundle for the monitoring of intracapillary hemoglobin oxygenation (oxy-deoxy Hb) using the EMPHO-I device (Frank et al., 1984a, b, 1989).

As an animal model for the studies, we adopted the Mongolian gerbil (Meriones Unguiculatus) which provides an easy way to produce reversible partial or complete ischemia (Donadio et al., 1982; Levine & Payan, 1966; Levy & Brierley, 1974; Mayevsky, 1978b; Mayevsky & Breuer, 1990). In the same animal, we also tested the effects of graded hypoxia as well as CSD (Leão, 1944a, b, 1947; Marshall, 1959). The effect of CSD wave on brain oxygenation as evaluated by oxy-deoxy Hb spectral analysis together with NADH redox state is described here for the first time.

Multiprobe Assembly The basic concepts of the MPA as previously described (Friedli et al., 1982; Mayevsky et al., 1985; Mayevsky & Zarchin, 1987) were adopted and modified to accommodate more probes in the same surface area of the brain. As shown in Fig. 2.10A, the center hole was used for the NADH light guide which was developed very recently to include fibers for relative CBF measurement by laser Doppler flowmeter (Mayevsky et al., 1991) in addition to NADH fluorescence. In the present study, two light guides were used, namely the center one (L) for NADH and CBF measurements and a microlight guide used for HbO_2 monitoring (Frank et al., 1989; Mayevsky et al., 1992). In order to prevent interference

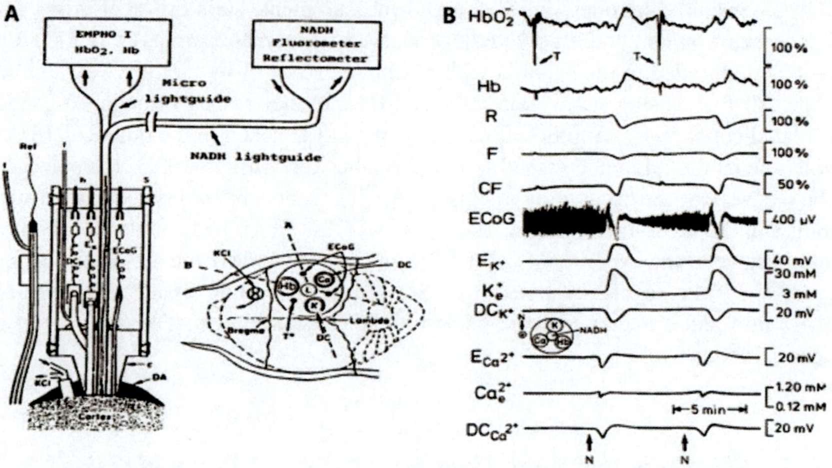

Fig. 2.10 (**A**) Schematic presentation of the multiprobe assembly (MPA) used in the present study. The longitudinal section, at the left, shows the two light guides connected to the EMPHO-I (microlight guide) and the NADH fluorometer reflectometer (NADH light guide). At the right side of the figure, the various probes are shown as located on the gerbil cerebral hemisphere. (*L* NADH light guide, *Hb* EMPHO-I light guide, *K, Ca* extracellular potassium and calcium electrodes, *DC* steady potential electrodes, *T°* temperature probe, *ECoG* electrocorticography electrodes, *KCl* push-pull cannula for topical application of KCl solution; *Ref* reference electrode, DC_k, E_k, *ECoG* cables connecting the various electrodes to the cable holder, *f* filling tubes connected to the reference and the DC electrodes, *C* plexiglass cannula, *DA* dental acrylic cement, *h, s* plexiglass cable holder and sleeve) (Mayevsky et al., 1992). (**B**) Metabolic ionic and electrical responses to CSD induced by needle insertion two millimeters anterior to the MPA. (Mayevsky et al., 1990b)

between the measurements of HbO_2 and CBF, we used them alternatively with the NADH measurements. The longitudinal section of the MPA, shown on the left side of the figure, corresponds to the cross-sectional line A shown on the right side. In order to see the two light guides in the same longitudinal section, a different sectioning line would need to be used (B). Extracellular K^+ and Ca^{2+} electrodes were made by WPI Inc. (Sarasota, FL, USA) using 1.2-mm-diameter PVC tubing (Cole Flex, W, Babylon USA).

The electrodes were assembled in the holder by the same basic approach described previously for the connection of the DC steady potential electrode (Friedli et al., 1982). For the DC electrode, we used the electrode holder (containing Ag/AgCl pellet) provided by WPI Inc. and polyethylene tubing was connected to it and inserted into the holder. The center light guide (NADH) was covered by a rigid brass tube and was glued by a 5-minute epoxy so that after cementation to the cannula (screwed and glued) it served as an axis to hold the cannula (c) and the connector holder (h) at a fixed convenient distance.

NADH Fluorescence Measurement In the present study, we used a single-channel DC fluorometer reflectometer (Mayevsky, 1984). The source for the 366 nm excitation light was a 100 W air-cooled mercury arc passing via an excitation fiber-optic bundle to the brain (240 μ diameter quartz fibers). The emitted light from the brain transmitted through another fiber-optic bundle was split into 90:10 ratios in order to measure the fluorescent light (F: 90%) and the reflected light (R: 10%). The calibration of the signals as well as other details were published previously (Mayevsky, 1984; Mayevsky & Chance, 1982).

Hemoglobin Spectrophotometry In order to evaluate the level of hemoglobin oxygenation, we used the Erlangen microlight guide spectrophotometer EMPHO-I (Frank et al., 1989). The light source was a water-cooler xenon high-pressure arc lamp (XBO 75 W/2 Osram) operated by a starting unit and a power supply (Muller, Moosinning, Germany). The light (400–1200 nm) passing through a heat absorption filter (KGI, Schott), located between two focusing lenses, entered a central 250 μ diameter microlight guide fiber closely surrounded by a hexagon of six detecting fibers (70 μ diameter) and was transmitted to the brain surface. The emitted light from the brain was passed through a rotating interference band pass filter disk (Anders, Nabburg) which works in the spectral domain from 502–650 nm. All other details regarding data collection, calibration, and calculations were published recently (Frank et al., 1989).

The CSD waves, seen in Fig. 2.10B, were induced by insertion of a needle into the cortical area next to the MPA and the propagated wave was monitored (Mayevsky et al., 1990b). The Ca^{2+} electrode was the first to detect the wave due to the MPA positioning on the brain in relation to the CSD initiation spot. The ionic hemostasis disturbances due to the CSD waves, including the increase in K^{+}, and NADH response, had started about the same time as can be seen also from the probe location scheme. The main new results are the HbO_2 responses to the CSD wave and its correlation to the NADH kinetic changes. The main change seen in the HbO_2 trace is the increase in oxygenation although a small short initial deoxygenation could be detected. In parallel, the Hb volume was increased and it correlated well to the decrease in the R signal. Those changes in the HbO_2 oxygenation and volume are secondary to the primary oxidation of the NADH as seen by the decrease in the CF signal. Those metabolic recorded in tissue oxygenation as well as the oxidation of the mitochondrial redox state correlated very well to the increase in energy demand by the Na^{+}-K^{+}- ATPase and other active transport systems. The short delay in the HbO_2 recorded responses is due to location of the HbO_2 probe in relation to the needle insertion site.

2.2.8 *Simultaneous Real-Time Monitoring of Brain NADH, CBF, ECoG, DC Potential, and Extracellular K^{+} and Ca^{2+}*

The interrelation between cerebral blood flow and metabolism is an important factor in understanding brain functions under normal and pathological situations. Since few of the situations, such as CSD and epilepsy, are transient and are not in steady state, it became necessary to monitor the various parameters under observation in real time and in a continuous mode. Until recently, CBF monitoring was limited to a single time point per animal using mapping technique or to a 10–15 minutes' interval between single measurements. The development of the laser Doppler flowmeter (LDF) opened up the possibility of monitoring CBF in a real-time continuous mode (Stern et al., 1977). As shown by a few groups, the CBF monitored by the LDF technique was significantly correlated to the flow measured by other quantitative techniques (Dirnagl et al., 1989; Haberl et al., 1989). Since CF is only one of the important parameters to be measured, we incorporated a laser Doppler probe into the multiprobe assembly developed by us a few years ago (Friedli et al., 1982; Mayevsky, 1983). Since all the probes used in the MPA are surface probes, we also monitored the CBF from the surface of the brain.

In order to compare the responses obtained under various brain perturbations, we used the Mongolian gerbil allowing an easy way to induce partial or near-complete ischemia (Mayevsky, 1990; Mayevsky & Breuer, 1990). Using this model, we were able to expose the same animal to all three types of brain perturbations, namely systemic hypoxia, ischemia, and CSD. In order to monitor the flow-metabolism coupling and correlations under those conditions, we have developed a combined fiber-optic probe for the simultaneous measurement of the relative CBF and mitochondrial redox state (NADH fluorescence). Incorporation of this metaflow probe (metabolism-flow) into the multiprobe assembly (Mayevsky, 1990) provides us with information regarding the metabolic, ionic, and electrical activities. In the present study, we monitored the extracellular levels of K^+ and Ca^{2+} using ion-selective minielectrodes as well as the DC steady potential and the ECoG. Preliminary results were described recently (Mayevsky et al., 1991).

Multiprobe Assembly This approach was developed by us several years ago (Friedli et al., 1982; Mayevsky, 1983, 1990) and has been used successfully in rats and gerbils. The multiprobe assembly (MPA) used in the present study has been modified (Fig. 2.11A) and contained the new optical probe named metaflow probe (MFP) (Mayevsky, 1992). The optical fibers of the two instruments, namely, the fluorometer and laser Doppler flowmeter, are mixed in the common part located on the surface of the brain. The laser Doppler flowmeter used in the present study was made by TSI Inc. (Laserflo BPM) as described recently (Borgos, 1988), but similar results were obtained by using the instrument made by Perimed Inc. being used in our laboratory for many years.

As shown in Fig. 2.11A in the present study, we used 2 potassium electrodes and one calcium electrode. All details regarding the construction of the metaflow probe as well as the preparation of the minielectrodes and the MPA were published very recently (Mayevsky et al., 1991).

Ten adult male Mongolian gerbils (60–80 gr) were used in the present study. The animal was anesthetized by Equithesin (each ml contains: pentobarbital 9.72 mg,

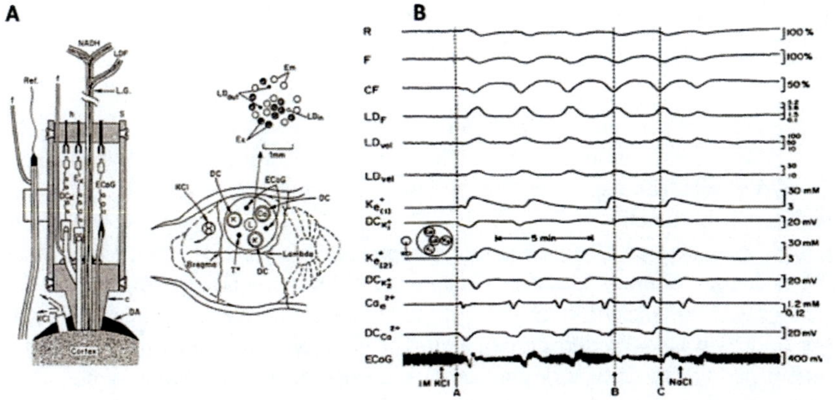

Fig. 2.11 (**A**) Schematic presentation of the experimental setup used. (*LG* Light guide monitoring the NADH redox state and relative CBF (LDF), *ECoG* electrocorticography electrodes, *K, Ca* minielectrodes for extracellular potassium and calcium monitoring, *T°* temperature thermistor probe, *DC* DC steady potential electrodes, *DA* dental acrylic cement, *Ref* reference electrode, *KCl* push–pull cannula for KCl application (CSD), *Ex, Em* excitation and emission fibers of the fluorometer, LD_{in}, LD_{out} fibers connected to the laser Doppler flowmeter, *C* plexiglass probe holder, *h* connectors holder, *s* plexiglass sleeve, *f* feeling tube of reference an X electrode) (Mayevsky, 1992). (**B**) Metabolic, hemodynamic ionic, and electrical responses to CSD induced by epidural application of KCl solution. (*R* 366 nm reflectance, *F* 450 nm fluorescence, *CF* NADH-corrected fluorescence, *LDF' LDvol' LDvel* laser Doppler flow, volume, and velocity $K^+_e(1)$, K^+_e (2), Ca^{2+}_e extracellular potassium (two electrodes) and calcium electrode, $DCK^+(1)$, $DCK^+(2)$, $DCCa^{2+}$ DC steady potential measured concentric to the three electrodes, *ECoG* electrocorticogram, (Mayevsky, 1992))

chloral hydrate 42.51 mg, magnesium sulfate 21.25 mg, propylene glycol 44.34% w/v, alcohol 11.5%, and water) injected IP (0.3 ml/100 gr). Body temperature was maintained at 36.5 ± 0.5 °C and Equithesin was added during the monitoring period (0.03 ml per gerbil every 30 minutes).

The effects of repetitive cycles of CSD, induced by 1 M KCl, on the hemodynamic metabolic and ionic activities are shown in Fig. 2.11B (Mayevsky, 1992). The application site of KCl is seen in the scheme shown inside the figure. The front of the propagating CSD wave affected the various electrodes at slightly different timing. As seen (Line a), the sequence was that the Ca^{2+} and $K^+_{(1)}$ were the first site to be affected followed by the light guide area and the $K^+_{(2)}$ was the last one to respond. Lines B and C show that the peak of the NADH oxidation precedes the peak of the maximal CBF recorded (LDF). Since the 1 M KCl was left in the washing cannula, repetitive cycles were recorded. Due to the depolarization wave developed as seen by the negative shift in the DC potential and the increase in the extracellular K^+, the energy metabolism was activated.

2.2.9 Simultaneous Monitoring of Brain NADH, CBF, ECoG, DC Potential, and Extracellular K^+, Ca^{+2}, H^+

Our approach was to monitor the brain on the "tissue level" rather than the "cellular level." We believe that the various component elements of the brain, namely, neurons, glia cells, and capillaries, are acting as an integrated system. Therefore, we monitored the functions of the brain by using "miniprobes" rather than "microprobes" that are normally used to monitor ionic homeostasis (Hansen, 1985). Also, the evaluation of other parameters such as cerebral blood flow (CBF) and NADH redox state was adapted to the tissue level rather than to the cellular level. Hence, our strategy was to develop a multiparametric monitoring assembly in which all the probes had the same type of contact with a small area of the brain. None of the probes penetrated the brain itself, thus avoiding severe damage to the brain or the formation of an artificial environment around the sensor that can be created around a penetrating microelectrode.

Multiprobe Assembly (MPA)
To monitor all the required parameters from the same cortical tissue, we used a multiprobe assembly (Mayevsky et al., 1995) that we have developed during the last 10 year. Hemodynamic responses were monitored with laser Doppler flowmetry (LDF), which allows real-time and continuous monitoring (Dirnagl et al., 1989; Haberl et al., 1989; Wadhwani et al., 1990). The metabolic state of the brain was evaluated by monitoring the intramitochondrial redox state by using NADH surface fluorometry/reflectometry (Chance & Williams, 1955; Mayevsky, 1984; Mayevsky & Chance, 1975, 1982). Ionic homeostasis was measured by real-time monitoring of extracellular levels of K^+, Ca^{2+} and Na^+, or H^+. The electrical status of the cortex was measured by electrocorticography (ECoG) and direct current (DC) steady potential.

Local Cerebral Blood Flow (CBF)
To measure in real time, the CBF from the same cortical area as the MPA location, we used the LDF technique (Dirnagl et al., 1989; Haberl et al., 1989; Wadhwani et al., 1990). The LDF measures relative flow changes, and readings have been shown to correlate with the relative changes in CBF measured by the two other quantitative approaches (for a review see Ref. Wadhwani et al., 1990) The principle of the LDF is to utilize the Doppler shift, namely, the frequency change that light undergoes when reflected by moving red blood cells. A beam of low-power or diode laser light is transmitted by an optical fiber to the tissue. After the multiple scattering of the light, another optical fiber picks up the reflected light that is recorded by a photodetector. The run signal is analyzed by a complicated algorithm developed by the manufacturers, and the results are presented in percentage of a full scale (0–100%), thereby providing arbitrary relative flow values. The change in the total back-scattered light is an indirect measure of the blood volume in the sampled tissue. To quantitate and normalize CBF values, we defined the reading value after

death as 0 CBF. The 100% value was defined as percent CBF read on the LDF scale during the control period.

Ion-Selective Electrodes
To monitor the extracellular levels of K^+, Ca^{2+}, Na^+, and H^+, we used specially designed minielectrodes made by World Precision Instruments (WPI; Sarasota, FL). A flexible tubing made of polyvinyl chloride was sealed at one end with a membrane sensitive to a specific ion. The tube was filled with the appropriate solution and connected to an electrode holder with a salt bridge between the membrane and an Ag-AgCl pellet located inside the holder. The interface between the polyvinyl chloride tubing and the holder was glued with epoxy, and such electrodes were usable for a few weeks. The sensitivity of the electrodes to the specific ion was close to the Nernstian value, namely, 50–60 mV/decade for K^+, Na^+, and H^+ or 25–30 mV/decade for Ca^{2+}.

Figure 2.12B (Mayevsky et al., 1995) shows the response of an awake rat exposed for 60 min to 1000 ppm of CO. The main change recorded under 1000 ppm CO was the increase in relative CBF. The CF was unchanged, suggesting no change in NADH redox state. The R and cerebral blood volume (CBV) were also very stable. The ionic homeostasis and DC potential were unaffected, as measured by the various electrodes

2.2.10 *Multiparametric Monitoring of Neurosurgical Patients*

The new era in monitoring of mitochondrial function in patients started in 1990, when our team developed a unique multiparametric monitoring system that included the measurement of NADH fluorescence, using a light-guide-based device (Mayevsky et al., 1991). This system was initially applied to monitor neurosurgical patients undergoing brain surgery. In the second half of the 1990s (Mayevsky et al., 1996, 1998b; Meilin et al., 1999), we were able to monitor the brain of comatose patients in the ICU and described, for the first time, the development of CSD-like responses, including marked NADH responses that were correlated to the microcirculatory blood flow. A modified multiparametric monitoring system was applied for NADH measurement during neurosurgical procedures in the operating room (Mayevsky et al., 1998a, 1999, 2001, 2002).

Based on the well-developed MultiProbe Assembly (MPA) used for animal experiments (Mayevsky et al., 1995), a new MPA was developed and applied to patients monitored in the neurosurgical operation rooms and ICU. This device allowed us to monitor, in real-time, the hemodynamic, metabolic, ionic, and electrical activities in the brain of comatose patients. All details regarding the technology and the clinical setup appear in our published papers (Mayevsky et al., 1996) and (Mayevsky & Barbiro-Michaely, 2013b). Figure 8.7A (in this book] shows the longitudinal section of the MPA (measuring 8 parameters) used in the neurosurgical operating room and ICU.

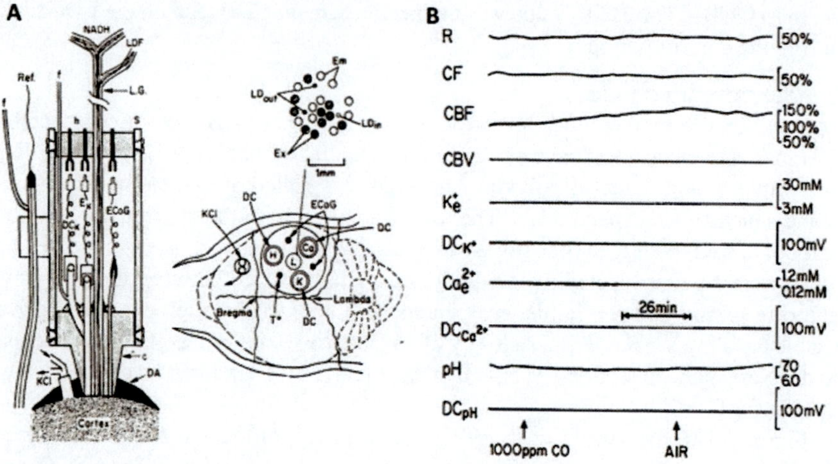

Fig. 2.12 (**A**) Multiprobe assembly used in awake rat to monitor hemodynamic, metabolic, ionic, and electrical activities from brain. **Left**: longitudinal section. **Right**: location of multiprobe assembly and skull. *Ref* reference electrode, *f* feeling tube of reference or other direct current electrodes, *h* connectors holder, *NADH* 2 arms of NADH monitoring light guide, *LDF* laser Doppler flowmeter, *L* light guide, s aluminum sleeve, *DCk and Ek* direct current and ion-specific electrodes, respectively, for K, *ECoG*, electrocortical electrodes, *c* Plexiglas probe holder, *DA* dental acrylic cement, *KCl* push–pull cannula for KCl, *K, Ca, and H* ion-specific electrodes, *DC* area of direct current steady potential monitoring, *Ex and Em* excitation and emission fibers, respectively, for NADH monitoring, LD_{in} *and* LD_{out} optical fibers for monitoring blood flow and volume, respectively, *T°* thermistor for local temperature measurement) (Mayevsky et al., 1995). (**B**) Responses to 60 min of 1000 ppm of carbon monoxide (CO). R and CF, reflectance and corrected NADH fluorescence, respectively; CBF and CBV, cerebral blood flow and cerebral blood volume, respectively; Ke^+ and Cae^{2+}, extracellular levels of K^+ and Ca^{2+}, respectively; pH, extracellular level of H^+ ions expressed in standard units; DCK^+, $DCCa^{2+}$, and DCpH, DC steady potential measured concentric to K^+, Ca^{2+}, and H^+ electrodes, respectively. CBV signal is presented without an ordinate due to nature of output of this signal. Laser Doppler flowmeter does not allow this signal to be calibrated. (Mayevsky et al., 1995)

The concept behind the development was that with more parameters monitored, a better diagnosis of the situation will be as presented (see Fig. 2.13A). The envisioned system was to monitor 8 parameters from the brain (right side) in addition to the various systemic parameters monitored in every patient hospitalized (left side). The long-term vision was that all parameters monitored by the 2 monitoring systems will be integrated in the same data bank and an expert system will be developed for a better diagnosis for the patients. The translation of this concept into a practical tool and monitoring device is presented in Figs. 8.6 and 8.7A. The patient who developed CSD cycles (presented in Chap. 8) showed signs of brain deterioration leading to brain death. The tracing presented in Fig. 2.13B demonstrates that metabolic and ionic changes preceded brain death by several hours. The ECoG became isoelectric 3 h prior to the beginning of the recording in Fig. 2.13B (68 hours from beginning of monitoring). A gradual increase in ICP up to the level of 100 mmHg was measured, followed by a sharp decrease in cerebral blood flow

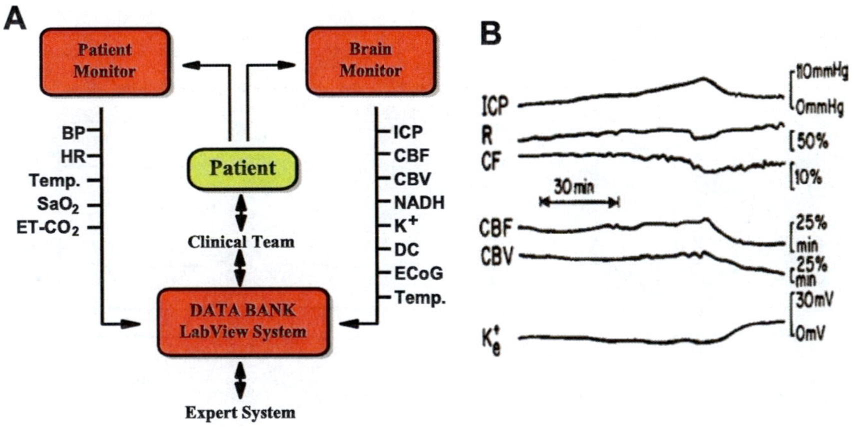

Fig. 2.13 (**A**) Neurosurgical patient monitoring includes usually systemic parameters (Left). However, we suggest the monitoring of specific brain parameters (Right). One Data bank (LabView system) will concentrate this data which will be analyzed by the clinical team. (*BP* blood pressure, *HR* heart rate, *Temp* body temperature, SaO_2 oxygen saturation, *ET* CO_2 end-tidal CO_2, *ICP* intracranial pressure, *CBF, CBV* cerebral blood flow or volume, *NADH* mitochondrial NADH redox state, K^+ extracellular potassium level, *DC* steady DC potential, *ECoG* electrocorticography, *Temp.* brain temperature) (Mayevsky & Barbiro-Michaely, 2013a). (**B**) Real-time monitoring of the brain death response in a head-injured patients (Mayevsky et al., 1996). (*ICP* intracranial pressure, *R* 366 nm reflectance, *CF* 450 nm corrected fluorescence, *CBF, CBV* cerebral blood flow and volume measured by laser Doppler flowmeter, K^+_e extracellular potassium levels)

(CBF) and volume (CBV) (Fig. 2.13B). During the terminal ischemic phase, a large increase in the extracellular K^+ level was recorded.

2.2.11 Monitoring of the Mechanism of CSD Propagation

The aim of this study was to identify the sequence of responses developed under the CSD phenomenon, initiated during various levels of brain oxygenation: normoxia, hypoxia, and partial ischemia (Sonn & Mayevsky, 2000). Increasing oxygen demand during these states, by inducing CSD waves, will cause different changes in the monitored parameters during CSD propagation. Back et al. (1994) showed no changes in cerebral blood flow (CBF) and a decrease in tissue pO_2 during spontaneous CSD waves in ischemic brain tissue, whereas induction of CSD waves (by KCl solution) in normal tissue showed a significant increase in CBF and in tissue pO_2. Since the CSD wave propagation is a multifactorial event, our strategy was to construct a MultiProbe assembly (MPA) containing probes to evaluate most of the parameters involved during CSD wave propagation, from its front line at the same time. This will verify the parallel changes in the various parameters after CSD initiation and will help us to identify which parameter/s start the chain responses and the reaction of the later ones. The MPA measured: extracellular K^+ concentration

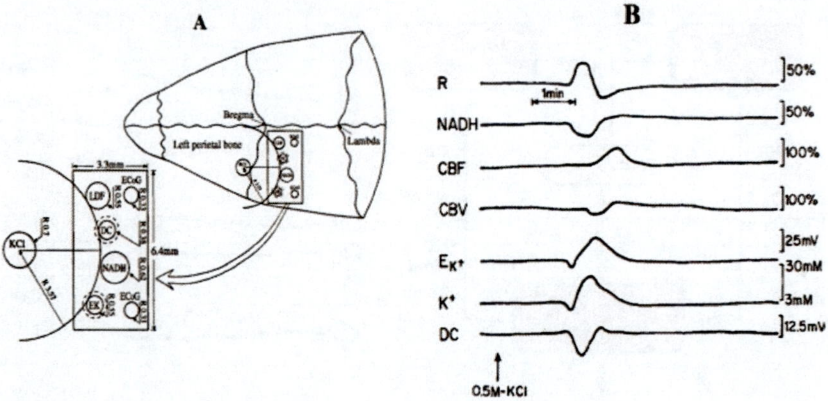

Fig. 2.14 (A) Schematic presentation of the special multiprobe assembly (MPA) developed in order to study the timing of the responses to CSD. The right side shows the MPA located on the cerebral cortex. The localization of the various sensors is shown on the left side of the Figure (**A**). (*LDF* light guide connected to the laser Doppler flowmeter, *NADH* light guide connected to the fluorometer/reflectometer for NADH fluorescence and 366 nm reflectance, *EK* selective K^+ electrode, *DC* direct current, *KCl* push–pull cannula, *ECoG* electrocorticography electrodes). The R values represent the radius of the particular probes (Sonn & Mayevsky, 2000). (**B**) Typical responses of the normoxic rat brain to CSD. (*R, NADH* – 366 nm reflectance and corrected NADH fluorescence, *CBF, CBV* cerebral blood flow and volume measured by the laser Doppler flowmeter; E_K^+, K^+ potential of the K^+ electrode and the DC corrected level of extracellular K^+ (mM), *DC* direct current steady potential). CSD was induced by 0.5 M KCl solution (arrow). (Sonn & Mayevsky, 2000)

$[K^+]e$, mitochondrial NADH/NAD ratio, cerebral blood flow and volume (CBF and CBV), steady DC potential, and bipolar electrocorticography (ECoG).

The probes were embedded in a Delrin rectangular cannula (6.4 × 3.3 mm) (Fig. 2.14A). It was found that the probes should be arranged on a circumference where the KCl cannula (CSD initiation area) is located in the center of the circle. This enabled the probes to be arranged according to the shape of the CSD wave determined and allowed the CSD wave to be monitored from its front line (by all the detected parameters) until complete recovery.

The MPA contained

surface minielectrodes for measuring $[K^+]_e$ and DC steady potential, fiber-optic probes that were used to monitor intramitochondrial NADH redox state (NADH fluorescence) and reflected light, as well as relative CBF and CBV by laser Doppler flowmetry and spontaneous bipolar ECoG via two platinum electrodes. All details regarding the various probe construction and calibration were published previously (Mayevsky, 1992, 1993).

The typical response to CSD shown in Fig. 2.14B presents the relationship between the electrical (DC), ionic, metabolic, and hemodynamic responses during the passage of a single wave. The negative shift seen in DC was also noticed in the K^+ trace as an initial decrease in potassium. A clear oxidation in the NADH redox state (decreased signal) was accompanied by a large increase in CBF. It is important to note that the edge of the various probes was positioned on the same front line.

2.2.12 *Multisite Monitoring of NADH and DC Potential in Rats Under Various Conditions*

The aim of this study was to investigate the changes in the metabolic, hemodynamic, and the electrical activities of CSD wave propagation through the entire ipsilateral cortex in real time, and the effects of ethanol administration on these processes (Sonn & Mayevsky, 2001). To this end, we constructed and applied a multisite assembly (MSA) to investigate the mechanism of propagation of the CSD phenomenon continuously and simultaneously, from four parasagittally arranged adjacent areas through the entire rat hemisphere, during a series of CSD cycles. This approach enabled us to determine whether the brain's metabolic, hemodynamic, and electrical responses to CSD stimulation, spread homogeneously from its source to the other cortical areas, and helped us determine the rate of CSD wave propagation before and after ethanol administration in the awake rat.

The MSA shown in Fig. 2.15A was constructed in order to monitor mitochondrial NADH redox state and DC steady potential from four subsequent areas in the same cerebral hemisphere (Mayevsky & Chance, 1982; Sonn & Mayevsky, 2001).

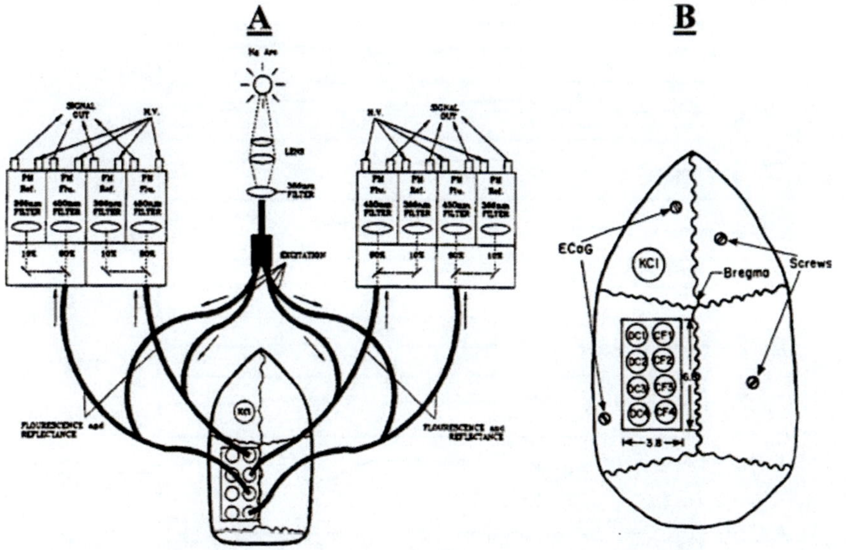

Fig. 2.15 **(A + B)** Schematic illustration of the multisite and multiprobe assembly system (MSMP). **(A)** A four-channel DC fluorometer/ reflectometer was connected to the rat brain using flexible fiber-optic bundles for detecting mitochondrial NADH and light reflectance (changes from four parasagittally adjacently arranged areas). The system contains one mercury lamp (Hg Arc) light source and 366 nm filter for excitation, and 366 nm [for reflectance (Ref.)] and 450 nm [for fluorescence (Flu.)] filters to measure emitted light, and photomultipliers (PM). H.V. high voltage. **(B)** The MSA system contains four light guides (CF1–CF4) and four DC electrodes (DC1–DC4). *KCl* Push–pull cannula, *ECoG* for bipolar ECoG (electrocorticography) recording, *screws* for MSA attachment to the skull. (Sonn & Mayevsky, 2001)

It contains four bundles of optical fibers (1.26 mm O.D.31 mm I.D.) connected to a four-channel fluorometer reflectometer providing information on the mitochondrial NADH redox state (surface fluorometry) as well as on tissue absorption properties (reflected light) from the four parasagittally arranged adjacent points (Fig. 2.15B). DC steady potential was monitored from four points located parallel to the four light guides, via Ag/AgCl electrodes (1.27 mm O.D.30.86 mm I.D.) relative to an Ag/AgCl reference electrode (Mayevsky et al., 1990a, 1993).

Propagation of CSD waves on the cerebral cortex (of a control untreated animal) showed typical biphasic waves in the reflected light, an oxidation–reduction waves in NADH fluorescence, a negative shift in DC potential, and a depression in ECoG (Fig. 2.16). It can be seen that the CSD wave propagated from the first channel (closest to initiation area in the frontal lobe) toward the fourth, with a time delay of

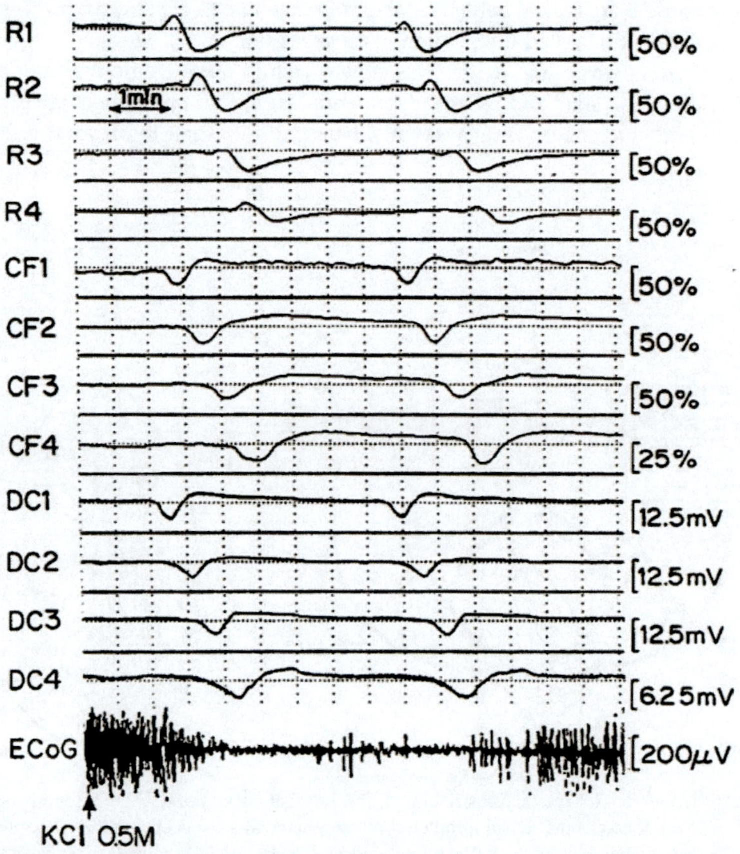

Fig. 2.16 Typical analog signals of reflected light 366 nm (R1–R4) and corrected fluorescence (CF1–CF4), DC steady potential (DC1–DC4) and ECoG (electrocorticography) measured from the rat cortex exposed to two CSD cycles elicited by 0.5 M KCl. Note the suppression in ECoG amplitude during CSD propagation and that the wave propagates from the first monitored channel (R1, CF1, DC1) to the rest of the hemisphere. (Sonn & Mayevsky, 2001)

approximately 20–30 s between two adjacent channels. (Mayevsky & Chance, 1982; Sonn & Mayevsky, 2001).

2.2.13 *Multisite Multiparametric Monitoring of NADH, CBF, and DC Potential in Rats Under Experimental Conditions*

In the present study, we show the typical results of the use of the multisite multiparametric monitoring approach during CSD in rats (Manor et al., 2008). To achieve this, we placed four probes at four different locations on the cerebral cortex. Two were placed on the left hemisphere, one on the parietal lobe and one on the frontal lobe and two on the contralateral hemisphere at the same locations.

The Multisite Multiparametric Monitoring System The monitoring of the cerebral cortex in four locations on the cerebral cortex was performed using the multisite multiparametric monitoring system (MSMP) (Fig. 2.17). Each probe of the MSMP (1–4), 3 mm in diameter, contains optical fibers for NADH redox state fluorometry, a laser Doppler probe, and a DC potential electrode. The position of each probe on the cerebral cortex was similar in all the experiments (Fig. 2.17B). The principle of NADH monitoring from the surface of the tissue (1 mm depth) is that excitation light (366 nm) passes from the fluorometer, which contains appropriate filters, to the brain via a bundle of optical fibers made of quartz. The emitted light (450 nm), together with the reflected light at the excitation wavelength (366 nm), is transferred to the fluorometer via another bundle of optical fibers (Mayevsky, 1993; Mayevsky & Rogatsky, 2007). Changes in the reflected light are correlated to changes in tissue blood volume and therefore serve to correct for hemodynamic artifacts in NADH monitoring (Mayevsky, 1984). The corrected fluorescence (NADH) is obtained by subtracting the reflectance from the fluorescence signal at a 1: 1 ratio (Mayevsky, 1984). The fluorometer is calibrated in such a way that a range of 500 mV is equal to a change of 100% in the monitored signal. Real-time monitoring of CBF from the same cortical area was achieved using the laser Doppler flowmetry technique which is based on the Doppler shift reflecting the flow of red blood cells in the microcirculation at a depth of 1–2 mm (Arvidsson et al., 1988). For CBF, the flow at death was considered as 0% flow and the range between the basal level and the flow in death was considered as 100%. The DC steady potential was measured by highly stable Ag/AgCl electrodes connected via a salt bridge. A reference electrode was located below the skin of the animal's neck. All of the signals monitored during the experiment were digitized and transmitted to a multichanneled computerized data acquisition and recording system (Labview A/D software, National Instruments Co., USA) for further analysis.

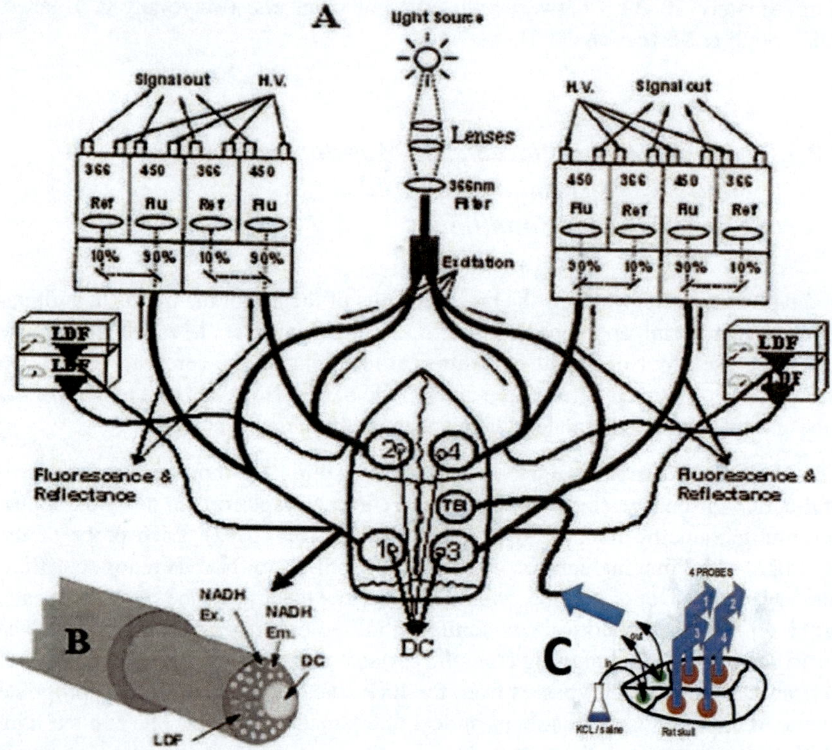

Fig. 2.17 (**A**) Schematic representation of the MSMP and the location of the various probes on the cerebral cortex. The MSMP includes four fluorometers units, optical fibers to conduct the excitation light from the light source to the cerebral tissue, and the emission light (fluorescence and reflectance) from the tissue to the photo-detectors. For CBF monitoring, four laser Doppler flowmeters (LDF) were used. (**B**) The tip of each probe includes optical fibers for NADH monitoring (Ex.: excitation, Em.: emission), for blood flow monitoring (LDF), and a dc electrode for dc steady potential monitoring. (**C**) The location of the MSMP probes on the cerebral cortex in experiments on CSD induction. (Manor et al., 2008)

To test the ability of the monitoring system to reliably represent the hemodynamic and metabolic state of the brain, not only in worsened cerebral conditions such as anoxia, but also when the brain is highly activated, we used the CSD model. As seen in Fig. 2.18, in probes 1 and 2, three CSD waves were developed; whereas in probes 3 and 4, only two waves were generated (Manor et al., 2008). The waves in each hemisphere resemble one another, clearly showing the spreading character of the CSD waves, as probes 1 and 3 show the waves before probes 2 and 4. The reflectance showed a biphasic response with an increase of 25–50% followed by a decrease back to the basal levels. CBF increased by 100%, mitochondrial NADH was oxidized by 25–50% and DC potential decreased by 6–10 mV.

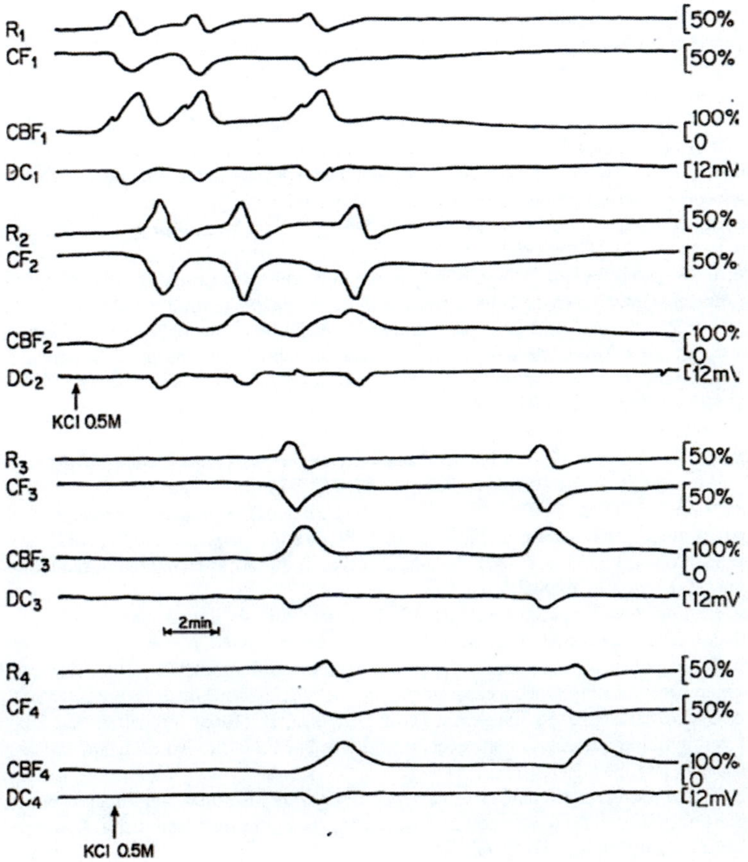

Fig. 2.18 Typical responses to CSD induced by 0.5 M KCl solution. (*R* reflectance, *CF* corrected NADH fluorescence, *CBF* cerebral blood flow, *DC* DC steady potential (Manor et al., 2008))

References

Acad, B., Sonn, J., Furman, E., & Kedem, J. (1986). Variations in left and right ventricular oxygen balance produced by paired electrical stimulations. *Archives Internationales de Physiologie et de Biochimie, 94*, 37–43.

Anderson, R. E. (1975). Instrumentation for in vivo cerebral NADH studies in squirrel monkey. *IEEE Transactions on Biomedical Engineering, BME-22*, 220–224.

Anderson, R. E. (1978). Comparison of dark-field and bright-field incident illumination for in vivo measurements of reduced pyridine nucleotides. *Analytical Biochemistry, 91*, 496–508.

Arvidsson, D., Svensson, H., & Haglund, U. (1988). Laser-Doppler flowmetry for estimating liver blood flow. *The American Journal of Physiology, 254*, G471–G476.

Back, T., Kohno, K., & Hossmann, K. A. (1994). Cortical negative DC deflections following middle cerebral artery occlusion and KCl-induced spreading depression: effect on blood flow, tissue oxygenation, and electroencephalogram. *Journal of Cerebral Blood Flow and Metabolism, 14*, 12–19.

Borgos, J. A. (1988). TSI's LDV blood flowmeter. In A. P. Shepher & P. A. Oberg (Eds.), *Laser doppler blood flowmetry* (pp. 73–91). KLuwer Academic Publishers.

Bures, J., & Krivanek, J. (1960). Ionic movements in the brain as studied with the aid of washing the cortical surface with an epidural cannula. *Physiologia Bohemoslovenica, 9*, 488–493.

Bures, J., Buresova, O., & Krivanek, J. (1974). *The mechanism and applications of Leão's spreading depression of electroencephalographic activity*. Academic.

Chance, B. (1959). The response of mitochondria to muscular contraction. *Annals of the New York Academy of Sciences, 81*, 477–489.

Chance, B. (1966). The identification and control of metabolic states. *Genootschap ter Bevordering van Natuur-, Genees-, en Heelkunde te Amsterdam*, 5–37.

Chance, B., & Jobsis, F. (1959). Changes in fluorescence in a frog sartorius muscle following a twitch. *Nature, 184*, 195–196.

Chance, B., & Legallias, V. (1959). Differential microfluorimeter for the localization of reduced pyridine nucleotide in living cells. *Review of Scientific Instruments, 30*, 732–735.

Chance, B., & Schoener, B. (1962). Correlation of oxidation-reduction changes of intracellular reduced pyridine nucleotide and changes in electro-encephalogram of the rat in anoxia. *Nature, 195*, 956–958.

Chance, B., & Schoener, B. (1965). A correlation of absorption and fluorescence changes in ischemia of the rat liver, in vivo. *Biochemische Zeitschrift, 341*, 340–345.

Chance, B., & Thorell, B. (1959). Fluorescence measurements of mitochondrial pyridine nucleotide in aerobiosis and anaerobiosis. *Nature, 184*, 931–934.

Chance, B., & Williams, G. R. (1955). Respiratory enzymes in oxidative phosphorylation (I-Kinetics of oxygen utilization). *The Journal of Biological Chemistry, 217*, 383–393.

Chance, B., Cohen, P., Jobsis, F., & Schoener, B. (1962a). Intracellular oxidation-reduction states in vivo. *Science, 137*, 499–508.

Chance, B., Legallias, V., & Schoener, B. (1962b). Metabolically linked changes in fluorescence emission spectra of cortex of rat brain, kidney and adrenal gland. *Nature, 195*, 1073–1075.

Chance, B., Legallias, V., & Schoener, B. (1963). Combined fluorometer and double-beam spectrophotometer for reflectance measurements. *Review of Scientific Instruments, 34*, 1307–1311.

Chance, B., Williamson, J. R., Jamieson, D., & Schoener, B. (1965). Properties and kinetics of reduced pyridine nucleotide fluorescence of the isolated and in vivo rat heart. *Biochemische Zeitschrift, 341*, 357–377.

Chance, B., Mayer, D., & Rossini, L. (1970). A time-sharing instrument for direct readout of oxidation- reduction states in intracellular compartments of cardiac tissue. *IEEE Transactions on Biomedical Engineering, BME-17*, 118–121.

Chance, B., Graham, N., & Mayer, D. (1971). A time sharing fluorometer for the readout of intracellular oxidation-reduction states of NADH and flavoprotein. *Review of Scientific Instruments, 42*, 951–957.

Chance, B., Oshino, N., Sugano, T., & Mayevsky, A. (1973). Basic principles of tissue oxygen determination from mitochondrial signals. Oxygen Transport to Tissue. *Advances in Experimental Medicine and Biology, 37A*, 277–292.

Chance, B., Legallias, V., Sorge, J., & Graham, N. (1975). A versatile time-sharing multichannel spectrophotometer reflectometer and fluorometer. *Analytical Biochemistry, 66*, 498–514.

Crowe, W., Mayevsky, A., & Mela, L. (1977). Application of a solid membrane ion selective electrode to in vivo measurements. *The American Journal of Physiology, 233*, C56–C60.

Dirnagl, U., Kaplan, B., Jacewicz, M., & Pulsinelli, W. (1989). Continuous measurement of cerebral cortical blood flow by laser-Doppler flowmetry in a rat stroke model. *Journal of Cerebral Blood Flow & Metabolism, 9*, 589–596.

Donadio, M. F., Kozlowski, P. B., Kaplan, H., Wisniewski, H. M., & Majkowski, J. (1982). Brain vasculature and induced ischemia in seizure-prone and non-seizure-prone gerbils. *Brain Research, 234*, 263–273.

Dora, E. (1984). A simple cranial window technique for optical monitoring of cerebrocortical microcirculation and NAD/NADH redox state. Effect of mitochondrial electron transport inhibitors and anoxic anoxia. *Journal of Neurochemistry, 42*, 101–108.

Dora, E., & Kovach, A. G. B. (1982). Effect of acute arterial Hypo- and Hypertension on cerebrocortical NAD/NADH redox state and vascular volume. *Journal of Cerebral Blood Flow & Metabolism, 2*, 209–219.

Duysens, L. N. M., & Amesz, J. (1957). Fluorescence spectrophotometry of reduced phosphopyridine nucleotide in intact cells in the near-ultraviolet and visible region. *Biochimica et Biophysica Acta, 24*, 19–26.

Frank, K. H., Rettig, V., Friedl, A., Brunner, M., Ellermann, R., Kerl, G., & Hoper, J. (1984a). Measurements of intracapillary haemoglobin spectra in the beating heart, skeletal muscle and the liver, using the Erlangen micro-lightguide spectrophotometer. *Pfluger Archiv, 400*(Suppl), 219.

Frank, K. H., Schabert, A., Friedl, A., Brunner, M., Hoper, J., Kerl, G., & Kessler, M. (1984b). Correlation between tissue PO_2 and intracapillary Hb spectra. *Advances in Experimental Medicine and Biology, 169*, 811–819.

Frank, K. H., Kessler, M., Appelbaum, K., & Dummler, W. (1989). The Erlangen micro-lightguide spectrophotometer EMPHO I. *Physics in Medicine and Biology, 34*, 1883–1900.

Friedli, C. M., Sclarsky, D. S., & Mayevsky, A. (1982). Multiprobe monitoring of ionic, metabolic, and electrical activities in the awake brain. *The American Journal of Physiology, 243*, R462–R469.

Ginsberg, M. D., Reivich, M., Frinak, S., & Harbig, K. (1976). Pyridine nucleotide redox state and blood flow of the cerebral cortex following middle cerebral artery occlusion in the cat. *Stroke, 7*, 125–131.

Gosalvez, M., Thurman, R. G., Chance, B., & Reinhold, H. (1972a). Mammary tumours in vivo demonstrated by fluorescence of pyridine nucleotide. *The British Journal of Radiology, 45*, 510–514.

Gosalvez, M., Thurman, R. G., Chance, B., & Reinhold, H. S. (1972b). Indication of hypoxic areas in tumours from in vivo NADH fluorescence. *European Journal of Cancer, 8*, 267–269.

Gyulai, L., Dora, E., & Kovach, A. G. B. (1982). NAD/NADH: redox state changes on cat brain cortex during stimulation and hypercapnia. *The American Journal of Physiology, 243*, H619–H627.

Haberl, R. L., Heizer, M. L., Marmarou, A., & Ellis, E. F. (1989). Laser-Doppler assessment of brain microcirculation: effect of systemic alterations. *American Journal of Physiology. Heart and Circulatory Physiology, 256*, H1247–H1254.

Hansen, A. J. (1985). Effect of anoxia on ion distribution in the brain. *Physiological Reviews, 65*, 101–148.

Harbig, K., Chance, B., Kovach, A. G. B., & Reivich, M. (1976). In vivo measurement of pyridine nucleotide fluorescence from cat brain cortex. *Journal of Applied Physiology, 41*, 480–488.

Jobsis, F. F. (1964). Basic processes in cellular respiration. In R. Fenn (Ed.), *Handbook of physiology - respiration* (1st ed., pp. 63–124).

Jobsis, F. F., & Duffield, J. C. (1967). Oxidative and glycolytic recovery metabolism in muscle. Fluorometric observations on their relative contributions. *The Journal of General Physiology, 50*, 1009–1047.

Jobsis, F. F., & Stainsby, W. N. (1968). Oxidation of NADH during contractions of circulated mammalian skeletal muscle. *Respiration Physiology, 4*, 292–300.

Jobsis, F., Legallias, V., & O'Connor, M. (1966). A regulated differential fluorometer for the assay of oxidative metabolism in intact tissues. *IEEE Transactions on Biomedical Engineering, BME-13*, 93–99.

Jobsis, F. F., O'Connor, M., Vitale, A., & Vreman, H. (1971a). Intracellular redox changes in functioning cerebral cortex. I. Metabolic effects of epileptiform activity. *Journal of Neurophysiology, 3465*, 735–749.

Jobsis, F. F., O'Connor, M. J., Rosenthal, M., & Van Buren, J. M. (1971b). Fluorometric monitoring of metabolic activity in the intact cerebral cortex. In G. G. Somjen (Ed.), *Excerpta medica international congress series, NO. 253* (Vol. 20–22, pp. 18–26).

Kedem, J., Mayevsky, A., Sonn, J., & Acad, B. (1981). An experimental approach for evaluation of the O_2 balance in local myocardial regions in vivo. *Quarterly Journal of Experimental Physiology, 66*, 501–514.

Kobayashi, S., Kaede, K., Nishiki, K., & Ogata, E. (1971). Microfluorometry of oxidation-reduction state of the rat kidney in situ. *Journal of Applied Physiology, 31*, 693–696.

Leão, A. A. P. (1944a). Pial circulation and spreading depression of activity in the cerebral cortex. *Journal of Neurophysiology, 7*, 391–396.

Leão, A. A. P. (1944b). Spreading depression of activity in cerebral cortex. *Journal of Neurophysiology, 7*, 359–390.

Leão, A. A. P. (1947). Further observations on the spreading depression of activity in the cerebral cortex. *Journal of Neurophysiology, 10*, 409–414.

Leão, A. A. P., & Morison, R. (1945). Propagation of spreading cortical depression. *Journal of Neurophysiology, 8*, 33–45.

Levine, S., & Payan, H. (1966). Effects of ischemia and other procedures on the brain and retina of the gerbil (*Meriones unguiculatus*). *Experimental Neurology, 16*, 255–262.

Levy, D. E., & Brierley, J. B. (1974). Communications between vertebro-basilar and carotid arterial circulations in the gerbil. *Experimental Neurology, 45*, 503–508.

Lukyanov, L. D., & Bures, J. (1967). Changes in pO_2 due to spreading depression in cortex and nucleus caudatus of rat. *Physiologia Bohemoslovaca, 16*, 449–455.

Manor, T., Barbiro-Michaely, E., Rogatsky, G., & Mayevsky, A. (2008). Real-time multi-site multi-parametric monitoring of rat brain subjected to traumatic brain injury. *Neurological Research, 30*, 1075–1083.

Marshall, W. H. (1959). Spreading cortical depression of Leão. *Physiological Reviews, 39*, 239–279.

Mayer, D., Chance, B., & Legallias, V. (1971). Time-sharing in spectrophotometry and fluorometry. Probes of structure and function of macromolecules and membranes. In *Probes and membranes function* (pp. 527–534). Academic Press.

Mayevsky, A. (1978a). Ischemia in the brain: The effects of carotid artery ligation and decapitation on the energy state of the awake and anesthetized rat. *Brain Research, 140*, 217–230.

Mayevsky, A. (1978b). Pyridine nucleotide oxidation reduction state of the cerebral cortex in the awake gerbil. *Journal of Neuroscience Research, 3*, 369–374.

Mayevsky, A. (1978c). Shedding light on the awake brain. In P. L. Dutton, J. Leigh, & A. Scarpa (Eds.), *Frontiers in bioenergetics: From electrons to tissues* (II ed., pp. 1467–1476). Academic Press.

Mayevsky, A. (1983). Multiparameter monitoring of the awake brain under hyperbaric oxygenation. *Journal of Applied Physiology, 54*, 740–748.

Mayevsky, A. (1984). Brain NADH redox state monitored in vivo by fiber optic surface fluorometry. *Brain Research Reviews, 7*, 49–68.

Mayevsky, A. (1990). Level of ischemia and brain functions in the Mongolian gerbil in vivo. *Brain Research, 524*, 1–9.

Mayevsky, A. (1992). Cerebral blood flow and brain mitochondrial redox state responses to various perturbations in gerbils. *Advances in Experimental Medicine and Biology, 317*, 707–716.

Mayevsky, A. (1993). Biochemical and physiological activities of the brain as in vivo markers of brain pathology. In E. F. Bernstein, A. D. Callow, A. N. Nicolaides, & E. G. Shifrin (Eds.), *Cerebral revascularization* (pp. 51–69). Med-Orion Pub.

Mayevsky, A. (2015). *Mitochondrial function in vivo evaluated by NADH fluorescence*. Springer.

Mayevsky, A., & Barbiro-Michaely, E. (2013a). Shedding light on mitochondrial function by real time monitoring of NADH fluorescence: II: human studies. *Journal of Clinical Monitoring and Computing, 27*, 125–145.

Mayevsky, A., & Barbiro-Michaely, E. (2013b). Shedding light on mitochondrial function by real time monitoring of NADH fluorescence: II: Human studies. *Journal of Clinical Monitoring and Computing, 27*, 125–145.

Mayevsky, A., & Bar-Sagie, D. (1978). The interrelation between CBF, energy metabolism and ECoG in a new awake brain model. *Advances in Experimental Medicine and Biology, 92*, 761–768.

Mayevsky, A., & Breuer, Z. (1990). In A. Schurr & B. M. Rigor (Eds.), *The Mongolian gerbil as a model for cerebral ischemia* (pp. 27–46). CRC Press.

Mayevsky, A., Chance, B., 1973a. Monitoring of brain NADH redox state using a fiber optic based fluorometer/reflectometer, The Annual Meeting of the International Society of Oxygen Transport to Tissue - ISOTT, .

Mayevsky, A., & Chance, B. (1973b). In H. I. Bicher & D. F. Bruley (Eds.), *A new long-term method for the measurement of NADH fluorescence in intact rat brain with implanted cannula* (pp. 239–244). Plenum Press.

Mayevsky, A., & Chance, B. (1974). Repetitive patterns of metabolic changes during cortical spreading depression of the awake rat. *Brain Research, 65*, 529–533.

Mayevsky, A., & Chance, B. (1975). Metabolic responses of the awake cerebral cortex to anoxia hypoxia spreading depression and epileptiform activity. *Brain Research, 98*, 149–165.

Mayevsky, A., & Chance, B. (1976). The effect of decapitation on the oxidation-reduction state of NADH and ECoG in the brain of the awake rat. *Advances in Experimental Medicine and Biology, 75*, 307–312.

Mayevsky, A., & Chance, B. (1982). Intracellular oxidation-reduction state measured in situ by a multichannel fiber-optic surface fluorometer. *Science, 217*, 537–540.

Mayevsky, A., & Rogatsky, G. (2007). Mitochondrial function in vivo evaluated by NADH fluorescence: From animal models to human studies. *American Journal of Physiology. Cell Physiology, 292*, C615–C640.

Mayevsky, A., & Sclarsky, D. S. (1983). Correlation of brain NADH redox state, K^+, PO_2 and electrical activity during hypoxia, ischemia and spreading depression. *Advances in Experimental Medicine and Biology, 159*, 129–141.

Mayevsky, A., & Shaya, B. (1980). Factors affecting the development of hyperbaric oxygen toxicity in the awake rat brain. *Journal of Applied Physiology, 49*, 700–707.

Mayevsky, A., & Zarchin, N. (1987). Metabolic ionic and electrical activities during and after incomplete or complete cerebral ischemia in the Mongolian gerbil. *Advances in Experimental Medicine and Biology, 215*, 265–273.

Mayevsky, A., Jamieson, D., & Chance, B. (1974a). Oxygen poisoning in the unanesthetized brain: correlation of the oxidation-reduction state of pyridine nucleotide with electrical activity. *Brain Research, 76*, 481–491.

Mayevsky, A., Zeuthen, T., & Chance, B. (1974b). Measurements of extracellular potassium, ECoG and pyridine nucleotide levels during cortical spreading depression in rats. *Brain Research, 76*, 347–349.

Mayevsky, A., Crowe, W., & Mela, L. (1980a). The interrelation between brain oxidative metabolism and extracellular potassium in the unanesthetized gerbil. *Neurological Research, 1*, 213–226.

Mayevsky, A., Lebourdais, S., & Chance, B. (1980b). The interrelation between brain PO_2 and NADH oxidation- reduction state in the gerbil. *Journal of Neuroscience Research, 5*, 173–182.

Mayevsky, A., Friedli, C. M., & Reivich, M. (1985). Metabolic, ionic and electrical responses of the gerbil brain to ischemia. *The American Journal of Physiology, 248*, R99–R107.

Mayevsky, A., Zarchin, N., Yoles, E., & Tannenbaum, B. (1986). Oxygen supply to the brain in hypoxic and hyperoxic conditions. In C. Nicolau (Ed.), *Advances in the biosciences. "O2-Transport in Red Blood Cells"* (pp. 119–132). Pergamon Press.

Mayevsky, A., Duckrow, R. B., Yoles, E., Zarchin, N., & Kanshansky, D. (1990a). Brain mitochondrial redox state, tissue hemodynamic and extracellular ion responses to four-vessel occlusion and spreading depression in the rat. *Neurological Research, 12*, 243–248.

Mayevsky, A., Frank, K. H., Nioka, S., Kessler, M., & Chance, B. (1990b). Oxygen supply and brain function in vivo: a multiparametric monitoring approach in the Mongolian gerbil. In J. Piper, T. K. Goldstick, & M. Meyer (Eds.), *Oxygen transport to tissue XII* (pp. 303–313). Plenum Pub.

Mayevsky, A., Flamm, E. S., Pennie, W., & Chance, B. (1991). A fiber optic based multiprobes system for intraoperative monitoring of brain functions. *SPIE, 1431*, 303–313.

Mayevsky, A., Frank, K., Muck, M., Nioka, S., Kessler, M., & Chance, B. (1992). Multiparametric evaluation of brain functions in the Mongolian gerbil in vivo. *Journal of Basic and Clinical Physiology and Pharmacology, 3*, 323–342.

Mayevsky, A., Zarchin, N., & Sonn, J. (1993). Brain redox state and O_2 balance in experimental spreading depression and ischemia. In A. Lehmenkuhler, K.-H. Grotemeyer, & F. Tegtmeier (Eds.), *Migraine - basic mechanisms and treatment urban & schwarzenberg* (pp. 379–393). Munchen-Wier.

Mayevsky, A., Meilin, S., Rogatsky, G. G., Zarchin, N., & Sonn, J. (1995). Multiparametric monitoring of the awake brain exposed to carbon monoxide. *Journal of Applied Physiology, 78*, 1188–1196.

Mayevsky, A., Doron, A., Manor, T., Meilin, S., Zarchin, N., & Ouaknine, G. E. (1996). Cortical spreading depression recorded from the human brain using a multiparametric monitoring system. *Brain Research, 740*, 268–274.

Mayevsky, A., Manor, T., Meilin, S., Doron, A., & Ouaknine, G. E. (1998a). Real-time multiparametric monitoring of the injured human cerebral cortex - a new approach. In R. Marmarou & R. Bullock (Eds.), *Acta Neurochir* (pp. 78–81). Suppl (Wien.) ICPX.

Mayevsky, A., Meilin, S., Manor, T., Ornstein, E., Zarchin, N., & Sonn, J. (1998b). Multiparametric monitoring of brain oxygen balance under experimental and clinical conditions. *Neurological Research, 20*, S76–S80.

Mayevsky, A., Doron, A., Meilin, S., Manor, T., Ornstein, E., & Ouaknine, G. E. (1999). Brain viability and function analyzer: multiparametric real-time monitoring in neurosurgical patients. *Acta Neurochirurgica. Supplementum (Wien), 75*, 63–66.

Mayevsky, A., Manor, T., Meilin, S., Razon, N., & Ouaknine, G. E. (2001). Multiparametric monitoring of tissue vitality in clinical situations. *SPIE, 4255*, 33–39.

Mayevsky, A., Ornstein, E., Meilin, S., Razon, N., & Ouaknine, G. E. (2002). The evaluation of brain CBF and mitochondrial function by a fiber optic tissue spectroscope in neurosurgical patients. *Acta Neurochirurgica. Supplement, 81*, 367–371.

Meilin, S., Mendelman, A., Sonn, J., Manor, T., Zarchin, N., & Mayevsky, A. (1999). Metabolic and hemodynamic oscillations monitored optically in the brain exposed to various pathological states. *Advances in Experimental Medicine and Biology, 471*, 141–146.

Rogatsky, G. G., Sonn, J., Kamenir, Y., Zarchin, N., & Mayevsky, A. (2003). Relationship between intracranial pressure and cortical spreading depression following fluid percussion brain injury in rats. *Journal of Neurotrauma, 20*, 1315–1325.

Rosenthal, M., & Jobsis, F. F. (1971). Intracellular redox changes in functioning cerebral cortex. II. Effects of direct cortical stimulation. *Journal of Neurophysiology, 34*, 750–762.

Rosenthal, M., Martel, D., LaManna, J. C., & Jobsis, F. F. (1976). In situ studies of oxidative energy metabolism during transient cortical ischemia in cats. *Experimental Neurology, 50*, 477–494.

Schuette, W. H., Lewis, D. V., O'Connor, M., & Van Buren, J. M. (1976). The design and operation of a dual-beam long-focal-length fluorometer for monitoring the oxidative metabolism in vivo. *Medical & Biological Engineering, 14*, 235–238.

Sonn, J., & Mayevsky, A. (2000). Effects of brain oxygenation on metabolic, hemodynamic, ionic and electrical responses to spreading depression in the rat. *Brain Research, 882*, 212–216.

Sonn, J., & Mayevsky, A. (2001). The effect of ethanol on metabolic, hemodynamic and electrical responses to cortical spreading depression. *Brain Research, 908*, 174–186.

Stern, M. D., Lappe, D. L., Bowen, P. D., Chimosky, J. E., Holloway, G. A., Jr., Keiser, H. R., & Bowman, R. L. (1977). Continuous measurement of tissue blood flow by laser-Doppler spectroscopy. *The American Journal of Physiology, 232*, H441–H448.

Sundt, T. M., & Anderson, R. E. (1975). Reduced nicotinamide adenine dinucleotide fluorescence and cortical blood flow in ischemic and nonischemic squirrel monkey cortex. I. Animal preparation, instrumentation, and validity of model. *Stroke, 6*, 270–278.

Van Harreveld, A., & Stamm, J. S. (1952). Vascular concomitants of spreading cortical depression. *Journal of Neurophysiology, 15*, 487–496.

Vyskocil, F., Kritz, N., & Bures, J. (1972). Potassium-selective microelectrodes used for measuring the extracellular brain potassium during spreading depression and anoxic depolarization in rats. *Brain Research, 39*, 255–259.

Wadhwani, K. C., Rapoport, S. I., Shepherd, A. P., & Oberg, P. A. (1990). Blood flow in the central and peripheral nervous systems. In A. P. Shephrd & P. A. Oberg (Eds.), *Laser doppler blood flowmetry* (pp. 265–304). Kluwer Academic Pub.

Walker, J. A. (1971). Ion specific liquid ion exchanger microelectrodes. *Analytical Chemistry, 43*, 89A–93A.

Zeuthen, T. (1971). A method to fill glass microelectrodes by local heating. *Acta Physiologica Scandinavica, 81*, 141–142.

Zeuthen, T., Hiam, R. C., & Silver, I. A. (1974). Microelectrode recording of ion activity. In H. J. Berman & N. C. Hebert (Eds.), *Brain. Ion-selective microelectrodes. Advances in experimental medicine and biology* (pp. 145–156). Springer.

Zeuthen, T., Dora, E., Silver, I. A., Chance, B., & Kovach, A. G. B. (1979). Mechanism of the cerebrocortical vasodilatation during anoxia. *Acta Physiologica Academiae Scientiarum Hungaricae, 54*, 305–318.

Chapter 3
Single Point and Multisite Monitoring of Brain NADH Fluorescence Under CSD

Abstract In this chapter, we will present our early studies on cortical spreading depression—CSD performed in the 1970s starting in 1972. After the development of the fiber-optic-based technology to monitor brain NADH redox state, we exposed the rat brain to CSD. Our monitoring system allowed us to study the effect of CSD on mitochondrial function as well as other physiological parameters in the *unanesthetized* small animal model using the flexible fiber-optic probe cemented to the surface of the brain using a special cannula cemented to the skull for few hours. In order to bridge the gap between the metabolic states of mitochondria in vitro and brain in vivo, we performed a set of studies in which the brain of rats was exposed to various perturbations while monitoring mitochondrial function.

Keywords Single site NADH monitoring · Two and four monitoring sites · Propagation of CSD wave · Dynamic range of NADH responses · Uncoupler effects in vivo

3.1 Introduction

Figure 1.27 (published in 1973), presented the first direct indication that mitochondrial NADH redox state, is affected by an increase in brain energy utilization induced by CSD. At the same year, we were able to repeat this type of study in a rat brain model as shown in Fig. 1.27. It was clear that the CSD activated mitochondrial function locally in the ipsilateral hemisphere, and the systemic blood pressure was not affected. Also, the ECoG was not affected in the contralateral hemisphere as presented in Fig. 1.27. The goal was to find the range of NADH redox state between the maximal level induced by anoxia or complete ischemia and the minimal level induced by stimulation of energy utilization using various perturbation. Part of

A. Mayevsky, J. Sonn, *Cortical Spreading Depression of Leao*,
https://doi.org/10.1007/978-3-031-08068-5_3

these experiments were performed using the time-sharing fluorometer shown in Fig.1.24B_2 and later on by the new device—the DC fluorometer seen in Chap. 2 Fig. 2.2A.

3.2 Single Point Monitoring of NADH

The first study was published in the years 1973 (Fig. 1.27) and 1974. The fluorometer and the flexible Y-shape light guide (2 mm effective diameter) were connected to the cannula above the rat brain. This system enables the animal to have a limited amount of free movement while contained in a small cage. The cannula described earlier (Mayevsky & Chance, 1973) was modified Fig. 2.2A_1B to permit KCI application near to the NADH fluorescence measurement area, via the two pairs of stainless-steel tubes, leading to the bottom of the cannula and thus onto the surface of the intact dura. These tubes were used also for measurement of the EEG. The light guide was held in a constant contact relationship to the brain by a step of 0.2 mm in the cannula. Male Wistar rats were used (200 g) and were operated under Equithesin anesthesia. The measurements on the NADH fluorescence and reflectance of 366 nm light, as well as EEG measurement, were started 30 min after the operation. The animal regained consciousness and remained alert in the small cage for 2-3 h.

In order to initiate a wave of CSD, a KCl solution was passed through one of the pairs of tubes leading to the dura. At threshold (~0.1 M KCl), only one wave of CSD developed, while at maximum (~0.4 M KCl), many cycles were observed. Figure 3.1 represents the latter condition, in which 6 CSD waves developed by 0.4 M KCl were terminated by washing with 0.9% NaCl (Mayevsky & Chance, 1974). Attention is drawn to the second trace from the top, which shows a series of NADH fluorescence decreases and increases during each wave of CSD, corresponding to states 4–3–4

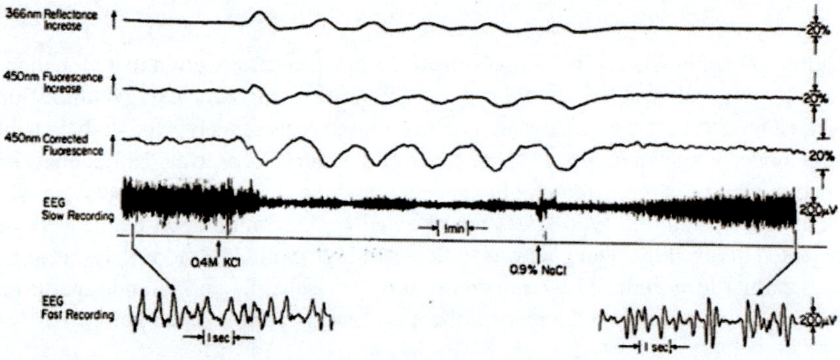

Fig. 3.1 The repetitive response of the brain to CSD evoked by application of 0.4 M KCI on the dura. The third trace is on an expanded amplitude scale. The arrow direction shows an increase in the optical signals. Time proceeds from left to right. (Mayevsky & Chance, 1974)

transitions, and may be correlated with varying energy demand during the CSD cycle. Five of the 6 cycles are identical and are closely synchronized with changes of electrical activity. The initial portion of the first cycle is obscured. In order to clarify this obscuration, the top trace records 366 nm reflectance, as has been proposed by Jobsis et al. (1971) and taken up in detail by Kobayashi et al. (1971a) and Harbig et al. (1976). Subtraction at a 1:1 ratio of the reflectance and fluorescence traces (first and second traces from the top) gives the third, corrected trace of 6 identical cycles. Our rationale for choosing a 1:1 ratio depends upon the observation that this ratio gives for the corrected trace symmetrical normoxic–anoxic cycles in rat brain, in spite of large reflectance changes, as is observed in isolated mitochondria (Chance & Williams, 1955). Kobayashi et al. (1971b) and Harbig et al. (1976) have employed replacement of blood with saline to cause a reflectance change without a change in the NADH redox state in the area under observation. Subtraction of the fluorescence and reflectance traces at a 1:1 ratio compensates correctly. If we assume that the saline flushes cause reflectance changes similar to those observed in CSD, we can apply three lines of argument in favor of the compensation: First, in Fig. 3.1, the compensation renders all 6 cycles identical and thus, comparable to the identical electrical changes observed in the 6 cycles of CSD; second, appropriate compensation for reflectance changes in normoxic-anoxic cycles is obtained; third, appropriate compensation for reflectance changes in saline flushes is obtained.

Application of 0.3 M KCI produces 3 separate waves of CSD (Fig. 3.2) which are again stopped by NaCl washing. In these 3 cycles, the initial phase of each response is obscured by the reflectance change (second trace from top), which through subtraction of the reflectance trace at a 1:1 ratio gives in the third trace from the top 3 symmetrical cycles of NADH oxidation and reduction (Mayevsky & Chance, 1974). While it is possible that there are geometrical problems about the interpretation of the exact timing of the electrical waves with respect to the optical data, it is of interest to note that the maximal mitochondrial activity occurs near the nodes of the CSD waves.

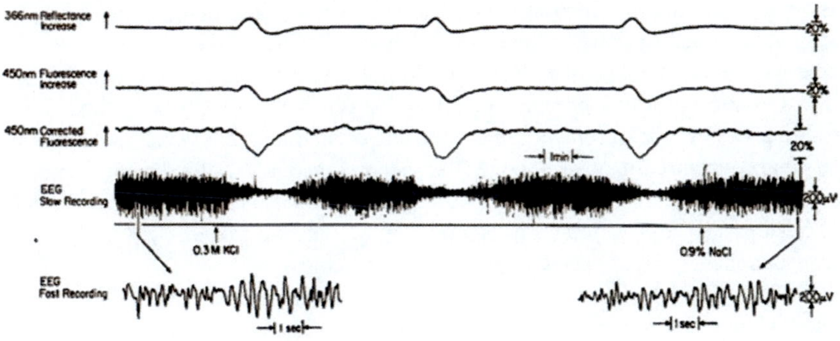

Fig. 3.2 The response of the brain to the application of 0.3 M KCI solution, epidurally. Three cycles of oxidation of NADH were developed. The arrow direction shows an increase in the optical signals. Note that the third trace is on an expanded scale. Time proceeds from left to right. (Mayevsky & Chance, 1974)

These two figures show that each cycle of the CSD involves a state 4–3-4 metabolic transition at the mitochondrial level. The extent of the metabolic transient can be evaluated by comparing the fluorescence decreases in an CSD cycle with the fluorescence increases in a normoxic–anoxic cycle (40 ± 10%) with respect to the normoxic level (Mayevsky & Chance, 1973, 1974). The fluorescence decrease in CSD is 20 ± 10%. It is characteristic of the state 4 to 3 transition in isolated mitochondria that the fluorescence decrease is equal to or greater than the increase in normoxic–anoxic transition. In CSD, we observe 20% decrease as compared with a 40% increase in anoxia. The difference between these values and those expected for a state 4–3 transition is presumably due to (1) the brain is not in state 4, but between states 4 and 3; and (2) fixed fluorescence signals from the brain (due to collagen components. Etc.) that do not respond metabolically would diminish the fractional change observed in CSD; (3) state 3 is not reached in CSD. The rapid responses of the fluorescence technique now permit much better resolution of the metabolic consequences of CSD than was possible in earlier studies (Krivanek, 1961; Lukyanova & Bures, 1967). However, better geometric resolution, i.e., smaller light guides, and better isolated application of the KC1 stimuli may be required to afford more precise time and space relationships of the metabolic and electrical parameters of CSD. Our present observation of maximal metabolic and presumably ionic transport activity at or near the node of the CSD cycle is thus of considerable interest.

In a new set of experiments, we compared the differential effects of CSD when the application site of the KCI was near or far away from the site of measurement of NADH. (These experiments were designed according to the suggestions of Dr. A. Van Hereveld). As an approximate indicator of CSD, the EEG was measured from 3 areas of the brain as seen in Fig. 3.3. One site was inside the light-guide holder (points 1–2 or 3–4) The second site was outside the light-guide holder in the same right hemisphere and the third monitoring site of ECoG was in the contralateral hemisphere (left)). As seen in part A of Fig. 3.3, the distance between the application site of KCl and NADH measurement area in the right hemisphere (the large circle) was about 2 mm (Mayevsky & Chance, 1975). When the KCI was applied inside the cannula, 4 oxidation cycles of NADH were recorded in Fig. 3.4 (right side) and stopped by NaCl (Mayevsky & Chance, 1975). These cycles showed the same pattern of changes in reflectance, fluorescence, corrected fluorescence, and EEG as were described previously (Fig. 3.1). When the KCI was applied 5 mm from the NADH measurement area, only one oxidation cycle was obtained using the same concentration of KCI (0.6 M) as seen in Fig. 3.4 left side. The first changes in EEG (depression) were recorded from the electrodes named "B" (near the KCl application site) and only later in electrodes "A." The control EEG, "C" (from the left hemisphere), did not show any significant changes.

We assumed that available oxygen or metabolic energy is a critical condition in order to keep the brain in a healthy condition. Depletion of oxygen may lead to the development of CSD and also may disturb the recovery from CSD situation. In Fig. 2.3, two exposures to 100% nitrogen are presented. In part A, the length of the nitrogen cycle was 70 seconds and the recovery from this event was fast (including the recovery of the ECoG) immediately after oxygen became available (Fig. 2.3A).

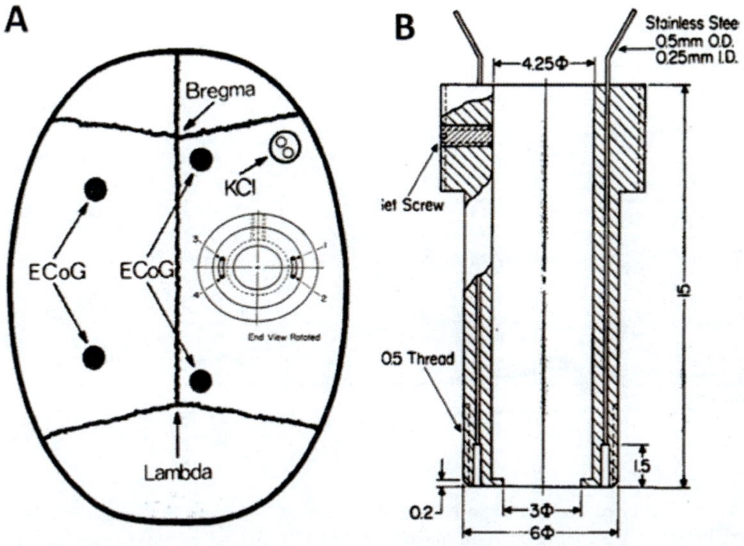

Fig. 3.3 (**A**) Schematic representation of location of cannulas and ECoG electrodes on rat's skull. (**B**) Light-guide holders (cannulas) for measurement of NADH fluorescence from the surface of the brain of an awake rat. (Mayevsky & Chance, 1975)

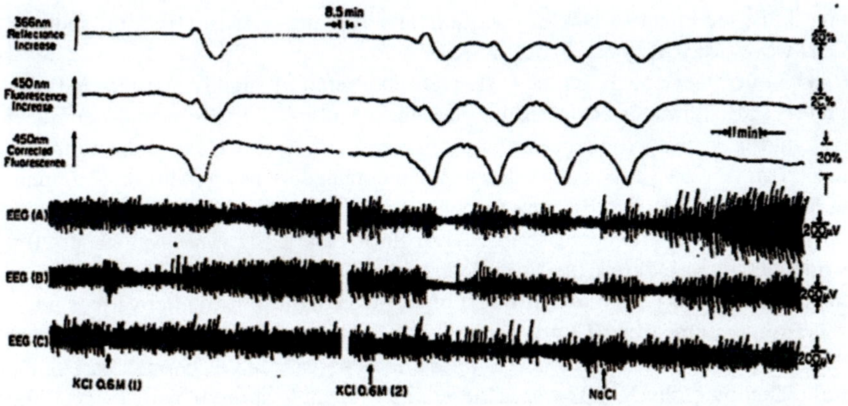

Fig. 3.4 The effects of CSD (elicited by KCl) on the NADH fluorescence 366 nm reflectance and electrical activity. The location of EEG electrode was as follows: (**A**), in the wall of the cannula; (**B**), outside the cannula; (**C**), contralateral hemisphere. (Mayevsky & Chance, 1975)

In part B, the rat was exposed to 100% nitrogen for 100 seconds as presented here also in Fig. 3.5A (Mayevsky & Chance, 1975). We assumed that during this period more potassium was leaking from the intracellular to the extracellular space and this was the stimulus that initiated a wave of CSD as seen in the bilateral ECoG signals as well as oxidation of NADH measured only in the right hemisphere. In

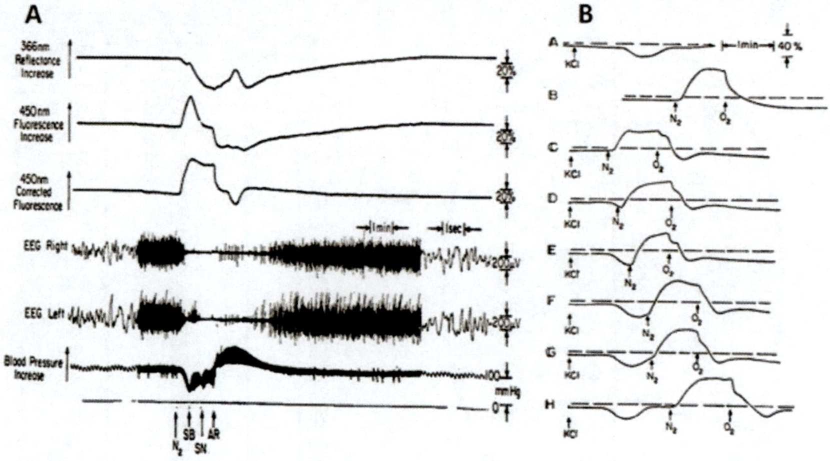

Fig. 3.5 (**A**) The effects of anoxia on NADH fluorescence, 366 nm reflectance, EEG, and blood pressure. In part (**A**), the duration of the anoxia was 100 sec. (*SB* the animal stopped breathing, *SN* stop nitrogen, *AR* short artificial respiration (Mayevsky & Chance, 1975)). (**B**) The response of the corrected fluorescence signal to CSD, N_2–0_2 breathing cycle, and the interaction between the two. The broken line represents the reference line for comparison between the different experiments (**A–H**). (Mayevsky & Chance, 1975)

Fig. 3.5B, the effect of lack of energy is avoiding the recovery from the induced CSD wave (Mayevsky & Chance, 1975).

The effects of anoxia on the CSD were measured in slightly anesthetized rats. Figure 3.5B shows the corrected fluorescence of a typical experiment. The broken straight line shows a reference arbitrary line which was very close to the normoxic level. Part A shows a typical oxidation cycle obtained by using a threshold amount of KCl (0.1–0.2 M). Part B shows the typical response to N_2-O_2 breathing cycle with no KCl addition. The decrease in NADH during the CSD cycle was about 20% while the increase during the N_2 cycle was about 45% in comparison with the normoxic level. Parts C-H show the effect of N_2 given to the animal after various intervals from the time that KCl was applied. Part H shows the N_2 cycle separated from the CSD cycle which, in practice, is the same as parts A and B. Even in part G, the full reduction of the NADH was achieved after N_2 application. In parts F, E, D, and C, in which the N_2 was applied at different stages of the CSD cycle, the level of NADH reduction was about 30–40% lower in comparison with parts B, G, and H. In parts F, E, and D, only a portion of the CSD cycle was finished before the N_2 was applied. In part C, the oxidation cycle was blocked completely by the nitrogen.

In order to analyze the results in more detail, calculation based on the corrected fluorescence data was done. In a group of 10 rats which were exposed to N_2 cycle as well as to CSD cycle, the average increase in fluorescence was 32.3 ± 6.1% and the average decrease in fluorescence with respect to the normoxic level was 17.3 ± 2.8%. The duration of the oxidation cycle was 127.1 ± 24.2 sec. The table seen in Fig. 3.6 shows the results in detail (Mayevsky & Chance, 1975).

Animal no.	*% reduction under N_2*	*% Oxidation in SD*	*Sum*	*% due to SD*	*Duration of SD cycle (sec)*
1	30	20	50	40	120
2	30	20	50	40	120
3	25	15	40	38	108
4	26	14	40	35	108
5	30	14	44	32	135
6	30	20	50	40	128
7	40	19	59	32	168
8	42	17	59	29	168
9	30	14	44	32	120
10	40	20	60	33	96
Mean ± S.D.	32.3 ± 6.1	17.3 ± 2.8	49.6 ± 7.7	35.1 ± 4.1	127.1 ± 24.2

Fig. 3.6 The effect of nitrogen and CSD on the reduction and oxidation of NADH in the brain cortex. (Mayevsky & Chance, 1975)

To determine the metabolic state of a tissue in vivo, the possibility of changing various factors, as has been done in isolated mitochondria (Chance et al., 1973), is required. In these experiments, the brain of an awake rat was exposed to various situations which caused an oxidation or reduction of NADH. The purpose of our investigation was to establish the range between the maximum increase and the maximum decrease in NADH, as compared to the normoxic level.

For the experiments during which the brain was exposed to uncoupler (pentachlorophenol), a small plastic cannula was implanted above the left side of the lateral ventricles (2 mm lateral, 2 mm posterior to bregma), to enable the injection to be made directly into the CSF (Mayevsky et al., 1977). Figure 3.7 shows the response of the brain to uncoupler injected into the lateral ventricle (Mayevsky, 1976). The left area in part A shows a typical response to N_2 exposure before the injection of the uncoupler. After injection of pentachlorophenol (PCP; 10 microliter saturated in ethanol solution), the ECoG (labeled EEG in Fig. 3.7) was affected as well as NADH, and both were similar to the response observed under CSD. The corrected fluorescence trace shows a gradual decrease (oxidation of NADH). In order to be certain that this decrease was in fact due to PCP, the animal was exposed to a N_2 cycle every few minutes. After each N_2 cycle, which causes a large increase in fluorescence, an oxidation cycle was recorded. The results show that PCP injection increases the range between maximal and minimal levels of NADH. Before PCP injection, the range was 55–60%, while after the injection it was 72–74%.

Figure 3.8 summarizes the effects on brain NADH oxidation–reduction state of various treatments given to the awake rat (Mayevsky & Chance, 1975). Nitrogen atmosphere led to a 40% increase in NADH while CSD was accompanied by a 25% decrease. Under hyperbaric oxygenation (HPO) conditions three levels of NADH oxidation were recorded (Mayevsky et al., 1974a). During phase I, the preconvulsive period, a 15% decrease in NADH was recorded. The convulsions (phase II) led to another 10% decrease, and during the third phase (probably CSD), the total oxidation of NADH was 35% as compared to the normoxic level. Uncoupler led to 10%

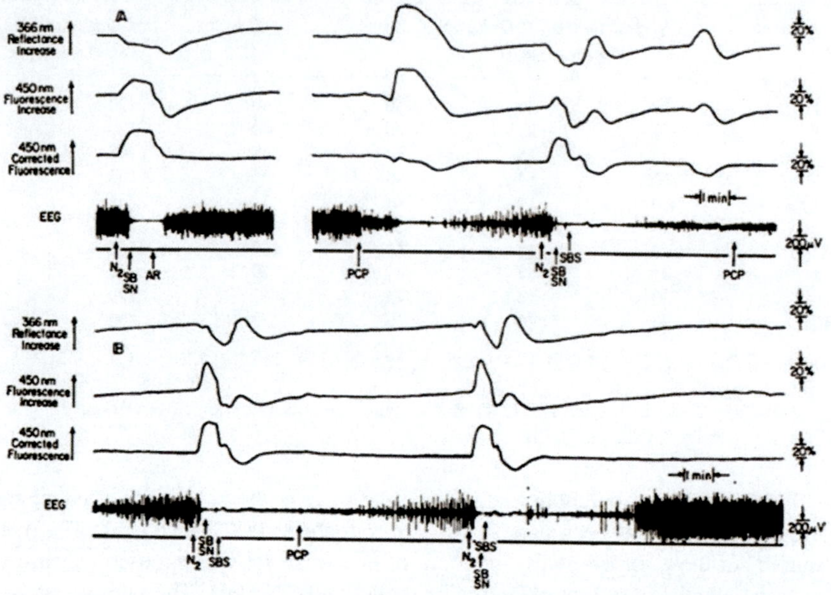

Fig. 3.7 Metabolic response of the brain to uncoupler (PCP) injected to the lateral ventricle. Part (**B**) is a continuation of (**A**). (*N_2* application of nitrogen, *SB* the animal stopped breathing, *SN* stop of nitrogen application, *AR* short artificial respiration, *SBS* the animal started breathing spontaneously. (Mayevsky, 1976))

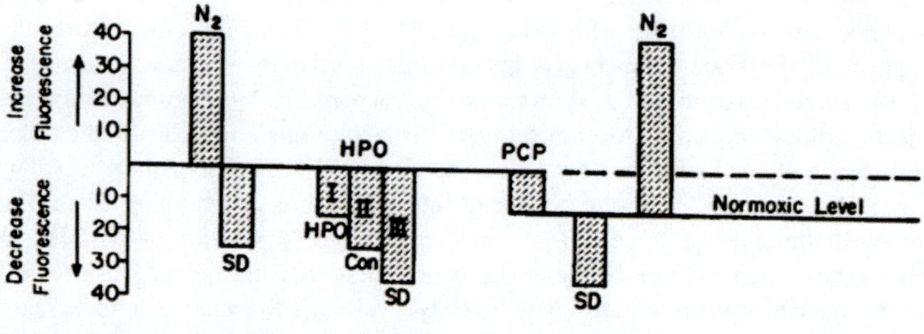

Fig. 3.8 The effect of various treatments on the oxidation–reduction state of NADH in the brain. (*N_2* nitrogen, *Con* convulsions, *CSD* cortical spreading depression, *PCP* pentachlorophenol (uncoupler), *HPO* hyperbaric oxygenation (Mayevsky, 1976))

oxidation of NADH. Therefore, under our experimental conditions, the maximal increase of brain NADH was 40%, while the maximal decrease was about 35%.

The identification of the metabolic state of an awake cerebral cortex was done by exposing our model to various treatments which caused an increase or decrease of NADH levels. The large decrease in NADH caused by CSD, HPO, and uncoupler can be interpreted as a state 4 to state 3 transition. In their study, Chance et al. (1965, 1966) were unable to show an effect of uncoupler on the NADH levels and concluded that this was due to the fact that the brain studied was in the active state (state 3). In the studies described (Fig. 3.7), the effect of uncoupler was successfully demonstrated, leading us to conclude that the awake cerebral cortex has an oxidation–reduction state of NADH between states 3 and 4, closer to 4.

3.3 Two Points Monitoring of NADH

A two-channel DC fluorometer/reflectometer was build according to the principles described earlier (Chance et al., 1975) and is shown in Fig. 3.9A (Mayevsky & Bar-Sagie, 1978). The excitation light (366 nm) from a 100-watt Hg arc is divided equally by a special light guide (Schott Co., Mainz, Germany) into the two spots of the measurements (2 mm diameter active fibers). The emitted light from each side of the brain was split by a special mirror so that 90% was used for fluorescence measurements and 10% for reflectance. Appropriate filters were located in front of the photomultipliers. Measurements were recorded by a multichannel Grass polygraph.

The measurement of intramitochondrial NADH oxidation–reduction state from the brain of an awake animal is a well-established technique and is in use for the last few years. We had used this animal model in studying the effects of hypoxia, CSD, epileptiform activity (Mayevsky & Chance, 1975), or ischemia (Mayevsky, 1978) on the metabolic activity. In all of the studies made by our group or other groups, only one site or side of the brain was monitored fluorometrically. In the ischemia studies, we found that under certain experimental conditions it is very important to

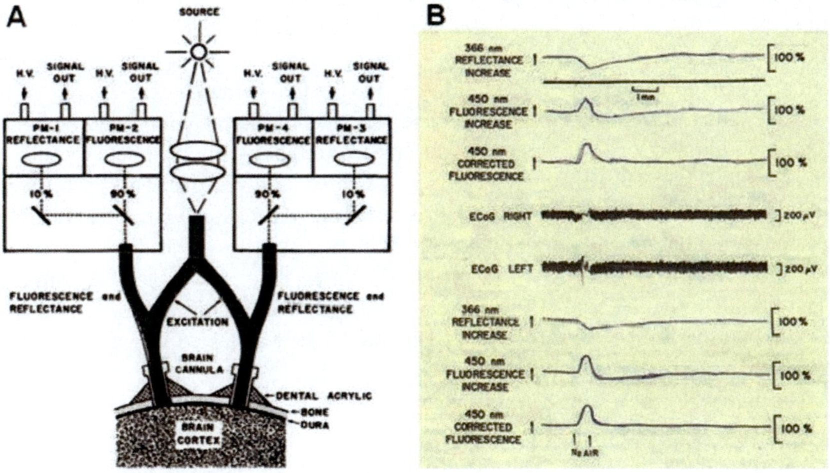

Fig. 3.9 (A) Two-channel DC fluorometer reflectometer for simultaneous measurement of NADH from two points of an awake brain. (B) The response of the brain NADH to 1 min anoxia measured from the right and left hemispheres. (Mayevsky & Bar-Sagie, 1978)

measure metabolic activity from the "experimental" site as compared to the "control" site or side simultaneously. In our awake model, we used a topical application technique to study the effects of drugs on the cortical tissue without any effects on the rest of the brain. This model also requires a bilateral measurement technique. We present in this presentation a new type of fluorometer/reflectometer by which NADH fluorescence, as well as reflected light, can be measured simultaneously from two points of an awake brain model. This model was exposed to anoxia, CSD, or ischemia and the results are presented. In another group of rats, the effects of topical application of 2-deoxyglucose were tested.

Wister male rats were used in our studies (100–200 gr.). After anesthesia with Equithesin, the animal was mounted to a head holder during the operation. Two holes of 5 mm diameter were drilled in the middle of the right and left parietal bone. In each hole a Plexiglas cannula was screwed epidurally. Two pairs of stainless-steel screws were implanted for ECoG measurements from both hemispheres. Two push–pull cannulas were also implanted for topical application of KCl solution in order to elicit CSD. The cannulas and screws were cemented to the skull by dental acrylic. The common parts of the light guide were inserted through the cannulas and measurements were started 30 min after the operation.

In order to test the new instrument, we used the N_2–0_2 breathing cycle and a typical result is shown in Fig. 3.9B (Mayevsky & Bar-Sagie, 1978). The upper three traces were measured from the right hemisphere and the lower three from the left hemisphere. The response to nitrogen was identical from the two hemispheres as one expected and was similar to those described previously using the time-sharing technique (Mayevsky & Chance, 1975).

Figure 3.10 shows a typical response of the brain to CSD elicited by O.5 M KCl solution. (Mayevsky & Bar-Sagie, 1978). One can see that the two hemispheres show the typical oxidation cycles due to increase in energy utilization in order to pump the large amount of K^+ ions into the cells (Mayevsky et al., 1974b). The frequency of the cycles was higher in the left hemisphere and this is probably due to difference in the washing of KCl solution above the two hemispheres.

After obtaining the metabolic response to CSD, the light guide was removed from the brain and 2-DG (65 mg/ml) was applied to the right hemisphere by filling the cannula. The left hemisphere served as a control and glucose (65 mg/ml) was applied. After 165 min, the two solutions were removed from the brain and NADH was measured before and after elicitation of CSD.

Figure 3.11 (Mayevsky & Bar-Sagie, 1978) shows the results obtained from the same brain shown in Fig. 3.10. It is very clear that 2-DG inhibited the metabolic response of the right hemisphere to CSD. This effect was tested in many animals and found to be very repeatable.

The inhibition of glucose metabolism by 2-DG was suggested previously by many investigators (Brown, 1962; Miselis & Epstein, 1975; Sakata et al., 1963) and our study shows that it is possible to obtain it in the cortical tissue by local application.

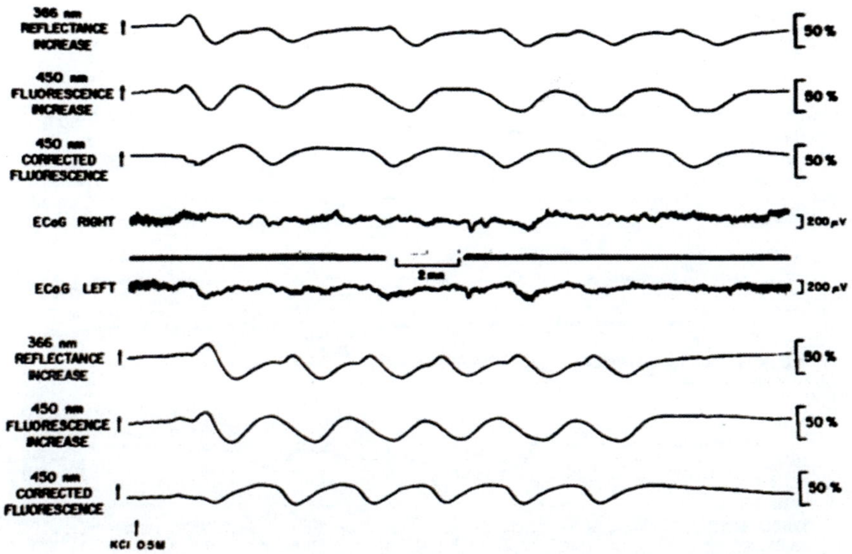

Fig. 3.10 The effects of CSD (elicited by 0.5 M KCl solution applied bilaterally) on the metabolic response of an awake brain. (Mayevsky & Bar-Sagie, 1978)

3.4 Four Points Monitoring of NADH

In order to monitor NADH redox state from various sites in the awake brain, we developed a new 4-channel DC fluorometer reflectometer connected to 4 points on the brain (Mayevsky & Breuer, 1992). The source for the excitation light was a 100-Watt water-cooled Mercury lamp split into 4 branches of the light guide after passing the lens and 366-nm filter (Corning 5860, 7–37 plus 9782, 4–96). The emitted light from the brain, transmit through the other bundle of fibers, is split in a 90:10 ratios in front of the two photomultipliers (RCA 93IB). The fluorescence and reflectance signals were amplified and calibrated as described previously (Chance et al., 1974; Mayevsky, 1976). The signals obtained from the normoxic brain after the operation served as a 100% of the signals and all changes were calculated as percentage change as compared to the normoxic level. Two types of light guides were used in the present study; a 2-mm active diameter of fibers was used in studies where the light guides were placed as shown in Fig. 3.12. In other experiments, we used the small diameter light guides (0. 8 mm active diameter).

Figure 3.13 shows typical responses to anoxia (A) and CSD (B) measured from a gerbil brain (Mayevsky & Chance, 1983). From each measuring site, only two channels were recorded namely the reflectance (R) and the corrected fluorescence

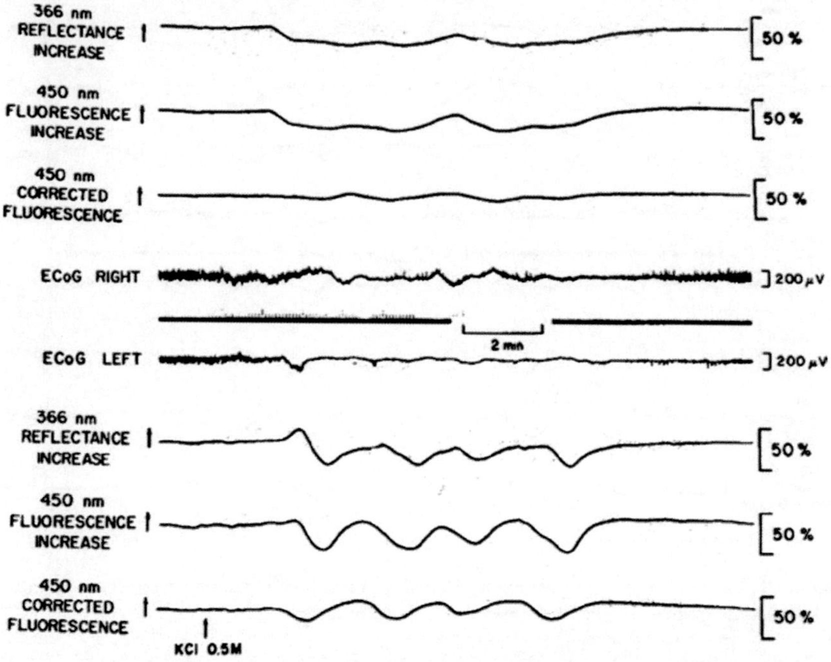

Fig. 3.11 The metabolic response of an awake brain to CSD after application of glucose (lower part) and 2-deoxyglucose (upper part) for 165 min. (Mayevsky & Bar-Sagie, 1978)

(CF). The uncorrected fluorescence was not recorded due to limitation in the recorder. Immediately after breathing the gerbil with N_2, the CF showed an increase of about 50%. A full recovery was achieved after returning the gerbil to air mixture. The reflectance trace showed a typical two-step decrease as described previously (Mayevsky, 1976; Mayevsky & Chance, 1973). This decrease is due to the increase in blood volume in the measuring site and this artifact was subtracted from the fluorescence signal to obtain the net corrected NADH fluorescence change (CF). We used the 1:1 correction factor in all animals. The variation between the four measuring sites is very small in respect to the kinetics of the changes as well as the magnitude of them. The recovery of the reflectance traces to the baseline was much slower as compared to the CF traces due to the longer time needed for the blood vessels to constrict back to the original diameter. The propagation of a CSD wave is shown in Fig. 3.13B (Mayevsky & Chance, 1983). The wave was initiated by washing the dura with a 1 M KCl solution via the push–pull cannula (labeled KCl in the center scheme). All four sites measured showed an oxidation cycle as a response to CSD (decrease in CF). The reflectance (R) trace showed also a biphasic response namely an initial increase and later on a long-lasting decrease until a complete recovery to the baseline levels. The vertical dashed line is a continuation of the time when the wave was under site 1 and could serve as a reference point to calculate the speed of the CSD wave propagation. For example, the distance between site 1 and 3 is about

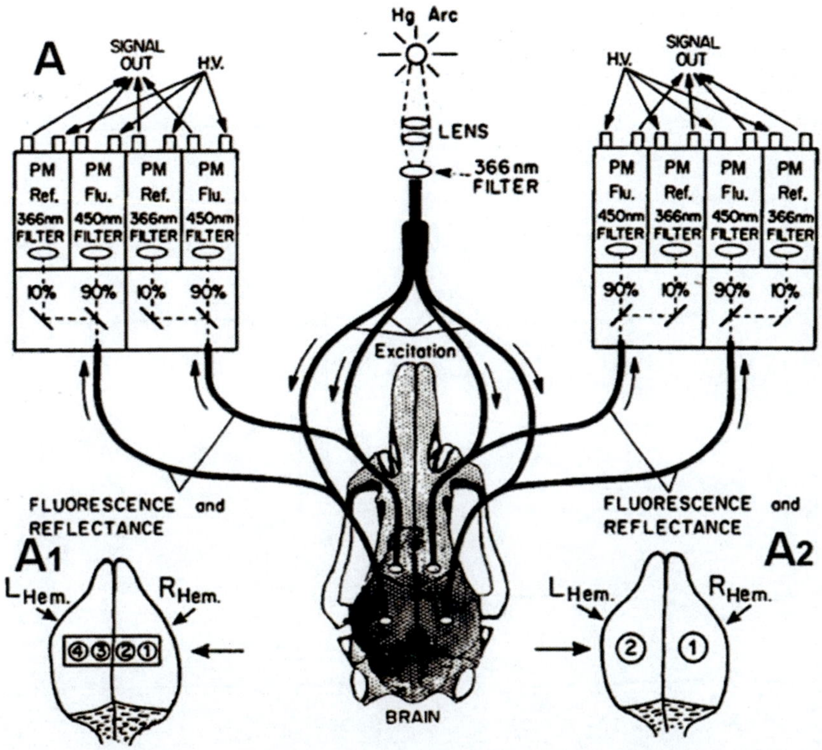

Fig. 3.12 Schematic presentation of the 4-channel DC fluorometer reflectometer used in the present study. Each channel contained a 366 nm as well as 450 nm measurement for reflected light and fluorescence, respectively. (*PM* photomultiplier, *HV* high voltage (Mayevsky & Breuer, 1992))

2.5 mm and the difference in the response was 50 sec so the speed of the propagation of the response to CSD was 3 mm/min as was calculated previously by monitoring electrical events (Leão & Morison, 1945).

Figure 3.14 shows a typical response of the gerbil brain to anoxia and CSD (Mayevsky & Chance, 1982). These responses were found in all normoxic gerbil brains (more than 300 gerbils tested during the last 3 years). The animal was breathing air spontaneously, and when it was exposed to 100 percent nitrogen, a typical two-step decrease in reflectance was recorded together with a large increase in corrected fluorescence, which reached a plateau shortly thereafter. When air breathing was restored, a fast reoxidation of NADH was recorded. In this gerbil, the anoxia induced a secondary response characterized by an oxidation cycle of NADH appearing first in site 1, then propagated to site 4 through sites 2 and 3. This phenomenon was a CSD initiated during the anoxia by the intrinsic elevated extracellular K^+ ions (Mayevsky et al., 1980). The same response to CSD was monitored after initiation of a wave by application of KCl (Fig. 3.14B) (Mayevsky & Chance, 1982). When the wave reached site 1 (the one closest to the KCl application site), a typical

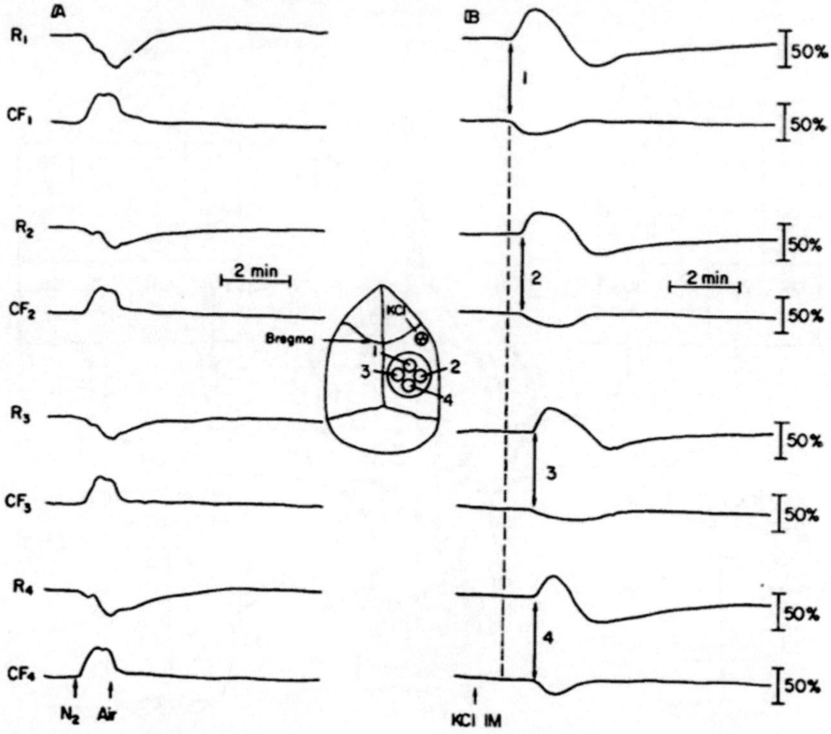

Fig. 3.13 The effects of anoxia (**A**) and CSD (**B**) on the reflectance (R) and corrected fluorescence (CF) measured from 4 sites in the gerbil brain (1–4 in the center schematic). (Mayevsky & Chance, 1983)

biphasic change in the reflected light was recorded. The corrected fluorescence shows an oxidation cycle lasting between 1 and 2 min, depending on the measuring site in relation to the propagation front of the CSD. The reflectance response had a short increase phase, followed by a very long decrease below base line and then a recovery phase. Four sources of error, which may affect the measurement of NADH fluorescence in vivo, are (1) movement artifacts; (2) changes in the oxygenation level of the blood; (3) changes in the absorption properties of the tissue monitored; and (4) blood volume changes in the tissue under observation. We found that when a good contact is made between the bundle of fibers and the brain (or any other organ), all movement artifacts are eliminated or avoided. The oxy-deoxy hemoglobin absorption changes are minimal in our system (Chance et al., 1973; Mayevsky & Chance, 1973; Mayevsky & Chance, 1975). Little is known about the changes in the absorption properties of the tissue during the various perturbations of the observed measuring site, since it is difficult to separate these changes from other factors affecting the NADH fluorescence reading. We have shown indirectly that under physiological or pathological conditions where there is movement of ions and water between the intracellular and the extracellular space, changes in the

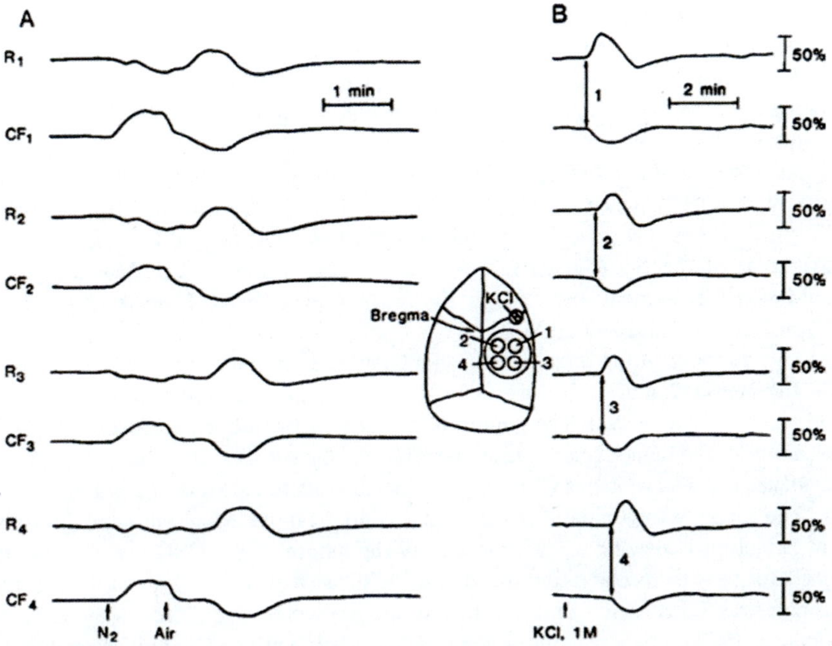

Fig. 3.14 Metabolic responses of a gerbil brain to (**A**) anoxia and (**B**) CSD in four different locations in one hemisphere. Arrows 1 to 4 in (**B**) show the propagation of the CSD throughout the measured area. The schematic drawing between (**A**) and (**B**) shows the locations of the four light. Guides above the parietal cortex area. (Reprinted with permission from AAAS. (Mayevsky & Chance, 1982))

absorption properties of the tissue being monitored will have an effect on NADH fluorescence measurements and we have corrected for this effect by the subtraction technique. Various investigators have attempted to correct for changes in the blood volume of the tissue under observation. Aubert et al. (1964), using slices of the electric organ of the electric fish, monitored the reflected light. Jobsis et al. (1971) used the reflected light at 366 nm and found that 1:1 subtraction was suitable for many brain preparations, although not for all. Using a fiber-optic fluorometer, we found in 1973 that 1:1 is a good correction factor for most of the brains tested by fast injection of saline into the common carotid artery (Mayevsky et al., 1984; Mayevsky & Ziv, 1991). This approach is today the dominant one for surface fluorometry in vivo (Mayevsky, 1990; Mayevsky & Chance, 1982). Dora and Kovach (1979) described the correction factor problem in monitoring cat brain NADH in vivo; they are using the saline flush technique to check and use a variable correction factor. Kramer and Pearlstein (1979), using an isosbestic microfluorometer rather than the 366 nm reflectance fluorometer, reported that the typical response to anoxia was "unexpected initial early oxidation," although an explanation for this finding was not given. Under anoxia, the kinetics of the corrected fluorescence would be expected to be similar to that of mitochondria, and this has been observed by many investigators. We believe that the correction factor, as well as the exact mathematical equation to be used for the correction, is dependent on the optical system used in the fluorometer and on the relation between the optical system used in the fluorometer and on the relation between the optical system and the tissue being monitored.

Activation of the mitochondria by adenosine diphosphate may result in an oxidation of NADH (Chance & Williams, 1955), and indeed, most investigators have shown that direct cortical stimulation in vivo, convulsions, and CSD lead to the oxidation of the brain NADH (Jobsis et al., 1971; Jobsis & LaManna, 1978; LaManna & Rosenthal, 1975; Mayevsky & Bar-Sagie, 1978; Mayevsky & Chance, 1975; Rosenthal & Jobsis, 1971).

The response of the brain to CSD can be used as an indicator for the intactness of the hemodynamic and metabolic compensation mechanisms in the brain (Mayevsky et al., 1982). The pattern of changes of the reflectance trace is a good indicator of the hemodynamic responses. Hence, the use of a short anoxia cycle (30 to 40 seconds) as well as a CSD can disclose the functional state of a brain.

The shape of the oxidation cycle was not identical in all four sites and a further investigation is needed in order to clarify the nature of the CSD wave shape. In order to study the propagation speed of the CSD wave, a second type of light-guide holder was constructed and the results are shown in Fig. 3.15 (Mayevsky & Chance, 1983). The light guides were organized along the KCl application site. In

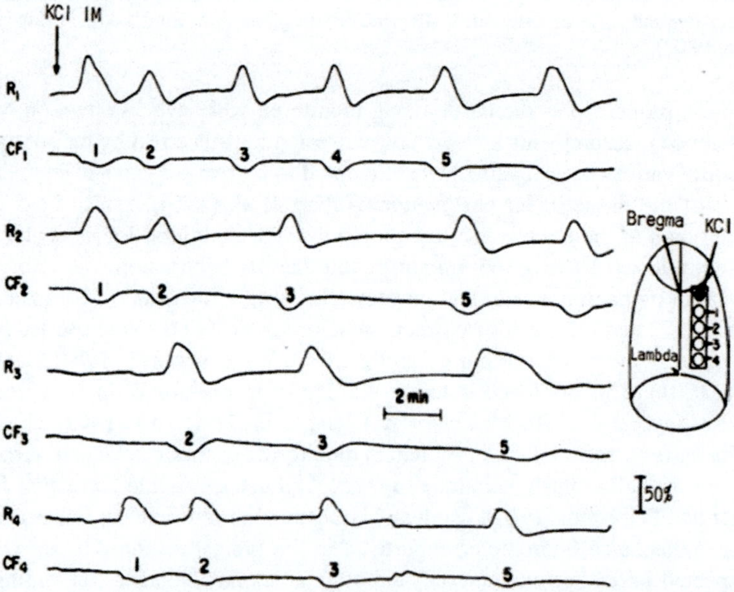

Fig. 3.15 Repetitive metabolic responses to CSD measured from 4 sites in the awake gerbil brain. (Mayevsky & Chance, 1983)

this specific gerbil, the response in site 3 wave was not as expected, namely, that only three waves were recorded while 4 waves were measured in site 4. We showed in the past (Mayevsky et al., 1981) that not all the metabolic waves started at one point in the brain will propagate through the entire cortex. As can be seen in Fig. 3.15, in site 1, five cycles were recorded and cycle number 4 did not propagate throughout the cortex. The distance between sites 1 and 4 was 5 mm, and the difference in the responses was about 130 seconds so the speed in this gerbil was 2. 3 mm/min.

References

Aubert, X., Chance, B., & Keynes, R. D. (1964). Optical studies of biochemical events in the electric organ of Electrophorus. *Proceedings of the Royal Society A, 160*, 211–245.

Brown, J. (1962). Effects of 2-deoxyglucose on carbohydrate metabolism: review of the literature and studies in the rat. *Metabolism, 11*, 1098–1112.

Chance, B., & Williams, G. R. (1955). Respiratory enzymes in oxidative phosphorylation (III- The steady state). *The Journal of Biological Chemistry, 217*, 409–427.

Chance, B., Jamieson, D., & Coles, H. (1965). Energy-linked pyridine nucleotide reduction: Inhibitory effects of hyperbaric oxygen in vitro and in vivo. *Nature, 4981*, 257–263.

Chance, B., Jamieson, D., & Williamson, J. R. (1966). *Control of the oxidation-reduction state of reduced pyridine nucleotides in vivo and in vitro by hyperbaric oxygen, Third international conference on hyperbaric medicine* (pp. 15–41). National Academy of Sciences.

Chance, B., Oshino, N., Sugano, T., & Mayevsky, A. (1973). Basic principles of tissue oxygen determination from mitochondrial signals. *Oxygen Transport to Tissue, Advances in Experimental Medicine and Biology, 37A*, 277–292.

Chance, B., Mayevsky, A., Goodwin, C., & Mela, L. (1974). Factors in oxygen delivery to tissue. *Microvascular Research, 8*, 276–282.

Chance, B., Legallias, V., Sorge, J., & Graham, N. (1975). A versatile time-sharing multichannel spectrophotometer reflectometer and fluorometer. *Analytical Biochemistry, 66*, 498–514.

Dora, E., & Kovach, A. G. B. (1979). Reactivity of the cerebrocortical vasculature and energy metabolism to direct cortical stimulation in haemorrhagic shock. *Acta Physiologica Academiae Scientiarum Hungaricae, 54*, 347–361.

Harbig, K., Chance, B., Kovach, A. G. B., & Reivich, M. (1976). In vivo measurement of pyridine nucleotide fluorescence from cat brain cortex. *Journal of Applied Physiology, 41*, 480–488.

Jobsis, F. F., & LaManna, J. C. (1978). Kinetic aspects of intracellular redox reactions. In vivo effects during and after hypoxia and ischemia. In E. Robin (Ed.), *Extrapulmonary manifestations of respiratory disease* (pp. 63–106). Marcel Dekker.

Jobsis, F. F., O'Connor, M., Vitale, A., & Vreman, H. (1971). Intracellular redox changes in functioning cerebral cortex. I. Metabolic effects of epileptiform activity. *Journal of Neurophysiology, 3465*, 735–749.

Kobayashi, S., Kaede, K., Nishiki, K., & Ogata, E. (1971a). Microfluorometry of oxidation-reduction state of the rat kidney in situ. *Journal of Applied Physiology, 31*, 693–696.

Kobayashi, S., Nishiki, K., Kaede, K., & Ogata, E. (1971b). Optical consequences of blood substitution on tissue oxidation- reduction state microfluorometry. *Journal of Applied Physiology, 31*, 93–96.

Kramer, R. S., & Pearlstein, R. D. (1979). Cerebral cortical microfluorometry at isosbestic wavelengths for correction of vascular artifact. *Science, 205*, 693–696.

Krivanek, J. (1961). Some metabolic changes accompanying Leão's spreading cortical depression in the rat. *Journal of Neurochemistry, 6*, 183–189.

LaManna, J. C., & Rosenthal, M. (1975). Effect of ouabain and phenobarbital on oxidative metabolic activity associated with spreading cortical depression in cats. *Brain Research, 88*, 145–149.

Leão, A. A. P., & Morison, R. (1945). Propagation of spreading cortical depression. *Journal of Neurophysiology, 8*, 33–45.

Lukyanova, L. D., & Bures, J. (1967). Changes in pO_2 due to spreading depression in cortex and nucleus caudatus of rat. *Physiologia Bohemoslovaca, 16*, 449–455.

Mayevsky, A. (1976). Brain energy metabolism of the conscious rat exposed to various physiological and pathological situations. *Brain Research, 113*, 327–338.

Mayevsky, A. (1978). Ischemia in the brain: The effects of carotid artery ligation and decapitation on the energy state of the awake and anesthetized rat. *Brain Research, 140*, 217–230.

Mayevsky, A. (1990). Level of ischemia and brain functions in the Mongolian gerbil in vivo. *Brain Research, 524*, 1–9.

Mayevsky, A., & Bar-Sagie, D. (1978). The interrelation between CBF, energy metabolism and ECoG in a new awake brain model. *Advances in Experimental Medicine and Biology, 92*, 761–768.

Mayevsky, A., & Breuer, Z. (1992). Brain vasculature and mitochondrial responses to ischemia in gerbils. I. Basic anatomical patterns and biochemical correlates. *Brain Research, 58*, 242–250.

Mayevsky, A., & Chance, B. (1973). In H. I. Bicher & D. F. Bruley (Eds.), *A new long-term method for the measurement of NADH fluorescence in intact rat brain with implanted cannula* (pp. 239–244). Plenum Press.

Mayevsky, A., & Chance, B. (1974). Repetitive patterns of metabolic changes during cortical spreading depression of the awake rat. *Brain Research, 65*, 529–533.

Mayevsky, A., & Chance, B. (1975). Metabolic responses of the awake cerebral cortex to anoxia hypoxia spreading depression and epileptiform activity. *Brain Research, 98*, 149–165.

Mayevsky, A., & Chance, B. (1982). Intracellular oxidation-reduction state measured in situ by a multichannel fiber-optic surface fluorometer. *Science, 217*, 537–540.

Mayevsky, A., & Chance, B. (1983). Multisite measurements of NADH redox state from cerebral cortex of the awake animal. *Advances in Experimental Medicine and Biology, 159*, 143–155.

Mayevsky, A., & Ziv, I. (1991). Oscillations of cortical oxidative metabolism and microcirculation in the ischaemic brain. *Neurological Research, 13*, 39–47.

Mayevsky, A., Jamieson, D., & Chance, B. (1974a). Oxygen poisoning in the unanesthetized brain: correlation of the oxidation-reduction state of pyridine nucleotide with electrical activity. *Brain Research, 76*, 481–491.

Mayevsky, A., Zeuthen, T., & Chance, B. (1974b). Measurements of extracellular potassium, ECoG and pyridine nucleotide levels during cortical spreading depression in rats. *Brain Research, 76*, 347–349.

Mayevsky, A., Gassner, S., & Samuel, D. (1977). A permanent plastic cannula for intraventricular injections into rat brain. *Journal of Neuroscience Research, 3*, 81–83.

Mayevsky, A., Crowe, W., & Mela, L. (1980). The interrelation between brain oxidative metabolism and extracellular potassium in the unanesthetized gerbil. *Neurological Research, 1*, 213–226.

Mayevsky, A., Mizawa, I., & Sloviter, H. A. (1981). Surface fluorometry and electrical activity of the isolated rat brain perfused with artificial blood. *Neurological Research, 3*, 307–316.

Mayevsky, A., Zarchin, N., & Friedli, C. M. (1982). Factors affecting the oxygen balance in the awake cerebral cortex exposed to spreading depression. *Brain Research, 236*, 93–105.

Mayevsky, A., Ziv, I., Ventura, V., & Levy, Y. (1984). *Multiparametric oscillations in the rat brain in vivo, Oscillations in Physiological Systems: Dynamics and Control. The Institute of Measurement and Control* (pp. 205–210).

Miselis, R. R., & Epstein, A. N. (1975). Feeding induced by intracerebroventricular 2-deoxy-D-glucose in the rat. *The American Journal of Physiology, 229*, 1438–1447.
Rosenthal, M., & Jobsis, F. F. (1971). Intracellular redox changes in functioning cerebral cortex. II. Effects of direct cortical stimulation. *Journal of Neurophysiology, 34*, 750–762.
Sakata, K., Hayano, S., & Sloviter, H. A. (1963). Effect on blood glucose concentration of changes in availability of glucose to the brain. *The American Journal of Physiology, 204*, 1127–1132.

Chapter 4
Sequence of Physiological Responses Recorded During Induced CSD Event

Abstract Since cortical spreading depression—CSD—is an event that led to increase in oxygen consumption, it was necessary to identify the mechanism behind the increase of oxygen supply to the mitochondria. The possible two options were tested, and from the results presented in this chapter, it was concluded that the extra oxygen needed for the recovery from the CSD wave was provided to the brain by the doubling of the CBF and not by downloading of oxygen from the oxygenated blood provided to the brain. This finding was the first and the sole direct proof of the very significant and important mechanism that regulates oxygen supply to the brain.

Keywords Multiparametric responses to CSD · Oxygen consumption during CSD · Hemoglobin oxygenation in CSD · Sequence of responses during CSD · CBF changes in CSD

4.1 Introduction

The first 30 years of CSD research (early 1940s to early 1970s) were summarized intensively in the book published in 1974 by Bures et al. (1974). During those 30 years, the physiological and biochemical effects of CSD were described in the published book. Chapter 2 covered the "Electrical properties of Spreading Depression." In Chap. 6, they discussed the "Chemical Aspects of Spreading Depression," and The "Mechanism of Spreading Depression," and the various hypothetical explanations were presented in Chap. 8 of this unique book. Chapters 2, 6, and part of 7 are presented in the appendix of our current new book.

As described in previous chapters, the discovery of brain CSD (Leão, 1944b) was done by Aristides Leão in 1944 (Fig. 1.5). After cortical stimulation, he showed that a wave of EEG depression propagated from the stimulation point to the rest of the stimulated hemisphere. In parallel to the EEG depression, the steady electrical potential shows a negative shift as seen in Fig. 1.9 (Leao, 1944).

A. Mayevsky, J. Sonn, *Cortical Spreading Depression of Leao*,
https://doi.org/10.1007/978-3-031-08068-5_4

It has become clear that the wave of depolarization passing through the tissue increases energy consumption (Leão, 1944b). Figures 1.6, 1.7, and 1.8 show the effects of CSD wave on the EEG activity and the diameter of the blood vessels on the surface (Leão, 1944a). The involvement of the microcirculation in the response to CSD was described by Van Harreveld and colleagues few years after the discovery of the phenomenon (Van Harreveld & Ochs, 1957a; Van Harreveld & Stamm, 1952). In other studies, the measurements of pO_2 in the brain had indicated that demand for energy and oxygen consumption increased during the CSD wave (Bures, 1956; Krivanek, 1961; Lukyanova & Bures, 1967). In order to understand the reason for the large increase in oxygen consumption during CSD, Bures and Krivanek (1960) used a special approach to check ion shift in the rat cortex and showed that extracellular potassium is elevated (Krivanek & Bures, 1960). In the early 1970s, Vyskocil et al. used a potassium-selective microelectrode and showed clearly that extracellular K^+ was elevated significantly (Vyskocil et al., 1972b).

It is very important to use the appropriate terminology when dealing with CSD and hypoxia or ischemia. The discovery of the CSD event happened when rabbits' brain was exposed to electrical stimulation in order to elicit "experimental epilepsy." The rabbits were breathing normally and oxygen supply to the brain was normal. Under ischemia, the EEG is also depressed as soon as the oxygen supply to the brain is limited but the mechanism behind the EEG depression is different from that operated in CSD. We will discuss this issue later in the book.

The effects of CSD on brain mitochondrial NADH were first described in 1973, by Rosenthal and Somjen, where a clear oxidation wave was recorded (Rosenthal & Somjen, 1973). Figure 1.26 shows the response of the cat brain to a wave of CSD (Rosenthal & Somjen, 1973).

In the same year (Mayevsky & Chance, 1973), we showed the same NADH oxidation response to CSD initiated in the rat brain (see Fig. 1.27).

A year later, we demonstrated the coupling between the elevated extracellular K^+ induced by CSD and the greater ATP requirement (NADH oxidation) needed to recover the normal ionic homeostasis (Mayevsky et al., 1974). In the same year, we studied in detail the nature of the NADH response to CSD as seen in Figs. 3.1 and 3.2 (Mayevsky & Chance, 1974).

In order to initiate a wave of CSD, KCl solution was passed through one of the pairs of tubes located above the Dura mater. At threshold (~0.1 M KCl), only one wave of CSD developed, while at maximum (~0.4 M KCl) many cycles were observed. Figure 3.1 represents the latter condition, in which six CSD waves developed by 0.4 M KCl and were terminated by washing the brain surface with 0.9% NaCl. Attention is drawn to the second trace from the top, which shows a series of NADH fluorescence decreases and increases during each wave of CSD, corresponding to states 4-3-4 transitions, and may be correlated with varying energy demand during the CSD cycle. Five of the 6 cycles are identical and are closely synchronized with changes in electrical activity.

Application of 0.3 M KCl produces three separate waves of CSD (Fig. 3.2) which are again stopped by NaCl washing. In these 3 cycles, the initial phase of each response shows a short increase in the fluorescence signal. The subtraction of the reflectance trace from the fluorescence at a 1:1 ratio gives in the third trace from the top 3 symmetrical cycles of NADH oxidation and reduction. While it is possible that there are geometrical problems about the interpretation of the exact timing of the electrical waves with respect to the optical data, it is of interest to note that the maximal mitochondrial activity occurs near the nodes of the CSD waves.

In order to test the hypothesis that CSD wave recovery is dependent on the availability of oxygen, another experiment was designed and the results appear in Fig. 3.5B (Mayevsky & Chance, 1975). The effects of anoxia on the CSD were measured in slightly anesthetized rats. The broken straight line shows a reference arbitrary line which was very close to the normoxic level. Part A shows a typical oxidation cycle obtained by using a threshold amount of KCl (0.1–0.2 M). Part B shows the typical response to N_2-O_2 breathing cycle with no KCl addition. The decrease in NADH during the CSD cycle was about 20% while the increase during the N_2 cycle was about 45% in comparison with the normoxic level. Parts C-H show the effect of N_2 given to the animal after various intervals from the time that KCl was applied. Part H shows the N_2 cycle separated from the CSD cycle which, in practice, is the same as parts A and B. Even in part G, the reduction of the NADH (increased signal) was achieved after N_2 application. In parts F, E, D, and C, in which the N_2 was applied at different stages of the CSD cycle, the level of NADH reduction was about 30–40% lower in comparison with parts B, G, and H. In parts F, E, and D, only a portion of the CSD cycle was finished before the N_2 was applied. In part C, the oxidation cycle was blocked completely by the nitrogen.

The recovery of the brain after an episode of CSD is dependent upon oxygen and energy availability. If energy production is in the normal range, the pumping of potassium into the cells starts immediately and its level recovers back to 3 mM. When energy production (ATP) is inhibited, the pumping of K^+ is prevented, as shown in Fig. 4.1 (Mayevsky et al., 1986) . In this animal, waves of CSD developed spontaneously at a certain stage of the experiment (after a few hypoxic episodes) and the left side of the figure shows the regular response to CSD. The efflux of K^+ reached a level of 14–15 mM and was then pumped back into the cells and the NADH (CF) shows a clear oxidation cycle as shown also in Fig. 4.5. In the second CSD event shown in Fig. 4.1, hypoxia was induced at the beginning of the recovery period and the recovery process was inhibited until 100% O_2 was given again to the rat.

In order to understand better the recovery phase from CSD, it was necessary to monitor CBF in parallel to the other parameters monitored. Indeed, in the early 1990s, we added the laser Doppler probe to our multiparametric monitoring system as shown in Fig. 2.11. Another record is presented. Figure 4.2 shows the use of a multiparametric system that includes a laser Doppler flowmeter to record microcirculatory blood flow (Mayevsky et al., 1992). As seen, two sets of responses to CSD were recorded after 0.1 M KCl application to the surface of the brain 2 mm anterior to the MPA. A clear correlation between the oxidation of NADH (CF) and large increase in relative CBF can be seen during the six cycles recorded. The changes in relative CBF are calculated in relation to the control values before the waves were initiated, and in this animal, it was 4–5 times higher. The cycles of CSD were not so clear on the volume trace, and we suspect it was a reading problem, since in other

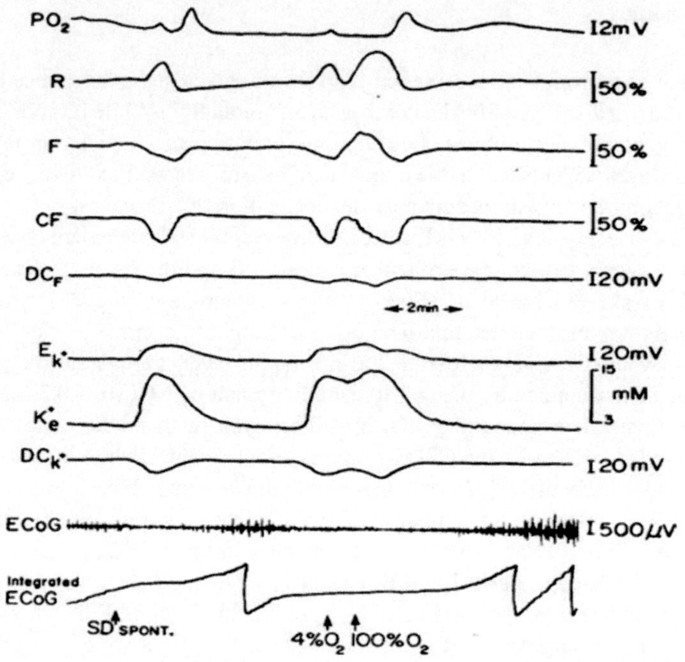

Fig. 4.1 The effects of hypoxia (4% O_2) on the metabolic, ionic, and electrical responses to CSD developed spontaneously. (Mayevsky et al., 1986)

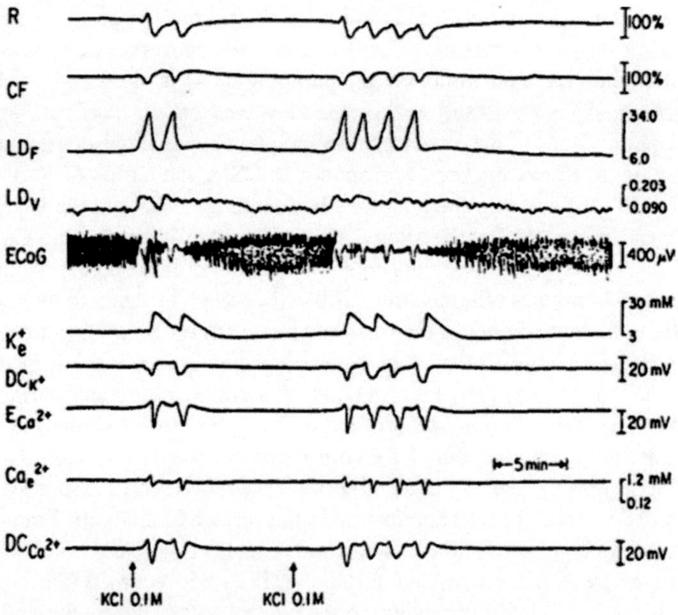

Fig. 4.2 Responses of the gerbil brain to CSD induced by 0.1 M KCI application, 2 mm anterior to the MPA. (*LDF, LDV* laser Doppler flow and volume monitored by a light guide inserted into the MPA. All other abbreviations are as in Fig. 2.11. (Mayevsky et al., 1992))

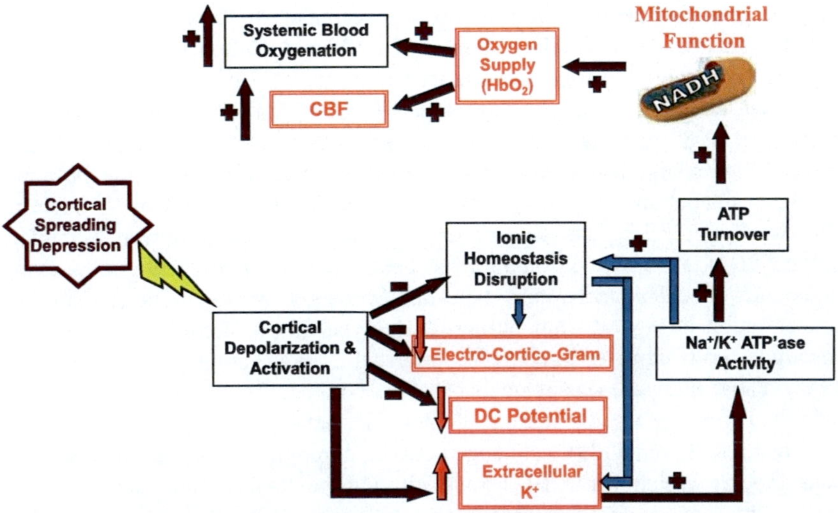

Fig. 4.3 Schematic representation of the effects of CSD on the energetic-mitochondrial function, ionic, and hemodynamic state of the cerebral cortex. (details are in the text)

studies a clear connection between flow and volume was found using the same laser Doppler flowmeter.

In order to analyze the sequence of physiological events developed during a wave of CSD, we constructed a scheme that summarized the possible changes in the measured parameters. Figure 4.3 presents, in a schematic way, the relationship between various physiological parameters monitored in vivo under CSD. This scheme is part of the scheme (Fig. 7) presented in the Frontmatter of the book. The CSD affects the processes in the brain such as CBF, NADH, and extracellular levels of K^+ but in the direction as seen in the scheme. When the brain is activated by CSD, the events propagate toward the activation of the brain and energy consumption is increased (purple arrows).

4.2 Effects of CSD on Cerebral Blood Flow and Brain Oxygen Consumption

4.2.1 Introduction

The depolarization caused by CSD is a response of the brain to various stimuli imposed as well as secondary metabolic and hemodynamic effects. Leão (1944a) found in rabbits a marked dilation and increased cerebral blood flow in pial vessels.

He concluded that the vascular response was secondary to the local change in the activity of neural elements. The responses of cerebral blood flow (CBF) before, during, and after the CSD wave were described by various investigators (Buresova, 1957; Hansen et al., 1980; Lauritzen et al., 1982; Lauritzen & Diemer, 1986; Mies & Paschen, 1984; Van Harreveld & Ochs, 1957b; Van Harreveld & Stamm, 1952). In all studies, a large increase in cerebral blood flow was recorded during the wave. Lauritzen and collaborators described a post-spreading depression wave hypoperfusion, while a preceding vasoconstriction was not established or proved. The effect of the CSD wave on O_2 consumption has been reported by very few groups owing to the lack of reliable techniques. Leão and Morison (1945) found a decreased O_2 concentration in sagittal sinus during CSD. A fall in $p0_2$ during CSD wave was described (Lukyanova & Bures, 1967; Marshall, 1959; Van Harreveld & Stamm, 1952). Those indirect measurements of increased O_2 consumption implied that the increased pumping activity stimulated mitochondrial activity to produce more ATP (Erecinska & Silver, 1989). A linkage between pumping activity and increased O_2 consumption was shown in 1973 by monitoring the NADH redox state using the surface fluorometry approach (Mayevsky & Chance, 1973; Rosenthal & Somjen, 1973). Mayevsky et al. (1974) were able to combine the measurements of extracellular K^+ with the NADH redox state showing an increased energy utilization during the passage of a CSD wave. To the best of our knowledge, no one has reported a direct quantitative measurement of cortical O_2 consumption and cerebral blood flow changes in the same brain during CSD. The aim of the present study (Mayevsky & Weiss, 1991) was to determine the local changes in O_2 supply, O_2 consumption, and oxygenation during the passage of a wave of CSD. We combined in vivo monitoring techniques (NADH redox state, extracellular K^+, and DC potential) with single-point measurements (cerebral blood flow and O_2 consumption) to assess the effect of CSD waves in anesthetized rats. We continuously monitored the NADH redox state as well as extracellular K^+ and DC potential from three cortical regions. After identification of the CSD wave, we determined cerebral blood flow. We used a microspectrophotometric technique to measure regional cerebral arterial and venous O_2 saturation. Using this approach, we were able to calculate O_2 consumption and cerebral blood flow in different brain regions representing the before, during, and after CSD wave areas and compared them with a contralateral control area. The in vivo monitoring was used to localize the CSD wave in real time throughout the cerebral cortex.

4.2.2 Methods

Male Long-Evans rats (250–300 g) were used in this study. They were anesthetized with sodium pentobarbital (50 mg/kg) and the femoral artery and vein were catheterized with polyethylene catheters. The femoral artery catheter was attached to a Statham P23Db pressure transducer connected to a Beckman R-411 recorder to obtain blood pressure. The venous catheter was used for the administration of

[^{14}C]-iodoantipyrine and supplemental anesthetic. Eleven rats were studied, although in only eight were complete data obtained. After the animal was connected to a head holder, the skull was exposed by a midline incision of the skin and removal of the connective tissue. A 5 × 11-mm hole was drilled in the frontoparietal bone area to accommodate the modified (described below) multiprobe assembly (MPA) shown in Fig. 4.4A. A small push–pull cannula was placed epidurally, anterior to the MPA, to elicit waves of CSD to propagate posteriorly (Fig. 4.4B). The dura mater was partially removed from the exposed cortex to measure K^+. Each animal was exposed to a set of cycles induced by application of 0.5 M KCI through the

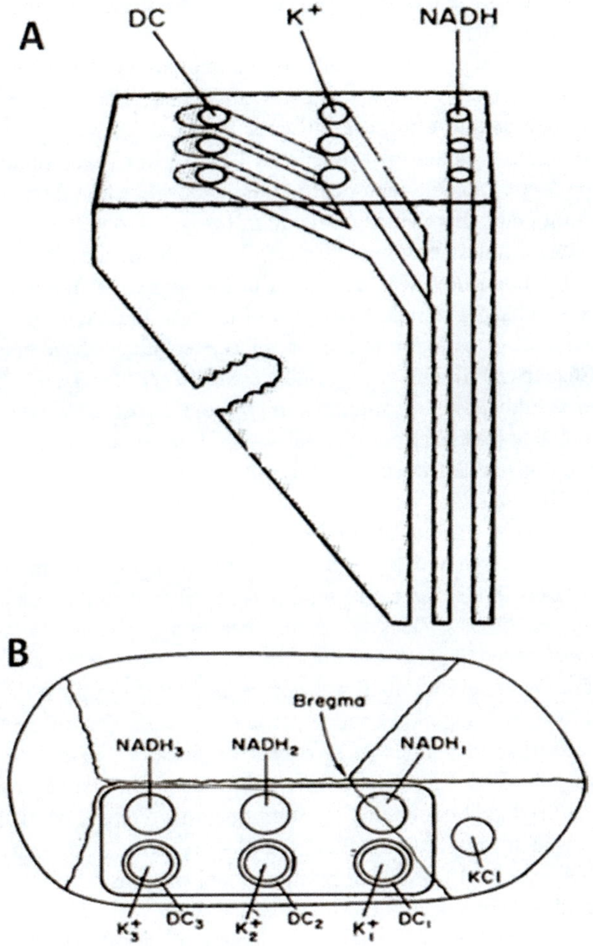

Fig. 4.4 Schematic presentation of the multiprobe assembly (**A**) and its location on the rats' brain during the monitoring phase (**B**), (*NADH$_1$-NADH$_3$* three light guides for NADH monitoring, *K^+_1-K^+_3* minielectrodes for extracellular K^+, *DC$_1$-DC$_3$* DC steady potential electrodes, *KCI* application of KCI solution for CSD initiation. (Mayevsky & Weiss, 1991))

push–pull cannula. The brain was allowed to recover (K^+ back to normal and NADH to the prewave steady state) for 20 min. After a new CSD wave reached, the second K^+ electrode, i.e., at the beginning of the steep rise of K^+, the rat was infused (venous catheter) with 50 micro Ci of 4-iodo-[N-methyl-^{14}C] antipyrine (Amersham) for the determination of cerebral blood flow. When the isotope entered the venous circulation, the arterial catheter was cut to a length of 20 mm to minimize smearing in the sampling catheter. Twenty-microliter blood samples were obtained from the arterial catheter every 3 sec during the next 30 sec. At the moment, the last sample was obtained, the animal was decapitated, and its head was frozen in liquid N_2. While frozen, the brain was sampled for four cortical regions representing areas from before the CSD wave, during the wave, after recovery, and the middle of the contralateral cerebral hemisphere (control). The aim of the MPA monitoring in the present study was to improve the tissue sampling-wave location in relationship to each animal. This helped to make the variation between animals smaller. Blood and tissue samples were then placed in a tissue solubilizer (Soluene; Packard), and 24 h later, they were put in a counting fluid (Dimiscint; National Diagnostic) and agitated. These samples were counted on a Beckman LS-230 liquid scintillation counter. Quench curves were prepared using carbon tetrachloride while the isotope counts were adjusted for color and quench correction. Regional blood flow determinations were made using a computer program; for details, see Reference (Buchweitz-Milton & Weiss, 1990). Blood flow was expressed in milliliters per minute per hundred grams. A second adjacent sample from each of the four brain regions was used for the determination of O_2 saturation. Details of these techniques have been published previously (Chance & Williams, 1954; Sinha et al., 1977). Briefly, the regions were mounted with an embedding medium in a microtome-cryostat. Twenty-micron sections were obtained on a rotary microtome at −30 °C under an N_2 atmosphere. They were transferred to precooled glass slides and covered with degassed silicone oil and a coverslip. These slides were placed on a Zeiss microspectrophotometer fitted with a N_2-flushed cold stage to obtain readings of optical density at 568, 560, and 506 nm. This three-wavelength method corrects for the light scattering in the frozen blood. Only vessels in the transverse section were studied so that the path of light traversed only the blood. The size of the measuring spot was 8 micrometers. Readings were obtained to determine O_2 saturation on 7–10 arteries and veins found in a region. The O_2 content of blood was determined by multiplying the percentage O_2 saturation by the hemoglobin concentration times 1.36. The difference between the average arterial and venous O_2 contents (regional O_2 extraction) was then obtained. Using the Fick principle, the paired product of local O_2 extraction and blood flow was used to determine O_2 consumption on a regional basis (Weiss et al., 1978). To monitor in vivo, in real time, the NADH redox state, extracellular K^+ activity, and DC potential from three brain regions, we constructed a special MPA shown schematically in Fig. 4.4A. The Plexiglas holder contained the following holes: three holes of 1.2 mm diameter for the common part of the light guide; three holes of 1.5 mm diameter for the 1.2 mm diameter K^+ electrodes. The clearance between the K^+ electrode and the hole contained physiological saline and was connected to the DC potential electrodes. The K^+ and DC electrodes were connected to an electrode holder containing an Ag/AgCl pellet (WPI, New Haven, CT, USA) and were sealed with epoxy glue. The electrodes as well as the light guides were cemented to the Plexiglas holder with epoxy glue. The testing and calibration of the electrodes were similar to those published previously (Crowe et al., 1977; Friedli et al., 1982). A factorial analysis of variance was employed to assess differences between the various regions examined (before, during, and after the wave as well as contralateral cortex) for cerebral blood flow, O_2 extraction, and O_2 consumption. The statistical significance of differences was determined by Duncan's procedure.

4.2.3 Results

Figure 4.5 shows the typical responses of the brain to waves of CSD passing over the right cerebral hemisphere of an anesthetized rat. Only two K^+ and two DC potential traces are shown. The upper six channels represent the three monitored areas for the NADH redox state. Each measuring point provided two signals: namely, 366 nm reflectance (R) and uncorrected NADH fluorescence (F). To

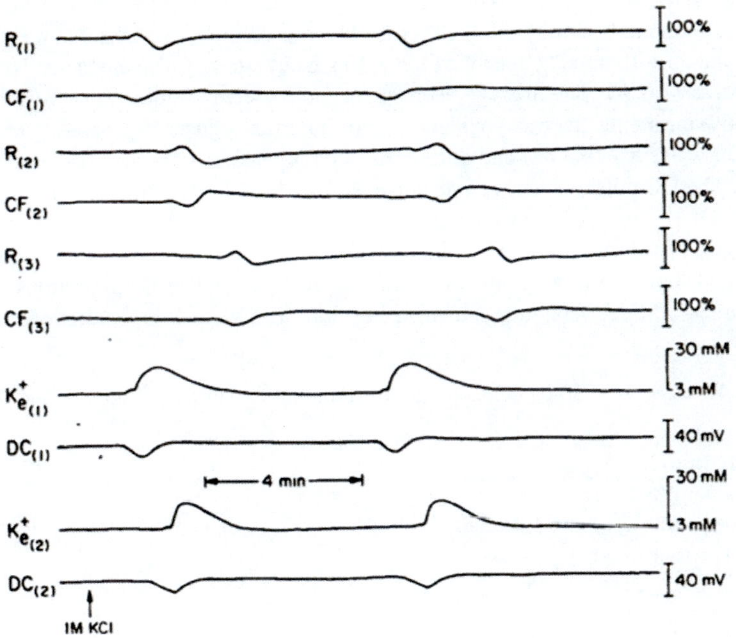

Fig. 4.5 The response of the brain to a wave of CSD induced by application of 1 M KCl to the frontal cortex. ($R_{(1)}$-$R_{(3)}$, 366-nm reflectance, $CF_{(1)}$-$CF_{(3)}$ 450-nm corrected NADH fluorescence, $K^+_{e(1)}$ *and* $K^+_{e(2)}$ extracellular K^+ levels, $DC_{(1)}$ *and* $DC_{(2)}$ DC steady potential. (Mayevsky & Weiss, 1991))

eliminate hemodynamic artifacts, we calculated the corrected NADH fluorescence by subtracting the reflectance from the fluorescence signal by 1:1 ratio (Mayevsky, 1984). This technique of correction provides the best approximation of the signal to the real mitochondrial NADH oxidation–reduction response. For each site, we present the reflectance signal and the corrected NADH fluorescence signal. In response to epidural application of 0.5 M KCI anterior to the MPA, two sets of waves were recorded. The reflectance signal shows the typical biphasic response, namely, an initial increase followed by a longer decrease that recovered to the baseline level. The corrected NADH signal showed a clear oxidation (decrease) cycle having a peak of 20% change as compared with the normoxic control value. The same kind of response was monitored in the various sites where NADH was monitored (R2, CF2, R3, and CF3 in Fig. 4.5). The lower four traces present the changes in extracellular K^+ (K^+_e) and the DC potential measured from two sites on the cortex parallel to sites 1 and 2 of the monitored NADH. The ring of the DC electrode shows the initial negative shift response followed by a large increase in K^+_e. The circular configuration of the DC electrode around the K^+ electrode led to some inaccuracy in the K^+ measurements. Since in the present study, we used the K^+ measurements as an indicator of the passage of the CSD wave, this factor was not critical in the determination of the wave location. To calculate the speed of the wave propagation, we used the difference in the DC charge appearance at sites 1 (range 2.5–5.0 mm/min). No change in the systemic blood pressure was observed during the waves of CSD, and for the 10 animals monitored, the average blood pressure was 113 ± 8 mm Hg systolic and 95 ± 10 mm Hg diastolic (Table 4.1), heart rate was also unchanged during the wave of CSD. The average decrease in the corrected fluorescence signal was 31 ± 6% during the maximal oxidation point (minimal signal). The baseline level of the K^+_e was 2.8 ± 0.7 mM and during the CSD it peaked at 16 ± 6 mM for the rats used. The DC potential shows a negative shift of 16 ± 3 mV during the peak of the wave. Of 11 rats used we were able to analyze the cerebral blood flow, O_2 extraction, and O_2 consumption in 8 animals. Table 4.2 shows the mean values for the 8 rats, and there were no significant regional differences. The O_2 saturations of all examined cortical veins are shown in Fig. 4.6. The venous O_2 saturations of all

Table 4.1 Effects of CSD on various parameters measured in 10 rats

	Mean ± SD
CF decrease (%)	31 ± 8
DC shift (mV)	−16 ± 3
Speed of CSD propagation (mm/min)	3.7 ± 0.8
Blood pressure (mm hg) systolic diastolic	113 ± 8 95 ± 10
K^+_e (mM) Before CSD During CSD	2.8 ± 0.7 16 ± 6

CF corrected NADH fluorescence, *CSD* cortical spreading depression, K^+_e extracellular potassium level

Table 4.2 Cerebral blood flow, arterial and venous oxygen saturation, cerebral oxygen extraction, and oxygen consumption in four areas of cerebral cortex exposed to CSD

			O_2 saturation		
Brain area	O_2 consumption (ml O_2/min/100 g)	Flow (ml/min/100 g)	Arterial (%)	Venous (%)	O_2 extraction (ml O_2/100 ml blood)
Before CSD wave	3.6 ± 1.1	66.1 ± 38.2	90.3 ± 3.6	60.1 ± 6.8	6.8 ± 1.3
During CSD wave	7.4 ± 3.7[a]	119.7 ± 50.5[a]	90.8 ± 2.0	60.8 ± 6.1	6.8 ± 1.7
After CSD wave	4.1 ± 1.7	65.3 ± 40.2	91.8 ± 2.4	57.6 ± 5.0	7.7 ± 1.1
Contralateral control	5.1 ± 2.6	69.3 ± 27.9	90.4 ± 2.1	57.4 ± 4.3	7.5 ± 1.7

Values are means ± SD for eight rats. *CSD* cortical spreading depression. [a]$p < 0.01$ compared with all other groups

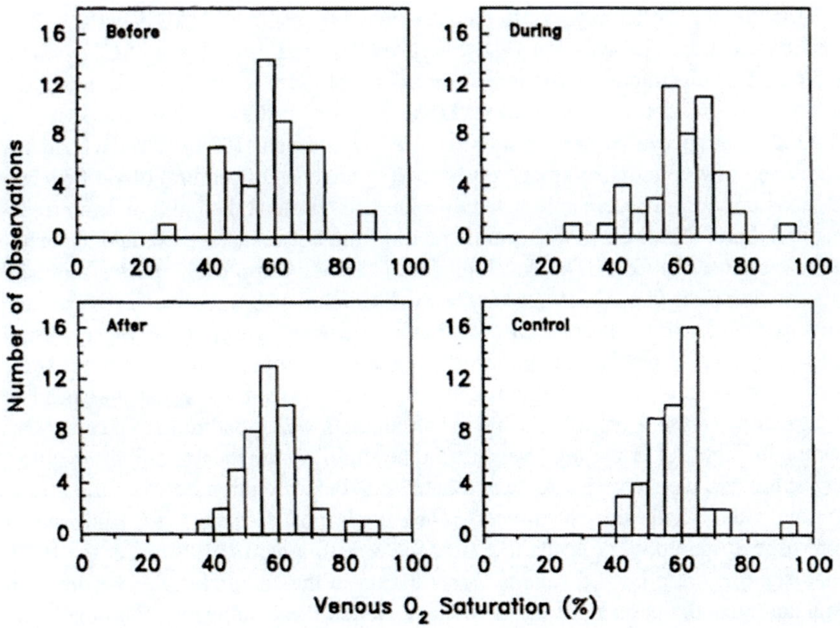

Fig. 4.6 Histograms of O_2 saturations from examined veins in cortical areas before, during, and after the passage of a wave of CSD as well as from the contralateral cortex (control). Note the lack of significant differences between the histograms. (Mayevsky & Weiss, 1991)

regions followed a normal distribution curve and were not different from each other. Oxygen extraction was not significantly different during the wave of CSD (6.8 ± 1.7 ml O_2/100 ml blood) compared with the contralateral cortex (7.5 ± 1.7 ml O_2/100 ml blood) and other regions (Table 4.2). Regional cerebral O_2 consumption was significantly elevated (+45%) in the area in which the wave of CSD was occurring compared with the contralateral cortex (Table 4.2). Regional O_2 consumption was not different from values in the contralateral cortex in cortical areas before or after the wave of CSD passed.

4.2.4 Discussion

The purpose of the present study was to examine the relationship between O_2 supply and O_2 consumption in the rat brain during the event of CSD. To identify the location of the wave in the hemisphere during the experimental procedure, we monitored in real time and continuously the extracellular K^+ level and DC potential as well as the metabolic response to the CSD wave by using an NADH surface fluorometry technique (Mayevsky, 1984). We hypothesized that a wave of CSD would increase O_2 consumption by increasing cerebral blood flow. We also examined the initial vasoconstriction prior to the CSD wave as well as the possible hypoperfusion state after the wave as previously suggested (Lauritzen, 1984, 1987). Using the combined in vivo monitoring and post-freezing analysis of cerebral blood flow and O_2 consumption, we were able to localize the CSD wave at the point of brain freezing and therefore have a more accurate relationship between the wave and the metabolic responses to CSD. The development of the MPA approach and the multisite fluorometer (Friedli et al., 1982; Mayevsky & Chance, 1982) enabled us to use such an approach for the first time. Cerebral blood flow was measured previously during CSD, but it was dependent on the assumption that the propagation speed of the wave was constant in all animals (Lauritzen, 1987). This assumption that the speed of propagation of the wave is the same for all animals was found not to be correct, as shown in Table 4.1. The range was 2.5–5.0 mm/min. Using the three-point monitoring approach, we were able to have a better spatial correlation between the propagating wave and the parameters measured post-freezing. We utilized a microspectrophotometric technique (Chance & Williams, 1954; Sinha et al., 1977) that examined arterial and venous O_2 saturation in the cerebral cortex before, during, and after the passage of the CSD wave. When these data were combined with regional blood flow measurements, regional O_2 consumption was determined as their product (Weiss et al., 1978). Our values for cerebral blood flow and O_2 consumption are in the same range as found by us previously (Buchweitz-Milton & Weiss, 1990), suggesting that the experimental procedure did not affect the results (brain and dura exposure). Also, the resting levels of K^+_e as well as the NADH response to CSD confirmed that the exposed brain was in the normal range (Hansen, 1985; Mayevsky, 1984) during our experimental procedure. The effects of the CSD wave on cerebral blood flow as well as other metabolic responses have been studied in two types of models, namely, after repetitive cycles (Mayevsky & Chance, 1973) and after a single wave (Lauritzen & Diemer, 1986). There were different responses under these two conditions, but it was not clear why. In our present study, we pre-tested the responses of the brain in vivo to a set of CSD waves, and after complete recovery of the in vivo parameters, we induced a new wave and froze the brain when the wave reached the second measurement point. This protocol provides three distinct cortical areas: a before-wave area, the area under the maximal effect of the wave, and an area in which recovery from the wave was nearly complete. We also sampled a contralateral control area that was never exposed to CSD. Differences between the contralateral cortex and the before and after areas were insignificant. This may suggest that we cannot support the finding of (Lauritzen et al., 1982) regarding the hypoperfusion in the post-wave period. Since we did expose the brain to repetitive cycles before the last one (at which the cerebral tissue was obtained), in the range of 25 ± 3 min/cycle, the flow values at the before-wave area were also in a post-wave period as suggested (Lauritzen et al., 1982). Therefore, it was not possible for us to see the decrease in cerebral blood flow prior to the wave as suggested by others (Van Harreveld & Ochs, 1957b; Van Harreveld & Stamm, 1952), but perhaps a better technique with higher time resolution was needed. Using a laser Doppler flowmeter (Mayevsky et al., 1990), we did not find a preceding vasoconstriction in most gerbils. Only in animals with a bad physiological status, it was observed. Even if such a short phase of vasoconstriction existed, it was not reflected in the intracellular oxygenation since we did not see any preceding NADH reduction prior to the large oxidation cycle (Mayevsky et al., 1982). The data presented here support the previously known fact that during the wave of CSD, a large increase in cerebral blood flow is observed (Mies & Paschen, 1984). This increased flow occurred initially and after repetitive cycles (Hansen, 1985; Hansen et al., 1980). In the present study, it rapidly (about 2 min) returned to control values in the area after the wave passed. Previously, no data existed as to the degree of coupling between this flow increase and cerebral metabolism in the same animal. Since K^+ is released, the flow increase could have been related to direct K^+-induced vasodilation. Others have reported increased metabolic rate or high blood flow due to the CSD wave. This study indicates that there is a good match between oxygen consumption and relative cerebral blood flow during CSD in the same animal. If the vasodilation caused by CSD were not metabolically induced, then one would expect decreased cerebral O_2 extractions and increased venous O_2 saturations at the site of the greatest flow. The fact that O_2 extractions were not different in all areas suggested that the brain was using the extra cerebral blood flow to meet an increased need for oxygen. This indicated a strong coupling between O_2 supply and demand. In addition, the lack of change in the distribution of venous O_2 saturation seen in Table 4.2 indicated that this coupling occurred at the level of areas drained by 20- to 40-micron veins. This is the first report of a direct measurement of O_2 consumption performed during the passage of CSD wave. It supports the previous assumption that cerebral O_2 consumptions were increased during the CSD wave (Lauritzen et al., 1990). This increase in O_2 consumption appeared related to the known stimulated state of mitochondrial activity during a CSD wave (Rosenthal & Sick, 1988; Rosenthal &

Somjen, 1973). The ability of the mitochondria to be shifted toward state 3 (Chance & Williams, 1955) is dependent on the excessive supply of 0_2 to the intracellular space. Lauritzen et al. (1987) studied the effects of CSD on energy change and ATP levels and confirmed the concept of increased energy utilization during CSD waves. The changes in the mitochondrial NADH redox state calculated in this study were comparable with those published before (Mayevsky et al., 1980; Mayevsky & Chance, 1973, 1975) using the same light guide surface fluorometry. As described in methods, the removal of the dura mater was not complete and we measured smaller changes in K^+_e during the CSD as compared with our other studies using surface electrodes [565] or levels found by other groups using microelectrodes (Hansen, 1985; Vyskocil et al., 1972a). One interesting comparison from the present study is that while venous O_2 saturation was not affected by CSD, the mitochondrial NADH redox state shifted to oxidation. This indicated that the delivery of O_2, which is usually limiting, was not an important factor in the cerebral metabolic response to CSD.

4.3 Sequence of Events Recorded During the CSD Cycle

4.3.1 The Use of the MPA System

Figure 2.14 shown in Chap. 2 presents the multiparametric monitoring system developed in order to identify the sequence of responses recorded after the initiation of CSD wave. The aim was to use this monitoring system under various experimental conditions of brain oxygenation including hypoxia, anesthesia, and partial ischemia. It was necessary to construct the MPA in such a way that all sensors will be positioned in the same front line facing the propagating wave of CSD. Due to the fact that the sensors are not very small, we performed a preliminary study before the construction of the system seen in Fig. 2.14. We built a multisite monitoring assembly (MSA) seen in Fig. 4.7A and B. Four DC steady potential electrodes were positioned inside a Plexiglas cannula as seen in parts A and B. The CSD waves were initiated in 2 distances from the monitoring cannula (KCl in a circle), one was implanted in left parietal bone and the second one in the frontal bone. Typical responses to CSD are presented in Fig. 4.7C. As seen, the minimal values of the DC steady potential recorded by electrodes DC_2 and DC_3 were recorded before those minimal values recorded by electrodes DC_1 and DC_4. At the end of each experiment, the distance between the two, KCl application cannulas was measured. This enabled us the calculate the value of the distance (C_1–C_4), between the KCl canula to the center of every DC electrode (DC1–DC4), seen in Fig. 4.7B.

Figure 4.8 presents the connection between the distance of the center of the KCl application site and the center of the monitoring probe, and the time (measured in seconds) to reach the minimal level of the DC response. The correlation value shows a significant value. ($N = 28$, $r = 0.96$, $p < 0.005$).

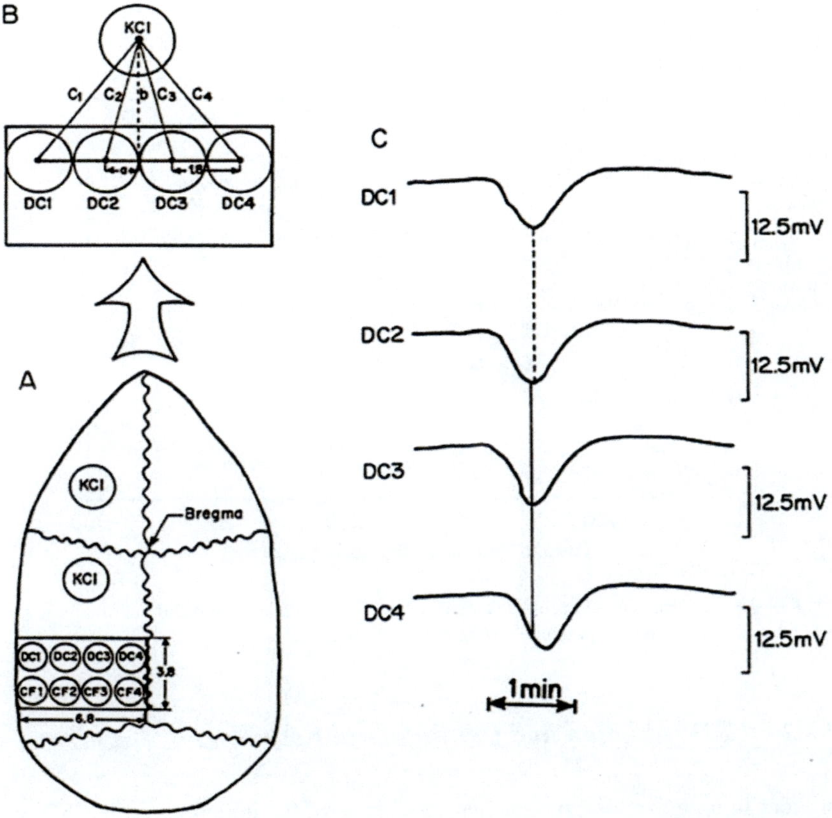

Fig. 4.7 (**A**) Schematic presentation of the multisite monitoring system enabled the monitoring of DC potential in 4 locations exposed at the same time to a CSD wave. The wave was elicited by applying KCl in two locations in the same hemisphere (KCl in circle). (**B**) The calculation of the distances between the application site of KCl and the measurement points. The distance between the centers of the DC electrodes was 1.8 mm. (**C**) Typical record of the DC potential measured in response to CSD wave. The dashed and the solid vertical lines demonstrate the time lapse to reach the minimal level of the DC potential

In order to achieve the same distance between the 4 sensors and the application site of the KCl, they were located on an arch as seen in Fig. 4.9B and C. Using this, MPA system provide a typical response to CSD seen in Fig. 4.10. The 2 vertical blue lines presents the initial response to CSD in changes of the DC potential and the extracellular potassium followed by the metabolic responses, namely the changes in NADH redox state and the CBF and CBV. The average values of the sequence of responses to CSD recorded in 52 experiments are presented in Fig. 4.11. The shortest response time was recorded in the changes of the DC potential (53 sec after the stimulation). The extracellular potassium responded after 70 sec while the metabolic responses appeared later. The reflectance responded after 83 sec and the NADH after 85 sec. The CBF responded after 88 sec and the CBV after 94 sec.

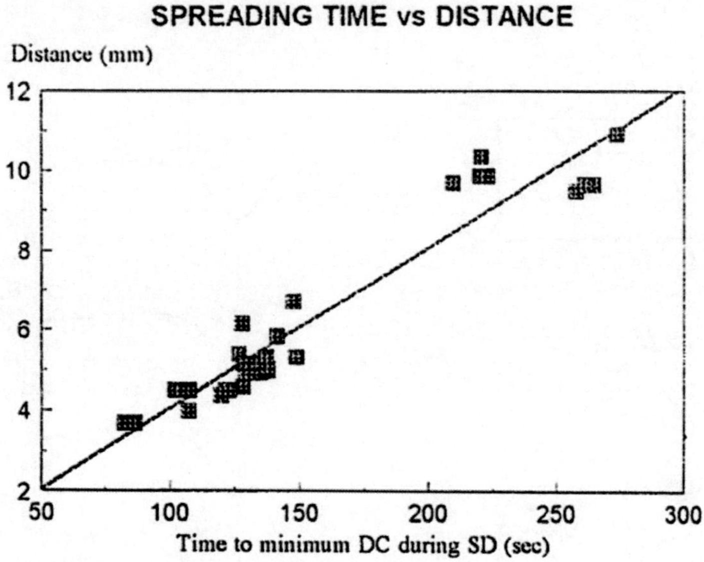

Fig. 4.8 The relationship between the distance of the KCl application site and the time to record the minimal level of the DC potential during the CSD wave

4.3.2 CSD Initiation and Contralateral Responses

In order to optimize the experimental procedure for the initiation of the CSD, we performed the following study:

The MPA system shown in Fig. 4.9 was connected to the left hemisphere and a second laser Doppler probe was connected to the right hemisphere in order to check whether or not the CSD stimulus is affecting the contralateral hemisphere. Also, we compared 2 types of stimulation in order to initiate the CSD response, namely by KCl solution or by needle puncture (NS). Figure 4.12 shows the results of a typical experiment. In part A, the CSD was initiated by a needle, and in part B, the KCl was used as a stimulus. Detailed analysis of the responses recorded under the 2 stimuli to initiate the CSD wave showed that no significant differences were found in the various parameters. Table 4.3 shows the mean and the standard error of the mean (SEM) calculates in the time parameters calculated in the two types of initiation of CSD. Only 2 parameters in the table were statistically different (Red oval line).

Figure 4.13 shows the statistical analysis of experiments performed in 10 animals. We compared the responses of the CBF and the CBV measured in the two hemispheres. The results show that a response to CSD induced either by KCl or by needle was recorded only in the left hemisphere. No response what so ever was recorded in the contralateral hemisphere—the right one.

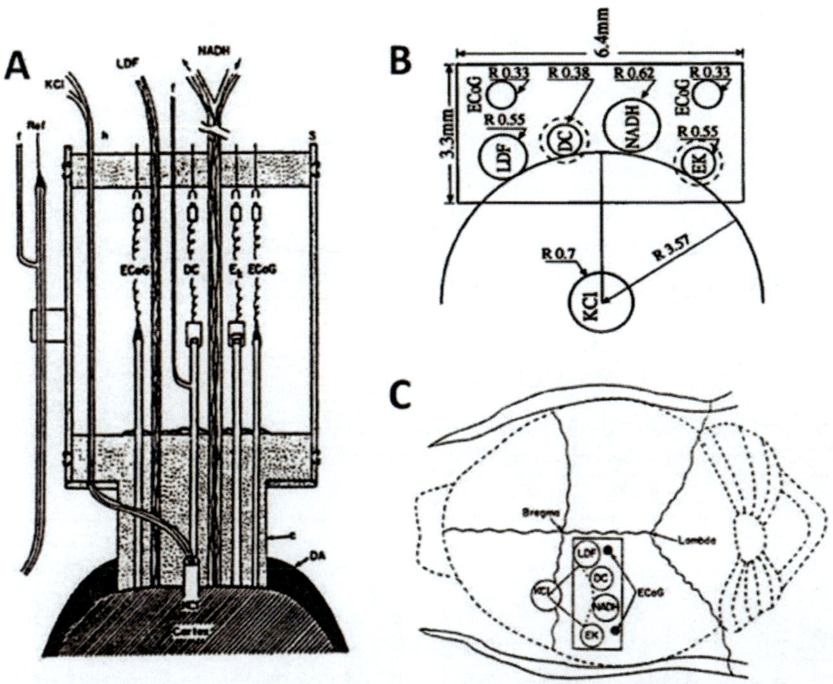

Fig. 4.9 (**A**) The multiparametric assembly used in order to study the sequence of the responses recorded after the initiation of CSD wave. (**B**) The location of the various sensors and the distances in relation to the KCl application point. (*LDF* the sensor measured the CBF using laser Doppler flowmeter, *DC* the DC steady potential electrode, *NADH* the fiber-optic sensor connected to the NADH fluorometer, *EK* minielectrode that measure the extracellular potassium levels, *ECoG* two electrodes measured the EEG, *R* The radius of the cannula that contain the KCl solution). (**C**) the location of the monitoring sensors on the cerebral cortex

4.3.3 Discussion

The results accumulated in our many studies on CSD and brain functions led us to formulate a possible explanation to the events that occurred during the CSD wave as seen in Fig. 4.14.

CSD is a phenomenon initiated by a rapid depolarization of neuronal tissue in the gray matter of the cerebral cortex (Fig. 4.14A, lower left side) detected by the depression of the EEG and a negative shift in the DC steady potential. A massive redistribution of ions between the intracellular and extracellular spaces led to a large increase in extracellular potassium and other ions shift (Fig. 4.14B). This activated the Na-K-ATPase in order to pump the potassium back to the intracellular space. The increase in ADP level shifts the mitochondrial state to state 3 and NADH becomes more oxidized. The increase in oxygen consumption by the mitochondria is creating an elevation in CBF. In most cases, the microcirculatory hemoglobin

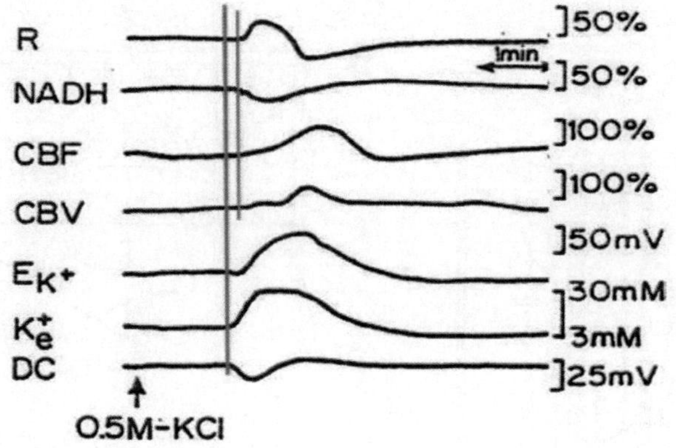

Fig. 4.10 Typical record demonstrating the MPA responses to CSD. The blue vertical lines enable to see the sequence of responses recorded after exposing the brain to CSD

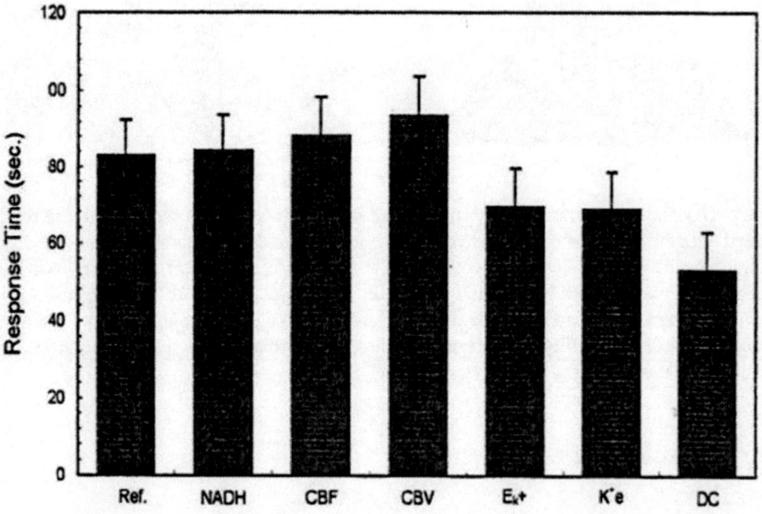

	Ref.	NADH	CBF	CBV	E_{K^+}	K^+_e	DC
Ref.		n.s.	*	**	**	**	**
NADH			n.s.	**	**	**	**
CBF				*	**	**	**
CBV					**	**	**
E_{K^+}						n.s.	**
K^+_e							**

Fig. 4.11 The average response time and SEM of 5 control cycles of CSD measured in total 52 animals. (*$p < 0.05$, **$p < 0.005$, *n.s.* not significant)

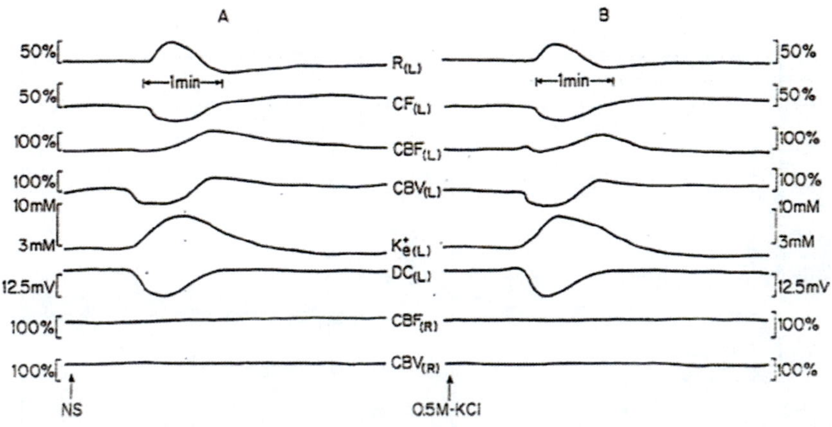

Fig. 4.12 The comparison between the responses to CSD initiated by insertion of a needle (NS) to the cortex (**A**) and the exposure to KCl solution (**B**). The stimulus and the responses were induced and recorded in the left hemisphere. (*CBF and CBV* were monitored simultaneously in the right hemisphere as well, $R_{(L)}$, $CF_{(L)}$, $CBF_{(L)}$, $CBV_{(L)}$ reflectance, *NADH* cerebral blood flow, and volume measured in the left hemisphere, $K^{+}_{e(L)}$, $DC_{(L)}$ extracellular potassium level and steady potential, $CBF_{(R)}$- $CBV_{(R)}$ cerebral blood flow and volume measured in the right hemisphere

Table 4.3 The mean time and SEM of the various parameters calculated during the time of CSD induced either by KCl or Needle puncture

Time	Reflectance		NADH		CBF		CBV		K^{+}_{e}		DC	
(min)	NS	KCl	NS	KCl	NS	KCl	NS	KCl	NS	KCl	NS	KCl
T1	2.1±0.03	2.3±0.02	2.0±0.03	2.3±0.02	1.5±0.02	1.9±0.02	1.8±0.03	2.2±0.02	1.8±0.03	2.1±0.01	1.8±0.04	2.0±0.02
Tmax	2.4±0.03	2.6±0.01	—	—	2.9±0.04	3.3±0.02	2.7±0.04	3.1±0.03	2.3±0.03	2.6±0.01	2.4±0.17	3.1±0.07
Tmin	3.2±0.04	3.4±0.02	2.5±0.04	2.8±0.02	1.7±0.03	2.3±0.02	2.2±0.05	2.5±0.02	—	—	2.1±0.03	2.6±0.02
Tqr	4.0±0.04	4.2±0.03	4.0±0.05	4.0±0.03	3.7±0.03	4.0±0.03	3.5±0.03	3.8±0.02	3.4±0.03	3.6±0.03	3.2±0.04	3.5±0.03
Tbr	4.9±0.05	4.9±0.03	5.9±0.06	5.2±0.06	4.0±0.03	4.3±0.03	4.0±0.03	4.4±0.02	4.0±0.03	4.2±0.03	4.0±0.06	4.0±0.03
Tr	7.6±0.08	6.5±0.07	9.2±0.06	6.8±0.09***	4.8±0.04	4.8±0.03	5.9±0.07	5.4±0.04	7.2±0.08	6.4±0.11	5.8±0.08	5.6±0.07

T1 the time to see the initial response, T_{max} the time to reach the maximal value(peak), T_{min} the time to reach the minimal value, *Tqr* time to the initial fast recovery, *Tbr* The time before the complete recovery, *Tr* Time to complete recovery
$*p < 0.05$; $**p < 0.005$; $n = 12$

saturation is elevated. The elevation of extracellular potassium may create a vasoconstriction response but the increase in NO release probably enabled the elevation of CBF.

Since the CSD wave propagation is a multifactorial event, our strategy was to construct a multiprobe assembly (MPA) containing probes to evaluate most of the parameters involved during CSD wave propagation, from its front line at the same time. This verified the parallel changes in the various parameters after CSD initiation and helped us to identify which parameter(s) start the chain responses and the

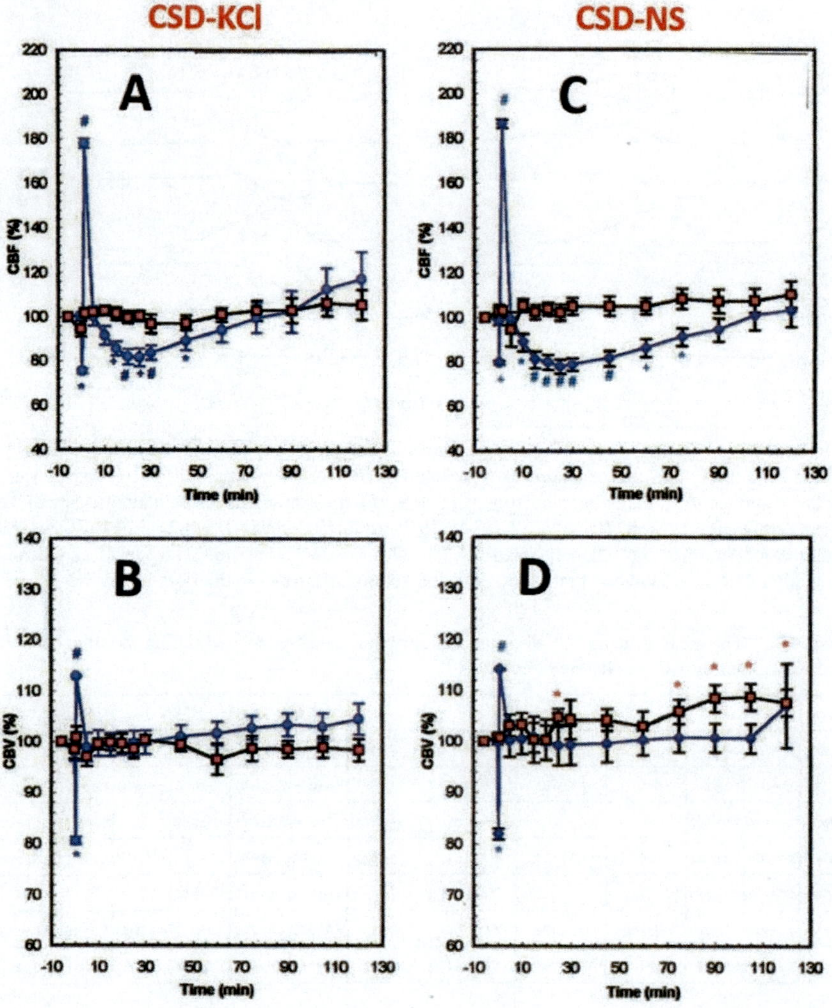

Fig. 4.13 Effects of CSD induced in the left hemisphere on the CBF and CBV measured simultaneously in the ipsilateral side (blue lines and symbols) and the contralateral hemisphere-right side (red lines and symbols). CSD was induced by KCl solution (**A**) and (**B**), or by Needle insertion (NS) (**C**) and (**D**). (**p* < 0.05; +*p* < 0.005; #*p* < 0.0005)

reaction of the later ones. The MPA measured: extracellular K concentration $[K^+]_e$ mitochondrial NADH/NAD ratio, cerebral blood flow and volume (CBF and CBV), steady DC potential, and bipolar electrocorticography (ECoG). The technical details were published in 2000 (Sonn & Mayevsky, 2000).

Using the special MPA in this study, we were able for the first time to prove the sequence of events during the changes in cortical blood flow and extracellular K^+ during spreading CSD wave propagation. The other aim was to present the

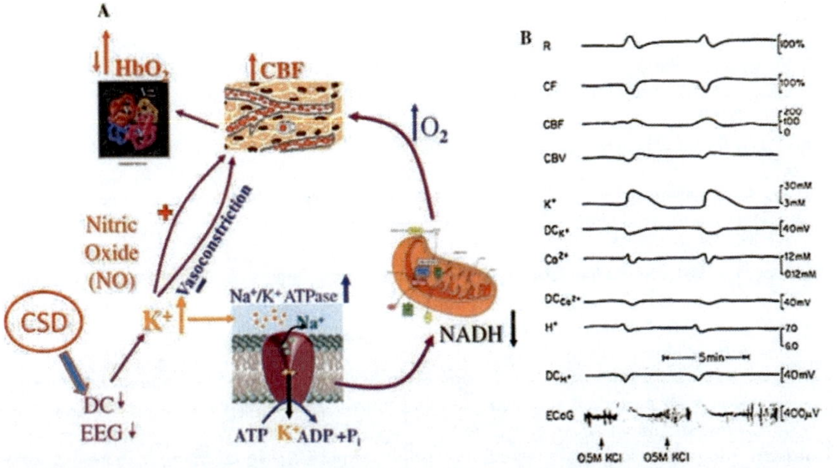

Fig. 4.14 (**A**) The cascade of events occurring after CSD. (**B**) Repetitive responses of the brain to CSD evoked by epidural application of 0.5 M KCl. (Mayevsky & Barbiro-Michaely, 2013)

interrelation between CBF and NADH during the various phases of its propagation. As found, the time to the decreased peak in CBF was initial to the increased peak in NADH during the first phase. The opposite interrelation can be seen in the later phase where parallel to the increase in CBF, a decrease in NADH was noticed. The simultaneous and opposite changes between NADH and CBF were found to be more interrelated when O_2 delivery was limited. These results may indicate an increase in the correlation between CBF and NADH under limited O_2 supply conditions, which is disclosed after increasing O_2 demand by CSD.

References

Buchweitz-Milton, E., & Weiss, H. R. (1990). Effect of salbutamol on regional cerebral oxygen consumption, flow and capillary and arteriolar perfusion. *Neurological Research, 12*, 169–175.

Bures, J. (1956). Some metabolic aspects of Leão's spreading depression. *Journal of Neurochemistry, 1*, 153–158.

Bures, J., & Krivanek, J. (1960). Ionic movements in the brain as studied with the aid of washing the cortical surface with an epidural cannula. *Physiologia Bohemoslovenica, 9*, 488–493.

Bures, J., Buresova, O., & Krivanek, J. (1974). *The mechanism and applications of Leão's spreading depression of electroencephalographic activity*. Academic.

Buresova, O. (1957). Changes in cerebral circulationin rats during spreading EEG depression. *Physiologia Bohemoslovenica, 6*, 1–11.

Chance, B., & Williams, G. R. (1954). Steady-state of reduced pyridine nucleotides in phosphorylating rat liver mitochondria. *American Society for Biochemistry and Molecular Biology, 13*, Abstract 633.

Chance, B., & Williams, G. R. (1955). Respiratory enzymes in oxidative phosphorylation (I-Kinetics of oxygen utilization). *The Journal of Biological Chemistry, 217*, 383–393.

Crowe, W., Mayevsky, A., & Mela, L. (1977). Application of a solid membrane ion selective electrode to in vivo measurements. *The American Journal of Physiology, 233*, C56–C60.

Erecinska, M., & Silver, I. A. (1989). ATP and brain function. *Journal of Cerebral Blood Flow and Metabolism, 9*, 2–19.

Friedli, C. M., Sclarsky, D. S., & Mayevsky, A. (1982). Multiprobe monitoring of ionic, metabolic, and electrical activities in the awake brain. *The American Journal of Physiology, 243*, R462–R469.

Hansen, A. J. (1985). Effect of anoxia on ion distribution in the brain. *Physiological Reviews, 65*, 101–148.

Hansen, A. J., Quistorff, B., & Gjedde, A. (1980). Relationship between local changes in cortical blood flow and extracellular K^+ during spreading depression. *Acta Physiologica Scandinavica, 109*, 1–6.

Krivanek, J. (1961). Some metabolic changes accompanying Leão's spreading cortical depression in the rat. *Journal of Neurochemistry, 6*, 183–189.

Krivanek, J., & Bures, J. (1960). Ion shifts during Leão spreading cortical depression. *Physiologia Bohemoslovenica, 9*, 494–503.

Lauritzen, M. (1984). Long-lasting reduction of cortical blood flow of the brain after spreading depression with preserved autoregulation and impaired CO2 response. *Journal of Cerebral Blood Flow and Metabolism, 4*, 546–554.

Lauritzen, M. (1987). Regional cerebral blood flow during CSD in rat brain: increased reactive hyperperfusion in low-flow states. *Acta Neurologica Scandinavica, 75*, 1–8.

Lauritzen, M., & Diemer, N. H. (1986). Uncoupling of cerebral blood flow and metabolism after single episode of CSD in the rat brain. *Brain Research, 370*, 405–408.

Lauritzen, M., Jorgensen, M. B., Diemer, N. H., Gjedde, A., & Hansen, A. J. (1982). Persistent oligemia of rat cerebral cortex in the wake of spreading depression. *Annals of Neurology, 12*, 469–474.

Lauritzen, M., Hansen, A. J., Kronborg, D., & Wieloch, T. (1990). Cortical spreading depression is associated with arachidonic acid accumulation and preservation of energy charge. *Journal of Cerebral Blood Flow and Metabolism, 10*, 115–122.

Leao, A. A. P. (1944). Further observations on the spreading depression of activity in the cerebral cortex. *Journal of Neurophysiology, 10*, 409–414.

Leão, A. A. P. (1944a). Pial circulation and spreading depression of activity in the cerebral cortex. *Journal of Neurophysiology, 7*, 391–396.

Leão, A. A. P. (1944b). Spreading depression of activity in cerebral cortex. *Journal of Neurophysiology, 7*, 359–390.

Leão, A. A. P., & Morison, R. (1945). Propagation of spreading cortical depression. *Journal of Neurophysiology, 8*, 33–45.

Lukyanova, L. D., & Bures, J. (1967). Changes in pO_2 due to spreading depression in cortex and nucleus caudatus of rat. *Physiologia Bohemoslovaca, 16*, 449–455.

Marshall, W. H. (1959). Spreading cortical depression of Leão. *Physiological Reviews, 39*, 239–279.

Mayevsky, A. (1984). Brain NADH redox state monitored in vivo by fiber optic surface fluorometry. *Brain Research Reviews, 7*, 49–68.

Mayevsky, A., & Barbiro-Michaely, E. (2013). Shedding light on mitochondrial function by real time monitoring of NADH fluorescence: I. Basic methodology and animal studies. *Journal of Clinical Monitoring and Computing, 27*, 1–34.

Mayevsky, A., & Chance, B. (1973). In H. I. Bicher & D. F. Bruley (Eds.), *A new long-term method for the measurement of NADH fluorescence in intact rat brain with implanted cannula* (pp. 239–244). Plenum Press.

Mayevsky, A., & Chance, B. (1974). Repetitive patterns of metabolic changes during cortical spreading depression of the awake rat. *Brain Research, 65*, 529–533.

Mayevsky, A., & Chance, B. (1975). Metabolic responses of the awake cerebral cortex to anoxia hypoxia spreading depression and epileptiform activity. *Brain Research, 98*, 149–165.

Mayevsky, A., & Chance, B. (1982). Intracellular oxidation-reduction state measured in situ by a multicannel fiber-optic surface fluorometer. *Science, 217*, 537–540.

Mayevsky, A., & Weiss, H. R. (1991). Cerebral blood flow and oxygen consumption in cortical spreading depression. *Journal of Cerebral Blood Flow & Metabolism, 11*, 829–836.

Mayevsky, A., Zeuthen, T., & Chance, B. (1974). Measurements of extracellular potassium, ECoG and pyridine nucleotide levels during cortical spreading depression in rats. *Brain Research, 76*, 347–349.

Mayevsky, A., Lebourdais, S., & Chance, B. (1980). The interrelation between brain PO_2 and NADH oxidation- reduction state in the gerbil. *Journal of Neuroscience Research, 5*, 173–182.

Mayevsky, A., Zarchin, N., & Friedli, C. M. (1982). Factors affecting the oxygen balance in the awake cerebral cortex exposed to spreading depression. *Brain Research, 236*, 93–105.

Mayevsky, A., Zarchin, N., Yoles, E., & Tannenbaum, B. (1986). Oxygen supply to the brain in hypoxic and hyperoxic conditions. In C. Nicolau (Ed.), *Advances in the biosciences. "O2-Transport in Red Blood Cells"* (pp. 119–132). Pergamon Press.

Mayevsky, A., Yoles, E., Zarchin, N., & Kaushansky, D. (1990). Brain vascular ionic and metabolic responses to ischemia in the Mongolian gerbil. *Journal of Basic and Clinical Physiology and Pharmacology, 1*, 207–220.

Mayevsky, A., Frank, K., Muck, M., Nioka, S., Kessler, M., & Chance, B. (1992). Multiparametric evaluation of brain functions in the Mongolian gerbil in vivo. *Journal of Basic and Clinical Physiology and Pharmacology, 3*, 323–342.

Mies, G., & Paschen, W. (1984). Regional changes of blood flow, glucose, and ATP content determined on brain sections during a single passage of spreading depression in rat brain cortex. *Experimental Neurology, 84*, 249–258.

Rosenthal, M., & Sick, T. J. (1988). Measurement of metabolic activity associated with ion shifts. *Neuronal Microenvironment*, 187–246.

Rosenthal, M., & Somjen, G. (1973). Spreading depression, sustained potential shifts, and metabolic activity of cerebral cortex of cats. *Journal of Neurophysiology, 36*, 739–749.

Sinha, A. K., Neubauer, J. A., Lipp, J. A., & Weiss, H. R. (1977). Blood O_2 saturation determination in frozen tissue. *Microvascular Research, 14*, 133–144.

Sonn, J., & Mayevsky, A. (2000). Effects of brain oxygenation on metabolic, hemodynamic, ionic and electrical responses to spreading depression in the rat. *Brain Research, 882*, 212–216.

Van Harreveld, A., & Ochs, S. (1957a). Electrical and vascular concomitants of spreading depression. *The American Journal of Physiology, 189*, 180–192.

Van Harreveld, A., & Ochs, S. (1957b). Electrical and vascular concomitants of spreading depression. *The American Journal of Physiology, 189*, 159–166.

Van Harreveld, A., & Stamm, J. S. (1952). Vascular concomitants of spreading cortical depression. *Journal of Neurophysiology, 15*, 487–496.

Vyskocil, F., Bures, J., & Kriz, N. (1972a). Potassium-selective microelectrodes used for measuring extracellular brain potassium during spreading depression and anoxic depolarization in rats. *Brain Research, 39*, 255–259.

Vyskocil, F., Kritz, N., & Bures, J. (1972b). Potassium-selective microelectrodes used for measuring the extracellular brain potassium during spreading depression and anoxic depolarization in rats. *Brain Research, 39*, 255–259.

Weiss, H. R., Neubauer, J. A., Lipp, J. A., & Sinha, A. K. (1978). Quantitative-determination of regional oxygen-consumption in dog heart. *Circulation Research, 42*, 394–401.

Chapter 5
Spontaneous Development of CSD Related to Brain Pathophysiological Conditions

Abstract In the current chapter, we will present the various pathophysiological that have the potential to induce CSD-like response. The mechanism of coupling between the availability of oxygen and the development of CSD may involve the leakage of potassium from the intracellular to the extracellular compartment in the cerebral cortex tissue. The interaction between the level of carbon dioxide in the breathing mixture in the hyperbaric chamber will be demonstrated. Also, the effects of various pharmacological agents on the development of CSD will be presented and discussed.

Keywords CSD and oxygen availability · Hyperoxia · Ischemia · Hypoxia · Hyperbaric hyperoxia · Anoxia

5.1 Introduction

Very early in our studies on CSD in 1973 and 1974, we found that in many experiments, exposing rats or gerbils to partial or complete lack of oxygen by anoxia, hypoxia, or ischemia may induce a CSD event during the recovery phase or immediately after the recovery from the pathological situation.

Actually, there are two options available in order to study the relationship between CSD and changes in oxygen availability:

1. To study the development of CSD as a result of an event developed during the exposure of the experimental animal to the availability of oxygen. This option will be presented and discussed in this chapter.
2. To expose the animal to a defined situation such as partial ischemia or hypoxia and then to induce CSD by application on KCl to the cerebral cortex and follow the responses of the brain to the CSD. This second option will be presented in the next chapter.

A clear demonstration of such a coupling between the anoxic event and the CSD wave is presented in Figs. 2.3 and 3.5A.

A. Mayevsky, J. Sonn, *Cortical Spreading Depression of Leao*,
https://doi.org/10.1007/978-3-031-08068-5_5

5.2 During Anoxia, Hypoxia, and Ischemia

In these experiments, the rats were slightly anesthetized by Equithesin. The nitrogen was applied via a nasal mask. In Fig. 2.3A, the exposure to nitrogen was ended after 70 s and the details of the responses were described in Chap. 2. Figure 5.1 shows the effect of exposure to 100 s N_2 on NADH fluorescence, reflectance, EEG, and systemic blood pressure (line A). The top trace shows the reflectance which in all animals decreases during the initial stage of the N_2 cycle. This decrease of reflectance was in two phases. The first decrease (blue circle) was small (in comparison with the second one) and occurred while the animal was breathing spontaneously. A second decrease occurred after the animal stopped breathing (SB).

The main difference between the two nitrogen cycles shown in Fig. 2.3A and 2.3B seen also here in Fig. 5.1 is in the recovery phase from the nitrogen exposure. After the N_2 administration had been discontinued (SN), artificial respiration (AR) was applied to induce spontaneous breathing. After the animal started breathing, a fast decrease of NADH was observed in the uncorrected fluorescence as well as in the corrected trace (line B). The recovery of the NADH level to the baseline is very fast in comparison with the recovery of the reflectance. The EEG of both hemispheres reaches a very low amplitude when the NADH level reaches 80–90% of the maximum increase during the N_2 cycle. The EEG responses of the two hemispheres were identical. The recovery of the EEG follows the NADH recovery to the normoxic level. The main differences between the two cycles are that after the recovery of the NADH to the normoxic level a further decrease in NADH occurred (third trace), and at this time, the EEG was depressed and recovered to normal only later.

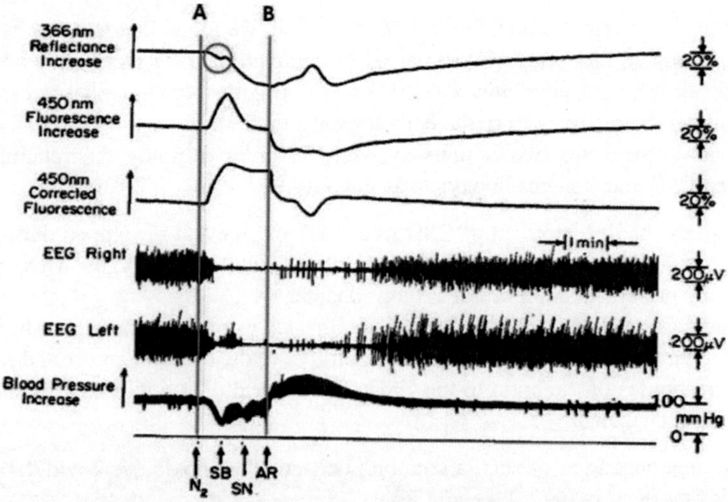

Fig. 5.1 Effects of anoxia on mitochondrial NADH redox state, ECoG, and systemic blood pressure. (Mayevsky & Chance, 1975)

This cycle of oxidation of NADH following the N_2 cycle was observed in most animals after exposure to a relative long N_2 cycle.

By using K^+-sensitive microelectrode, Zeuthen et al. (1979, 1974) and Vyskocil et al. (1972) described two groups of animals which showed different levels of K^+ accumulation in the extracellular space during anoxic depolarization. In one group of rats, the extracellular K^+ was ten-fold higher than in the second one, suggesting different sensitivity to anoxia. Marshall (1959), in his review, also suggested that anoxia can lead to the initiation of a CSD wave. Bures (1957) and Bures and Buresova (1960) reported that the threshold of the CSD is reduced by a short period of asphyxia or anoxia. On the basis of the information available, we can assume that during anoxia an increase of extracellular K^+ can reach the threshold level for the initiation of CSD wave in the most sensitive area in the cortex. The results presented in Fig. 5.1 can be explained in the same way. The late recovery of the EEG (Fig. 5.1) is probably also a result of the CSD. This change in the EEG was found in both hemispheres so that the mechanism of the initiation is a general one. In a few animals, only one hemisphere was under CSD cycle after the N_2-O_2 breathing cycle which can be related to the different levels of extracellular K^+. It seems that in the cycle shown in Fig. 2.3A, the amount of K^+ accumulated in the extracellular space was low so that when energy is available this K^+ is pumped into the cells and the amount of ATP needed to pump the K^+ was very high so the high ADP level may stimulate the respiratory activity, and thus, as a result, an oxidation of NADH was recorded.

In another group of gerbils, unilateral occlusion diminished blood supply to the ipsilateral hemisphere. Figure 5.2 shows the results obtained from an awake gerbil (Mayevsky, 1978b). The occlusion of the left carotid artery caused a transient increase in fluorescence in the contralateral hemisphere, which recovered to normal while the artery was occluded. The occlusion of the right artery (when the left was occluded) led to an increase of 25% in NADH, and this high level remained until the reopening of both carotid arteries (R, Fig. 5.2). The recovery of the brain from the ischemic condition was accompanied by an oxidation–reduction cycle of NADH, and at the same time, ECoG was depressed. The transient increase in NADH, when the left artery was occluded, was probably due to a sudden change in blood pressure which affects the distribution of blood between the two hemispheres. When the right artery was occluded alone, the same increase in NADH was found (not shown in Fig. 5.2). From these results, we concluded that in the gerbil (experimental results on which are shown in Fig. 5.2), the blood vessel connections between the left and right hemispheres were not minimal.

The CSD-like response which follows the ischemia episode seems to be identical to the connection between anoxia episode and oxidation–reduction cycle we described in the rat (Mayevsky & Chance, 1975). We can assume that K^+ is accumulated in the extracellular space during the ischemia, and when blood containing oxygen is restored, the K^+ ions are pumped back into the cells by an active process which consumes energy (Mayevsky & Chance, 1974, 1975; Mayevsky et al., 1974b). The gerbil has the advantage of having a peculiar blood supply to the brain; ischemia can be investigated in more detail in a small awake animal.

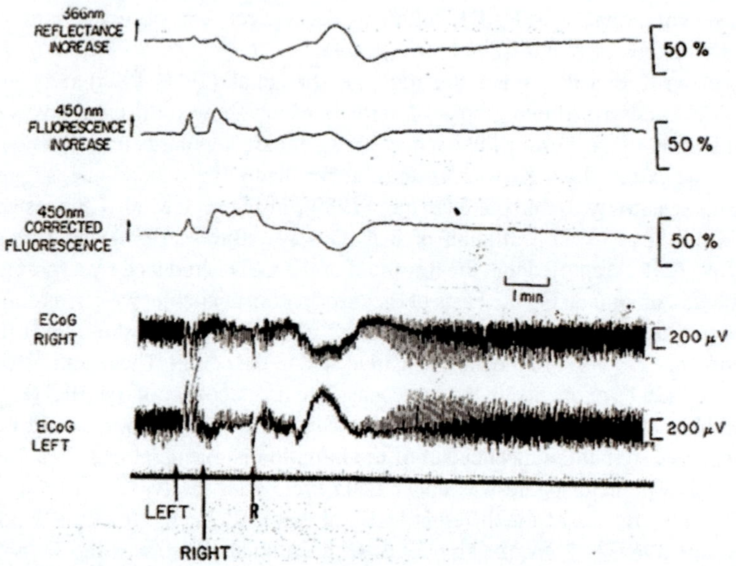

Fig. 5.2 The effect of ischemia (carotid artery occlusion) on brain NADH oxidation–reduction state and ECoG of slightly anesthetized gerbil. (John Wiley and Sons, reprinted with permission. (Mayevsky, 1978b))

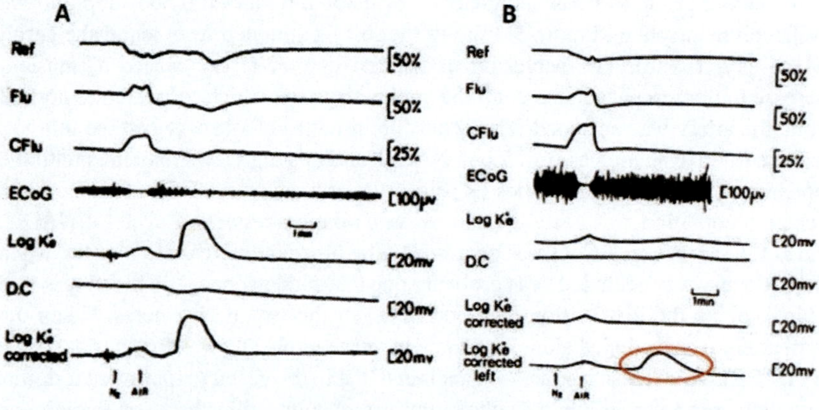

Fig. 5.3 Effects of anoxia on mitochondrial NADH, ECoG, extracellular potassium, and DC steady potential in the gerbil brain. (Mayevsky et al., 1980b)

Figure 5.3A shows a response to a 1-min anoxia induced by nitrogen respiration. Within 20 s, extracellular potassium (K^+_e) concentration started to change as the log of K^+ activity increased gradually. K^+_e returned to baseline level very shortly after room air was reintroduced. At the same time, NADH levels (third trace from top) responded as previously described and reached an elevated level under anoxic conditions. The typical decrease of the reflected light is recorded in the upper trace. The

ECoG activity became severely depressed during complete anoxia. Very shortly after full recovery of K^+_e and NADH to their initial levels, a steep increase of K^+_e, which peaked at 91 meg/l, was measured, while the surface potential showed a wave whose negative shift reached 8.5 mV. The late response of the K^+_e is identical to that recorded during CSD.

As shown in Fig. 5.3, anoxia induced changes in K^+_e levels as well as in the oxidation state of the brain (Mayevsky et al., 1980b). The initial effect of anoxia is a small increase of K^+_e due to the lack of available energy (as shown in the increase of NADH levels) with depolarization of a large population of neurons. This elevated level of K^+_e leads to varied responses that depend upon the subject; differing responses were even obtained in two hemispheres of the same subject. This point is made clear by comparing Fig. 5.3A with Fig. 5.3B.

Two nitrogen cycles measured in another gerbil are shown in Fig. 5.3B. In this animal, K^+_e was measured bilaterally while NADH level and ECoG were monitored only from the right hemisphere. Of the eight traces shown in Fig. 5.3B, only the one on the bottom was measured from the left hemisphere. As can be seen, the responses of the two hemispheres to N_2 anoxia are markedly different. As a response to anoxia, NADH level increased (CFlu) about 30% as compared to the normoxic level on the right side. ECoG became isoelectric when NADH level reached 80–90% of its maximum reduction level. The response of K^+_e was very small. K^+_e recovery after air was reintroduced revealed some differences between the two hemispheres of the same animal. In the right hemisphere, NADH was oxidized to its baseline levels rapidly. ECoG and K^+_e returned to their baseline levels in less than 1 min. The response of the right hemisphere was similar in the two N_2 cycles. The responses of the left hemisphere measured during the same N_2 cycles varied from those of the right side.

This variation in the responses to anoxia was shown earlier when only NADH was measured in the rat brain (Mayevsky & Chance, 1975). At that time, we suggested that the explanation might be found in individual variations of brain sensitivity to potassium leakage. If Fig. 5.3A is examined carefully, it is seen that on our recordings the oxidation cycle that appears following the N_2 cycle is simultaneous with the K^+_e recovery phase. We believe this occurred because the combination electrode and the light-guide fibers were necessarily slightly displaced from one another so that the two components were not measuring from exactly the same field of tissue.

Figure 5.4 shows the effects of anoxia in the anesthetized brain while measuring also the local pO_2 (Mayevsky et al., 1980a). Parts A and B show the responses to two nitrogen cycles. The reflectance and fluorescence response are typical and NADH became fully reduced within less than one minute. The decrease in pO_2 had faster kinetics and stayed at the zero level until rebreathing started. The main difference between the two cycles can be seen in the recovery phase from anoxia. In the first cycle (A), an overshoot of the pO_2 was higher than the normoxic level and showed a lower level followed by a small overshoot. At the same time, the NADH showed an oxidation cycle typical to the CSD response. The ECoG trace showed a typical anoxic depression in Parts A and B and a secondary depression in Part B concomitant to the metabolic and hemodynamic responses.

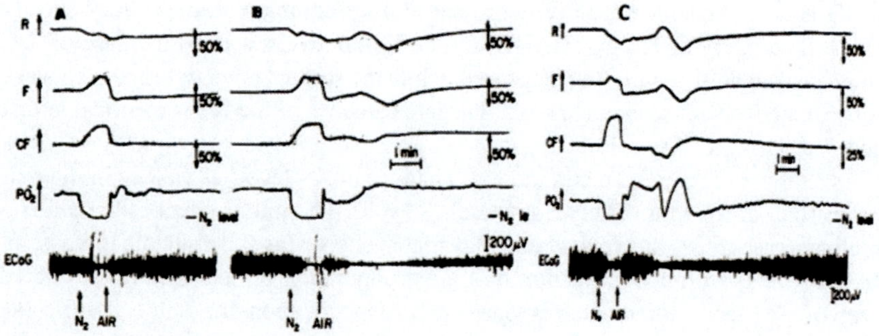

Fig. 5.4 The effects of anoxia (A, B, C) on NADH, pO_2 and ECoG in the gerbil brain. (John Wiley and sons, reprinted with permission. (Mayevsky et al., 1980a))

Figure 5.4C presents results obtained from a different gerbil. The response to anoxic episode is basically similar to the one shown in Fig. 5.4A and B, but at the recovery phase, the pO_2 response to CSD has a different nature. One minute after breathing was started, a biphasic response was recorded inversely related to the reflected light changes, namely, that when the reflectance showed an increase, the pO_2 trace showed a parallel decrease; and when the reflectance showed a large decrease (increased blood volume), the pO_2 showed three-to-four-fold increase and then gradually recovered to the normoxic level in parallel to the gradual increase in the reflected light.

The biphasic response in the pO_2 trace is in correlation to the response of the NADH fluorescence trace and we assume that it is the typical response to CSD, as described previously (Mayevsky & Chance, 1975) after anoxic cycle.

Figure 5.5 presents the effects of unilateral or bilateral carotid artery occlusion (Mayevsky et al., 1980a). In this animal, the occlusion of the contralateral carotid (L_{occ}) had a very small effect on the NADH or the pO_2. Occlusion of the right carotid (while the left one was closed) induced a large fall in pO_2 and an increase of NADH with a similar kinetics. The recovery after bilateral recirculation was fast, without any significant overshoot in the pO_2. In the second, longer, ischemia episode, when N_2 was applied to the ischemic brain, a large overshoot in the pO_2 was recorded after recirculation. Here again, as in Fig. 5.4, the large increase in pO_2 was parallel to the large decrease in reflectance with the same kinetics toward the normoxic level. The NADH shows an oxidation cycle after recirculation due to CSD, as can be seen in the ECoG trace.

Figure 5.6 shows the response to anoxia. In the two hemispheres, a spontaneous CSD was initiated during the anoxia, and as a response, oxidation cycles were recorded bilaterally after the recovery of the brain to the normoxic state (Mayevsky et al., 1982). The same type of response was described previously (Mayevsky & Chance, 1975; Mayevsky et al., 1980b).

Figure 5.7 shows a typical response of the gerbil brain to anoxia (left) and bilateral carotid occlusion (right) (Friedli et al., 1982). The initial response to anoxia is an increase of the NADH level as shown in the corrected fluorescence (CF) trace

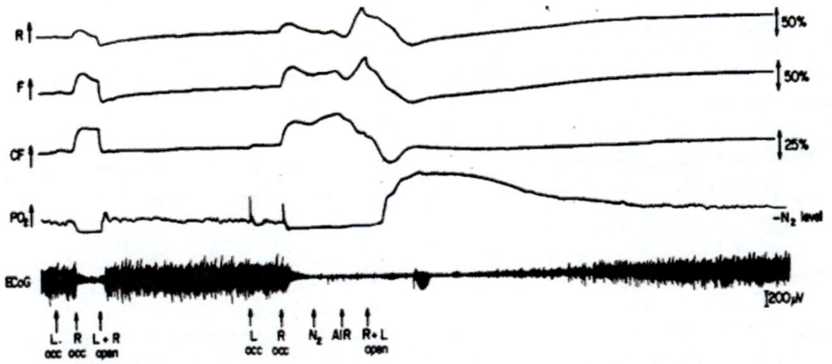

Fig. 5.5 The effects of carotid arteries occlusion alone (left side), and in combination with anoxia on the metabolic and electrical activity of the slightly anesthetized gerbil brain. (John Wiley and sons, reprinted with permission. (Mayevsky et al., 1980a))

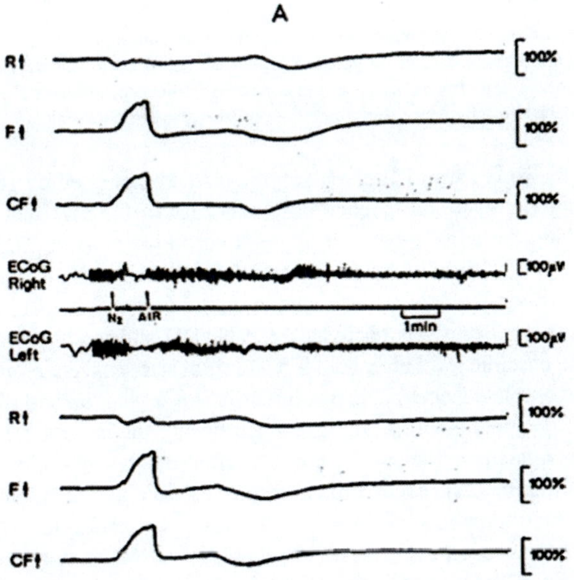

Fig. 5.6 Effects of anoxia on NADH and ECoG recorded from the two hemispheres of the brain. (Mayevsky et al., 1982)

(30% increase compared with the normoxic level). As a result, the extracellular level of K^+ increased slightly and returned to the baseline level after restoration of air to the breathing mixture. The pH showed an increased response under anoxia. The DC potential measured in all three electrodes site (DC_{H+}, DC_{K+}, and DC_F) did not show any change during the anoxic cycle. The ECoG showed a typical depression during the anoxic episode. The effect of ischemia was tested in two steps. After

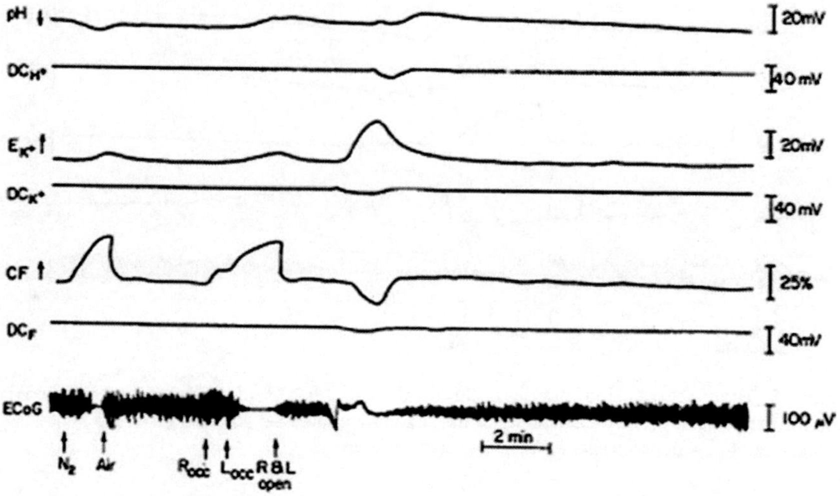

Fig. 5.7 Effects of anoxia (left) and ischemia (center) on metabolic, ionic, and electrical signals measured from a slightly anesthetized gerbil. Spontaneous CSD cycle was developed after ischemic episode. In this experiment, we recorded corrected fluorescence alone (CF) without uncorrected fluorescence and reflectance. For abbreviations, see text. (Friedli et al., 1982)

occlusion of the right carotid artery (R_{occ}), the NADH increased by 10% compared with the normoxic level and reached only one-third of the maximal NADH level measured during the anoxic cycle or bilateral carotid occlusion. The partial ischemia induced by R_{occ} did not cause any other significant changes during the short period of occlusion. After complete ischemia induced by occlusion of the left artery (L_{occ}; while the right one was occluded), the NADH showed a 30% increase and returned to the baseline level after both left and right arteries were reopened. During the period of complete ischemia, extracellular K^+ level went up and pH went down (more acidic). A spontaneous CSD developed during the complete ischemia, and the wave propagated to the measuring site after the two carotid arteries were reopened. During the CSD event, typical responses were recorded; namely, K^+ was elevated and then pumped back into the cells, and as a result, the NADH became more oxidized (oxidation cycle). The pH showed an acidification response to the CSD. As expected, the three DC potential signals showed a transient negative shift during CSD. We found that the resting level of extracellular K^+ was 3 ± 1 mM in all undamaged brains. When the initial levels were higher, the animal was not included.

Figure 5.8 shows the response of a gerbil brain to anoxia (Mayevsky & Sclarsky, 1983). The animal was breathing 100% O_2 and then under complete anoxia, pO_2 dropped down to its minimal level simultaneously with an increase in the NADH levels (CF). The anoxic cycle was very short (30 sec), and after rebreathing the animal with 100% O_2, the NADH returned to the nontoxic level while the pO_2 showed a large overshoot due to the increase in blood flow during the anoxia. During the anoxic cycle, the K^+ accumulated in the extracellular space and later on it

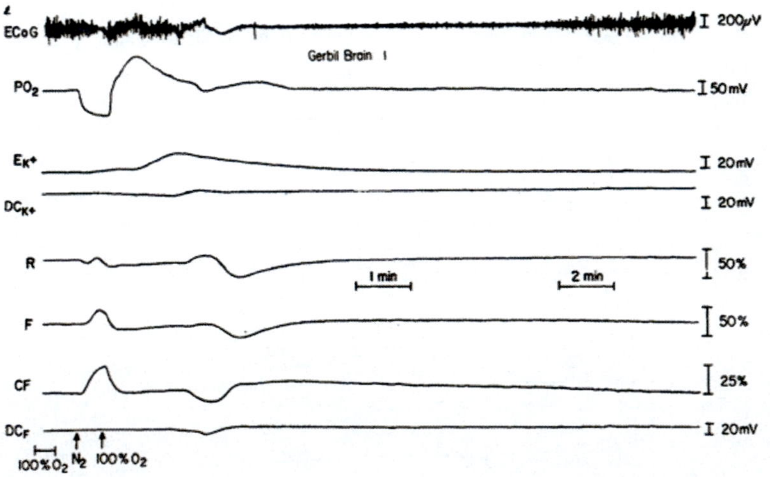

Fig. 5.8 The effects of anoxia on the metabolic, ionic, and electrical activities in the gerbil brain. (Mayevsky & Sclarsky, 1983)

initiated a spontaneous CSD cycle leading to a further increase in K^+. The wave of CSD propagated in the cortex and reached also the light-guide area and an oxidation cycle of NADH was recorded together with a negative shift in the DC potential near the light guide (DC_F). During the CSD wave, the pO_2 showed a biphasic change starting with a short decrease followed by a long-lasting increase. These changes in pO_2 were parallel with the biphasic changes in the reflected light as was also shown previously when NADH was monitored with pO_2 in the gerbil brain (Mayevsky et al., 1980a). In the present study, we verified those results and were able to correlate them with the changes in extracellular K^+ and DC potential. The ECoG showed a long-lasting depression during the CSD and the complete recovery took a few more minutes after all other parameters reached the baseline levels.

In the same gerbil, we exposed the brain to ischemia by occlusion of the carotid arteries (Fig. 5.9). When the contralateral carotid artery was occluded (L_{occ}), very small transient changes were recorded in the pO_2 and NADH (Mayevsky & Sclarsky, 1983). Only after inducing complete ischemia (R_{occ}), pO_2 dropped to 0 and NADH increased to its maximal level. As a result, a spontaneous CSD cycle was initiated and the type of responses recorded were similar to those described in Fig. 5.8 (after anoxia). These results confirm again our previous data about the oxidation of NADH when an extra pumping of K^+ is stimulated by CSD (Mayevsky et al., 1980b).

Figure 5.10 shows typical responses to unilateral (R_{occl}) and bilateral (L_{occl}) in CBF (LDF) and an increase in NADH levels (CF) (Mayevsky, 1992). During the period of ischemia, accumulation of K^+ in the extracellular space was recorded (K_1^+, K_2^+) but the DC steady potential and the Ca^{2+} levels remained unchanged during the occlusion period.

The ECoG reached the isoelectric level very soon after the second occlusion. During the reopening of the carotid arteries, a fast reperfusion was recorded together

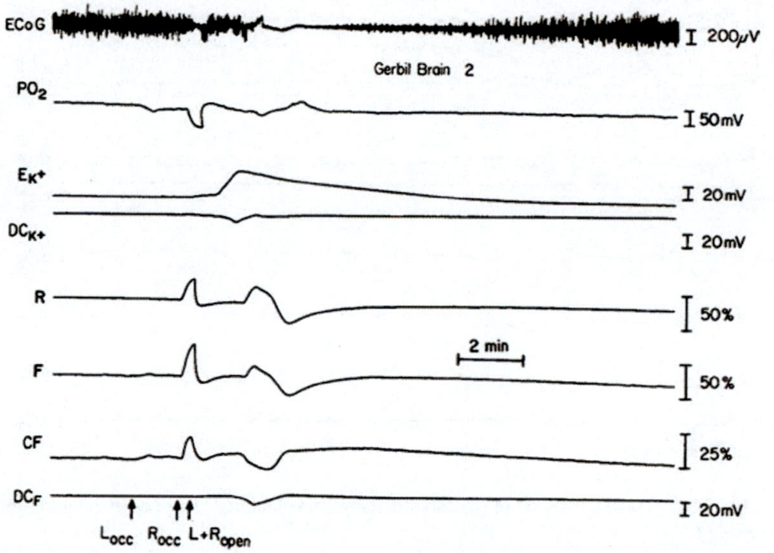

Fig. 5.9 Responses to unilateral and bilateral carotid occlusion in the gerbil brain. (Mayevsky & Sclarsky, 1983)

with the oxidation of NADH. A spontaneous wave of CSD (spreading depression) was developed during the recovery phase, characterized by a large increase in K^+_e and a decrease in Ca^{2+}_e together with a negative shift in the DC steady potential. During the recovery from the CSD wave, a large increase in CBF (300%) was recorded accompanied by an oxidation wave of the NADH (decrease CF).

Exposing the rat to hypoxia (2% O_2) exhibits very similar responses, as shown in Fig. 5.11 (Mayevsky et al., 1986a). The response of the pO_2 electrode was very small, probably due to low sensitivity, but the NADH redox state shows a large increase simultaneously with a large decrease in the reflectance signal (R). When O_2 was supplied to the rat, the pO_2 showed the expected overshoot simultaneously with the oxidation of NADH. The extracellular K^+ accumulated during the hypoxia and very soon after O_2 breathing recovered to the pre-hypoxic level. During the recovery period, a secondary response to the hypoxia was recorded, namely, a CSD wave propagated in the cortex (Leão, 1944; Mayevsky & Chance, 1974, 1975; Mayevsky & Sclarsky, 1983; Rosenthal & Somjen, 1973). The initiation of the wave occurs during hypoxia at a certain point in the cortex and is then propagated to the entire hemisphere, including the area below the multiprobe assembly. During the CSD, a large increase in extracellular K^+ was recorded (from below 3 mM to about 14–15 mM) together with the negative shift in the DC steady potential near the K^+ electrode ($DC_K{}^+$) and the light guide (DC). The ECoG remained depressed after the hypoxia due to the concomitant CSD which occurred. The recovery period from the CSD includes the active pumping of the K^+ into the cells which led to greater oxidation of NADH (CF) in order to supply more ATP needed.

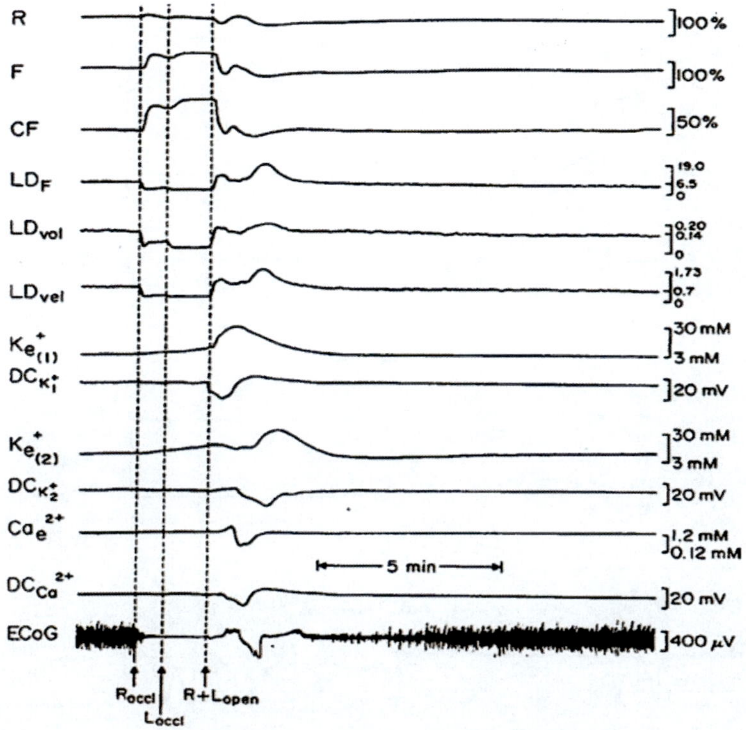

Fig. 5.10 The effects of unilateral (R_{occl}) and bilateral (L_{occl}) carotid occlusion on the metabolic hemodynamic, ionic, and electrical activities in the gerbil brain. *R* 366 nm reflectance, *F* 450 nm fluorescence, *CF* NADH-corrected fluorescence LD_F LD_{vol} LD_{vel} laser Doppler flow, volume, and velocity, $K^+_e(1)$, $K^+_e(2)$, Ca^{2+}_e extracellular potassium (two electrodes) and calcium electrodes, *DCK+(1) DCK(2)*, $DCCa^{2+}$ DC steady potential measured concentric to the three electrodes, *ECoG* electrocorticogram. (Mayevsky, 1992)

Figure 5.12 shows a typical response to graded hypoxia covering the range from 100% to 0% oxygen in the breathing mixture and then a recovery to 100% O_2 for another 5 min (Mayevsky et al., 1990b). As seen, a gradual decrease in hemoglobin saturation (HbO_2) was recorded simultaneously with the increase in the intramitochondrial redox state (CF). In parallel, changes were recorded in the reflected light intensity (R) and the Hb concentration representing the autoregulation response due to the hypoxia. The first change in the ECoG was noted at FiO_2 level below 9% and was severely depressed at 4.8% and became isoelectric under N_2 breathing. The resting level of the extracellular K^+ level was 2.5 mM and increased gradually to a maximum of 3.9 mM during the short complete anoxic period. The resting level of Ca^{2+} was quite low as compared to other brains monitored (due to a film of blood below the electrode) and was decreased from 0.6 mM (normoxia) to 0.5 mM during the anoxia. The DC steady potential around the K^+ and Ca^{2+} electrodes showed a 1–2 mV-positive change during the hypoxia but we suspect it is because of a small

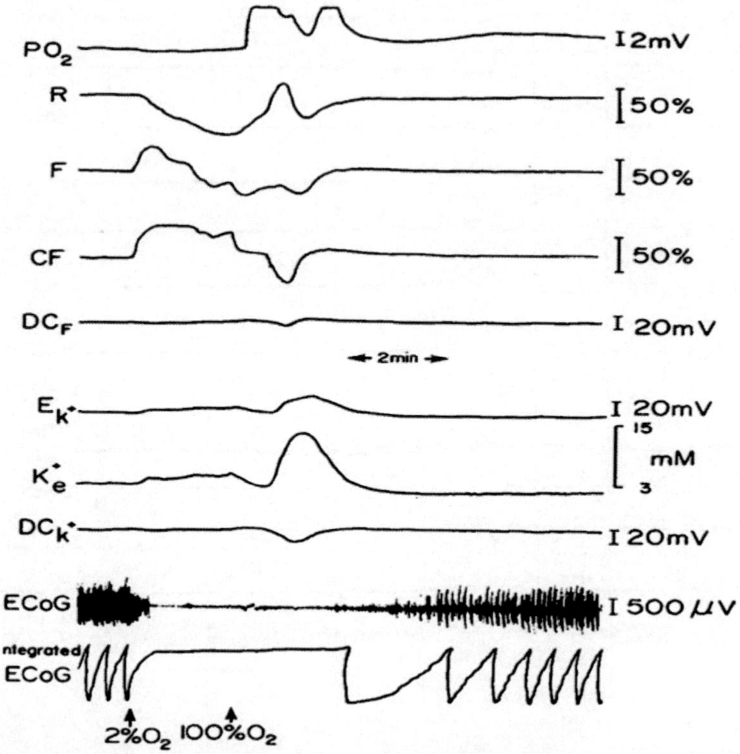

Fig. 5.11 Effects of 2-min hypoxia (2%) on the functional state of the brain. A spontaneous CSD developed at the recovery phase. (Mayevsky et al., 1986a)

instability of the reference electrode since it did not recover during the normoxic recovery period. The recovery period presents a short period of reoxygenation followed by a wave of CSD developed probably by the increased K^+ level during the anoxic period. As seen, the brain became hyperoxygenated due to the vasodilated state (at the end of the anoxia) and the breathing of 100% oxygen. In order to describe the exact relationship between the various parameters, we analyzed the analog signals.

5.3 During Exposure to Hyperbaric Hyperoxia (HBO)

In the introduction to the first paper published by Leão in 1944 (1944), he described the aim of the study as cited here:

> This study originated in an attempt to secure more data for the understanding of the cortical electrogram which occurs in 'experimental epilepsy', and of the conditions in which it is brought forth by electrical stimulation.

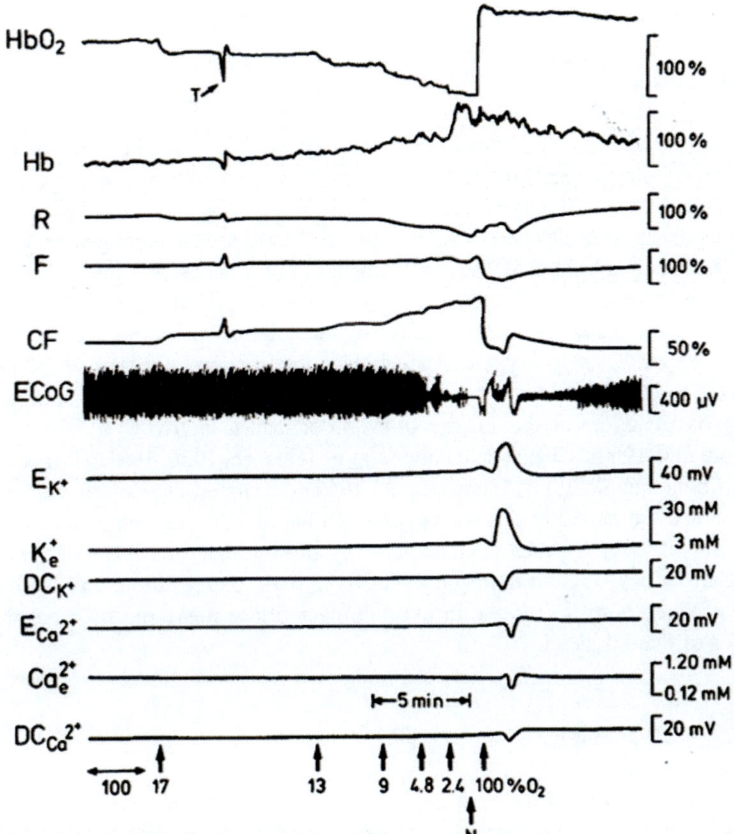

Fig. 5.12 Metabolic, ionic, and electrical responses to graded hypoxia induced in the gerbil brain. (Mayevsky et al., 1990b)

In the Summary part of the paper, he concluded that:

> The depression and 'tonic-clonic' activity of experimental cortical epilepsy seems to be closely related phenomena. The spread of tonic-clonic responses is probably mediated by the same cortical elements which are involved in the spread of depression. The two processes are mainly or exclusively cortical, i.e., they do not require a contribution from subcortical centers.

5.3.1 *First Study—HBO Effects on Mitochondrial Function*

5.3.1.1 Introduction

In the early studies in our laboratory, we investigated the effect of hyperbaric hyperoxia on the metabolic state of the brain evaluated by the monitoring of mitochondrial NADH and ECoG in the un-anesthetized rat. It was well documented that providing animals or man with elevated oxygenation will lead to the development of "oxygen toxicity." The time needed for the development of this toxic event is inversely proportional to the level of oxygenation, namely the higher the pO_2, the shorter the time. On the other hand, providing more O_2 may be beneficial in conditions such as carbon monoxide toxicity, body oxygenation pathology (heart or lung problems), and severe trauma. Therefore, it became necessary to understand the relationship between the level of oxygenation and the function of the mitochondria in vivo.

In the mid-1960s, Chance and collaborators (Chance, 1966, 1967; Chance et al., 1965; Chance et al., 1966; Chance et al., 1969) developed the experimental setup that enabled the exposure of various types of mitochondria as well as of the entire small animal to the hyperbaric chamber. They showed that the NADH of the brain, liver, and kidney became oxidized under hyperbaric oxygenation, and this effect was correlated with a decrease in pyridine nucleotides measured by biochemical analysis of fixed tissue.

5.3.1.2 Methodology

After the introduction of the light-guide-based fluorometry, we were able to expose an awake brain to hyperbaric oxygenation conditions as seen in Fig. 5.13A (Mayevsky, 1978c). Figure 5.13B (Mayevsky & Shaya, 1980) illustrates the location of the light-guide holder, seen in Fig. 5.13C, on the right hemisphere. The ECoG electrodes were inserted and measured in the two hemispheres (Mayevsky, 1975).

Male albino Wistar rats (180–220 g) were used in all experiments. The animals were anesthetized with Equithesin (Jensen-Salsbery Laboratories, U.S.A.) using a minimal level of anesthetic (0.5–0.6 ml) to enable the operation to be performed. The skull was exposed and a 6-mm-diameter hole drilled through the right parietal bone. A specialized combined cannula and light pipe holder, fully described in a previous communication (Mayevsky & Chance, 1973), was then screwed into the skull to maintain constant contact with the dura without exerting any pressure upon it. Three stainless-steel screws were fixed into the skull around the cannula, which was then cemented to the skull with dental cement. This process ensured that even with gross movement of the animal, the cannula remained in place. The EEG was recorded from electrodes passing inside the wall of the cannula and onto the surface of the dura. Recordings were taken with two Offner "Dynograph" recorders connected in parallel so that both fast and slow time scales could be used

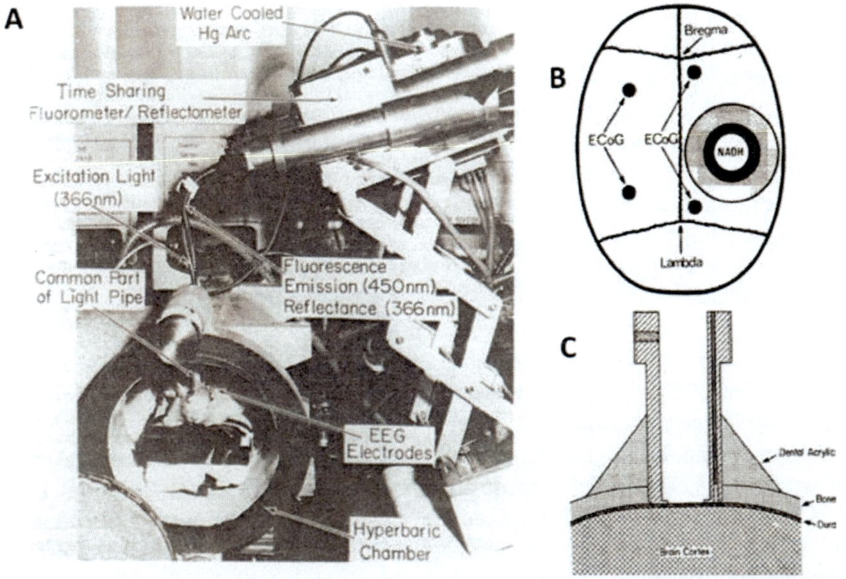

Fig. 5.13 (**A**) Time-sharing fluorometer/reflectometer, attached to the hyperbaric oxygen chamber enables the measurement of NADH from the cortex of the awake rat exposed to HBO (Mayevsky, 1978c). (**B**) Topical view of the rat's skull prepared for the monitoring inside the hyperbaric chamber (Mayevsky & Shaya, 1980). (**C**) The light-guide holder located epidurally on the rat brain and cemented by dental acrylic. (John Wiley and sons, reprinted with permission. (Mayevsky, 1975))

simultaneously. The animal was placed in a small hyperbaric chamber (Bethlehem Corp. FM21-A) and connected to the light guide. The indwelling cannula accepts the bifurcated light guide which exits through the wall of the chamber onto the time-sharing fluorometer. This has two channels, one for excitation and one for reflectance measurements. At the same time, a 90% reflecting, 10% transmitting mirror directs the emitted light toward the 3 measuring photomultipliers. The majority of the signal returning from the rat brain is directed onto a photomultiplier using an appropriate filter for the measurement of reduced pyridine nucleotide fluorescence at 450 nm. The photomultiplier gives sets of pulses, depending upon the position of the rotor of the air turbine; for example, with 366 nm excitation transmitted into the light guide, 366 nm reflectance signals, and 450 nm fluorescence signals are recorded.

The first study on the effects of hyperbaric oxygenation (HBO) was described in detail in our study (Mayevsky et al., 1974a). Recordings were begun immediately, but the animals were left for 2 or 3 h before compression was begun. At that time, the animal was fully awake, as seen by gross movement and the "awake" EEG response. The chamber door was closed and oxygen was flushed through the chamber for 5 min. The animal was then compressed to 75 psi (6 ATA) in 3 min. With the use of light guides, previous difficulties encountered with fluorescence measurements taken through the windows of the pressure chamber (Chance et al., 1966) were eliminated. Recordings were taken continuously until the animal died, as seen by the cessation of respiration and the flattening of the EEG. A continual leak from the chamber was allowed in order to maintain constant CO_2 tension. A second series of experiments was performed in which a gas mixture of 1.5% CO_2, –98.5% O_2 was used for compression. A third series of experiments was undertaken to determine the effect of succinate on the pattern of the responses. The timing and method of administration exactly followed that described by Sanders et al. (1965).

5.3.1.3 Results

The general pattern of responses obtained is illustrated in Fig. 5.14 (Mayevsky et al., 1974a). The upper trace shows the reflectance at 366 nm. In this experiment, very little change was observed during compression but a small increase (8%) occurred 3 min later. This increase corresponds to a change in hemodynamic response in the direction of vasoconstriction and reduced blood flow (Harbig et al., 1976; Jobsis et al., 1971). Fifteen minutes after compression, a large decrease in reflectance (25%) occurred. In this animal, the reflectance returned to the pre-compression level after a further 18 min and remained at this level until respiration ceased. However, in the majority of preparations the post-compression decrease in reflectance, which is probably due to vasodilatation (Bean & Leatherman, 1969; Bean et al., 1972) remained at the lower level throughout the remainder of the experiment. During compression, a decrease of NADH fluorescence of 4–5% was observed. However, as the maximum reflectance change occurred just after this point, the uncorrected fluorescence appeared to return to the baseline. This observation indicates that the 450 nm fluorescence emission is affected by the intensity of the light absorbed by the blood and scattered by the tissue, as monitored by the

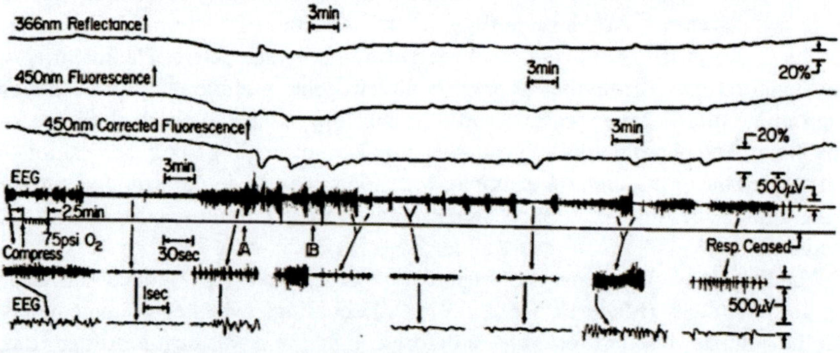

Fig. 5.14 The response of the cortex of an unanesthetized rat exposed to hyperbaric oxygen at (75 psi) 6 ATA until cessation of respiration and flattening of the EEG signal. Reflectance and fluorescence calibrations are indicated as a percentage of the normoxic signal. Calibrations for both these traces are identical. The two lower traces are sections from the EEG recording, taken on faster time bases than in trace 4, with a recorder connected in parallel. Full description in Results. (Mayevsky et al., 1974a)

reflectance measurements (Barlow & Chance, 1976; Harbig et al., 1976; Jobsis et al., 1971; Kobayashi et al., 1971; Mayevsky & Chance, 1973, 1974). Thus, some of the changes of the NADH redox state are obscured in the uncorrected fluorescence trace. Consequently, the third trace, showing the corrected fluorescence, gives a more accurate indication of the true oxidation state of pyridine nucleotides. An 8% oxidation of NADH started immediately upon compression and reached a steady level 5 min after 6 ATA oxygen was reached. Ten minutes later, a further oxidation occurred, concomitant with the decreased reflectance signal and with the onset of convulsions, as shown in the EEG trace. Six minutes after the onset of the seizures, a series of oxidation–reduction cycles of PN was observed. Six such cycles were observed during this experiment. At the peak of such an oxidation cycle, the maximum fluorescence decrease was to 30% of the normoxic level. After cessation of respiration, a slow, continual increase in PN fluorescence was seen. The fourth trace of Fig. 5.14 shows the EEG on a very slow time scale. Prior to compression, the EEG was composed of mixed high and low amplitude waves. Within 4 min of reaching 6 ATA oxygen, the EEG changed to a low amplitude, high-frequency response, indicative of the hyper alert state. Visual observation of the rat during this period also showed the typical hyperactive behavior of rats exposed to HBO. Sixteen minutes after compression, the first signs of seizures were noted on the EEG. These early signs consisted of isolated spikes which gradually increased in intensity and number to give a typical grand mal convulsive pattern. It should be pointed out that the oxidation–reduction cycles of trace 3, mentioned above, occurred between the bursts of convulsion. Such PN cycles continued until the eventual cessation of respiration and EEG occurred. These cycles, which are superimposed upon the hyperbaric oxidized level and occur between bursts of convulsive activity, have not been described previously in HBO, and therefore, a more detailed record of the portion of trace 3, Fig. 5.14, marked "A-B" was obtained. This trace was taken on a much faster time scale to show the EEG changes accompanying such cycles. These cycles in the redox state of PN, which are concurrent with changes in reflected light and depression of the EEG, are characteristic of the effects seen following application of KCl solution to the cortex of the unanesthetized rat, and are believed to be CSDs (Mayevsky & Chance, 1974). Traces 5 and 6 of Fig. 5.14 show portions of the EEG recording taken on a faster time base to show details of the frequency and amplitude of the EEG.

Effect of CO_2 An example of the signals of reflectance, PN fluorescence, and EEG from a rat compressed to 6 ATA in 1.5% CO_2–98.5% O_2 is shown in Fig. 5.15 (Mayevsky et al., 1974a). The principal differences seen with this gas mixture, in comparison with oxygen alone, are in the type of response of the reflectance trace and in the shorter time scale of the whole experiment. When rats were compressed with the CO_2-O_2 mixture, the reflectance almost always decreased, probably due to vasodilatation, rather than showing the customary increase followed by a late decrease, seen in rats exposed to HBO alone. In this case, the uncorrected fluorescence signal also shows a large decrease, but again we use the corrected fluorescence trace for more detailed measurements and here a NADH oxidation of

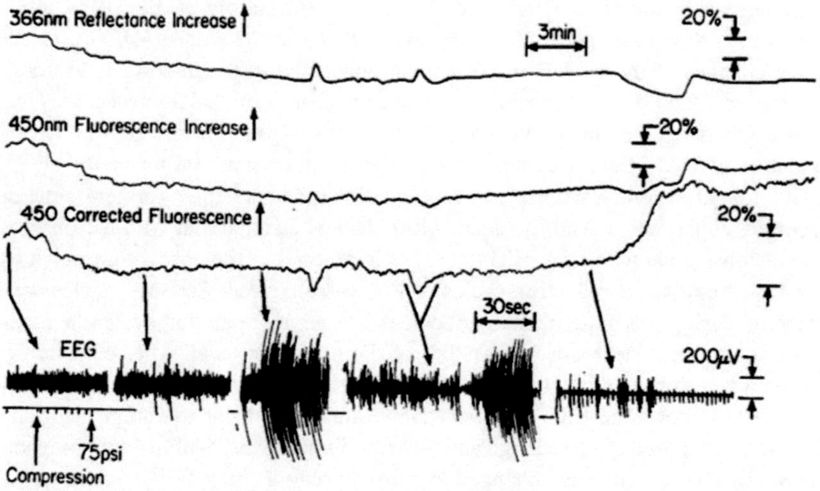

Fig. 5.15 Fluorescence, reflectance, and EEG signals from cortex of an unanesthetized rat exposed to 1.5% CO_2 + 98.5% O_2 at 6 ATA (75 psi) until respiration and EEG cease. Calibrations of fluorescence and reflectance are based on the normoxic signal. Note faster time scale on the EEG trace. (Mayevsky et al., 1974a)

approximately 16% occurs, reaching its maximal level within 1 min at 6 ATA. As in rats compressed with HBO alone, cycles of NADH oxidation can be seen superimposed upon the hyperbaric oxidized state. The fourth trace of Fig. 5.15 shows sections of the EEG taken on the same time scale as that of trace 5 of Fig. 5.14 and indicating the pattern of the EEG at various points in the experiment. Isolated spikes can be seen within 17 min of the completion of compression, and then, large seizures occur spasmodically with flattening of the EEG during the oxidation cycle response.

Effect of Succinate The pattern of responses seen after succinate injection is very similar to that in the untreated animal exposed to HBO, but all events occur on a slower time scale. There is an increase in the time to maximal oxidation, to onset of convulsions, to onset of oxidation waves, and in the total survival time. However, the degree of oxidation of PN observed was unchanged from that in control rats (see Table 5.1).

Table 5.1 indicates that the initial oxidation of NADH prior to convulsions is only slightly different in the 3 conditions studied, i.e., HBO alone, HBO with CO_2 addition, or after pretreatment with succinate. However, the time course of the events varies considerably, and the principal correlations are shown in the histogram of Fig. 5.16 (Mayevsky et al., 1974a). The time from compression to the onset of convulsions is lengthened in the presence of succinate and very greatly decreased in HBO + CO_2. Under these different conditions, a comparable pattern may be observed for the onset of oxidation waves and survival time as well. Thus, there

Table 5.1 Fluorescence and Reflectance Changes in Rats Exposed to Hyperbaric Oxygen

Treatment (no. rats)	Max. PN* oxidation at 6 ATA prior to seizures (% initial)**	PN oxidation of a.c. (% initial)**	% Reflectance increase (A)	Time (min.) to reach (A)	% Max reflectance decrease (B)	Time (min) to reach (B)	Survival time (min)
HPO (16)	14.2±1.6	10.1±1 .3	20.2±2.4	8.3±1.0	26.3±2.8	12.1±1.1	78.3±7.4
HPO+CO_2 (7)	19.3±2.9 N.S.	9.1±1.6 N.S.	-	-	15.0±2.2 P<0.01	8.5±1.9 N.S.	38.6±6.1 P<0.001
HPO+Succinate (5)	12.8±2.0 N.S.	13.5±2.2 N.S.	17.6±6.3 N.S.	10.2±4.4 N.S.	20.4±2.6 N.S.	12.6±4.9 N.S.	110.2±19.4 N.S.

*** PN oxidations calculated from the corrected fluorescence traces**

** The normoxic fluorescence and reflectance levels are taken as 100%, and the values in the Table are measured as a percentage of this normoxic signal

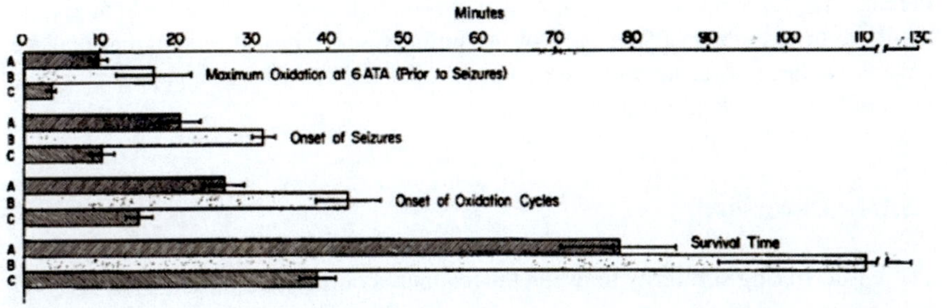

Fig. 5.16 The time course of events occurring following compression of animals to 6 ATA. (**A**) control group. (**B**) succinate pretreated rats. (**C**) rats compressed in 98.5% O_2–1.5% CO_2. Comparison of the time to reach maximum oxidation of PN, and survival time in control and succinate treated rats shows no significant difference with the number of animals used (as in Table 5.1). All other differences are highly significant ($P < 0.005$). (Mayevsky et al., 1974a)

appears to be good correlation between the time course of pyridine nucleotide oxidation and that of physiological events as indicated by the EEG. It should be pointed out that during the convulsive interval a slight further oxidation occurred (approximately 5%), probably due to the convulsions themselves as already noted by other investigators (Jobsis et al., 1971; O'Connor et al., 1972; 1973; 1977) and by ourselves (unpublished observations). Fig. 5.17 shows the level of pyridine nucleotide oxidation during the convulsive stage of HBO plotted against the additional increment of NADH oxidation due to the oxidation wave (probably a CSD wave) occurring between bursts of convulsive activity (Mayevsky et al., 1974a). It can be clearly seen that if the NADH oxidation caused by HBO plus convulsions is large, then the oxidation wave is comparatively lower in amplitude, and vice versa. The oxidation level reached at the peak of an oxidation wave in the hyperbaric state reaches a maximum of approximately 30% of the initial normoxic level. Previous data

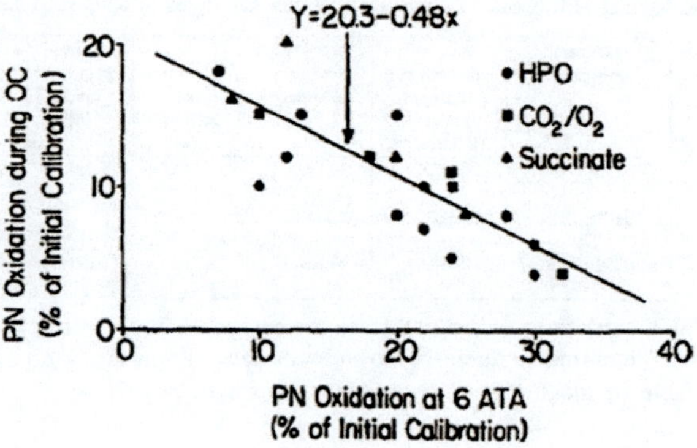

Fig. 5.17 The correlation of the maximum oxidation level of pyridine nucleotide during convulsive activity and between oxidation cycles, with the added percentage oxidation during the oxidation cycle itself. (Mayevsky et al., 1974a)

obtained by Mayevsky and Chance (Mayevsky & Chance, 1973, 1974) indicate that the maximal fluorescence increase due to NADH reduction during a cycle of anoxia attains approximately 40% ± 10% of the initial value. Thus, it is apparent that we are able in these experiments to reach a level of NADH oxidation approximating the level of NADH reduction caused by anoxia.

5.3.1.4 Discussion

There is a striking similarity between the responses of reflectance, fluorescence, and EEG in the oxidation waves seen between convulsive bursts and the responses seen when KCl is applied directly onto the surface of the brain cortex (Mayevsky & Chance, 1974) or injected into the brain or cerebral ventricles (Zuckermann & Glaser, 1968) where the extracellular K^+ concentration is related to the seizure activity. Other investigators (Zuckermann & Glaser, 1968) have reported seizures induced by K^+ injected into brain or cerebral ventricles, and related the degree of predisposition to seizure activity to the levels of extracellular K^+ ion (Izquierdo et al., 1970). We postulate that in the hyperbaric state, K^+ is released from cortical cells and the extracellular K^+ concentration could become high enough to produce the seizures observed in HBO, and the seizures themselves would then cause more K^+ accumulation in the interstitial spaces (Fertziger & Ranck Jr., 1970). Thus, levels of K^+ sufficient to produce CSD may readily arise. O'Connor et al. (O'Connor et al., 1973) have recently reported that when K^+ was topically applied to the hippocampus there was an oxidation of NADH, manifesting increased energy metabolism, possibly due to increased Na^+-K^+ ATPase activity, and that as the K^+ concentrations were increased, there were often resultant seizures. The greatest extent of pyridine nucleotide oxidation that had been obtained previously was with hyperbaric oxygen (Chance et al., 1966). The fact that we can here achieve further oxidation when CSD occurs gives a new viewpoint on the metabolic state of brain pyridine nucleotides. In order to interpret the NADH changes, it is necessary to determine the initial metabolic state of the brain cortex, i.e., whether it tends toward the resting state 4 condition, or is closer to the active state 3. If we consider that the total increase of

NADH oxidation from the normoxic condition is approximately equal to the total reduction obtained in anoxia, the normoxic state would lie at a ratio of oxidized to reduced NADH of approximately 1.0. A comparison of this oxidation–reduction ratio with that obtained in isolated mitochondria suggests that the normoxic state corresponds rather closely to the active state 3, while the state showing maximal NADH oxidation corresponds to state 2, characteristic of a combination of substrate depletion and an activation of electron flow by ADP and phosphate.

5.3.2 Second Study—Protection Against Oxygen Toxicity by Trimethadione

The phenomenon of "oxygen toxicity" has been under investigation since Bert described it in 1878. The basic mechanism by which this toxicity appears in man and animals is still unknown. The most dramatic effect of oxygen at high pressure, HBO (above 3 ATA), on the central nervous system is the induction of clonic–tonic convulsions. These convulsions are the first external sign of HBO toxicity, occurring within minutes after compression. The most rapid biochemical change which is observed in the anesthetized brain in vivo (as well as in other organs in vivo or in vitro) is the oxidation of pyridine nucleotides, NADH (Chance et al., 1965; Chance et al., 1966). We have recently measured the oxidation–reduction state of NADH and the EEG activity of the unanesthetized rat brain exposed to HBO (Mayevsky et al., 1974a). We found that by addition of carbon dioxide to the hyperbaric gas mixture (1.5% CO_2) or pretreatment of the animals with succinate, almost the same level of PN oxidation is observed as in a control group. The same sequence of events was observed in all three groups, but the time scale was shortened by CO_2 and lengthened by succinate. The succinate did not prevent the appearance of the clonic–tonic seizures but only prolonged the interval before their onset. Paton (1967) reported protection against oxygen toxicity by the use of trimethadione (TMO); however, these results were statistically insignificant. In the present study, the use of an anticonvulsant drug, trimethadione, as a protector against oxygen toxicity is described in detail. The correlation between the oxidation of NADH and the EEG activity of the conscious rat has been measured after injection of TMO.

Figure 5.18 shows a typical response of a control rat exposed to 75 psi oxygen. The control animals show the same response as we had found previously (Mayevsky et al., 1974a). The reflectance at 366 nm increases during the compression period, and a few minutes later, a large decrease of reflectance occurs. This pattern of reflectance changes was observed in all animals. The third trace from the top—the corrected fluorescence—represents the difference between the fluorescence emission at 450 nm and the reflectance at 366 nm. By this subtraction, one can eliminate absorbance changes due to hemodynamic effects which produce artifacts in the fluorescence measurements. During the compression, an oxidation of NADH of 10% of the normoxic fluorescence level is observed, which is maintained for

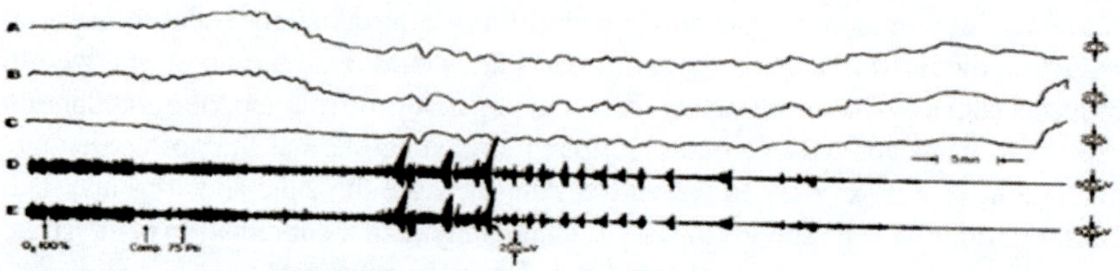

Fig. 5.18 The responses of awake brain cortex exposed to 75 psi pure oxygen (control animal). (**A**) 366 nm reflectance; (**B**) 450 nm fluorescence; (**C**) 450 nm corrected fluorescence; (**D**) EEG of right hemisphere; (**E**) EEG of left hemisphere. Note that the EEG amplitude (traces D and E) was changed during the experiment. Upward deflection of the pen represents an increase in the signal size. (John Wiley and sons, reprinted with permission. (Mayevsky, 1975))

15 min. A series of oxidation–reduction cycles of NADH then appears. Approximately 10 min before the animal stops breathing, a reduction of NADH is observed which increases to 50–60% at the end. The fourth and fifth traces of Fig. 5.18 show the EEG measured from the two hemispheres. In most animals, the two hemispheres of the cortex respond to HBO in the same way. A few minutes after compression, the EEG changes from the typical "awake" pattern to the activated pattern, and then, the convulsions appear. The number of bursts of convulsive activity differs between animals. The EEG becomes flat just before the animal stops breathing, and the reduction of NADH is seen. In Fig. 5.19, the typical response of an animal pretreated with TMO is shown (Mayevsky, 1975). The pattern of changes in reflectance and fluorescence is indistinguishable from that of the control animals. The main changes appear in the EEG traces. The TMO-treated animal does not show the bursts of high-amplitude activity. A higher amplitude appears in the EEG traces, but the EEC typical of tonic seizures is very rare. Table 5.2 summarizes this set of experiments. Some parameters were measured from the traces, and the averages and standard deviations were calculated. The average weight of the control animals was almost the same as in the TMO group. The degree of pyridine nucleotide oxidation, calculated from the corrected fluorescence traces, does not show a significant difference between control and TMO-treated groups, either on pressurization to 6 ATA or during the later oxidation–reduction cycles. However, the time to the onset of convulsions was more than twice as long in the TMO-treated group as in the controls, as was the time to the onset of the oxidation cycles. The survival time of the TMO-treated group was also longer. However, the number of tonic seizures and the number of oxidation cycles were greater in the control group than in the TMO-treated group.

5.3.3 *Third Study—Effects of CO_2 Under HBO*

The interrelations between convulsive activity, and CO_2 concentration under hyperbaric oxygen conditions were investigated by few investigators. It was shown in mice (Marshall & Lambertsen, 1961) and in cats (Taylor, 1949) that small elevation

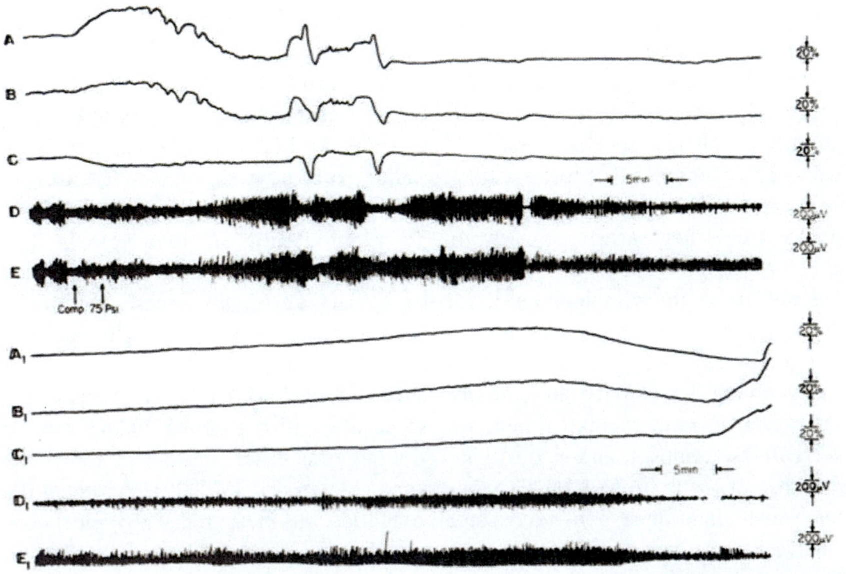

Fig. 5.19 The effect of trimethadione on the response of the awake brain cortex to 75 psi pure oxygen. *A, A_1* 366 nm reflectance, *B, B_1* 450 nm fluorescence, *C, C_1* 450 nm corrected fluorescence, *D, D_1* EEG of the right hemisphere, *E, E_1* EEG of the left hemisphere. Upward deflection of the pen represents an increase in the signal size. (John Wiley and sons, reprinted with permission. (Mayevsky, 1975))

Table 5.2 The effect of trimethadione (TMO) on different parameters measured during the exposure of rats to 75 psi

Treatment (no. rats)	Weight (gm)	Initial PN oxidation* at 6 ATA	Onset of convulsions (min)	Number of tonic convulsions	Onset of oxidation cycles (min)	Number of oxidation cycles	PN oxidation of oxidation cycles	Survival time (min)
HPO	200.2	15.6	9.4	11.4	12.7	4.8	17.8	69.6
(10)	±19.2	±5.1	±3.2	±4.2	±4.1	±1.3	±4.0	±31.7
HPO+TMO	203.8	18.0	24.2	0.5	31.4	2.7	18.2	102.0
(10)	±17.0	±4.3	±6.8	±0.8	±8.5	±1.3	±6.8	±23.2
Level of significance	N.S.†	N.S.	$p<0.01$	$p<0.01$	$p<0.01$	$p<0.01$	N.S.	$p<0.05$

***PN (pyridine nucleotides) oxidation calculated from the corrected fluorescence traces. The normoxic fluorescence levels are taken as 100%, and the values in the table are calculated as percentage of the normoxic level**

†N.S. **not significant

of CO_2 under HBO potentiated the toxicity phenomenon while higher levels of CO_2 inhibited convulsions. Other investigators (Levy & Richards, 1962) came to the conclusion that addition of CO_2 to the HBO chamber provided the animal with protection against toxicity. In our previous communication (Mayevsky et al., 1974a), we found that 1.5% CO_2 in oxygen potentiates the toxicity under pressure of 6 ATA. In order to understand the effects of CO_2 on the response of the brain to HBO, we measured the effects of various concentrations of CO_2 on the electrical activity as well as the metabolic activity of the awake brain.

A typical response of the awake cerebral cortex to oxygen was published previously (Mayevsky et al., 1974a). In the present study, we exposed a total number of 30 rats to HBO with various concentrations of CO_2 (as shown in Table 5.3). Figure 5.20 shows a record taken from a rat exposed to 2% CO_2 in pure oxygen (Mayevsky, 1978c). As the compression started, the reflected light (upper trace) decreased, probably due to vasodilatation of blood vessels. The uncorrected trace shows the same decrease, so the corrected trace of this animal did not show the classical oxidation of NADH as we described previously. Several minutes later, the electrical activity increased, namely, spiking activity was recorded.

Following this activity, an oxidation–reduction cycle of NADH was measured. The termination of the experiment was when respiration stopped, ECoG activity decreased to almost 0 and NADH showed a large increase. Figure 5.21 shows the response of the brain to 3.1% CO_2 in oxygen (Mayevsky, 1978c). Here, again the same vasodilatation response and a small oxidation was measured. The typical convulsive activity was not recorded and probably was blocked. Only one oxidation cycle was measured from this animal.

In order to understand the effect of CO_2 in detail, each record was analyzed and the results are shown in Table 5.3.

The initial oxidation of NADH was calculated (not shown in the Table) and appeared in 19 out of 30 rats used and was in the range of 5–15% of the initial fluorescence level in the normoxic brain. One can see from the Table that 4% CO_2 and 5% CO_2 blocked all the metabolic and electrical responses of the cortex to HPO. The survival time was shorter in all the CO_2 groups but was not statistically significant due to high variability between the animals. The number of tonic convulsions decreased as CO_2 went up and disappeared at 3.1% CO_2. The time to the onset of oxidation cycle was longer than that for the convulsion and it suggests that perhaps these cycles are the CSD type response after the convulsion, as was previously suggested (Mayevsky et al., 1974a).

Table 5.3 The effects of various CO_2 concentrations under HBO, on the various parameters calculated from the records

	Onset Convulsion (min.)	Number of tonic Convulsions	Onset of Oxidation Cycles (min)	Number of Oxidation Cycles	PN Oxidation of Oxidation Cycles	Survival Time (min.)
No CO_2	9.4±3.2	11.4±4.2	12.7±4.1	4.8±1.3	17.8±4.0	69.6±31.7
1% CO_2	7.4±1.8	3.8±1.6	11.6±2.1	2.6±0.9	17.2±2.6	41.8±12.3
2% CO_2	10.6±9.4	1.4±1.1	15.6±12.3	1.4±0.5	18.0±4.5	40.0±13.9
3.1% CO_2	10.2±2.2	0	-	0.4±0.5	-	48.0±19.8
4% CO_2	0	0	-	0.2±0.4	-	51.2±25.4
5% CO_2	0	0	0	0	0	41.8±11.4

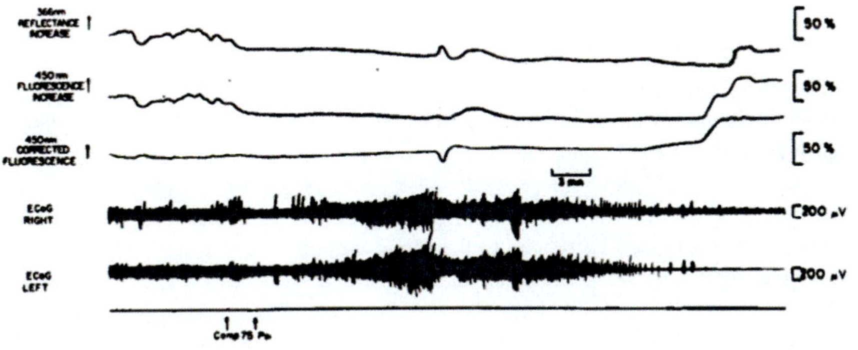

Fig. 5.20 The response of the awake brain to 2% CO_2 in pure oxygen under 6 ATA. (Mayevsky, 1978c)

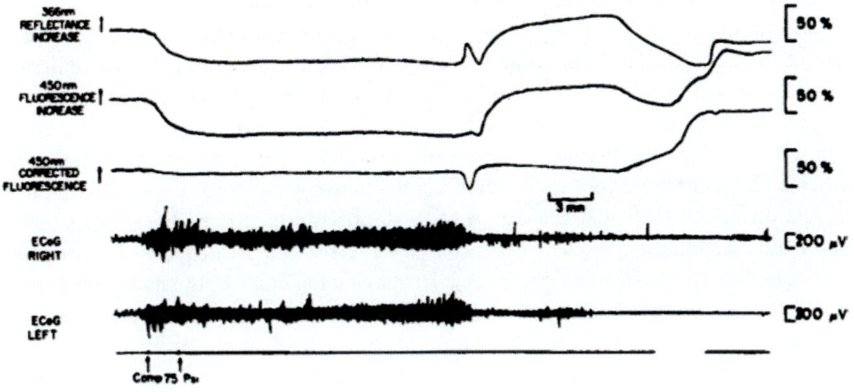

Fig. 5.21 The effects of 3.1% CO_2 on the metabolic and electrical response of an awake brain. (Mayevsky, 1978c)

The oxidation of NADH during the cycles was the same in all groups found to show this response, suggesting that whenever extracellular potassium elevates, energy utilization will go up. All the results show that increase in CO_2 concentration in oxygen toxicity may block the convulsive activity but does not protect the animal against the toxicity process and the animal may die earlier than the control animals (Chapin, 1955).

5.3.4 More Factors Affecting HBO Toxicity

To understand in detail, the responses to hyperbaric oxygenation, as evaluated from the NADH and electrocorticogram (ECoG) measurements, we exposed awake rats (200 g) to various levels of pressure and awake rats of various ages to 90 psi O_2. In a different set of experiments, we tested the effects of two protection agents, pentobarbital and diazepam, on the response under 60 and 90 psi O_2.

Effects of Pressure Level In the present set of experiments, 52 rats were used. A typical response of the brain to HBO was previously described (Mayevsky, 1978c; Mayevsky et al., 1974a) and can be seen in Fig. 5.22 (Mayevsky & Shaya, 1980). The upper trace (A), which represents the changes in the reflected light, shows the biphasic response, namely, an initial increase and a sharp decrease later on (at T_1). This parameter is well correlated to the blood volume in the tissue under observation (Harbig et al., 1976; Jobsis et al., 1971; Mayevsky, 1978c; Mayevsky et al., 1974a) and we use the term "hemodynamic parameter" when analyzing the reflectance trace. The second trace (B) is the measurement of the intensity of the NADH fluorescent light measured from the intramitochondrial space. The trace (C) showed a decrease of the signal interpreted as an oxidation of NADH. This initial oxidation reached a steady level and later showed oxidation–reduction cycles (at T_3). The corrected fluorescence signal showed a large increase (reduction of NADH) when the animal died (T_5). The oxidation–reduction cycles recorded during the epileptic activity period were very similar to those measured during CSD (Mayevsky, 1975; Mayevsky & Chance, 1975; Mayevsky et al., 1974a) and named "oxidation cycles" or "CSD cycles." The ECoG responses are identical in the two hemispheres; that is, after a short latent period, tonic–clonic convulsions were recorded until complete cessation of the electrical activity. A clear depression of the ECoG was recorded after the convulsion episode. The cessation of respiration occurred 5–10 min after the flattening of the ECoG signals (T_5). To analyze the analog signals in detail, several parameters were measured or calculated (Fig. 5.22), as well as analyzed and summarized in Table 5.4. The total of 52 rats were divided into seven pressure levels, and the mean for each group, as well as the standard error of the mean, is shown in Table 5.4. Three types of parameters are shown in Table 5.4 as follows:

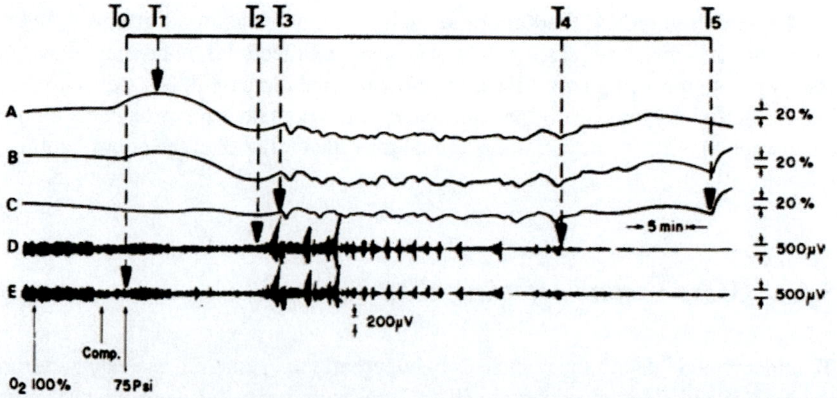

Fig. 5.22 Typical responses of awake rat brain to 6 ATA O_2 (see text for details). (Mayevsky & Shaya, 1980)

1. Changes in NADH levels. **a.** Corrected fluorescence (CF) increases during anoxia. (This parameter was measured at the beginning of each experiment.) **b.** Initial CF decreases when the pressure was stabilized in the chamber.
2. Time to occurrence of various events. **a.** Time-to-reflectance decrease (T_1), representing the second hemodynamic event that occurred after pressurization (the first one is the reflectance increase due to vasoconstriction during pressurization). **b.** Time to the onset of convulsion (T_2), representing the latent period until the fast tonic convulsion was measured. **c.** Time-to-oxidation cycles (T_3). This event always occurred after the first tonic convulsion. **d.** The duration of the convulsive period. This parameter was measured between the two points T_2, the flattening of the ECoG record. Survival time (T_5). This parameter was determined according to the cessation of respiration, as observed via the window of the chamber. All measurements were compared to To, time at which the needed pressure level was achieved (Fig. 5.22).
3. Number of various events. **a.** Number of tonic convulsions. **b.** Number of oxidation cycles.

From the results shown in Table 5.4, it is clear that the effect of the level of pressure on the NADH decrease during pressurization is very insignificant and without any clear direction. The lowest oxidation was recorded at 75 psi (18 ± 1%) and the highest at 120 psi (34 ± 7%). The general pattern of oxidation of NADH and the oxidation cycles were measured in all animals at various pressures. For this reason, the NADH oxidation was measured as a parameter but was not used in the correlation study made between the various parameters. Table 5.5 shows the correlation coefficients of the parameters belonging to type 2 and 3 groups mentioned above. For better understanding, we plotted a few of the significant correlations as shown in Fig. 5.23 (Mayevsky & Shaya, 1980). All the correlations between the parameters of timing are statistically significant, but the level of significance differs. It seems that between the three parameters of time: time-to-reflectance decrease (A), time to convulsions (B), and time-to-oxidation cycle (C), the correlations are highly

Table 5.4 Effects of various levels of hyperbaric pressure on different parameters measured from the awake brain

Pressure, psi	CF Increase during Anoxia, %	Initial CF Decrease at Pressure, %	Time-to-Reflectance Decrease, min	Tonic Convulsions: Onset Time, min	Tonic Convulsions: No.	Duration of Convulsive Period, min	Oxidation Cycles: Onset Time, min	Oxidation Cycles: No.	Survival Time, min
150 (7)	53±3	27±5	6±2	8±2	2±0.3	14±5	10±3	2±0.3	42±5
120 (9)	48±3	34±7	6±1	12±2	3±0.5	8±1	13±1	3±0.3	38±2
90 (7)	57±4	26±3	24±7	28±9	8±1.7	24±2	28±8	4±0.6	65±12
75 (7)	58±3	18±1	31±7	36±5	9±1.0	29±7	37±5	3±0.6	113±17
60 (9)	52±3	27±3	34±6	38±6	18±2.3	92±13	43±7	14±2.0	243±34
45 (7)	48±3	27±6	154±23	166±24	11±2.4	68±14	168±24	6±1.0	336±22
30 (6)	58±3	21±4	230±9	240±9	3±0.3	57±23	242±10	2±0.2	420±38

Values are means ± SE. Number of rats are given in parenthesis. CF, corrected fluorescence

significant, but between those three parameters and the other two parameters (i.e., convulsive period and survival time), the correlations are significant but have lower r values. This point can be seen by comparing the distribution of the points in Fig. 5.23A and B, to that of Fig. 5.23C. The correlation between the convulsions and the number of oxidation cycles is very highly significant (r = 0.862). The correlation between the convulsive period and the other two connected parameters, namely, the number of convulsions and the number of oxidation cycles, is also highly significant. The survival time was shown to be in a significant correlation with all the parameters, except with the number of convulsions. The effects of pressure on the various parameters measured are demonstrated in Fig. 5.24 (Mayevsky & Shaya, 1980). The three parameters shown in Fig. 5.24A are probably connected to each other and in most conditions occurred in the same order (i.e., the change in reflectance is the first event, followed by the convulsive activity, and the oxidation cycles, which appeared later). Between 30 and 60 psi, the slopes of the changes of all three parameters are very sharp, whereas between 60 and 150, they are more moderate. Thus, the 60-psi pressure is a breaking point of the line. On the other hand, the other three parameters shown in Fig. 5.24B are affected differently by the pressure. The maximum effect was observed at 60 psi, and the curves had a bell shape. The differences between the 60-psi point and the 30- or 150-psi point (Fig. 5.24) are statistically significant ($P < 0.005$), as calculated by Student's t-test.

Effects of Age A total of 41 rats were used in this set of experiments and were divided into six groups, ranging from 14 to 90 days. Unfortunately, due to technical difficulties, it was impossible to use the same monitoring technique below 14 days. From the analog signals obtained, we calculated the various values of the various

Table 5.5 Correlation coefficient values and level of significance between various parameters measured during oxygen toxicity of awake brain

	Time to Convulsion	No. of Convulsions	Convulsive Period	Time-to-Oxidation Cycle	No. of Oxidation Cycles	Survival Time
Time-to-reflectance decrease	0.994#	NS	0.320*	0.994‡	NS	0.845‡
	Time to convulsion	NS	0.307*	0.999‡	NS	0.843‡
		No. of convulsions	0.655#	NS	0.862‡	NS
			Convulsive period	0.323*	0.684†	0.657†
				Time-to-oxidation cycle	NS	0.853‡
					No. of oxidation cycles	0.334*

NS—not significant. *$P < 0.05$, †$P < 0.01$, ‡ < 0.001

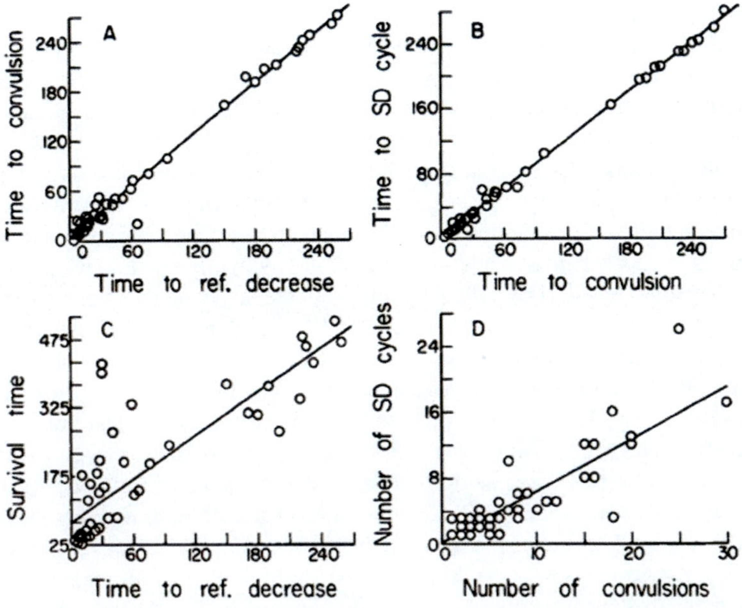

Fig. 5.23 Correlations between various parameters measured from the rat exposed to various levels of hyperbaric oxygenation. The correlation coefficient values (r) are presented in Table 6.2. (Mayevsky & Shaya, 1980)

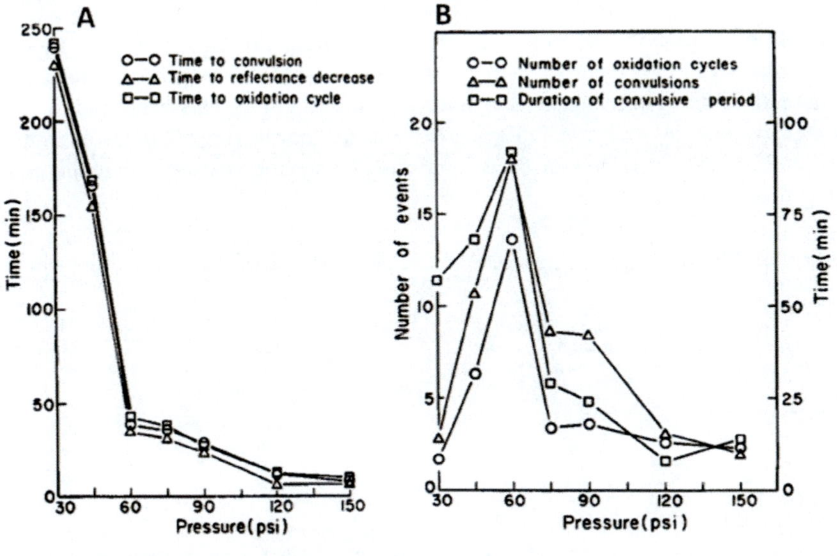

Fig. 5.24 (**A**) Effects of pressure level during hyperbaric oxygenation on hemodynamic, metabolic, and electrical activity of the brain. (**B**) Effects of pressure level during hyperbaric oxygenation on electrical activity and its concomitant phenomena of the brain. (Mayevsky & Shaya, 1980)

parameters as described in the preceding section. Table 5.6 presents the means and standard error of the mean of the different age groups. The results show that age had very small effects on the various parameters measured. The percent increase of NADH during anoxia and the percent decrease of NADH after pressurization to 90 psi were not affected by the age of the rats, although some variations were found between the groups. The response of the blood vessels to hyperbaric oxygenation, as evaluated from the 366 nm reflectance trace, showed that the younger the animal, the longer it took to reach the time-to-reflectance decrease. The differences between the groups are not significant due to the variability between the animals, leading to a high standard error. The effects of age on the appearance and duration of convulsions, as well as the oxidation cycles following the convulsions, were not consistent and in most cases were not statistically significant. It seems that the main changes were found in the 14-day-old rats; namely, at this young age, the toxicity process developed slower, as can be seen from the very long survival time. In this group, the appearance of the convulsion as well as the oxidation cycles was the longest, as compared to the rest of the groups. Although the variation between the groups was not so large and significant, the various parameters showed a wide range of values. Figure 5.25 shows the correlations between various parameters measured in the 41 rats used in this study (Mayevsky & Shaya, 1980). From the results, it is clear that the first event taking place after the oxidation of NADH is the decrease in the reflected light (A), followed by the appearance of the convulsion, and then, the oxidation cycles (CSD cycles) took place (B). The number of oxidation cycles (CSD cycles) was affected by the length of the convulsive period (C), as well as by the total number of the tonic convulsions (D).

Effects of Pentobarbital and Diazepam Two protection agents against hyperbaric oxygen toxicity were tested in our study under two levels of pressure. Table 5.7 summarizes the results obtained by analyzing the analog signals measured during the exposure to HPO. From Table 5.7, one can see the significant effects of pentobarbital as well as diazepam on the various parameters measured. Pentobarbital

Table 5.6 Effects of age on the metabolic and electrical responses of the awake brain to 90 psi 100% O_2

Age, days	Body wt, g.	CF increase during Anoxia, %	Initial CF Decrease at 90 psi, %	Time to Reflectance-Decrease, min	Tonic Convulsions: Onset Time, min	Tonic Convulsions: No.	Duration of Convulsive Period, min	Oxidation Cycles: Onset Time, min	Oxidation Cycles: No.	Survival Time, min
90	300 (4)	42±2	23±7	15±2	30±5	12±3	31±1	35±6	5±1	71±19
60	200 (10)	41±4	18±2	12±2	21±3	7±1	27±5	21±3	3±1	88±14
35	100 (7)	57±4	25±3	24±7	28±9	8±2	24±2	28±8	4±1	65±12
21	60 (5)	62±3	18±2	20±5	28±4	7±2	28±6	28±4	2±0.3	89±3
17	40 (8)	58±2	25±4	22±4	25±3	4±1	13±2	25±3	2±0.7	63±7
14	20 (7)	55±5	22±4	33±13	39±12	4±1	33±2	45±10	2±0.2	292±25

Values are means ± SE. Number of rats is given in parenthesis. CF, corrected fluorescence

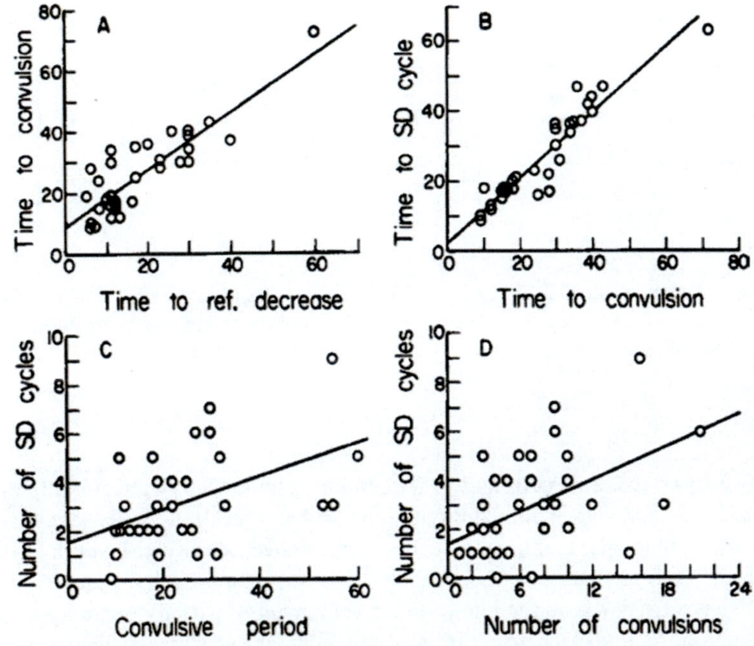

Fig. 5.25 Interrelation between few parameters measured from brain of rats at various ages. Correlation coefficients (r) were 0.863*, 0.941**, 0.465*, 0.532* for A, B, C, and D, respectively. *P < 0.01; **P < 0.001. (Mayevsky & Shaya, 1980)

completely inhibited the tonic convulsions under 60 and 90 psi. As a result, the oxidation cycles did not occur in most of the animals treated by pentobarbital. Diazepam also inhibited the convulsions and the oxidation cycles under 60 and 90 psi but not completely as was found with pentobarbital.

The main difference between the 60 and 90 psi can be seen in the survival time values. At 90 psi, the pentobarbital and diazepam significantly increased the survival time, whereas at 60 psi no significant difference between the three groups of animals was observed although the convulsions were almost completely inhibited.

Discussion We can summarize the various events found during exposure to hyperbaric oxygenation as follows. After pressurization, three different processes were recorded in relation to blood flow, electrical activity, and metabolism. Those three events were described previously, and their significance to the development of the oxygen toxicity is still unknown. Chance et al. (1966) discussed the toxicity phenomenon on the basis of the metabolic activity of the mitochondria. Due to the lack of a correlation between the pressure level and the rate of NADH oxidation, we doubt there is a direct primary relation between mitochondrial activity and the toxicity. It seems that NADH oxidation is only one of the responses of the brain to HPO but not the key one. The main factor in the toxicity, according to the data presented,

Table 5.7 Effects of pentobarbital and diazepam on the responses of the brain to 60 and 90 psi 100% O_2

Pressure, psi	Treatment	CF Increase during Anoxia, %	Initial CF, Decrease at Pressure, %	Time-to-Reflectance Decrease, min	Tonic Convulsions: Onset Time, min	Tonic Convulsions: No.	Oxidation Cycles: Onset Time, min	Oxidation Cycles: No.	Survival Time, min
90	Control (7)	57±4	25±3	24±7	28.4±8.6		27.6±7.6	3.6±0.6	65±12
	Pentobarbitol (7)	54±3	19±5	97†±15	∞ †	0‡	91‡	0.4†	118†±6
	Diazepam (6)	89*±6	24±4	65†±9	53†	0.3‡	55†	0.7†	121†±4
60	Control (9)	52±3	27±3	34±6	38.3±6.1	18±2.3	42.6±6.8	13.6±1.9	243±34
	Pentobarbitol (7)	52±2	16±2	204†±19	∞	0‡	218†	0.3†	297±15
	Diazepam (6)	86*±4	39±2	161†±22	164†	3†	109†	1†	283±29

Values are means ± SE. Number of rats is given in parentheses. CF, corrected fluorescence. Levels of significance in comparison to the control group: *P < 0.05; ‡P < 0.001

is the effect on the brain cells and its membrane system. One of the possible initial events is a change in the permeability to ions, and as a result, K^+ will accumulate in the extracellular space and may lead to convulsive activity (Mayevsky, 1975; Mayevsky et al., 1974a). From our results, it is not clear what the exact effects of convulsive activity are on the rate of the development of toxicity under hyperbaric oxygenation. As it is clear from Fig. 5.24, the number of tonic convulsions and the duration of the convulsive period are not correlated to the pressure level. The concomitant event of the convulsions (the oxidation cycles of NADH) is a good indication that the Na^+-K^+-ATPase system is working normally and shows the typical response to CSD, appearing after convulsions (Mayevsky & Chance, 1975).

5.3.5 *Fifth Study—Multiparametric Monitoring of CSD Under HBO*

5.3.5.1 Introduction

Due to technical limitations (in most of the studies published), only one to three physiological parameters were monitored simultaneously from the same site in the brain in vivo under HBO. Stein and Sonnenschein (1950) measured the electrical activity and O_2 partial pressure (pO_2). The same two parameters were measured recently in unanesthetized rabbits (Torbati et al., 1977). The first biochemical response to HB0 recorded in vivo is the oxidation of reduced pyridine nucleotide (NADH). Chance et al. (1965, 1966) described it in detail for the anesthetized rat brain, and later Mayevsky et al. (1974a, 1975, 1978c, 1980a, b) measured it from the brain of unanesthetized rats. In a recent report, we described in detail a proposed mechanism explaining the metabolic, ionic, and electrical activities during the development of brain O_2 toxicity (Mayevsky & Shaya, 1980). To elucidate the mechanism of brain O_2 toxicity, it is necessary to measure as many physiological parameters as possible from the same site. Our present study aimed to develop a new multiparameter monitoring system located in a hyperbaric chamber, using an awake animal. The basic features of the multiparameter assembly used under normal conditions were described previously (Friedli et al., 1982), and in the present study, we extended and adapted the system for the hyperbaric chamber. The multiparameter assembly can measure up to eight parameters representing the metabolic, ionic, and electrical activities on the cerebral cortex. Metabolic activity is evaluated by measuring tissue O_2 tension and by monitoring the intramitochondrial NADH oxidation–reduction state. The ionic state of the tissue is evaluated by monitoring extracellular potassium (K^+) activity. Electrical activity was monitored by bipolar electrocorticography (ECoG) and the DC steady potential. In addition, we monitored the surface temperature with a miniature thermistor.

5.3.5.2 Methods

The multiprobe assembly has the following main features. (1) All parameters have the same type of noninvasive surface contact with the cortex. (2) The multiprobe assembly can be removed without any damage from the skull very easily at the end of the experiment so that repetitive experiments can be performed in a short period of time. (3) All signals monitored by the multiprobe assembly show a very small sensitivity to movements of the animal while exposed to hyperbaric pressure. The aim of the present communication is to show the technical aspects of the system, as well as to describe the various responses measured during hyperbaric brain O_2 toxicity.

Figure 5.26 presents, in a schematic way, the connection between the multiprobe assembly and the cerebral cortex of the rat brain (A) and the location of the various probes in relation to the skull (B) (Mayevsky, 1983b). The various electrodes were held in the multiprobe assembly as described in detail elsewhere (Friedli et al., 1982; Mayevsky & Sclarsky, 1983). A general description follows. The probes were mounted in a Lucite holder (cannula) that can be cemented directly to the skull. The holes accommodating the specific sensing electrodes converge at the implantation site to occupy less space on the brain and diverge at the top of the cannula to leave maximal room for mounting and replacing the electrodes. Another hole was drilled obliquely from the top to intersect the first chamber approximately half way down. This hole houses an Ag-AgCl wire to record DC potential concentrically to the sensing electrodes. One more hole was drilled horizontally into the chamber of the DC wire for filling with saline and for cleaning extracellular fluid and blood from around the electrodes at the end of an experiment. A Lucite sleeve was placed over the entire assembly and was held in place with set screws. This sleeve holds the assembly together as one unit, protects the electrodes and wiring, and also provides shielding for the K^+ electrode's high impedance. The sensing electrodes, Ag-AgCl wires for DC measurement, stainless-steel wires for ECoG measurement, and a

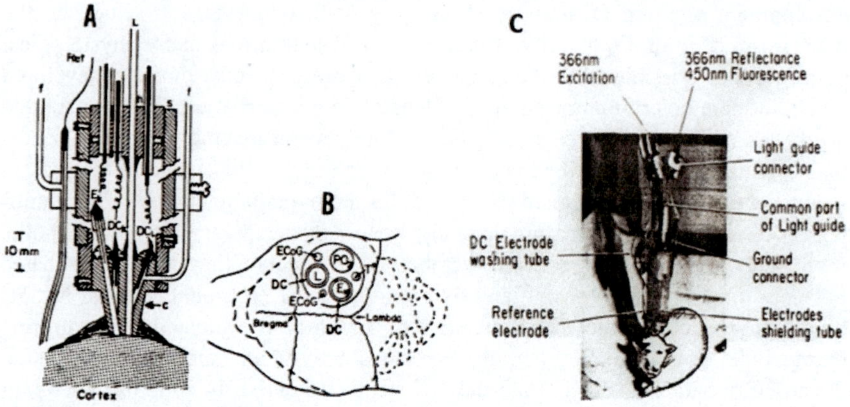

Fig. 5.26 Schematic presentation of multiprobe assembly (MPA) used in hyperbaric chamber. (**A**) Longitudinal section of MPA; (**B**) The relative location of various probes above brain. *Ref* reference, *Ag-AgCl* electrode, *f* refill tube for Ref or DC electrode, *c* Lucite cannula, *s* Plexiglas sleeve, *L* light guide, *h* cable holder, *Ek* K^+ electrode, *DCK, DCL, Ag* AgCl electrode, pO_2 O_2 electrode, *ECoG* electrocorticography electrodes. (**C**) Rat connected to multiprobe assembly inside hyperbaric chamber. (Mayevsky, 1983b)

polyethylene-sleeved temperature probe were held in and sealed to the cannula with 5-min epoxy or by melting wax around them. Another block of Lucite was attached to the outer shield that holds the reference assembly and the tubes for filling the DC chambers.

Surface Fluorometry The intramitochondrial NADH redox state was monitored by a light-guide DC fluorometer reflectometer (Mayevsky & Chance, 1975). The source for the 366 nm light was a 100-W mercury arc cooled by water. The emitted light was split in a 90–10 ratio for the measurement of the NADH fluorescence and reflectance, respectively. The common part of the light guide was cemented inside the multiprobe assembly in the same way as the other electrodes. A short piece of the excitation and emission bundle of fibers was connected to the common part. Two bundles of fibers were connected inside the wall of the hyperbaric chamber and used to transmit the light into and from the multiprobe assembly. Specially designed connectors were used when the animal was introduced into the chamber for measurements (Fig. 5.26C). The signals were calibrated as described previously (Mayevsky, 1976). More technical details are available in our previous publication (Mayevsky, 1983b).

Animal Preparation The rat was anesthetized with Equithesin (0.3 ml/ 100 g body wt ip; each ml contains 9.72 mg pentobarbital sodium, 42.51 mg chloral hydrate, 21.25 mg magnesium sulfate, 44.34% wt/vol propylene glycol, 11.5% alcohol, and water). After a midline incision of the skin, the skull was exposed and a 6.5-mm-diam hole was drilled in the right parietal bone. Four holes (1.5 mm diameter) were drilled in various points of the skull, and stainless-steel screws were used to hold the cement to the skull. The dura mater was removed very gently only from the area where the K^+ electrode was placed. The multiprobe assembly was held by a micromanipulator and located above the cortex without creating any extra pressure on the brain. The ground and reference electrodes were placed below the skin and were cemented to the skull together with the multiprobe assembly and the screws by dental acrylic cement. The rat was placed in the hyperbaric chamber (Bethlehem model 1836-HP) 10 min after the end of the operation and connected to the measuring system (located outside the chamber) at the connection points (C in Fig. 5.27). Figure 5.27 shows the interconnection between the rat and the monitoring system (Mayevsky, 1983b). The various cables and the light guide were fixed in the chamber wall with screw(s) by epoxy glue. The following parameters were monitored and recorded on a multichannel recorder (Beckman Dynograph): (1) ECoG; (2) tissue pO_2; (3) corrected K^+ (the uncorrected signal was not recorded in our experiments due to a limited number of recording channels; the actual measurements of K^+ electrode are presented in Figs. 5.28–5.32 as EK^+ representing a potential reading of the electrode; the meaning of the potential changes are expressed in terms of K^+_e; (4) DC potential from a narrow ring around the K^+ electrode ($DC_k{}^+$); (5) reflectance (R) at the excitation wavelength of the NADH (366 nm); (6) uncorrected 450 nm fluorescence (F); (7) corrected fluorescence obtained by 1:1 subtraction of the reflectance signal from the fluorescence (F-R); (8) DC potential from a narrow ring around the light guide (DC_L). Temperature measured from the brain, core, and chamber was recorded on a second Beckman Dynograph.

5.3.5.3 Results

The results presented here are preliminary, and only the qualitative aspects are discussed now. Each of the rats used in our study was exposed to a 1-min anoxic cycle to test the state of the brain monitored. Figure 5.28 shows a typical response to complete anoxia and recovery to the preanoxic state (Mayevsky, 1983b). The first response to the N_2 breathing is the drop in pO_2 to its minimal level (0), which recovered as soon as air was introduced into the breathing mixture. A large overshoot in the pO_2 was recorded followed by a slow recovery to the normoxic level. In parallel to the pO_2 changes, the intramitochondrial NADH (CF signal) also responded very quickly to anoxia.

The high steady level of and quantitative analysis will be described in the near NADH during the anoxia is in parallel to the very low level of the pO_2. During the recovery phase, the intramitochondrial NADH (CF) reached the normoxic state quickly, while the reflectance and the uncorrected fluorescence recovered more slowly in parallel to the slow recovery in the pO_2 trace. During anoxia, a small increase in K^+_e was recorded. As soon as energy was available again, the K^+ returned to the resting level after a small undershoot. The ECoG reached the isoelectric state about 30 s after breathing N_2 and was followed by a fast recovery after rebreathing air.

To demonstrate the effects of hyperbaric oxygenation, five sections were selected from a continuous recording of a rat exposed to 5 ATA 100% O_2 (60 psi). To avoid

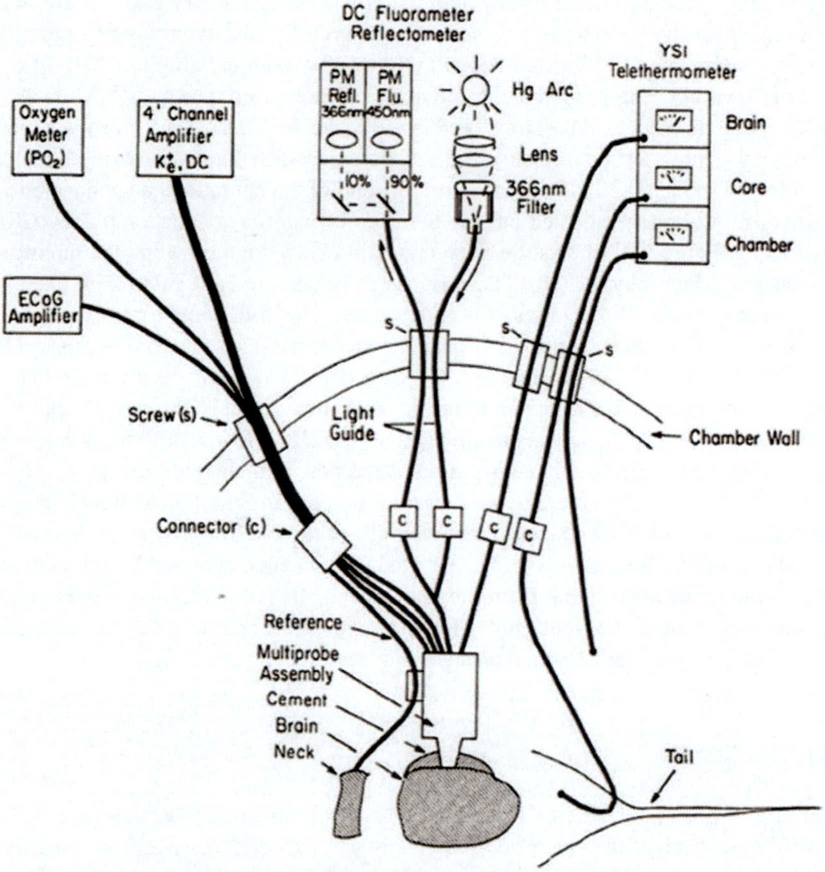

Fig. 5.27 Interconnection between multiprobe assembly connected to a rat inside chamber and monitoring system located outside hyperbaric chamber. (Mayevsky, 1983b)

artifacts in the K^+ measurement, a slow compression rate was found to be necessary. In Fig. 5.29 and 5.30, the compression phase and the first tonic–clonic convulsion are shown (Mayevsky, 1983b). After flushing the chamber with 100% O_2, the pressure was increased gradually and 5 ATA was reached after 28 min (Fig. 5.30). In response to the compression, tissue pO_2 was elevated very slowly due to the vasoconstrictive response of the blood vessels in the brain. A small jump in K^+ was recorded during the first minutes after starting the compression, and a small gradual increase was recorded until the first convulsion episode appeared. A few minutes before the seizure the ECoG trace showed a few spikes, which later developed into the seizure. The highly synchronized epileptic activity appeared 35 min after compression and was terminated by a silent period (postictal depression) followed by a "recovery" of the "regular" ECoG trace. The changes in K_e^+ during the epileptic activity can be explained as follows. During the high electrical activity, K^+_e starts to

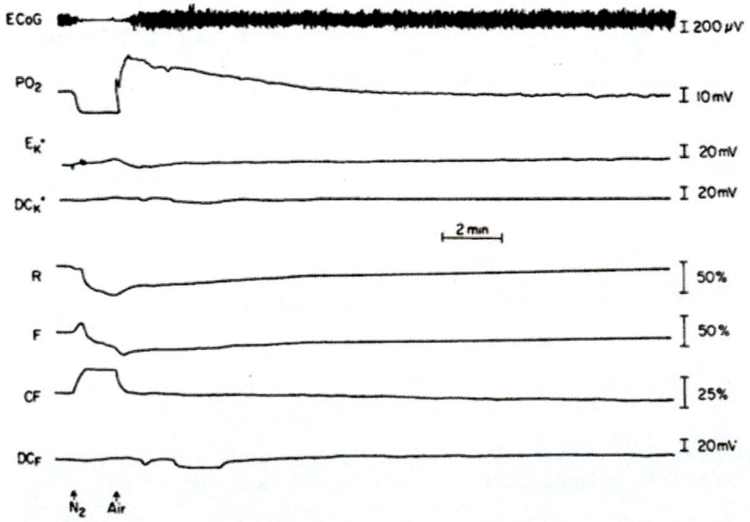

Fig. 5.28 Effects of anoxia (100% N_2) on metabolic ionic and electrical, activities in the rat brain. *ECoG* electrocorticogram, pO_2 partial pressure of O_2, *Ek* potential measured due to net changes in extracellular K^+, *DCK*$^+$ DC potential measured near K^+ electrode or light guide, *R* 366-nm reflectance, *F* 450-nm fluorescence, *CF* 450-nm corrected fluorescence, N_2 nitrogen breathing. (Mayevsky, 1983b)

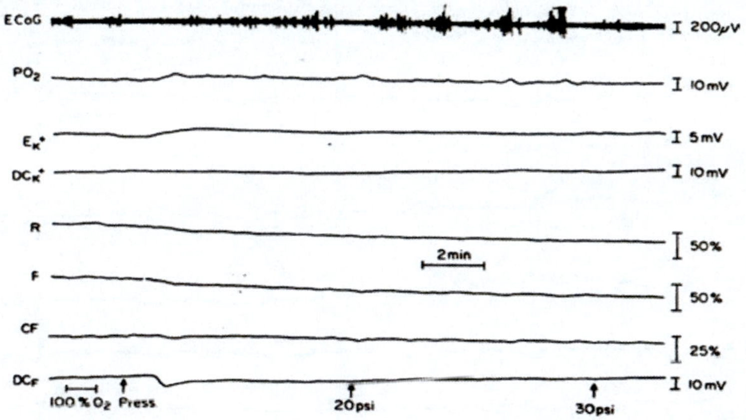

Fig. 5.29 Responses of various parameters measured from awake brain to slow compression with 100% O_2. Abbreviations as in Fig. 5.28. (Mayevsky, 1983b)

accumulate in the extracellular space followed by a short recovery to the resting level, but the high potassium level initiates a wave of CSD as can be identified by the large K^+ efflux and the negative shift in the DC potential. The recovery phase from the CSD is characterized by the large undershoot in the extracellular potassium level, which later recovers to the preconvulsive level. The metabolic responses

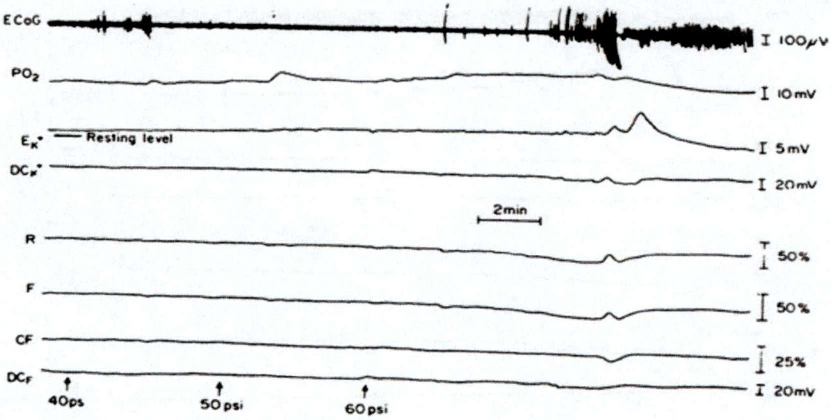

Fig. 5.30 A continuation in real time of Fig. 5.29. Pressure reached 60 psi (5 A T A) and first tonic–clonic episode is shown. Abbreviations as in Fig. 5.28. (Mayevsky, 1983b)

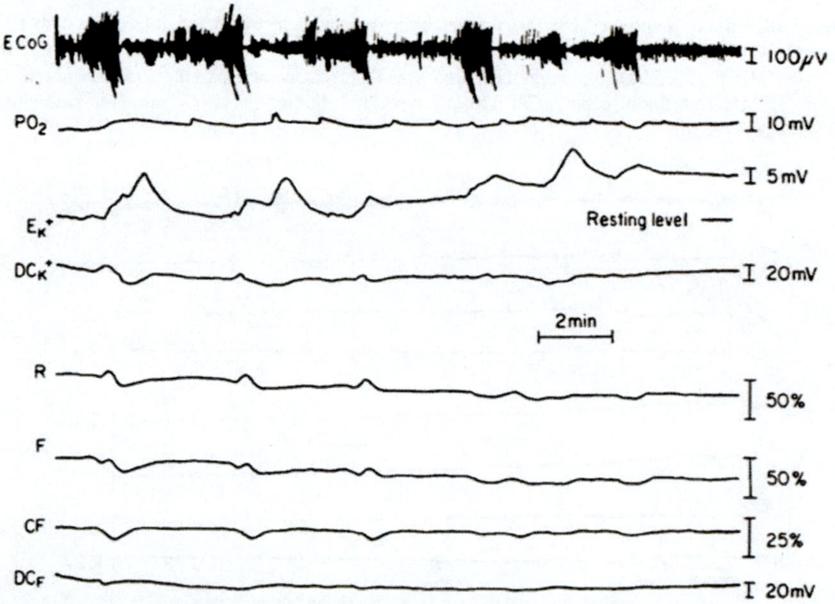

Fig. 5.31 Metabolic ionic and electrical responses to repetitive tonic-clonic episodes appeared under 5 ATA of 100% O_2. Abbreviations as in Fig. 5.28. (Mayevsky, 1983b)

to the first tonic–clonic event can be explained as follows. During the spiking activity, blood flow to the cortex was increased as indicated by the increase in the pO_2 level and the decrease in the reflectance signal. During the CSD episode, the NADH showed a clear oxidation cycle (CF trace). Figure 5.31, which is a continuation of Fig. 5.30, represents stage C in the same animal showing a few repetitive epileptic

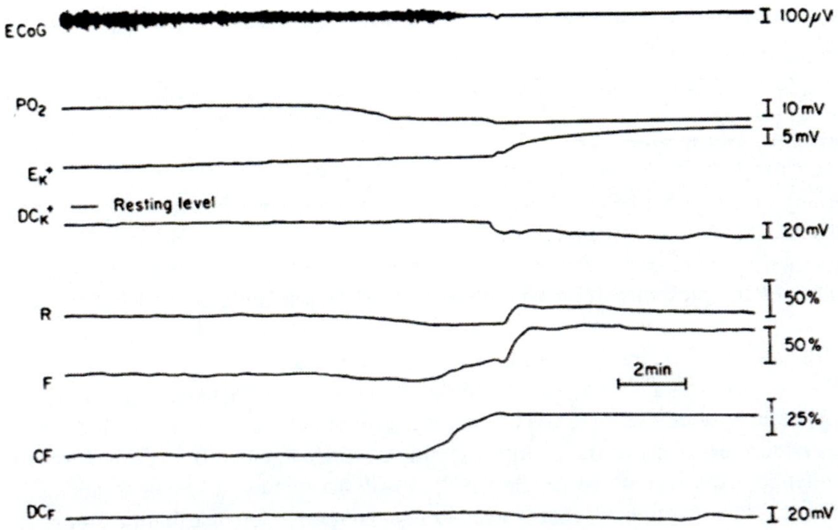

Fig. 5.32 Final period in response to 5 ATA of awake rat that represents cessation of respiration and the death of rat. Abbreviations as in Fig. 5.28. (Mayevsky, 1983b)

episodes (measured between min 40 and 58) (Mayevsky, 1983b). One can see a very clear correlation among the ionic, electrical, and metabolic activities during the epileptic behavior. In the first two episodes (at least), the termination of the seizure by a CSD wave was very clear. After the third episode, the K^+ remained high and accumulated above the resting level for the rest of the experiment. Figure 5.32 shows the terminal step in this experiment (stage E), starting with a decrease in pO_2 due to cessation of respiration (this record was taken 160 min after compression) (Mayevsky, 1983b). As a result, NADH showed an increase due to the lack of O_2 in the mitochondria. K^+ began accumulating simultaneously with the disappearance of the ECoG activity. Complete depolarization was recorded in all areas of the cortex monitored, as can be seen in the DC-negative shift and the large increase in K^+. The reflectance trace shows the large increase described previously by us as secondary reflectance increase (SRI) under the complete depolarization (Friedli et al., 1982; Mayevsky, 1978a; Mayevsky & Sclarsky, 1983). Preliminary results can be summarized as follows. Under 5 ATA (100% O_2), the temperatures were stable during the compression period (ca. 8–10 min). Several minutes later temperatures dropped gradually with different slopes in various animals. Toward the end, the decrease in temperatures was faster. In a typical rat, we obtained the following results. Before compression, core and brain temperatures were 38.2 and 36 °C, respectively; at 5 ATA, 38.4, and 36.1 °C, respectively; and during the epileptic activity (30 min), 36.5 and 34.4 °C, respectively. At the end of the seizure activity (70 min under 5 ATA), the temperatures were 36.3 and 34.0 °C, respectively, and before death (120 min), 33.6 and 31.0 °C, respectively. The results obtained from our experiments are similar to those obtained by Puglia et al. (Puglia et al., 1974). The temperature of the chamber was 22–23 °C with a very small increase (0.3–05 °C) due to compression.

5.3.5.4 Discussion

Many investigators believed that the key cause of the toxicity development at the cellular level is the intactness of the membranes. However, it is very difficult to measure it in vivo while the animal is exposed to the HPO condition. The development of the presently described system aimed to provide a tool for testing very indirectly an outcome of the status of membrane function in the awake brain under HBO toxicity. The level of potassium in the brain is affected by membrane permeability, energy availability, and the normal function of the adenosine triphosphatase (ATPase) system leading to the low resting level of potassium. If extracellular K^+ is elevated (due to any of the factors affecting it), brain excitability will increase and epileptic activity could be developed. It is well accepted that the prerequisites for normal brain function are energy availability and normal ion distribution around the cell membrane. Extracellular potassium is a good indicator for the functional state of the brain, as was shown by several groups (Fertziger & Ranck Jr., 1970; Heinemann & Lux, 1975; Somjen et al., 1976). The measurement of the DC steady potential, together with the ECoG, is the best way to evaluate the polarized state of the neuronal elements in the brain (Friedli et al., 1982; Somjen et al., 1976). The main breakthrough in the present study is that K^+_e was monitored continuously and simultaneously with the surface DC steady potential under hyperbaric conditions. In addition, for the first time so many parameters representing the metabolic, ionic, and electrical activities were correlated in the same site of the awake brain exposed to HB0 toxicity. For the convenience of the discussion, the exposure time to HB0 is divided into three periods. (1) Preconvulsive period. During this period, a number of events may occur and lead to the development of seizure activity. As we showed in the past (Mayevsky, 1978c; Mayevsky et al., 1974a; Mayevsky & Shaya, 1980), it is unlikely that the oxidation of NADH is the key biochemical event leading to the development of the convulsions. Also, the level of pO_2 during this period is one to three times greater than that of the air-breathing animal, as was found earlier (Torbati et al., 1977). The reason for the relatively low pO_2 is probably the vasoconstrictive response due to high O_2 pressure. Our hypothesis is that if at this stage of the exposure to HBO cellular membranes are affected by any of the toxicity processes, then the level of K^+_e will show some changes, as was shown in brain slices exposed to HPO (Kaplan & Stein, 1957). The results presented in Figs. 5.29 and 5.30 show that K^+_e did increase slightly above the resting level, but such a change must be proven to be statistically significant after also having a few controls such as pressure effects. In preliminary experiments, we compressed the two rats with air to 5 ATA and the K^+_e changes were undetectable, eliminating the effects of the pressure itself. (2) Convulsive period. The events occurring during this stage of exposure to HBO contribute to the better understanding of the toxicity process, as well as provide us with basic information regarding epileptic activity in general. The results shown in

Fig. 5.31, which represent the convulsive period, show a very clear correlation between ionic, metabolic, and electrical activities in the awake epileptic brain. We also have similar preliminary results obtained when epilepsy was induced by injection of pentylenetetrazol. The results shown here confirm our results published previously (Mayevsky, 1975; Mayevsky & Chance, 1975; Mayevsky & Shaya, 1980). The higher electrical activity, as detected by the ECoG, causes an increase in K^+_e level during the seizure episode. The high extracellular potassium level induced a CSD wave starting at the most sensitive area in the cortex and propagating to the measuring site. The evidence that we have a CSD wave comes from the large increase in K^+_e, as well as the negative shift in the DC potential, as described previously (Mayevsky & Chance, 1975; Mayevsky & Sclarsky, 1983). The same connection between epileptic activity and CSD was found in chicken brain exposed to pentylenetetrazole injection (Ookawa & Bures, 1975). It seems that the development of CSD may serve as a mechanism to terminate the epileptic seizure. According to Chance and Williams (1955), the oxidation of NADH during the higher ionic pumping activity is due to state 4-to-3 transition in mitochondrial function. Such results were found by other investigators (Jobsis et al., 1971; Lewis & Schuette, 1975; Somjen et al., 1976) as well. According to our results, we can conclude that during the convulsive period, energy is available for pumping ions after each of the depolarizing episodes.

5.4 During Drug-Induced Seizure Activity

In addition to the development of CSD after anoxia, hypoxia, and ischemia (described in Sect. 5.2) as well as after hyperbaric oxygenation (shown in Sect. 5.3), the effects of various epileptic and sensory stimulation are presented in this section.

The effects of locally applied Metrazol were tested in awake animals. The Metrazol (100 mg/ml) was applied epidurally using the cannula shown in Chap. 2, Fig. 2.2A_1. The typical response to Metrazol occurs 3–5 min after the administration of the drug, and the results are shown in Fig. 5.33A (Mayevsky & Chance, 1975). The Metrazol was applied to the right hemisphere and the left one served as a control. After application of the Metrazol, an increase in electrical activity was found and as a result an oxidation of NADH which was in the range of 5–10% of the normoxic level. After a period of a very high activity, the EEG became isoelectric while the NADH showed a very large oxidation cycle which then recovered to the normoxic level. In some animals, a different response was recorded and it is shown in Fig. 5.33B. In these cases, the small oxidation under Metrazol effect was recorded but the recovery to the normoxic level showed no large oxidation cycle. The reflectance changes were very small during the effect of Metrazol (before the development of the CSD).

Also, Metrazol was applied as above but under hypoxic conditions. Figure 5.34 shows the typical results (Mayevsky & Chance, 1975). The animal was exposed to air (A), 10% 0_2 (B), 7.5% O_2 (C), and 5% O_2 (D). The line N.L. represents the normoxic level of NADH. In all oxygen concentrations, an increase in activity followed by a flat EEG was recorded. The changes in NADH are the same as in Fig. 5.33A, namely, that two phases in the oxidation were found. By giving the animal lower concentrations of O_2, the baseline shifted to a more reduced level depending on the O_2 level.

The aim of this study was to use an awake rat brain model to investigate the effect of hypoxia on the response of the brain to convulsive activity induced by Metrazol. The question was whether or not brain oxidative metabolism will respond in the same way under normoxic or hypoxic conditions in the awake rat. In some animals, the hypoxia caused a failure of the brain to recover from a CSD cycle so that the recovery was very slow or even uncompleted during longer periods of time (Mayevsky & Chance, 1975). The main advantage of using the awake model is the elimination of the decrease in blood pressure due to the effect of anesthetics. By using the awake rat, the blood pressure was stable for a few hours and the deterioration of the animal model is negligible.

The effects of Metrazol on brain metabolism and its impact upon the oxidation–reduction state of NADH are of great interest. The first evidence that epileptiform activity of brain is accompanied by an oxidation of NADH described by Jobsis and

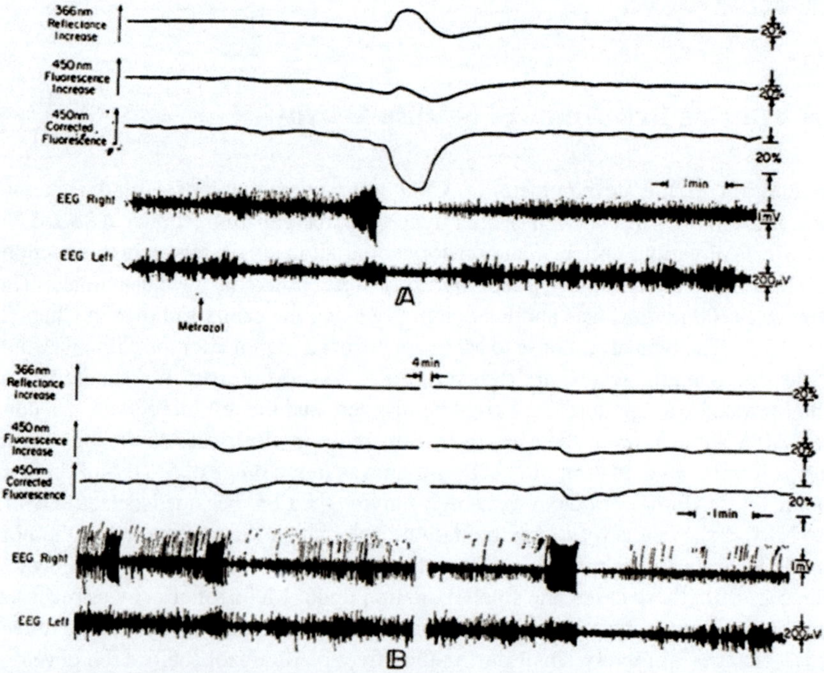

Fig. 5.33 The effects of Metrazol on brain NADH fluorescence, 366 nm reflectance, and the EEG of the two hemispheres. Note that the amplitude of the EEG calibration in the right hemisphere is 1 mV. The typical response is shown in part **A**. Part **B** shows another type of response to Metrazol applied 10 min before the record was taken. (Mayevsky & Chance, 1975)

his group (Jobsis et al., 1971). The same effects were described later by O'Connor et al. (1972) and Lewis et al. (1974). The results obtained by us confirm and extend these results to our awake rat brain model with slightly different results. In all our animals, Metrazol convulsions caused an oxidation of NADH (Fig. 5.33B), and in most animals, following this oxidation, another larger cycle of NADH oxidation was recorded (Figs. 5.33A and 5.34). The changes in reflectance, fluorescence, and corrected fluorescence have the same pattern as was found under CSD conditions, so that the oxidation cycle recorded after the oxidation of NADH induced by Metrazol is closely related to an CSD cycle. Various investigators showed that during epileptic activity, extracellular potassium was accumulated (Fertziger & Ranck Jr., 1970; Izquierdo et al., 1970; Moody et al., 1974). This increase in extracellular potassium is a potential source of local high potassium and may induce CSD to start in a low threshold area in the cortex. It was shown by Janebova (1971) that using threshold amounts of K^+ applied electrophoretically, one can initiate CSD. It seems that in our model in most animals, epileptiform activity was followed by an CSD cycle. In a few animals where CSD was not observed, it is possible that the threshold of extracellular K^+ was above the level of K^+ accumulated during the epileptic activity. It seems that hypoxic conditions have only a small effect on the epileptic activity and the following CSD cycles. The main changes are longer duration of the

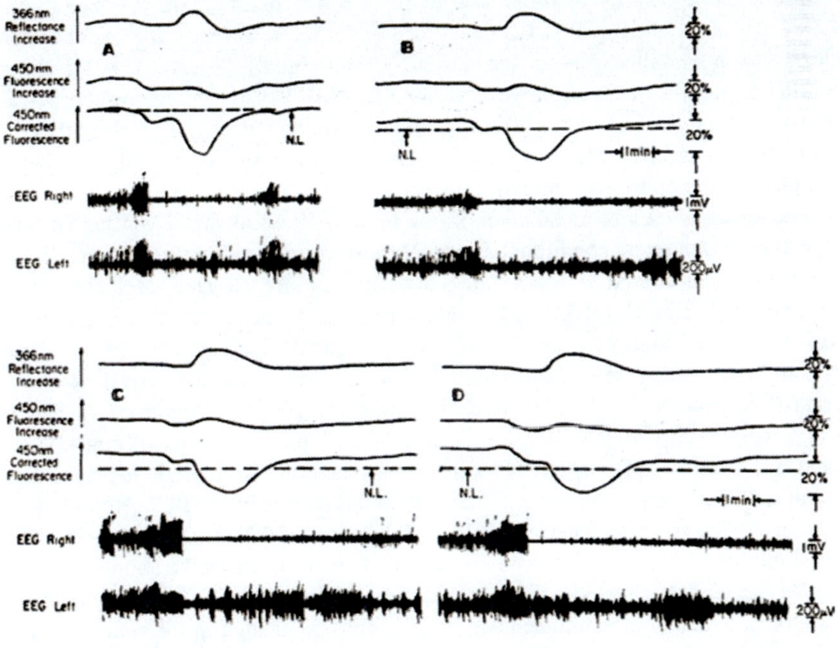

Fig. 5.34 The effects of graded hypoxia on the response of the awake brain to local application of Metrazol. The fluorescence responses occur 3–5 min after the application of Metrazol. The broken (N.L.) represents the normal oxidation–reduction state of the NADH. The levels of O_2 were in **A**, air; **B**, 10%; **C**, 7.5%; and **D**, 5%. (Mayevsky & Chance, 1975)

cycle and the decrease in amplitude of the maximum oxidation but no change in the rate of oxidation during the convulsive activity was obtained. Our conclusion from these experiments is that under our hands and preparation, no anoxic episode was followed after the epileptic activity. The same observation was made by Jobsis et al. (1971) in anesthetized paralyzed cats under artificial respiration. Furthermore, on the basis of our data it seems that the factor which may terminate the epileptiform burst is the CSD occurring after accumulation of K^+ in the extracellular space during the convulsive activity. This postulation is based also on results obtained when rats were exposed to hyperbaric oxygen toxicity (Mayevsky et al., 1974a) and convulsive activity was followed by oxidation cycles.

The contribution of the awake model to the understanding of brain energetics in pathological situations can be seen in Fig. 5.35 (Mayevsky, 1978d). We developed an epilepsy model by local application of Metrazol (Mayevsky & Chance, 1975) or penicillin in our present study. A solution of 500,000 units/ml was applied epidurally on the left hemisphere while NADH was measured from the right one. The ECoG of the left side shows the tonic-clonic pattern but a mirror focus was developed on the right side, as shown on the ECoG trace. The depression of the ECoG was followed by an oxidation cycle of the NADH, as was described previously in the Metrazol model (Mayevsky & Chance, 1975).

Figure 5.36 illustrates very clearly that CSD is developed as a secondary response to the epileptic activity recorded due to IP injection of Metrazol (Chance et al., 1978). At line A, the typical EEG developed under the seizure event was correlated clearly to the small oxidation (decreased NADH signal) of NADH. After about 1 min, at point B the very large decrease in NADH-corrected fluorescence was recorded at the same time that EEG was depressed dramatically. This second event is identifying as CSD. The two processes are more obvious in the third event demonstrated in the figure (A_1 and B_1).

Activation of energy metabolism in the brain may occur under various physiological or pathological conditions. If (under those conditions) oxygen is not a limiting factor, then the NADH redox state will shift to a more oxidized level, defined by Chance and Williams (1955) to be state 4 to state 3 transitions. The main factor which increases energy metabolism in vivo is the activation of the Na^+, K^+-ATPase system. This occurs when extracellular K^+ is elevated during the activation of the brain (Leniger-Follert et al., 1977; Lothman et al., 1975; O'Connor et al., 1972; 1973; Rosenthal & Jobsis, 1971; Somjen et al., 1976). The effects of convulsions and CSD (Leão, 1944) on NADH redox state in the brain were originally described in the artificially ventilated cat cerveau isole brain (Jobsis et al., 1971; Rosenthal & Somjen, 1973) and also later on by others (Tenny et al., 1980; Vern et al., 1976). The development of fiber-optic fluorometry enabled us to use an awake animal exposed to CSD or epileptic activity. In the awake rat exposed to CSD, an oxidation of NADH was recorded simultaneously with a biphasic change in the reflected light (Mayevsky & Chance, 1973). The repetitive nature of the CSD phenomenon and the metabolic responses to it were recorded when higher concentrations of KC1 (0.3–0.4 M) were used to stimulate the cortex (Mayevsky & Chance, 1974). The correlation between the behavior of the extracellular K^+ and the NADH redox state

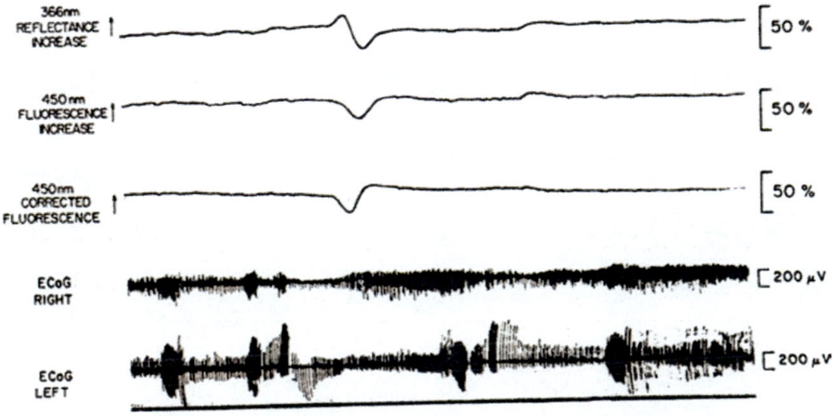

Fig. 5.35 The effects of mirror epilepsy focus (induced by penicillin) on the metabolic and electrical activity of the awake brain. (Mayevsky, 1978d)

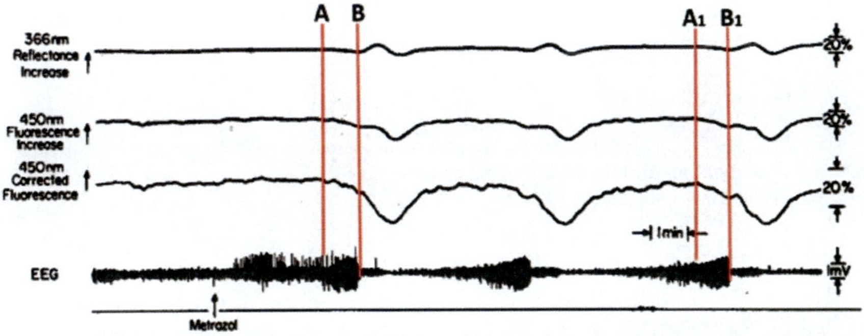

Fig. 5.36 NADH oscillations due to spatiotemporal metabolic heterogeneity represented by periodic electrical and metabolic activity introduced by an intraperitoneal metrazole injection in an anesthetized rat. (Chance et al., 1978)

was first using light-guide fluorometry (Mayevsky et al., 1974b) and later using the K^+ surface electrode (Mayevsky et al., 1980b). The effects of epileptic activity on brain NADH fluorescence were tested using a specially designed cannula by which Metrazol was applied directly to the cortex of the awake rat (Mayevsky & Chance, 1975). In most animals, the epileptic activity was followed by a CSD-like response so that the NADH showed a biphasic oxidation. Recently, by using the multiprobe assembly (Mayevsky, 1983b; Mayevsky & Sclarsky, 1983), we were able to induce convulsions by intraperitoneal injection of Metrazol. Figure 5.37 shows a typical metabolic, ionic, and electrical response to Metrazol injected IP in the awake rat (Mayevsky, 1984a). In the third episode obtained, the increase in extracellular K^+ was clearly in two steps. The first (A) was due to the epileptic activity and the second (B) due to a CSD wave. The same pattern of changes was recorded in the pO_2, NADH, and reflectance traces.

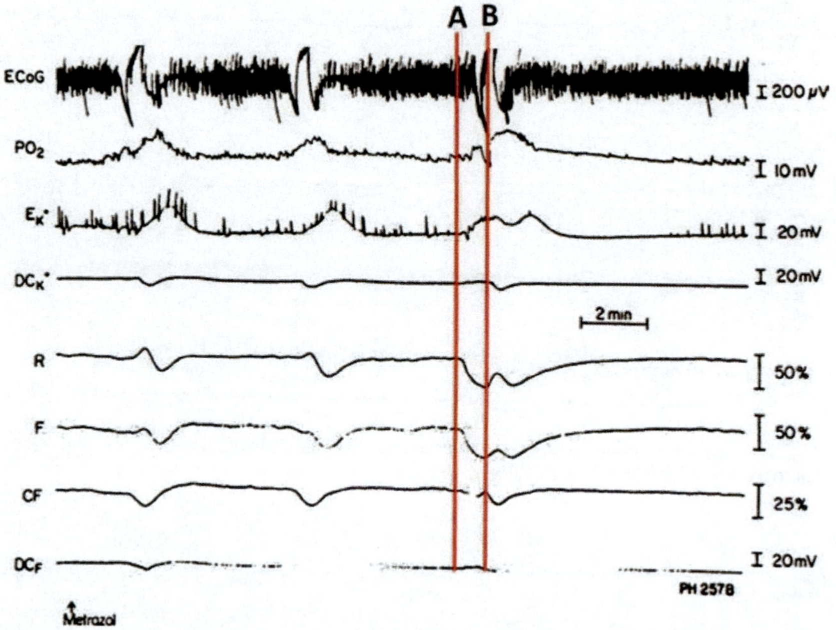

Fig. 5.37 Repetitive responses of the awake brain (rat) to epileptic activity initiated by intraperitoneal injection of Metrazol. In these specific measurements, movement noise was recorded in the pO_2 and EK^+ signals. (Mayevsky, 1984a)

The effects of partial brain ischemia on the responses to epileptic activity are shown in Fig. 5.38 (Mayevsky, 1983a). In this rat, the two carotid arteries were ligated 20 h prior to the experiment. During the discharge activity, K^+ increased in the extracellular space and then CSD developed. The main effects of the ischemia are on the recovery processes due to the slow oxygen delivery to the cells.

The PO_2 showed a decrease cycle, indicating that the compensatory mechanism of increased blood flow is impossible as measured also by the increased reflectance. The short oxidation cycle of NADH found in the normoxic brain (Fig. 5.37) is very unclear in the ischemic brain. Due to the slow supply K^+ into the cells was much slower as compared with normoxic brain. The electrical activity of the brain was also affected, as could be seen in the slow recovery of the ECoG and DC traces after the CSD episode.

Another response to epileptic activity was recorded (Fig. 5.39) at a secondary focus developed by topical application of penicillin on the contralateral hemisphere (Mayevsky, 1983a). The qualitative responses in this model were very similar to those obtained by systemic injection of Metrazol. The main difference is in the PO_2 response which was not clear in Fig. 5.39, probably due to instability of the

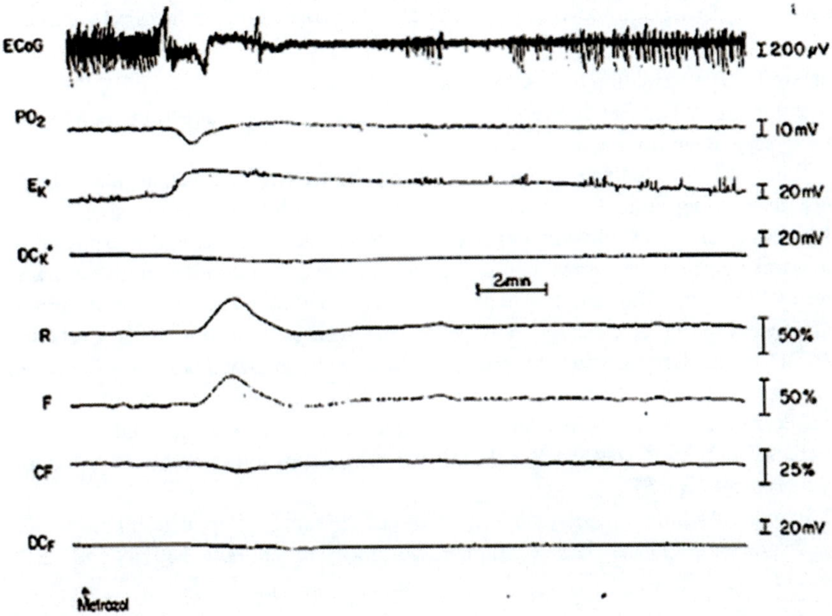

Fig. 5.38 Effects of partial ischemia (bilateral carotid artery occlusion) on the metabolic, ionic, and electrical responses to Metrazol injection in the awake rat. (John Libbey, reprinted with permission. (Mayevsky, 1983a))

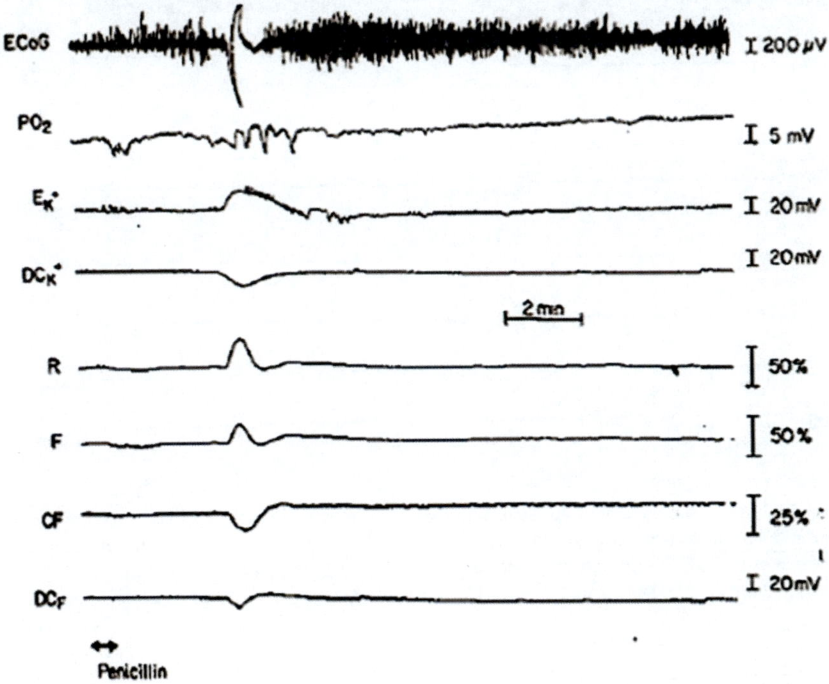

Fig. 5.39 Metabolic, ionic, and electrical activities measured in a secondary focus developed by topical application of penicillin in the contralateral hemisphere. (John Libbey, reprinted with permission. (Mayevsky, 1983a))

electrode. The oxidation of NADH was recorded simultaneously with the decrease in extracellular K^+. The third epileptic animal model was developed under a hyperbaric oxygen environment.

Figure 5.40 presents the progression of seizure activity in a special strain of gerbils, developing epileptic effects as a result of monotonic noise or other factors (Mayevsky et al., 1986b; Mayevsky & Rogatsky, 2007). As seen, the exposure of the awake gerbil to noise resulted in the development of epileptic activity followed by a CSD wave. The coupling between the two pathological events is clear and manifested by the electrical activity (ECoG) as well as extracellular K^+ levels and NADH redox state. During the epileptic stage, extracellular potassium increased and started to recover to the baseline, but then a larger elevation was recorded. There is a clear correlation between the various parameters during epilepsy and CSD. The decrease in NADH was smaller during the first stage, followed by an oxidation cycle typical for CSD.

Figure 5.41 shows the continuation record of the same epileptic gerbil developed the 2 interest responses. Here again, the two processes, epileptic activity and CSD, are very clear.

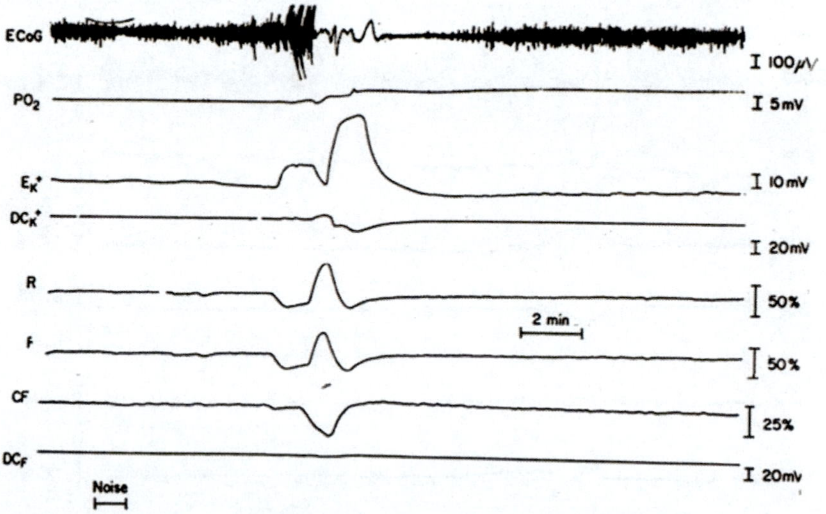

Fig. 5.40 The development of epileptic activity followed by CSD in the brain of a seizure-prone gerbil. *ECoG* electrocorticogram, *PO*$_2$ partial pressure of O_2 in the tissue, EK^+, DCK^+, extracellular potassium and DC steady potential measured around the K^+ electrode; R, F, and CF, reflectance, NADH fluorescence, and corrected NADH fluorescence (Mayevsky & Rogatsky, 2007)

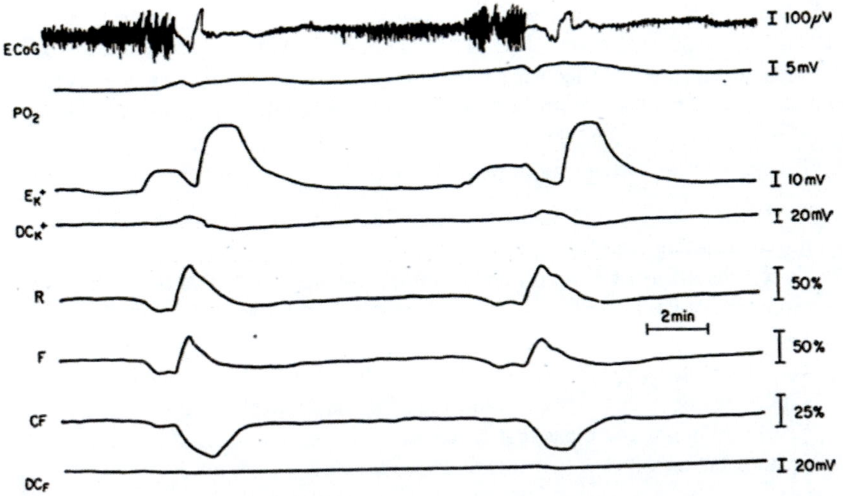

Fig. 5.41 The development of two episodes of epileptic activity followed by CSD in the brain of a seizure-prone gerbil. This figure is a continuation of Fig. 5.40 *ECoG* electrocorticogram, *PO_2* partial pressure of O_2 in the tissue, EK^+, DCK^+, extracellular potassium and DC steady potential measured around the K^+ electrode; R, F, and CF, reflectance, NADH fluorescence, and corrected NADH fluorescence

5.5 Development of CSD During and After Head Injury

Severe head injury can result in a high mortality rate or irreversible brain damage. One technique used to induce traumatic brain injury (TBI) is exposure of the brain to fluid-percussion pressure while monitoring the increase in intracranial pressure (ICP). Since brain injury is a multifactorial, pathological, time-dependent state, the multiparametric monitoring approach was adopted for studying fluid-percussion effects on the rat brain.

5.5.1 First Study—Development of the TBI Model

5.5.1.1 Introduction

We adopted the multiparametric monitoring approach (Friedli et al., 1982; Mayevsky, 1983a) to the fluid-percussion traumatic brain injury (TBI) model in the rat. Brain hemodynamics were measured by monitoring microcirculatory blood flow using laser Doppler flowmetry (Wadhwani et al., 1990). Metabolic activity was evaluated by the intramitochondrial redox state as measured by NADH fluorescence

(Mayevsky, 1984a). Ionic homeostasis was measured by ion-sensitive K^+, Ca^{2+}, and H^+ mini-electrodes (Crowe et al., 1977; Friedli et al., 1982; Mayevsky, 1983a; Mayevsky et al., 1991). Electrical activity was measured by the spontaneous ECoG and steady DC potential. Intracranial pressure was continuously measured using a fiber-optic probe (Crutchfield et al., 1990).

The aims of the present study were as follows:

1. To develop the rat TBI model for recording typical responses using the multi-probe assembly (MPA);
2. To test the effect of CSD (brain activation) induced after TBI on the metabolic responses of the brain;
3. To evaluate qualitatively the potential use of hyperbaric oxygen therapy in the TBI model using these monitoring approaches. It is assumed that our approach would only detect changes developed during the acute phase after the TBI (up to 12 h), although it is clear that secondary damage may also develop after this monitoring period.

5.5.1.2 Technological Aspects

The details regarding the multiparametric monitoring system used were presented in Chap. 3 and also in reference (Rogatsky et al., 1996). In addition, we monitored also the intracranial pressure (ICP) using the following technique.

In order to monitor parenchymal ICP, a fiber-optic ICP probe was used (Camino Lab., USA) as detailed in Fig. 5.42A (Rogatsky et al., 1996). This probe has recently been introduced into the MPA such that it has the same contact area with the brain surface as the other probes. The values measured by the optical probe were found to be well correlated with the standard ICP monitoring device (Crutchfield et al., 1990).

The second monitoring system used in the current study is presented in Fig. 5.44—left side (Rogatsky et al., 1996). In experiments in which the effects of CSD after TBI were tested or in which the animal was exposed to hyperbaric oxygenation (HBO), another monitoring setup was used. NADH was monitored either from one hemisphere (Fig. 5.44 left side,) or bilaterally (not shown). A push–pull cannula (for KCl application) was located anteriorly to the NADH cannula while the ICP probe was located posteriorly. TBI was induced in the contralateral hemisphere. One pair of ECoG electrodes was implanted in each hemisphere.

Fluid-Percussion Injury

Traumatic brain injury (TBI) was induced by the lateral fluid-percussion method (Dixon et al., 1987b; McIntosh et al., 1989; Sullivan et al., 1976). Hydraulic impact of moderate or severe magnitude (3.1–5.0 atm) was applied to the left parietal cortex (monitoring of all parameters was performed in the right hemisphere).

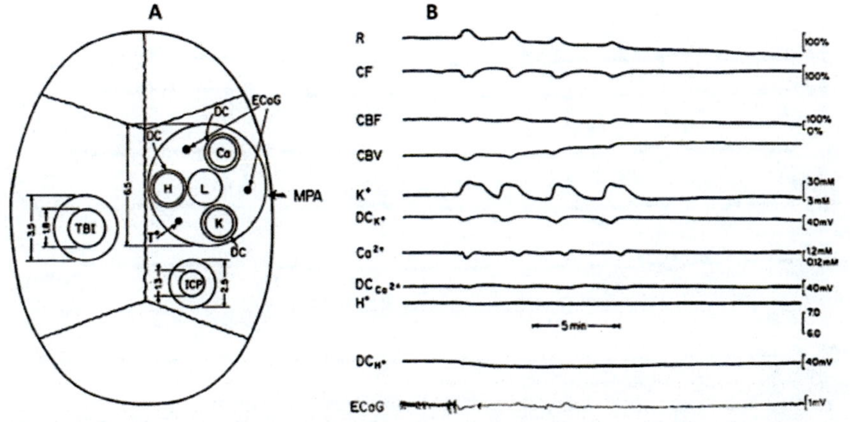

Fig. 5.42 The relative location of various sensing elements on the rat skull. (**A**) using the MPA. *L* light-guide monitoring NADH redox state and CBF, *ECoG* electrocorticography electrodes, *To* temperature thermistor probe, *K, Ca, H* mini-electrodes for monitoring extracellular potassium, calcium, and hydrogen ion concentrations, *DC* DC steady potential electrodes (dimensions, external, and internal diameters for cannula, in mm). (**B**) Effect of traumatic brain injury (3.0 atm) on the metabolic, hemodynamic, ionic, and electrical activities of the rat brain. *R* 366 nm reflectance, *CF* NADH-corrected fluorescence, *CBF* CBV cerebral blood flow and volume monitored by a laser Doppler flowmeter, K^+, Ca^{2+}, H^+ extracellular concentrations of potassium, calcium, and hydrogen ions, DC_{K+}, D_{Ca2+}, DC_{H+} DC steady potential measured concentric to the three electrodes, *ECoG* electrocorticogram, *ICP* intracranial pressure (Rogatsky et al., 1996)

5.5.1.3 Results

The results presented in this study are only preliminary (24 experiments), and their purpose is to illustrate the possibilities of modern monitoring in research on the pathophysiological post-concussion brain. CSD is one of the brain's reactions to mechanical trauma. A typical CSD response is presented in Fig. 5.42B (Rogatsky et al., 1996), which shows that directly after fluid-percussion impact, cyclic changes occur on a background of a severe decrease in brain electrical activity (ECoG), in parameters representing brain blood supply (R, CBF, and CBV), oxidative metabolism (CF), and ion transport (K^+, Ca^{2+}, H^+). After four waves of post-traumatic CSD, gradual normalization of the measured parameters was observed.

Spreading depression following brain injury was observed in 12 experiments. The duration of CSD periods in the various experiments varied from 3 to 42 min. In three rats, CSD episodes recurred for more than 10 h.

The development of post-concussion status epilepticus was recorded in five experiments. The dynamics of parameter changes in one of the experiments is illustrated in Fig. 5.43 (Rogatsky et al., 1996). Immediately after trauma, four typical CSD waves appeared on a background of decreased CBF (not shown in the figure). Two hours after the trauma, the first epileptiform episode appeared in the ECoG, followed by severe depression similar to CSD. This was accompanied by typical ion transport fluxes (increase in concentration of extracellular K^+ and H^+ and decrease

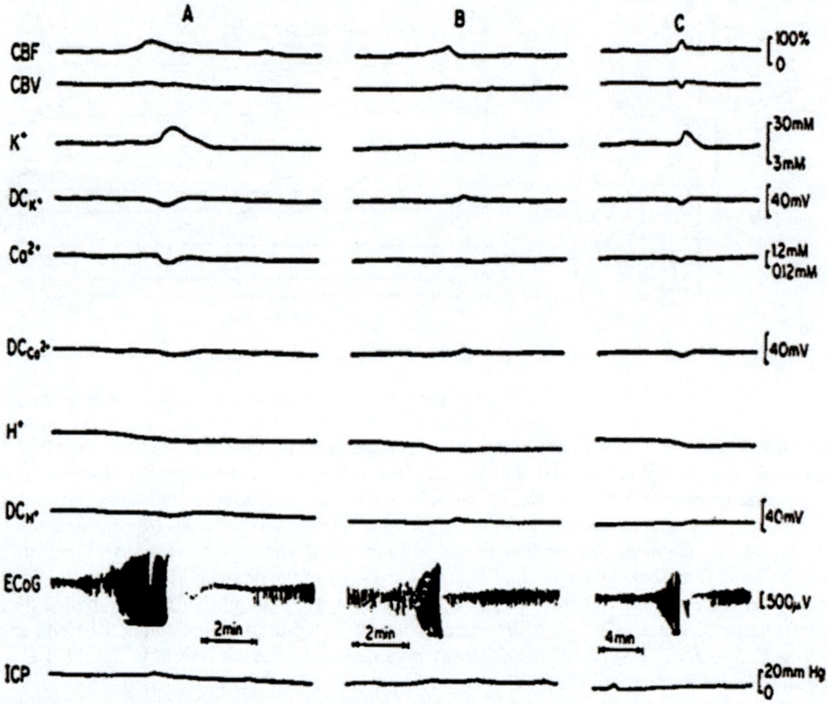

Fig. 5.43 Hemodynamic, ionic, and electrical responses to lateral fluid percussion (4.5–5 atm). The distinct epileptiform episodes as observed in the electrical activity (ECoG). A, B, and C were recorded 120, 160, and 260 min after induction of TBI, respectively. Abbreviations as in Fig. 5.42B (Rogatsky et al., 1996)

in extracellular Ca^{2+}) as well as a transient twofold increase in CBF (Fig. 5.43A). Another seizure event is shown in Fig. 5.43B. Termination of this seizure was not followed by a CSD wave. The increase in CBF was due to the seizure alone. In this case, the large increase in K^+ was not found and the ECoG depression was less pronounced and of shorter duration. The absence of a DC-negative shift was another indicator that CSD did not occur after this seizure. Figure 5.43C (recorded 260 min after TBI) shows events similar to those observed in A, i.e., a combination of seizure and CSD waves (note the change in time scale). Furthermore, the same episodes (with or without CSD) recurred many times either cyclically or at 8- to 12-min intervals. Fifteen episodes of this kind were recorded during a 1.5-h period. Most episodes were accompanied by changes in ion balance (spreading depression type) and a transient rise in the CBF level which did not, however, return to initial levels. The amplitude of the ICP changes was small during these cycles of seizure activity. Immediately after TBI, ICP reached the 20–21 mm Hg level and then decreased to 10 mm Hg or lower.

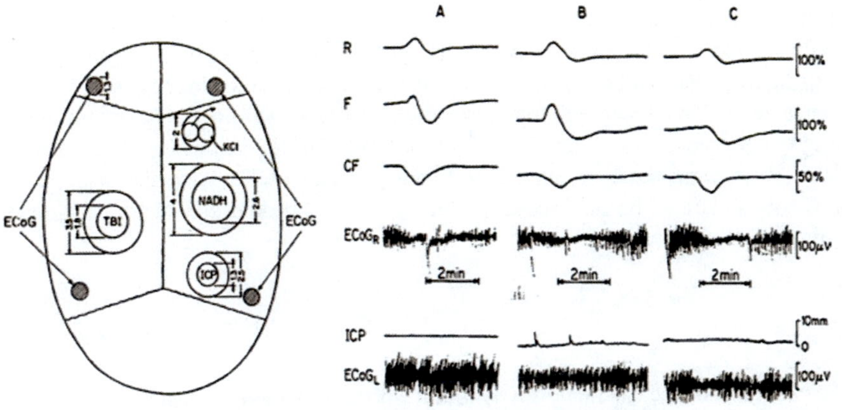

Fig. 5.44 Left side—The relative location of various sensing elements on the rat skull. *NADH* cannula for light-guide monitoring the NADH redox state, *KCl* push–pull cannula for inducing CSD, *ECoG* electrocorticography electrodes, *ICP* cannula for intracranial pressure sensor, *TBI* cannula for inducing brain injury (dimensions, external, and internal diameters for cannula, in mm). **Right side**—Metabolic, ICP and electrical responses to right-side cortical application of a 0.5 M solution of KCl. *R*- 366 nm reflectance, *F*- 450 nm fluorescence, *CF*- NADH-corrected fluorescence, *ECoG- R, L* electrocorticogram measured from the right and left hemispheres, *ICP*- intracranial pressure (in A, the ICP was not measured due to technical difficulties). A- was measured before the TBI, B and C- 30 and 60 min after TBI (4.5–5 atm) (Rogatsky et al., 1996)

The data presented above demonstrate that during the early post-traumatic period changes in electrical activity and ion balance correlated with changes in CBF and in oxidative metabolism adjustments.

A second experimental setup, shown in Fig. 5.44 (left side), was used for examining the functional state of the brain exposed to TBI (Rogatsky et al., 1996). NADH and ECoG were monitored from one or both hemispheres as well as from one ICP probe. The brain was exposed to CSD (epidural application of KCl) before and after the TBI event. Typical responses measured in 12 experiments are shown in Fig. 5.44A. The typical control response to CSD, which indicates an oxidation wave (decrease in CF) as well as ECoG depression in the ipsilateral hemisphere, is presented in Fig. 5.44A. Responses measured 30 min after TBI (Fig. 5.44B) show that the amplitude of the NADH wave decreased. 60 min after TBI the response to CSD was similar to the control response (Fig. 5.44C). The ICP level was very low after TBI and was not affected by the CSD waves.

Effects of HBO treatment after the exposure to severe TBI are presented in Fig. 5.45A (Rogatsky et al., 1996). In this animal, TBI led to cessation of spontaneous respiration. After 2 min of artificial respiration, the chamber was washed with 100% O_2 and compressed to 50 psi. NADH oxidation occurred during the compression period. NADH remained oxidized for the duration of the decompression period and recovery of ECoG was observed. When the rat was exposed to normal air, the NADH reduction occurred concomitant to an increase in blood volume as reflected by a decrease in the reflectance signal. This same rat was exposed to CSD (while breathing normal air) before TBI as well as after exposure to HBO. A typical normal response to CSD, namely, an NADH oxidation cycle and a biphasic response in reflectance, is presented in Fig. 5.45B. Thirty minutes after the induction of TBI and exposure to HBO, the amplitude of the oxidation cycle was only 50% of that occurring under control conditions (Fig. 5.45C). Furthermore, the changes in the reflectance trace were much smaller. Two hundred minutes after TBI, the response to CSD demonstrated further recovery, although the duration of the cycles was longer (Fig. 5.45D).

5.5.1.4 Discussion

The efficacy of the treatment given to a head injury patient is dependent upon the effective and early diagnosis of the pathological state of the brain. If treatment is immediate, irreversible damage, which is a critical factor in the prognosis, will be

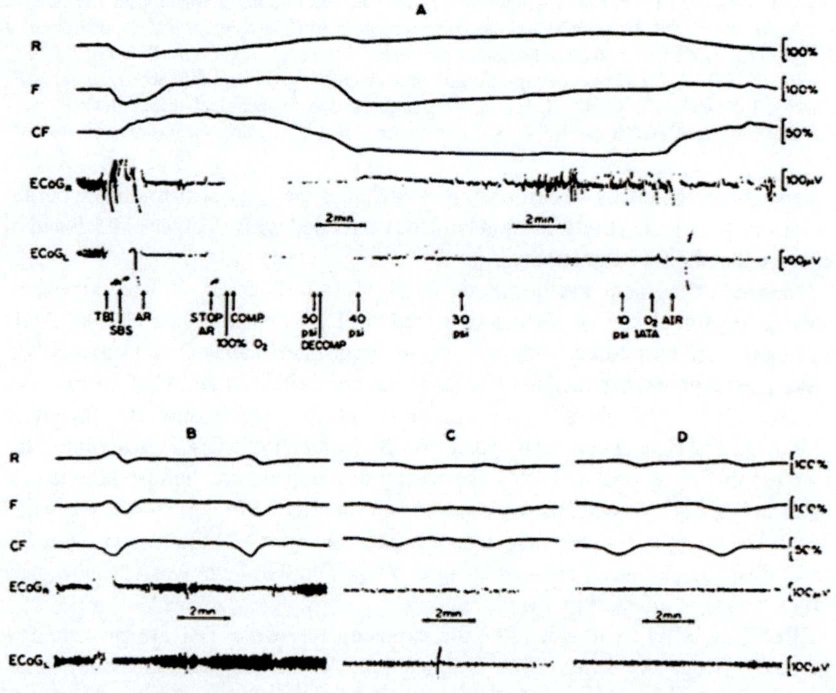

Fig. 5.45 Metabolic and electrical responses to CSD in the rat brain before and after severe lateral fluid percussion treated by hyperbaric oxygenation. **A.** shows the effects of TBI (4.5–5 atm) and recovery by hyperbaric oxygenation: *SBS* spontaneous breathing stopped, *AR* artificial ventilation, *stop AR* stop artificial ventilation, *Comp* compression with 100% O_2, *Decomp* initial decompression, *50, 40, 10 psi* the level of compression O_2 in chamber, *air* breathing of room air. **B, C,** and **D**: metabolic and electrical responses to CSD during control (**B**); **C** and **D**: 30 and 200 min after TBI. Other abbreviations as in Fig. 5.44 right side (Rogatsky et al., 1996)

minimal or prevented. In order to achieve significant progress in the treatment of TBI patients, it is necessary to establish a representative animal model in which as many physiological and biochemical activities as possible can be monitored simultaneously in real time. In order to achieve this goal, we adopted the multiparametric monitoring approach, developed by our group (Friedli et al., 1982; Mayevsky, 1983a), and applied it to the rat fluid-percussion model. The main advantage of the MPA approach is its easy application to patients after severe head injury (Mayevsky et al., 1996). Our development is a response to the challenge made by Wiebers et al. (1990) who declared that: "Ultimately, however, the answers to many of our questions regarding the underlying pathophysiology and treatment of stroke do not lie with continued attempts to model the human situation perfectly in animals, but rather with development of techniques to enable the study of more basic metabolism, pathophysiology and anatomical imaging detail in living humans." This statement was used by Wiebers for stroke and it can well be applied to TBI. The MPA sensors touch the surface of the brain after dura mater removal, without penetrating the brain tissue. Under such conditions, the brain itself remains very close to normal, as opposed to other methods in which penetration of the tissue by a microdialysis probe or K^+ electrode is necessary (Hubschmann & Kornhauser, 1983; Katayama et al., 1990; Nilsson et al., 1990; Persson & Hillered, 1992). Since most real-time probes sample a very small tissue volume, this brain region should be maintained in as normal a condition as possible. Indeed, the MPA has this advantage over other techniques. When TBI was induced in an animal model or developed in head-injured patients, the nature of the pathological state could be defined better using the multiparametric monitoring approach. We added two important parameters to the MPA, ICP and tissue temperature, which can identify other aspects of TBI. The extensive publications, as well as our preliminary experiments regarding TBI, led us to conclude that TBI induced by fluid percussion is a gradual dynamic process which can be divided into two intensity levels, severe and mild or moderate. The initial effect of TBI affects CBF autoregulation (severe) or induces disturbances in ionic and electrical homeostasis. These events may lead to the development of the following pathophysiological states: ischemia, CSD, or seizures. Each of these states is a very well-defined condition and can be described in terms of changes in various physiological and biochemical parameters. Ischemia, for example, affects brain activity primarily by changing the energy availability measured as a decrease in CBF (O_2 supply), an increase in the intramitochondrial redox state (NADH accumulation) and an inhibition of ion pumps. These changes in energy supply affect ionic homeostasis mechanisms resulting in an increase in extracellular K^+ levels while Ca^{2+} and Na^+ will enter the cellular compartments (Hansen, 1985). pH decreases due to the stimulation of glycolysis. The brain enters a phase of ischemic depolarization (ID) characterized by a negative shift in the DC potential as well as depression of spontaneous electrical activity (ECoG). CSD developing spontaneously by TBI (or induced experimentally by KCl application) is primarily a result of changes in the ion homeostasis of the brain monitored as changes in the extracellular levels of various ions. These changes stimulate the activity of various ion pumps (in order to restore ion homeostasis), and as a result, energy metabolism is enhanced (Mayevsky et al., 1982). This is characterized by the oxidation of mitochondrial NADH as well as by a large increase in CBF and O_2 consumption (Mayevsky & Weiss, 1991). Seizures, induced by epileptogenic drug injection (Mayevsky, 1983b) or by exposure to hyperbaric oxygenation (Mayevsky, 1984b), will usually affect electrical activity as well as ion homeostasis. Similar responses were measured when seizures developed after TBI. The metabolic responses were similar to those measured under CSD, although in some cases, ischemic effects overlap the seizure episode and the responses are different.

CSD developed after TBI without any significant change in ICP (Fig. 5.42B). Coupling between the ionic and metabolic events is similar to that observed in the response of the normal brain to CSD (Mayevsky, 1992; Mayevsky & Weiss, 1991). In some animals, CSD develops sometime after TBI and the nature of the response to CSD may serve as an indicator for the integrity of the microcirculation, mitochondrial activity, and ionic homeostasis. Therefore, we used the rat TBI model in which CSD was later induced (Fig. 5.42B), and indeed, in this specific animal, the response was very similar to that monitored before TBI. The same CSD test was applied to the rat shown in Fig. 5.45 which was exposed to severe TBI and treated by HBO. In some cases, the response to CSD is abnormal, as identified by an increase instead of a decrease in NADH (Mayevsky, 1978d; Mayevsky et al., 1982). Under these conditions, the CBF response will exhibit a large, typical, hyperemic response to CSD (Mayevsky, 1992; Mayevsky et al., 1990a). The seizure activity shown in Fig. 5.43, developed more than an hour after TBI, led to the typical response published by us and other groups (Jobsis et al., 1971; Kreisman et al., 1987; Mayevsky, 1983a, b). The combination of the MPA and the fiber-optic ICP sensor enables the study of the correlation between recorded values. This is important for understanding the mechanisms causing increased ICP after trauma as well as for elucidating the clinical relevance of ICP as a universal and integral parameter indicating the status of the injured brain (Bouma et al., 1991). The facts presented above together with information in the literature show that an immediate and severe energy deficiency resulting from CBF disorders may occur during the post-traumatic period. Therefore, attempts must be made to counteract the rapid reduction of intramitochondrial NADH by using the powerful modern means of improved tissue oxidation via hyperbaric oxygenation (HBO). Recently, clinical and experimental data have accumulated on the application of HBO in the treatment of acute functional disturbances caused by brain injury (Brown et al., 1988; Sukoff et al., 1968; Sukoff & Ragatz, 1982). Whether HBO treatment is beneficial for these conditions is still not known.

5.5.2 *Second Study—Level of ICP and CSD After TBI*

5.5.2.1 Introduction

The mechanism underlying the pathophysiology of cerebral contusion is complex and currently not well understood (Graham et al., 1978; Marmarou et al., 2000; McIntosh et al., 1998). However, in recent years (Strong et al., 2002), CSD cycles were found during clinical investigations following traumatic brain injury (TBI) (Mayevsky et al., 1996; Mayevsky et al., 1998; Strong et al., 2002) as well as in models of traumatic injury in the rat cortex (Koura et al., 1998; Nilsson et al., 1996; Nilsson et al., 1993; Ozawa et al., 1991; Rogatsky et al., 1996; Sunami et al., 1989).

CSD causes a disruption in ionic homeostasis (specifically, an increase in extracellular K^+ levels, intracellular Ca^{2+} and Na^+ levels, and acidosis), which augments ion pump activities, such as Na^+-K^+-ATPase, consequently increasing energy metabolism and oxygen consumption (Lauritzen, 1987; Mayevsky et al., 1993; Somjen, 2001; Sonn & Mayevsky, 2000). Thus, in the normal brain, where delivery of oxygen to the tissue is not disturbed, an increase in oxygen supply will compensate for the energetic demands, thus inducing ionic homeostasis. In contrast, under conditions where the tissue suffers from a lack of oxygen (such as ischemia and hypoxia), an increase in oxygen demand by CSD waves may cause decreased CBF and tissue pO_2, delaying the recovery phase of ionic homeostasis (Back et al., 1994; Mayevsky et al., 1999a; Sonn & Mayevsky, 2000). Further, the disruption in ionic homeostasis during ischemia, hypoxia, or head trauma, can stimulate the development of continuous CSD cycles, thereby inducing acidosis, and the release of noxious substances such as glutamate and other excitatory amino acids (EAAs), thus enhancing tissue damage (Benveniste et al., 1988; Bullock et al., 1998; Dietrich et al., 1994; Gido et al., 1994a, b; Koroleva & Bures, 1996; Koura et al., 1998; Mutch & Hansen, 1984; Otsuka et al., 2000; Somjen, 2001).

When occurring after traumatic brain injury, this cascade of events can adversely affect brain tissue, resulting in varied damage, linked to the degree of tissue oxygenation (Koroleva & Bures, 1996; Nedergaard & Hansen, 1993; Somjen, 2001; Sonn & Mayevsky, 2000). In a recent study, we found that the number of CSD cycles, together with the mean changes in ICP levels after severe FPBI, was the most significant parameters linked to morbidity when compared to the others measured by a multiparametric assembly (Manor et al., 2003). However, the relationship of CSD cycles to the severity of TBI was not clear. Therefore, the relationship between the level of post-traumatic intracranial hypertension and the number of CSD cycles generated, as well as the linkage between CSD and the severity of brain injury, is yet unknown.

To clarify these issues, the present work was carried out in rats in order to study the dynamic relationship between the level of intracranial pressure (ICP) and the number of CSD cycles during the early post-traumatic brain injury period under various levels of TBI severity.

5.5.2.2 Methods

Mayevsky et al. (1995) and in clinical research (Mayevsky et al., 1996, 1999a, b). We used the MPA for continuous, simultaneous online measurement of hemodynamic, metabolic, ionic, and electrical activity of the rat cerebral cortex. The measurements included NADH redox state (using surface fluorometry), CBF (using laser Doppler flowmetry), extracellular K^+, Ca^{2+}, and H^+ concentrations, DC steady

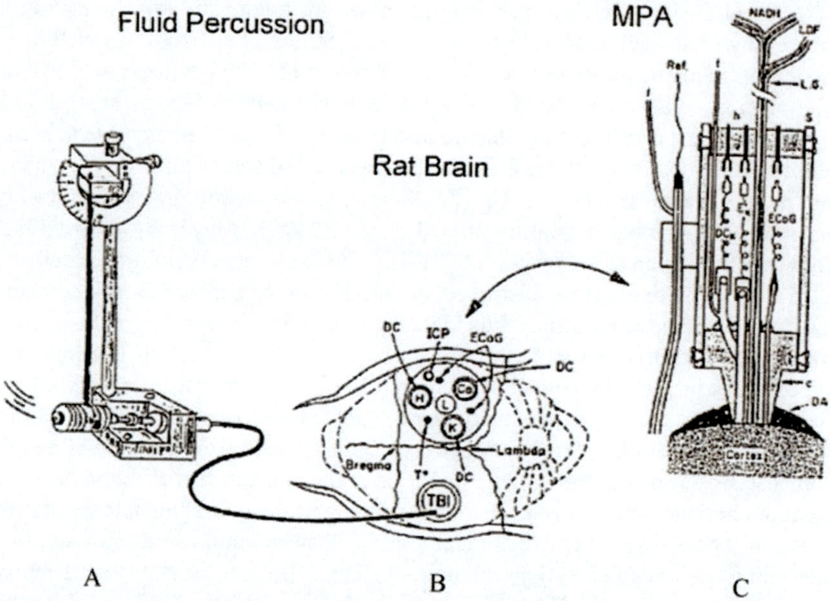

Fig. 5.46 Schematic presentation of the multiprobe assembly (MPA) used for monitoring the brain of head-injured rats. (**A**) Schematic presentation of the fluid-percussion impactor. (**B**) Location of the MPA in the skull. (**C**) Longitudinal section of the MPA. *DC* steady DC potential, *ECoG* electrocorticography, *ICP* intracranial pressure probe, *K, Ca, H* potassium, calcium, and hydrogen electrodes, *L* fiber-optic light-guide probe to measure local cerebral blood flow (CBF) and mitochondrial redox state (NADH), *TBI* cannula for inducing traumatic brain injury. (Mary Ann Liebert, reprinted with permission. (Rogatsky et al., 2003))

potential, as well as ECoG and ICP (Fig. 5.46B and C). The characteristic simultaneous increase in extracellular K^+ levels and the decrease in DC potential as well as depression in ECoG (if ECoG was not previously depressed by the FPBI) were used to define CSD initiation. Normalization of these parameters to the baseline levels showed that CSD phenomenon stopped.

From these data, the changes in ICP levels and the number of CSD cycles generated every 30 min under various severity levels of fluid-percussion brain injury (FBPI) were analyzed and calculated.

Animal Preparation
The experimental protocol was approved by the Institutional Animal Care and Use Committee of Bar-Ilan University. All experiments were performed with 27 adult male rats (180–220 g). Each rat was anesthetized by an intraperitoneal injection (IP) of Equithesin 0.3 mL/100 g (each ml contains 42.51 mg chloral hydrate, 9.72 mg pentobarbital, 21.25 mg magnesium sulfate, 44.43% w/v propylene glycol, 11.5% alcohol, and distilled water). First, the rat skull was exposed, and two holes were drilled: a 6-mm-diameter hole in the right parietal bone for the MPA, and a

3.5-mm-diameter hole in the left parietal bone for inducing TBI (Fig. 5.46B). Next the dura mater was removed and the MPA was located on the brain, avoiding any pressure on it. A special cannula was used for fixing the Leur–Lock syringe hub (1.8 mm inner diameter) and inserted into the second hole (3.5 mm diameter). Screws were inserted into the skull to secure it mechanically. Finally, the MPA, the cannula, and the screws were cemented to the skull with dental acrylic.

Fluid-Percussion Brain Injury

TBI was induced by the lateral fluid-percussion brain injury (FPBI) technique (Dixon et al., 1987a; McIntosh et al., 1998; Rogatsky et al., 1996; Sullivan et al., 1976). The magnitudes of the hydraulic impact with the Peak experimental groups of animals were 5 h. Each experimental group consisted of 4 rats in group A, 7 in B, 7 in C, and 9 in D. Observation of the highest trauma level group (D) continued for more than 5 h (up to 18 hours only for determining survival). Group D was divided into 2 subgroups: subgroup D_1—animals that died during monitoring (4 animals), and subgroup D_2—survivors (5 animals).

Statistical Analysis

A Pearson correlation analysis was performed initially to determine the most pronounced relationships between the investigated variables. Then, one-way analysis of variance (ANOVA) was used to test possible differences between the trauma level groups. Null hypotheses were rejected at the 5% level of significance. Linear regression analysis was applied to obtain quantitative evaluations of the most interesting trends. The computer program SPSS 10.0 for Windows (Norusis, 1999) was used for all statistical analyses. Data are presented in mean ± SE.

5.5.2.3 Results

The mean ICP levels of the four experimental groups, as a function of time, are presented in Fig. 5.47 (Rogatsky et al., 2003). FPBI caused a marked increase in ICP (240 ± 2.52, 250 ± 1.5, 250 ± 2.23, and 312 ± 0.95 mm Hg in groups A, B, C, and D, respectively). However, immediately afterward, a rapid decline was observed. Then, a slow secondary increase in ICP was seen (30–90 min after the traumatic impact). This was especially pronounced in groups C and D. Next, a trend of increased ICP was noted (the most striking trend was found in group D). At the end of the 5 h experiment, the mean ICP levels were 6.75 ± 2.87, 8.40 ± 2.70, 12.75 ± 4.03, 29.56 ± 9.25 for groups A, B, C, and D, respectively.

Using one-way ANOVA with Bonferroni correction for multiple comparisons, we found significant differences in the ICP levels ($p < 0.05$) affected by the PPP levels as a function of time, starting 30 min after trauma. These differences were conserved and tended to increase with time.

Significant differences in ICP were found between the experimental groups. ICP in group D was significantly higher than in all other groups during hours 1–5 post-injury and was higher than group A at 30 min post-injury. These differences were more significant at later periods. Thus, the statistical results support the graphical

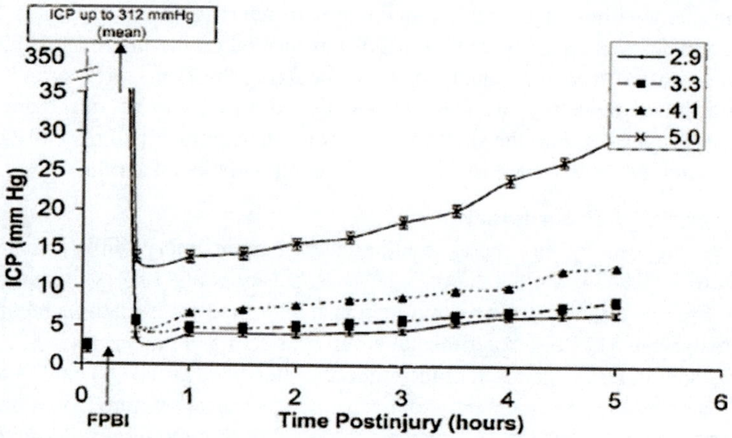

Fig. 5.47 The effect of FPBI levels (2.9, 3.3, 4.1 and 5.0 ATM, for groups A, B, C, D, respectively) on ICP levels (mean ± SE, mm Hg) in four experimental groups of animals [A (n = 4), B (n = 7), C (n = 7), and D (n = 9)]. The first mean ICP values in all experimental groups were under control conditions before FPBI. Lower arrow, FPBI induction; upper arrow, maximal ICP levels immediately after FPBI (Mary Ann Liebert, reprinted with permission. (Rogatsky et al., 2003))

presentation in Fig. 5.47 and show that group D differed from the other three groups during the monitored time after FPBI.

CSD cycles were observed in all experimental groups. Usually, one to four cycles were observed for each animal during the 10–15 min immediately after FPBI (Fig. 5.42B). Then, the total number of CSD cycles increased in all groups, except in group A. During the 300 min of monitoring, the mean total number of CSD cycles increased with the impact magnitude. The maximum number was found in group D and the minimum in group A. The mean cumulative number of CSD cycles found was 2.00 ± 1.41 in group A, 4.29 ± 4.23 in group B, 11.71 ± 13.29 in group C, and 20.11 ± 19.26 in group D. Regarding the CSD total count, the differences increased with time, but even for the latest moments they were not significant.

The highest number of CSD cycles was observed during a specific mode that we term "CSD status"—when CSD cycles repeated almost continuously, until the end of the monitoring period (Rogatsky et al., 2003). Here, the total number of CSD cycles increased to 53 during 300 min of observation. The "CSD status" was noted in 50% of the animals in group D and only in 14% of the animals in group C. In groups A and B, the "CSD status" was not found.

The intensity of CSD cycles generated was expressed by the number of CSD cycles observed every 30 min after brain trauma. Figure 5.48 shows the intensity of CSD cycles during the post-impact period (Rogatsky et al., 2003). After an initial high level (during the first 30 min following FPBI), the number of CSD cycles declined in all four groups for the next 60–90 min. Then, a new stage began, characterized by an increase in CSD cycle intensity (most pronounced in groups C and D). After this stage, the second CSD intensity decline took place in group C, with

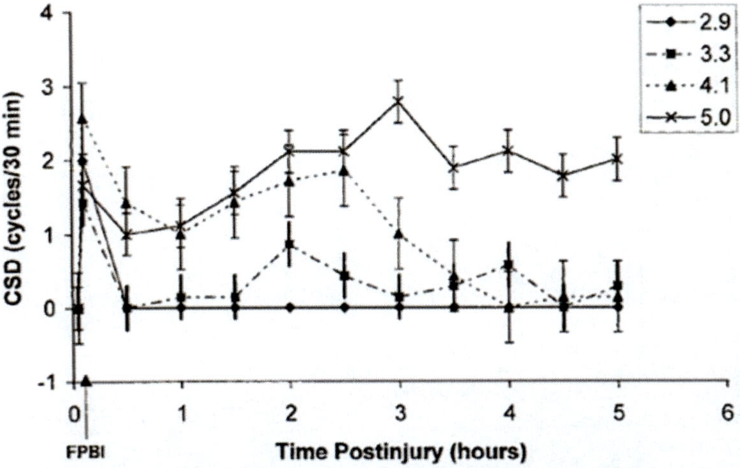

Fig. 5.48 Changes in the mean number of CSD cycles generated (per 30 min mean ± SE) during 5 h of monitoring after induction of different levels of FPBI: 2.9, 3.3, 4.1, and 5.0 ATM, respectively. CSD cycles started immediately after FPBI (lower arrow). Note that before FPBI, no CSD cycles were observed (0 cycles). (Mary Ann Liebert, reprinted with permission. (Rogatsky et al., 2003))

no clear tendency in group B. Only group D, with almost constant activity (up to 2 CSD cycles per 30 min), was maintained until the end of the experiment.

To estimate the ICP level (ICP_{xx}), the intensity of CSD cycle generation (CSD_{xx}), and the total number of CSD cycles (CSD_{tot}), approximation regression equations were obtained (Fig. 5.49, 5.50, and Table 5.8) (Rogatsky et al., 2003).

For all time points measured, the ICP regression equations were significant ($p < 0.001$; $n = 7$), while equations for CSD had p-level < 0.05 only beginning 180 min (t180) after traumatic impact. The coefficient of determination (r^2) and significance level for CSD equations were lower than the corresponding values for ICP equations. Using the mean levels for each PPP level ($n = 4$), a highly significant linear equation was obtained for the total number of CSD cycles: $CSD_{mean} = a + bX$, where X is the PPP level, $a = -23.35 \pm 2.05$, $b = 8.62 \pm 0.52$, $r^2 = 0.992$, $p = 0.004$.

These results indicate that ICP level is a much more reliable parameter for evaluating the trend of head trauma than the number of CSD cycles.

Correlation Between Parameters

The Pearson coefficient (r) of paired correlation between PPP (the level of traumatic impact) and ICP levels was calculated for all time points measured. A significant, strong correlation was found 30 min after trauma (t30) ($r = 0.644$, $p < 0.001$, $n = 27$), and then, the correlation coefficient increased to 0.706 (t60) and 0.777–0.841 (t90–t300).

The coefficient of correlation between PPP level and the intensity of CSD cycle generation were also calculated and became significant starting from t180. For time periods t180-t300, $r = 0.411–0.496$ ($p < 0.05$, $n = 27$).

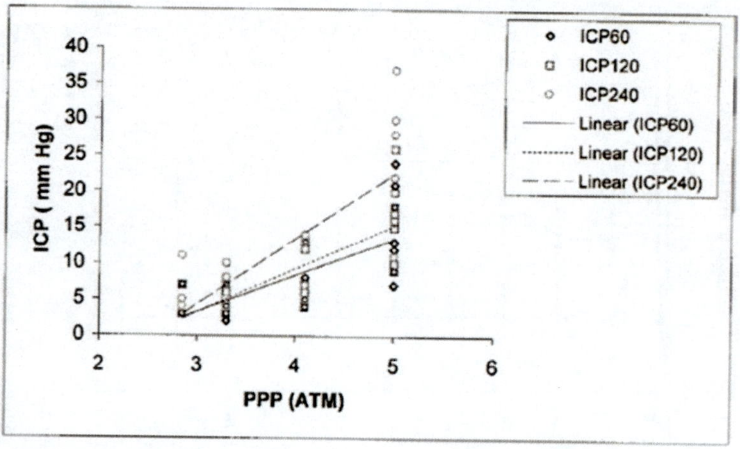

Fig. 5.49 Dependence of the ICP level on the peak pressure pulse (PPP). Parameters of the regression equations are given in Table 5.8. ICP_{xx}, CSD_{xx} mean estimates for time xx. Lines CS240 and CS300 overlap since they have very close parameters (Table 5.8). (Mary Ann Liebert, reprinted with permission. (Rogatsky et al., 2003))

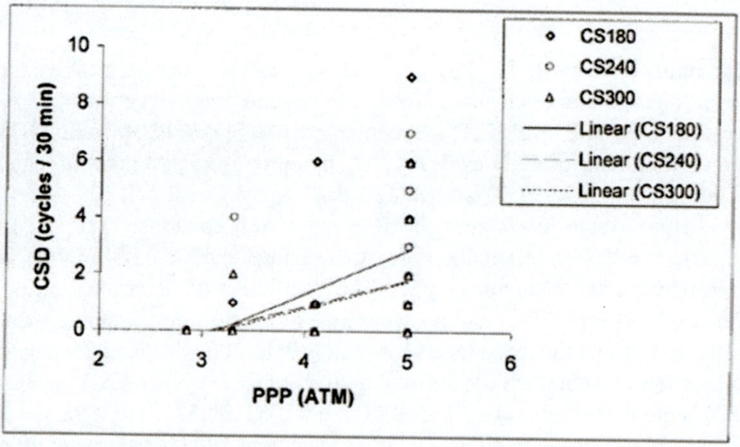

Fig. 5.50 Dependence of the number of CSD cycles, generated per 30 min, on the peak pressure pulse (PPP). Parameters of the regression equations are given in Table 5.8. ICP_{xx}, CSD_{xx} mean estimates for time xx. Lines CS240 and CS300 overlap since they have very close parameters (Table 5.8) (Mary Ann Liebert, reprinted with permission. (Rogatsky et al., 2003))

Thus, the growing effect of the impact factor on the level of ICP was observed during the whole experimental period, whereas the analogical effect on the intensity of CSD cycle generation was noted only during the second half of the experimental period (from 180 min after FPBI).

Table 5.8 Values of linear regression parameters (± SEM), the coefficient of determination (r^2), and significance levels (p) for the model. Y = a + bX

Dependent variable (Y)	Time (xx)	Intercept a (SEM)	Slope b (SEM)	r^2	p-level
ICP_{xx}	60	−11.12 (3.95)	4.82 (0.97)	0.499	<0.001
	120	−13.80 (3.79)	5.72 (0.92)	0.604	<0.001
	240	−22.28 (5.35)	8.89 (1.30)	0.649	<0.001
CSD_{xx}	180	−43.5 (2.09)	1.39 (0.51)	0.229	0.012
	240	−2.76 (1.63)	0.90 (0.040)	0.169	0.033
	300	−2.95 (1.33)	0.93 (0.32)	0.246	0.009
CSD_{tot}	300	−24.17 (12.49)	8.81 (3.05)	0.250	0.008

Mary Ann Liebert, reprinted with permission (Rogatsky et al., 2003)
Where x is the PPP level (ATM): ICP_{xx} is the intracranial pressure at time xx (min): CSD_{xx} is the intensity of CSD cycle at time xx (min); and CSD_{tot} is the number of CSD cycles generated during the experimental period. N = 27 for each regression

A significant coefficient of correlation between ICP level and the total number of CSD cycles was found for ICP measurements ($r = 0.47–0.63$, $p < 0.05$, $n = 27$) for each time point. For time periods t120–210, $p < 0.001$. Therefore, from the results of the correlation analysis, we can select t120–210 as the best period to analyze the relation between intracranial pressure and CSD.

5.5.2.4 Discussion

We found a high correlation between the magnitude of PPP and intracranial hypertension level in our experiments, showing that ICP levels can provide an estimate of the trauma severity. These results support those of Engelborghs et al. (1997), who concluded earlier that a significant correlation exists between the severity of the traumatic challenge and the resultant pressure rise. This dependence and the fact that the maximal correlation coefficients previously listed between PPP and ICP were noted in group D (i.e., at the maximal level of FPBI) demonstrate the diagnostic importance of ICP as a parameter used to estimate the risk level of the unfavorable development of head trauma (Bardt et al., 1998; Czosnyka et al., 1996; Lang & Chesnut, 1995; Stocchetti et al., 1991).

Recent clinical observations have shown that TBI outcome was often unfavorable when an ICP level > 20 mm Hg was reached (Miller et al., 1977; Pfenninger, 1984; Piper et al., 1990; Signorini et al., 1999). In our experiments, such a level of ICP was reached during 5 h of monitoring in seven of nine rats (almost 80%) in group D and only in this group. More than half (four of seven) of those animals died spontaneously with 2-3 h after the first 5 h of monitoring.

The ICP increase that persisted in group D may have been caused by the appearance of brain contusion in the contralateral hemisphere (Kochanek et al., 1995), with the development of significant brain edema (Kochanek et al., 1995; Marmarou et al., 2000) and traumatic subarachnoid hemorrhage. ICP (Thomas et al., 2000).

Results of linear regression analysis (Table 5.8) revealed that the slope coefficient (b) was positive and almost constant (0.9–1.39) for $t > 180$ for CSD; b was positive and grew with time (4.82–8.89) for ICP. Therefore, mean CSD generation intensity grew with the level of traumatic impact (PPP) and did not change with time. The ICP level also grew with PPP level, and these differences are more and more pronounced with time (Table 5.8 and Fig. 5.47).

A positive correlation between the total number of CSD cycles and the PPP levels ($r^2 = 0.250$, $n = 27$, $p < 0.01$) emphasizes the correlation between the intensity and duration of the CSD wave generation period and the severity level of TBI. It has been noted in clinical studies evaluating this problem that the CSD phenomenon was only noted in patients with well-developed forms of TBI (Mayevsky et al., 1996; Strong et al., 2002). As CSD data are heteroscedastic, separate t-tests were used to find significant differences between the groups, rather than the use of ANOVA for ICP differences. Significant differences ($p < 0.05$) were found between 5.0 and 3.3 ATM for t180, between 5.0 and 2.9 ATM for t120–t240, and t300.

Our results showed that in the groups with maximal injury, an increase in the mean number of CSD cycles was found together with the highest levels of ICP (subgroups D1 and D2). The growing number of CSD cycles found under these circumstances of steadily increasing ICP could be related to secondary hypoxic insult, which may result from pronounced metabolic disorders ongoing in the severely traumatized brain (Bergsneider et al., 2000; Ishige et al., 1987; Signoretti et al., 2001). In such cases, progressing post-traumatic intracranial hypertension is a factor of major importance in influencing ischemic brain damage (Graham et al., 1978; Graham et al., 1989; Miller et al., 1992; Obrist et al., 1984). Our results of high ICP levels accompanied by CSD cycles in the contralateral hemisphere may indicate the global and severe development of the secondary brain ischemic/hypoxic insult.

It is also known that tissue damage caused by brain trauma can induce marked disorders in ionic homeostasis characterized by increasing extracellular K^+ levels that can in turn initiate CSD cycles (Mayevsky et al., 1996; Somjen, 2001). The lack of an adequate oxygen supply under these circumstances can induce continuous CSD cycles without progressing to sustained depolarization. Further, the high ICP levels that develop after brain trauma may further disturb the availability of oxygen to the tissue, partially compromising ionic homeostasis, thus causing a continuation of the progression of CSD cycles and a further increase in tissue damage.

An indication of this hypoxic insult can be found in our results 60–90 min after FPBI in the groups of maximal impact level (5.0 ATM), where a tight coupling can be seen between the total number of CSD cycles, the increase in ICP levels, and the high mortality of rats in group D.

It can be presumed that in such cases, the cumulative damaging influences of many spontaneous and continuous CSD cycles can, at a specific stage of development of severe brain pathology, promote an increase in ICP levels. This raises the possibility that a vicious cycle can exist between increased ICP levels and the generation of numerous CSD cycles, especially during a period of CSD status. Such a relationship between these pathophysiological processes can take place under conditions induced by secondary hypoxic insult after severe TBI. Some additional support for this hypothesis can be found, as noted above, in the strengthening correlation between increased ICP levels and the generation of a vicious cycle of CSD intensity, beginning 2–3 h after the trauma and continuing until the end of the experiment. It, therefore, seems reasonable that during the secondary hypoxic insult, under conditions of increasing ICP, the increased number of CSD cycles becomes a more important factor on brain trauma pathogenesis.

As shown in Fig. 5.48, the diminishing intensity of CSD cycles after 2–3 h in groups C (especially in comparison with group D), paralleled favorable post-traumatic development. It can be assumed that the appearance and growth of the number of CSD cycles (even multiple cycles) in group C, non-confluent only with growing ICP, could be initiated by different mechanisms. For example, such phenomena, associated with only acute ischemia lesion, are described in studies dealing with CSD (Dietrich et al., 1994; Otsuka et al., 2000; Takano et al., 1996). Among the most probable causes of CSD cycle induction are those originating from local and regional CBF disorder, provoked by TBI. Theoretically, some such CSD generation zones can appear, and such CSD waves can be induced without any considerable elevation of ICP, as can be seen for group C. With low-level trauma, the number of CSD cycles can cease (groups B and A).

From our results, we conclude that repeating numerous CSD cycles are not rare but rather a typical phenomenon for all (including the moderate and heavy) forms of TBI. The possibility of CSD cycles appearing and the augmentation of their number grows with increased intracranial hypertension. The growing number of CSD cycles under conditions of ICP level ≥20 mm Hg demonstrates with high probability of the unfavorable development of post-traumatic syndrome. This increasing number of spontaneous repeating CSD cycles under conditions of growing intracranial hypertension has been interpreted as manifestation of progressive brain mitochondrial dysfunction. Probably, this paroxysmal character of generation of repeated CSD cycles, characterized by us "CSD status," may be a compensatory mechanism required for elimination local ionic disbalance and reactivation of transmembrane ionic transport. It is possible that this mechanism of adaptation toward mitochondrial insufficiency after severe traumatic brain mitochondrial insufficiency (Zauner et al., 2002). In contrast, a discontinuation or a slowdown in the generation of CSD cycles, even under persisting hypertension, can be considered a sign of recovery.

As was pointed out previously, the classical detection of the CSD phenomenon requires special multi-parametric monitoring. Nevertheless, an analysis of the clinical and experimental materials that we have collected makes it reasonable to assume that repeating numerous CSD cycles may be possibly identified using EEG monitoring alone. In combination with the monitoring of ICP and other usually measured parameters, the monitoring of the number of CSD cycles can provide important information for monitoring the progression of trauma and its overall implications for mortality and/or morbidity.

References

Back, T., Kohno, K., & Hossmann, K. A. (1994). Cortical negative DC deflections following middle cerebral artery occlusion and KCl-induced spreading depression: Effect on blood flow, tissue oxygenation, and electroencephalogram. *Journal of Cerebral Blood Flow and Metabolism, 14*, 12–19.

Bardt, T. F., Unterberg, A. W., Hartl, R., Kiening, K. L., Schneider, G. H., & Lanksch, W. R. (1998). Monitoring of brain tissue PO_2 in traumatic brain injury: Effect of cerebral hypoxia on outcome. *Acta Neurochirurgica. Supplement, 71*, 153–156.

Barlow, C. H., & Chance, B. (1976). Ischemic areas in perfused rat hearts: Measurement by NADH fluorescence photography. *Science, 193*, 909–910.

Bean, J. W., & Leatherman, N. E. (1969). Cerebral blood flow during convulsions. Alterations induced in animals by high pressure oxygen. *Archives of Neurology, 20*, 396–405.

Bean, J. W., Lignell, J., & Burgess, D. W. (1972). Cerebral O_2, CO_2, regional cerebral vascular control, and hyperbaric oxygenation. *Journal of Applied Physiology, 32*, 650–657.

Benveniste, H., Jorgensen, M. B., Diemer, N. H., & Hansen, A. J. (1988). Calcium accumulation by glutamate receptor activation is involved in hippocampal cell damage after ischemia. *Acta Neurologica Scandinavica, 78*, 529–536.

Bergsneider, M., Hovda, D. A., Lee, S. M., Kelly, D. F., McArthur, D. L., Vespa, P. M., Lee, J. H., Huang, S. C., Martin, N. A., Phelps, M. E., & Becker, D. P. (2000). Dissociation of cerebral glucose metabolism and level of consciousness during the period of metabolic depression following human traumatic brain injury. *Journal of Neurotrauma, 17*, 389–401.

Bouma, G. J., Muizelaar, J. P., Choi, S. C., Newlon, P. G., & Young, H. F. (1991). Cerebral circulation and metabolism after severe traumatic brain injury: The elusive role of ischemia. *Journal of Neurosurgery, 75*, 685–693.

Brown, J. A., Preul, M. C., & Taha, A. (1988). Hyperbaric oxygen in the treatment of elevated intracranial pressure after head injury. *Pediatric Neuroscience, 14*, 286–290.

Bullock, R., Zauner, A., Woodward, J. J., Myseros, J., Choi, S. C., Ward, J. D., Marmarou, A., & Young, H. F. (1998). Factors affecting excitatory amino acid release following severe human head injury. *Journal of Neurosurgery, 89*, 507–518.

Bures, J. (1957). The effect of anoxia and asphyxia on spreading EEG depression. *Physiologia Bohemoslovenica, 6*, 447–453.

Bures, J., & Buresova, O. (1960). Activation of latent foci of spreading cortical depression in rats. *Journal of Neurophysiology, 23*, 225–236.

Chance, B. (1966). The identification and control of metabolic states. *Genootschap ter Bevordering van Natuur-, Genees-, en Heelkunde te Amsterdam*, 5–37.

Chance, B. (1967). *Non-destructive readout of biochemical data*. Williams and Wilkins.

Chance, B., & Williams, G. R. (1955). Respiratory enzymes in oxidative phosphorylation (I- kinetics of oxygen utilization). *The Journal of Biological Chemistry, 217*, 383–393.

Chance, B., Jamieson, D., & Coles, H. (1965). Energy-linked pyridine nucleotide reduction: Inhibitory effects of hyperbaric oxygen *in vitro* and *in vivo*. *Nature, 4981*, 257–263.

Chance, B., Jamieson, D., & Williamson, J. R. (1966). *Control of the oxidation-reduction state of reduced pyridine nucleotides in vivo and in vitro by hyperbaric oxygen, Third international conference on hyperbaric medicine* (pp. 15–41). National Academy of Sciences.

Chance, B., Thurman, R. G., & Gosalvez, M. (1969). Oxygen affinities of cellular respiration. *Forvarsmedicin, 5*, 235–243.

Chance, B., Barlow, C., Haselgrove, J., Nakase, Y., Quistorff, B., Matschinsky, F., & Mayevsky, A. (1978). Microheterogeneities of redox states of perfused and intact organs. In P. Srere (Ed.), *Microenvironments and metabolic compartmentation* (pp. 131–148). Academic Press.

Chapin, J. L. (1955). Anticonvulsant threshold of CO_2 in oxygen under high pressure. *Proceedings of the Society for Experimental Biology and Medicine, 90*, 663–664.

Crowe, W., Mayevsky, A., & Mela, L. (1977). Application of a solid membrane ion selective electrode to *in vivo* measurements. *The American Journal of Physiology, 233*, C56–C60.

Crutchfield, J. S., Narayan, R. K., Robertson, C. S., & Michael, L. H. (1990). Evaluation of a fiber-optic intracranial pressure monitor. *Journal of Neurosurgery, 72*, 482–487.

Czosnyka, M., Guazzo, E., Whitehouse, M., Smielewski, P., Czosnyka, Z., Kirkpatrick, P., Piechnik, S., & Pickard, J. D. (1996). Significance of intracranial pressure waveform analysis after head injury. *Acta Neurochirurgica, 138*, 531–541. discussion 541-532.

Dietrich, W. D., Feng, Z. C., Leistra, H., Watson, B. D., & Rosenthal, M. (1994). Photothrombotic infarction triggers multiple episodes of cortical spreading depression in distant brain regions. *Journal of Cerebral Blood Flow and Metabolism, 14*, 20–28.

Dixon, C. E., Lyeth, B. G., Povlishock, J. T., Findling, R. L., Hamm, R. J., Marmarou, A., Young, H. F., & Hayes, R. L. (1987a). A fluid percussion model of experimental brain injury in the rat. *Journal of Neurosurgery, 67*, 110–119.

Dixon, C. E., Lyeth, B. G., Povlishock, J. T., Findling, R. L., Hamm, R. J., Marmarou, A., Young, H. F., & Hayes, R. L. (1987b). A fluid percussion model of experimental brain injury in the rat. *Journal of Neurosurgery, 67*, 110–119.

Engelborghs, K., Verlooy, J., Van Deuren, B., Van Reempts, J., & Borgers, M. (1997). Intracranial pressure in a modified experimental model of closed head injury. *Acta Neurochirurgica. Supplement, 70*, 123–125.

Fertziger, A. P., & Ranck, J. B., Jr. (1970). Potassium accumulation in interstitial space during epileptiform seizures. *Experimental Neurology, 26*, 571–585.

Friedli, C. M., Sclarsky, D. S., & Mayevsky, A. (1982). Multiprobe monitoring of ionic, metabolic, and electrical activities in the awake brain. *The American Journal of Physiology, 243*, R462–R469.

Gido, G., Kristian, T., Katsura, K., & Siesjo, B. K. (1994a). The influence of repeated spreading depression-induced calcium transients on neuronal viability in moderately hypoglycemic rats. *Experimental Brain Research, 97*, 397–403.

Gido, G., Kristian, T., & Siesjo, B. K. (1994b). Induced spreading depressions in energy-compromised neocortical tissue: Calcium transients and histopathological correlates. *Neurobiology of Disease, 1*, 31–41.

Graham, D. I., Adams, J. H., & Doyle, D. (1978). Ischaemic brain damage in fatal non-missile head injuries. *Journal of the Neurological Sciences, 39*, 213–234.

Graham, D. I., Ford, I., Adams, J. H., Doyle, D., Teasdale, G. M., Lawrence, A. E., & McLellan, D. R. (1989). Ischaemic brain damage is still common in fatal non-missile head injury. *Journal of Neurology, Neurosurgery, and Psychiatry, 52*, 346–350.

Hansen, A. J. (1985). Effect of anoxia on ion distribution in the brain. *Physiological Reviews, 65*, 101–148.

Harbig, K., Chance, B., Kovach, A. G. B., & Reivich, M. (1976). *In vivo* measurement of pyridine nucleotide fluorescence from cat brain cortex. *Journal of Applied Physiology, 41*, 480–488.

Heinemann, U., & Lux, H. D. (1975). Undershoots following stimulus-induced rises of extracellular potassium concentration in cerebral cortex of cat. *Brain Research, 93*, 63–76.

Hubschmann, O. R., & Kornhauser, D. (1983). Effects of intraparenchymal hemorrhage on extracellular cortical potassium in experimental head trauma. *Journal of Neurosurgery, 59*, 289–293.

Ishige, N., Pitts, L. H., Pogliani, L., Hashimoto, T., Nishimura, M. C., Bartkowski, H. M., & James, T. L. (1987). Effect of hypoxia on traumatic brain injury in rats: Part 2. Changes in high energy phosphate metabolism. *Neurosurgery, 20*, 854–858.

Izquierdo, I., Nasello, A. G., & Marichich, E. S. (1970). Effects of potassium on rat hippocampus: The dependence of hippocampal evoked and seizure activity on extracellular potassium levels. *Archives Internationales de Pharmacodynamie et de Thérapie, 187*, 318–328.

Janebova, M. (1971). Excitation of Leão's cortical spreading depression by threshold amounts of electrophoretically applied K^+ ions. *Physiologia Bohemoslovaca, 20*, 447–451.

Jobsis, F. F., O'Connor, M., Vitale, A., & Vreman, H. (1971). Intracellular redox changes in functioning cerebral cortex. I. Metabolic effects of epileptiform activity. *Journal of Neurophysiology, 3465*, 735–749.

Kaplan, S. A., & Stein, S. N. (1957). Effects of oxygen at high pressure on the transport of potassium, sodium and glutamate in Guinea pig brain cortex. *The American Journal of Physiology, 190*, 157–162.

Katayama, Y., Becker, D. P., Tamura, T., & Hovda, D. A. (1990). Massive increases in extracellular potassium and the indiscriminate release of glutamate following concussive brain injury. *Journal of Neurosurgery, 73*, 889–900.

Kobayashi, S., Nishiki, K., Kaede, K., & Ogata, E. (1971). Optical consequences of blood substitution on tissue oxidation- reduction state microfluorometry. *Journal of Applied Physiology, 31*, 93–96.

Kochanek, P. M., Marion, D. W., Zhang, W., Schiding, J. K., White, M., Palmer, A. M., Clark, R. S., O'Malley, M. E., Styren, S. D., Ho, C., et al. (1995). Severe controlled cortical impact in rats: Assessment of cerebral edema, blood flow, and contusion volume. *Journal of Neurotrauma, 12*, 1015–1025.

Koroleva, V. I., & Bures, J. (1996). The use of spreading depression waves for acute and long-term monitoring of the penumbra zone of focal ischemic damage in rats. *Proceedings. National Academy of Sciences. United States of America, 93*, 3710–3714.

Koura, S. S., Doppenberg, E. M., Marmarou, A., Choi, S., Young, H. F., & Bullock, R. (1998). Relationship between excitatory amino acid release and outcome after severe human head injury. *Acta Neurochirurgica. Supplement, 71*, 244–246.

Kreisman, N. R., Hodin, R. A., Brizzee, B. L., Rosenthal, M., Sick, T. J., Busto, R., & Ginsberg, M. D. (1987). Seizure-associated pulmonary edema and cerebral oxygenation in the rat. *Journal of Applied Physiology, 62*, 658–667.

Lang, E. W., & Chesnut, R. M. (1995). Intracranial pressure and cerebral perfusion pressure in severe head injury. *New Horizons, 3*, 400–409.

Lauritzen, M. (1987). Cerebral blood flow in migraine and cortical spreading depression. *Acta Neurologica Scandinavica. Supplementum, 113*, 1–40.

Leão, A. A. P. (1944). Spreading depression of activity in cerebral cortex. *Journal of Neurophysiology, 7*, 359–390.

Leniger-Follert, E., Urbanics, R., Harbig, K., & Lubbers, D. W. (1977). The behavior of local pH and NADH-fluorescence during and after direct activation of the brain cortex. In D. H. Ingvar & N. A. Lassen (Eds.), *Acta Neurologica Scandinavica* (56th ed., pp. 214–215). Munkgaard.

Levy, J. V., & Richards, V. (1962). Effect of oxygen at high pressure (OHP) on asphyxial survival time of rats. *Proceedings of the Society for Experimental Biology and Medicine, 109*, 941–944.

Lewis, D. V., & Schuette, W. H. (1975). NADH fluorescence and $[K^+]_o$ changes during hippocampal electrical stimulation. *Journal of Neurophysiology, 38*, 405–417.

Lewis, D. V., O'Connor, M. J., & Schuette, W. H. (1974). Oxidative metabolism during recurrent seizures in the penicillin treated hippocampus. *Electroencephalography and Clinical Neurophysiology, 36*, 347–356.

Lothman, E., LaManna, J., Cordingley, G., Rosenthal, M., & Somjen, G. (1975). Responses of electrical potential potassium levels, and oxidative metabolic activity of the cerebral neocortex of cats. *Brain Research, 88*, 15–36.

Manor, T., Rogatsky, G., Zarchin, N., & Mayevsky, A. (2003). Effects of fluid percussion injury on rat brain hemodynamics, ionic, electrical activity, and energy metabolism *in vivo*. *Advances in Experimental Medicine and Biology, 510*, 409–414.

Marmarou, A., Fatouros, P. P., Barzo, P., Portella, G., Yoshihara, M., Tsuji, O., Yamamoto, T., Laine, F., Signoretti, S., Ward, J. D., Bullock, M. R., & Young, H. F. (2000). Contribution of edema and cerebral blood volume to traumatic brain swelling in head-injured patients. *Journal of Neurosurgery, 93*, 183–193.

Marshall, W. H. (1959). Spreading cortical depression of Leão. *Physiological Reviews, 39*, 239–279.

Marshall, J. R., & Lambertsen, C. J. (1961). Interactions of increased pO_2 and pCO_2 effects in producing convulsions and death in mice. *Journal of Applied Physiology, 16*, 1–7.

Mayevsky, A. (1975). The effect of trimethadione on brain energy metabolism and EEG activity of the conscious rat exposed to HPO. *Journal of Neuroscience Research, 1*, 131–142.
Mayevsky, A. (1976). Brain energy metabolism of the conscious rat exposed to various physiological and pathological situations. *Brain Research, 113*, 327–338.
Mayevsky, A. (1978a). Ischemia in the brain: The effects of carotid artery ligation and decapitation on the energy state of the awake and anesthetized rat. *Brain Research, 140*, 217–230.
Mayevsky, A. (1978b). Pyridine nucleotide oxidation reduction state of the cerebral cortex in the awake gerbil. *Journal of Neuroscience Research, 3*, 369–374.
Mayevsky, A. (1978c). The responses of an awake brain to HPO under increased CO_2 concentration. *Advances in Experimental Medicine and Biology, 92*, 735–740.
Mayevsky, A. (1978d). Shedding light on the awake brain. In P. L. Dutton, J. Leigh, & A. Scarpa (Eds.), *Frontiers in bioenergetics: From electrons to tissues* (2nd ed., pp. 1467–1476). Academic Press.
Mayevsky, A. (1983a). Metabolic, ionic and electrical responses to experimental epilepsy in the awake rat. In M. Baldy, D. H. Moulinier, D. H. Ingvar, & B. S. Meldrum (Eds.), *Proceedings of the first international congress of cerebral blood flow, metabolism and epilepsy* (pp. 263–270). John Libbey.
Mayevsky, A. (1983b). Multiparameter monitoring of the awake brain under hyperbaric oxygenation. *Journal of Applied Physiology, 54*, 740–748.
Mayevsky, A. (1984a). Brain NADH redox state monitored *in vivo* by fiber optic surface fluorometry. *Brain Research Reviews, 7*, 49–68.
Mayevsky, A. (1984b). Brain oxygen toxicity. Invited review. In A. J. Bachrach & M. M. Matzen (Eds.), *Underwater physiology. 8th symposium* (pp. 69–89). Undersea Medical Society Inc.
Mayevsky, A. (1992). Cerebral blood flow and brain mitochondrial redox state responses to various perturbations in gerbils. *Advances in Experimental Medicine and Biology, 317*, 707–716.
Mayevsky, A., & Chance, B. (1973). A new long-term method for the measurement of NADH fluorescence in intact rat brain with implanted cannula. *Advances in Experimental Medicine and Biology, 37A*, 239–244.
Mayevsky, A., & Chance, B. (1974). Repetitive patterns of metabolic changes during cortical spreading depression of the awake rat. *Brain Research, 65*, 529–533.
Mayevsky, A., & Chance, B. (1975). Metabolic responses of the awake cerebral cortex to anoxia hypoxia spreading depression and epileptiform activity. *Brain Research, 98*, 149–165.
Mayevsky, A., & Rogatsky, G. (2007). Mitochondrial function *in vivo* evaluated by NADH fluorescence: From animal models to human studies. *American Journal of Physiology. Cell Physiology, 292*, C615–C640.
Mayevsky, A., & Sclarsky, D. S. (1983). Correlation of brain NADH redox state, K^+, PO_2 and electrical activity during hypoxia, ischemia and spreading depression. *Advances in Experimental Medicine and Biology, 159*, 129–141.
Mayevsky, A., & Shaya, B. (1980). Factors affecting the development of hyperbaric oxygen toxicity in the awake rat brain. *Journal of Applied Physiology, 49*, 700–707.
Mayevsky, A., & Weiss, H. R. (1991). Cerebral blood flow and oxygen consumption in cortical spreading depression. *Journal of Cerebral Blood Flow and Metabolism, 11*, 829–836.
Mayevsky, A., Jamieson, D., & Chance, B. (1974a). Oxygen poisoning in the unanesthetized brain: Correlation of the oxidation-reduction state of pyridine nucleotide with electrical activity. *Brain Research, 76*, 481–491.
Mayevsky, A., Zeuthen, T., & Chance, B. (1974b). Measurements of extracellular potassium, ECoG and pyridine nucleotide levels during cortical spreading depression in rats. *Brain Research, 76*, 347–349.
Mayevsky, A., Crowe, W., & Mela, L. (1980b). The interrelation between brain oxidative metabolism and extracellular potassium in the unanesthetized gerbil. *Neurological Research, 1*, 213–226.
Mayevsky, A., Lebourdais, S., & Chance, B. (1980a). The interrelation between brain PO_2 and NADH oxidation- reduction state in the gerbil. *Journal of Neuroscience Research, 5*, 173–182.

Mayevsky, A., Zarchin, N., & Friedli, C. M. (1982). Factors affecting the oxygen balance in the awake cerebral cortex exposed to spreading depression. *Brain Research, 236*, 93–105.

Mayevsky, A., Kaplan, H., Haveri, J., Haselgrove, J., & Chance, B. (1986b). Three-dimensional metabolic mapping of the freeze-trapped brain: Effects of ischemia on the Mongolian gerbil. *Brain Research, 367*, 63–72.

Mayevsky, A., Zarchin, N., Yoles, E., & Tannenbaum, B. (1986a). Oxygen supply to the brain in hypoxic and hyperoxic conditions. In C. Nicolau (Ed.), *Advances in the biosciences. "O_2-transport in red blood cells"* (pp. 119–132). Pergamon Press.

Mayevsky, A., Duckrow, R. B., Yoles, E., Zarchin, N., & Kanshansky, D. (1990a). Brain mitochondrial redox state, tissue hemodynamic and extracellular ion responses to four-vessel occlusion and spreading depression in the rat. *Neurological Research, 12*, 243–248.

Mayevsky, A., Frank, K. H., Nioka, S., Kessler, M., & Chance, B. (1990b). Oxygen supply and brain function *in vivo*: A multiparametric monitoring approach in the Mongolian gerbil. In J. Piper, T. K. Goldstick, & M. Meyer (Eds.), *Oxygen transport to tissue XII* (pp. 303–313). Plenum Pub.

Mayevsky, A., Flamm, E. S., Pennie, W., & Chance, B. (1991). A fiber optic based multiprobes system for intraoperative monitoring of brain functions. *SPIE, 1431*, 303–313.

Mayevsky, A., Zarchin, N., & Sonn, J. (1993). Brain redox state and O_2 balance in experimental spreading depression and ischemia. In A. Lehmenkuhler, K.-H. Grotemeyer, & F. Tegtmeier (Eds.), *Migraine - Basic Mechanisms and Treatment Urban & Schwarzenberg* (pp. 379–393). Munchen-Wier.

Mayevsky, A., Meilin, S., Rogatsky, G. G., Zarchin, N., & Sonn, J. (1995). Multiparametric monitoring of the awake brain exposed to carbon monoxide. *Journal of Applied Physiology, 78*, 1188–1196.

Mayevsky, A., Doron, A., Manor, T., Meilin, S., Zarchin, N., & Ouaknine, G. E. (1996). Cortical spreading depression recorded from the human brain using a multiparametric monitoring system. *Brain Research, 740*, 268–274.

Mayevsky, A., Manor, T., Meilin, S., Doron, A., & Ouaknine, G. E. (1998). Real-time multiparametric monitoring of the injured human cerebral cortex – A new approach. In R. Marmarou & R. Bullock (Eds.), *Acta neurochirurgica supplement (Wien.)* (pp. 78–81). ICPX.

Mayevsky, A., Doron, A., Meilin, S., Manor, T., Ornstein, E., & Ouaknine, G. E. (1999b). Brain viability and function analyzer: Multiparametric real-time monitoring in neurosurgical patients. *Acta Neurochirurgica. Supplementum (Wien), 75*, 63–66.

Mayevsky, A., Meilin, S., Manor, T., Zarchin, N., & Sonn, J. (1999a). Optical monitoring of NADH redox state and blood flow as indicators of brain energy balance. *Advances in Experimental Medicine and Biology, 471*, 133–140.

McIntosh, T. K., Vink, R., Noble, L., Yamakami, I., Fernyak, S., Soares, H., & Faden, A. L. (1989). Traumatic brain injury in the rat: Characterization of a lateral fluid-percussion model. *Neuroscience, 28*, 233–244.

McIntosh, T. K., Juhler, M., & Wieloch, T. (1998). Novel pharmacologic strategies in the treatment of experimental traumatic brain injury: 1998. *Journal of Neurotrauma, 15*, 731–769.

Miller, J. D., Becker, D. P., Ward, J. D., Sullivan, H. G., Adams, W. E., & Rosner, M. J. (1977). Significance of intracranial hypertension in severe head injury. *Journal of Neurosurgery, 47*, 503–516.

Miller, J. D., Dearden, N. M., Piper, I. R., & Chan, K. H. (1992). Control of intracranial pressure in patients with severe head injury. *Journal of Neurotrauma, 9*(Suppl 1), S317–S326.

Moody, W. J., Futamachi, K. J., & Prince, D. A. (1974). Extracellular potassium activity during epileptogenesis. *Experimental Neurology, 42*, 248–263.

Mutch, W. A., & Hansen, A. J. (1984). Extracellular pH changes during spreading depression and cerebral ischemia: Mechanisms of brain pH regulation. *Journal of Cerebral Blood Flow and Metabolism, 4*, 17–27.

Nedergaard, M., & Hansen, A. J. (1993). Characterization of cortical depolarizations evoked in focal cerebral ischemia. *Journal of Cerebral Blood Flow and Metabolism, 13*, 568–574.

Nilsson, P., Hillered, L., Ponten, U., & Ungerstedt, U. (1990). Changes in cortical extracellular levels of energy-related metabolites and amino acids following concussive brain injury in rats. *Journal of Cerebral Blood Flow and Metabolism, 10*, 631–637.

Nilsson, P., Hillered, L., Olsson, Y., Sheardown, M. J., & Hansen, A. J. (1993). Regional changes in interstitial K^+ and Ca^{2+} levels following cortical compression contusion trauma in rats. *Journal of Cerebral Blood Flow and Metabolism, 13*, 183–192.

Nilsson, P., Gazelius, B., Carlson, H., & Hillered, L. (1996). Continuous measurement of changes in regional cerebral blood flow following cortical compression contusion trauma in the rat. *Journal of Neurotrauma, 13*, 201–207.

Norusis, M. J. (1999). *Guide of data analysis*. Prentice-Hall.

Obrist, W. D., Langfitt, T. W., Jaggi, J. L., Cruz, J., & Gennarelli, T. A. (1984). Cerebral blood flow and metabolism in comatose patients with acute head injury. Relationship to intracranial hypertension. *Journal of Neurosurgery, 61*, 241–253.

O'Connor, M. J., Herman, C. J., Rosenthal, M., & Jobsis, F. (1972). Intracellular redox changes preceding onset of epileptiform activity in intact cat hippocampus. *Journal of Neurophysiology, 35*, 471–483.

O'Connor, M. J., Lewis, D. V., & Herman, C. J. (1973). Effects of potassium on oxidative metabolism and seizures. *Electroencephalography and Clinical Neurophysiology, 35*, 205–208.

O'Connor, M. J., Welsh, F., Komarnicky, L., Davis, L., Stevens, J., Lewis, D., & Herman, C. (1977). *Origin of labile NADH tissue fluorescence* (pp. 90–99). Colloquium on Oxygen and Physiological Function. Professional Information Library, Dallas.

Ookawa, T., & Bures, J. (1975). Extracellular potassium shifts accompanying epileptic discharge induced in chicken hyperstriatum by systemic injection of Metrazol. *Brain Research, 97*, 171–176.

Otsuka, H., Ueda, K., Heimann, A., & Kempski, O. (2000). Effects of cortical spreading depression on cortical blood flow, impedance, DC potential, and infarct size in a rat venous infarct model. *Experimental Neurology, 162*, 201–214.

Ozawa, Y., Nakamura, T., Sunami, K., Kubota, M., Ito, C., Murai, H., Yamaura, A., & Makino, H. (1991). Study of regional cerebral blood flow in experimental head injury: Changes following cerebral contusion and during spreading depression. *Neurologia Medico-Chirurgica (Tokyo), 31*, 685–690.

Paton, W. D. (1967). Experiments on the convulsant and anaesthetic effects of oxygen. *British Journal of Pharmacology and Chemotherapy, 29*, 350–366.

Persson, L., & Hillered, L. (1992). Chemical monitoring of neurosurgical intensive care patients using intracerebral microdialysis. *Journal of Neurosurgery, 76*, 72–80.

Pfenninger, J. (1984). Early prediction of outcome after severe head injury in children. *Zeitschrift für Kinderchirurgie, 39*, 223–228.

Piper, I. R., Miller, J. D., Dearden, N. M., Leggate, J. R., & Robertson, I. (1990). Systems analysis of cerebrovascular pressure transmission: An observational study in head-injured patients. *Journal of Neurosurgery, 73*, 871–880.

Puglia, C. D., Glauser, E. M., & Glauser, S. C. (1974). Core temperature response of rats during exposure to oxygen at high pressure. *Journal of Applied Physiology, 36*, 149–153.

Rogatsky, G. G., Mayevsky, A., Zarchin, N., & Doron, A. (1996). Continuous multiparametric monitoring of brain activities following fluid-percussion injury in rats: Preliminary results. *Journal of Basic and Clinical Physiology and Pharmacology, 7*, 23–43.

Rogatsky, G. G., Sonn, J., Kamenir, Y., Zarchin, N., & Mayevsky, A. (2003). Relationship between intracranial pressure and cortical spreading depression following fluid percussion brain injury in rats. *Journal of Neurotrauma, 20*, 1315–1325.

Rosenthal, M., & Jobsis, F. F. (1971). Intracellular redox changes in functioning cerebral cortex. II. Effects of direct cortical stimulation. *Journal of Neuropsychology, 34*, 750–762.

Rosenthal, M., & Somjen, G. (1973). Spreading depression, sustained potential shifts, and metabolic activity of cerebral cortex of cats. *Journal of Neurophysiology, 36*, 739–749.

Sanders, A. P., Hall, I. H., & Woodhall, B. (1965). Succinate: Protective agent against hyperbaric oxygen toxicity. *Science, 150*, 1830–1831.

Signoretti, S., Marmarou, A., Tavazzi, B., Lazzarino, G., Beaumont, A., & Vagnozzi, R. (2001). N-Acetylaspartate reduction as a measure of injury severity and mitochondrial dysfunction following diffuse traumatic brain injury. *Journal of Neurotrauma, 18*, 977–991.

Signorini, D. F., Andrews, P. J., Jones, P. A., Wardlaw, J. M., & Miller, J. D. (1999). Adding insult to injury: The prognostic value of early secondary insults for survival after traumatic brain injury. *Journal of Neurology, Neurosurgery, and Psychiatry, 66*, 26–31.

Somjen, G. G. (2001). Mechanisms of spreading depression and hypoxic spreading depression-like depolarization. *Physiological Reviews, 81*, 1065–1096.

Somjen, G. G., Rosenthal, M., Cordingley, G., LaManna, J., & Lothman, E. (1976). Potassium, neuroglia, and oxidative metabolism in central gray matter. *Federation Proceedings, 35*, 1266–1271.

Sonn, J., & Mayevsky, A. (2000). Effects of brain oxygenation on metabolic, hemodynamic, ionic and electrical responses to spreading depression in the rat. *Brain Research, 882*, 212–216.

Stein, S. N., & Sonnenschein, R. R. (1950). Electrical activity and oxygen tension of brain during hyperoxic convulsions. *The Journal of Aviation Medicine, 21*, 401–405.

Stocchetti, N., Barbagallo, M., Bellini, G. C., Furlan, A., Vezzani, A., & Nizzoli, V. (1991). Arterio-jugular difference of oxygen and intracranial pressure in comatose, head injured patients. II. Clinical correlations. *Minerva Anestesiologica, 57*, 327–334.

Strong, A. J., Fabricius, M., Boutelle, M. G., Hibbins, S. J., Hopwood, S. E., Jones, R., Parkin, M. C., & Lauritzen, M. (2002). Spreading and synchronous depressions of cortical activity in acutely injured human brain. *Stroke, 33*, 2738–2743.

Sukoff, M. H., & Ragatz, R. E. (1982). Hyperbaric oxygenation for the treatment of acute cerebral edema. *Neurosurgery, 10*, 29–38.

Sukoff, M. H., Hollin, S. A., Espinosa, O. E., & Jacobson, J. H., 2nd. (1968). The protective effect of hyperbaric oxygenation in experimental cerebral edema. *Journal of Neurosurgery, 29*, 236–241.

Sullivan, H. G., Martinez, J., Becker, D. P., Miller, J. D., Griffith, R., & Wist, A. O. (1976). Fluid-percussion model of mechanical brain injury in the cat. *Journal of Neurosurgery, 45*, 521–534.

Sunami, K., Nakamura, T., Ozawa, Y., Kubota, M., Namba, H., & Yamaura, A. (1989). Hypermetabolic state following experimental head injury. *Neurosurgical Review, 12*(Suppl 1), 400–411.

Takano, K., Latour, L. L., Formato, J. E., Carano, R. A., Helmer, K. G., Hasegawa, Y., Sotak, C. H., & Fisher, M. (1996). The role of spreading depression in focal ischemia evaluated by diffusion mapping. *Annals of Neurology, 39*, 308–318.

Taylor, H. J. (1949). The role of carbon dioxide in oxygen poisoning. *The Journal of Physiology, 109*, 272–280.

Tenny, R. T., Sharbrough, F. W., & Anderson, R. E. (1980). Correlation of intracellular redox states and pH with blood flow in primary and secondary seizure foci. *Annals of Neurology, 8*, 564–573.

Thomas, S., Tabibnia, F., Schuhmann, M. U., Brinker, T., & Samii, M. (2000). ICP and MABP following traumatic subarachnoid hemorrhage in the rat. *Acta Neurochirurgica. Supplement, 76*, 203–205.

Torbati, D., Parolla, D., & Lavy, S. (1977). Changes in local brain tissue PO_2 and electrocortical activity of unanesthetized rabbits under high oxygen pressure. *Aviation, Space, and Environmental Medicine, 48*, 247–250.

Vern, B., Schuette, W. H., Whitehouse, W. C., & Mutsuga, N. (1976). Cortical oxygen consumption and NADH fluorescence during metrazole seizures in normotensive and hypotensive cats. *Experimental Neurology, 52*, 82–99.

Vyskocil, F., Kritz, N., & Bures, J. (1972). Potassium-selective microelectrodes used for measuring the extracellular brain potassium during spreading depression and anoxic depolarization in rats. *Brain Research, 39*, 255–259.

Wadhwani, K. C., Rapoport, S. I., Shepherd, A. P., & Oberg, P. A. (1990). Blood flow in the central and peripheral nervous systems. In P. A. Oberg (Ed.), *Shephrd, a.P* (pp. 265–304). Laser Doppler Blood Flowmetry Kluwer Academic Pub.

Wiebers, D. O., Adams, H. P., Jr., & Whisnant, J. P. (1990). Animal models of stroke: Are they relevant to human disease? *Stroke, 21*, 1–3.

Zauner, A., Daugherty, W. P., Bullock, M. R., & Warner, D. S. (2002). Brain oxygenation and energy metabolism: Part I-biological function and pathophysiology. *Neurosurgery, 51*, 289–301. discussion 302.

Zeuthen, T., Hiam, R. C., & Silver, I. A. (1974). Microelectrode recording of ion activity in brain. In H. J. Berman & N. C. Hebert (Eds.), *Ion-selective microelectrodes. Advances in experimental medicine and biology* (pp. 145–156). Springer.

Zeuthen, T., Dora, E., Silver, I. A., Chance, B., & Kovach, A. G. B. (1979). Mechanism of the cerebrocortical vasodilatation during anoxia. *Acta Physiologica Academiae Scientiarum Hungaricae, 54*, 305–318.

Zuckermann, E. C., & Glaser, G. H. (1968). Hippocampal epileptic activity induced by localized ventricular perfusion with high-potassium cerebrospinal fluid. *Experimental Neurology, 20*, 87–110.

Chapter 6
The Effects of Brain Metabolic and Other Perturbations on the Responses to Induced CSD

Abstract In this chapter, we will test the effects of various metabolic perturbations on the responses to CSD induced by exposing the brain to solution of high level of potassium. The metabolic perturbations included oxygen availability (mild hypoxia, hyperoxia, and partial ischemia) hypothermia, anesthesia, and the age of the rats. In the studies described in this chapter, we used the multiparametric monitoring systems as well as multisite system measuring mitochondrial function and electrical activity of the brain in 4 locations.

Keywords Normobaric hyperoxia · Partial Ischemia · Anesthesia · Hypothermia · Aging rats · Induced CSD

6.1 Effects of Oxygen Availability

6.1.1 Introduction

The dependency of the recovery from CSD wave on the availability of oxygen, compensated by an increase in CBF, was discovered by Leão in his original studies in the early 1940s. In his Ph.D. thesis and in his 2nd published paper on CSD (Leão, 1944), Leão summarized his findings as appear in the next 2 paragraphs.

> In rabbits, under dial narcosis, a wave of marked dilatation of and increased blood flow in the pial vessels travels over the cerebral hemispheres concomitantly with the depression of electrical activity that is elicited by weak electrical or mechanical stimulation of the cerebral cortex.
>
> The changes in the pial circulation are similar whether only depression of the spontaneous electrical activity takes place or whether 'convulsive' activity, of any intensity, develops during the depression. The vascular response is apparently secondary to a local change in the activity of nervous elements. Whatever the mechanism of the vasodilation, the marked increase in blood flow probably influences in turn the activity of the cortical neurons.

A. Mayevsky, J. Sonn, *Cortical Spreading Depression of Leao*,
https://doi.org/10.1007/978-3-031-08068-5_6

The direct evidence that shows the critical need of oxygen for the recovery from CSD comes from the real-time monitoring of mitochondrial NADH redox state performed in the early 1970s (Mayevsky & Chance, 1973; Rosenthal & Somjen, 1973).

Using our multiparametric monitoring systems presented in Figs. 6.1 and 6.2, we were able to show for the first time the connection between mitochondrial function, ionic homeostasis, microcirculatory blood flow, and electrical activities, during the CSD event. The second conclusion from our studies is that, as long as CBF and oxygen supply are regulated to stay in the normoxic state, repeated CSD responses could persist.

In this chapter, the effects of oxygen availability on the recovery from CSD waves are presented. The effects of induced CSD are shown in Fig. 6.1, recorded from the gerbil brain (Mayevsky, 1992b). Solution of 1 M KCl was applied

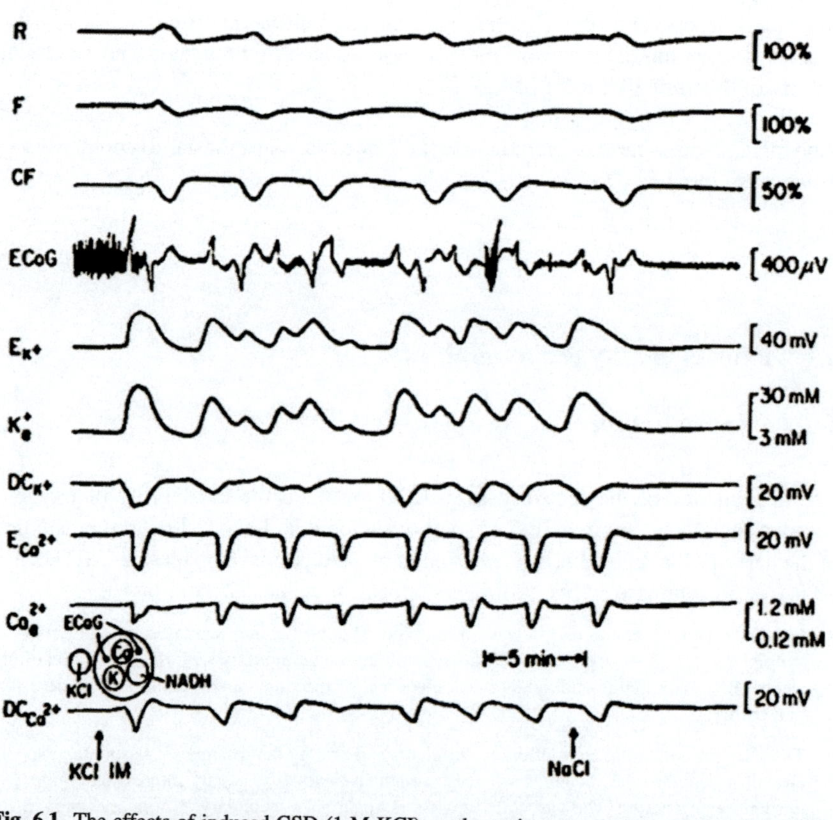

Fig. 6.1 The effects of induced CSD (1 M KCl) on the various parameters monitored from the gerbil brain. EK$^+$, ECa^{2+}—uncorrected potential measured by the two surface electrodes. K^+_e, Ca^{2+}—corrected extracellular potassium and calcium. DCK$^+$, DCCa^{2+}—steady potential measured around the two electrodes used. ECoG—electrocorticogram R, F, CF—reflectance, fluorescence, and corrected fluorescence, respectively (Mayevsky, 1992b)

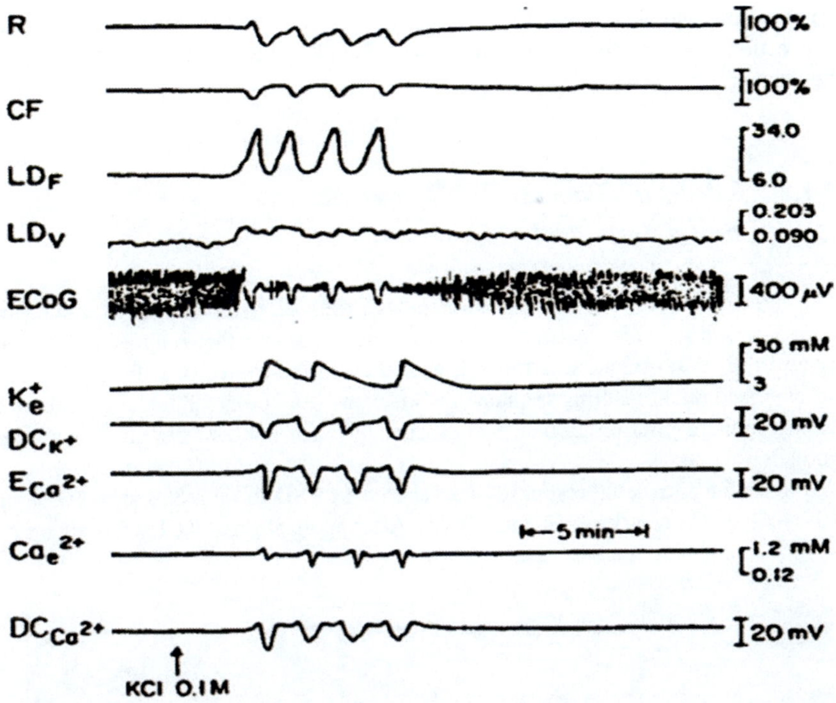

Fig. 6.2 Responses of the brain to CSD induced by KCl (0.1 M) application, 2 mm anterior to the MPA. LDF, LDV—laser Doppler flow and volume monitored by a light guide inserted into the MPA. R, CF—reflectance and corrected fluorescence of NADH; ECoG—electrocorticogram; ECa^{2+}—uncorrected calcium potential; K^{+}_{e}, Ca^{2+}—corrected extracellular potassium and calcium levels; DCK^{+}, $DCCa^{2+}$—DC steady potential measured around the potassium and calcium electrodes (Mayevsky et al., 1992)

epidurally 2 mm anteriorly to the MPA via a push–pull cannula. As can be seen, the wave reached the K^+ electrode first and moved toward the Ca^{2+} electrode and finally to the NADH monitoring area. The shifts of ions between extra- and intracellular spaces led to the stimulation of the pumps and the NADH became more oxidized. The cycles were quite regular and appeared one after the other in K^+ monitored area. Not all cycles were propagated to the light guide area; therefore, only 6 NADH cycles were recorded out of 7 or 8 measured in the K^+ and Ca^{2+} areas.

As seen in Fig. 6.2, a set of responses to CSD was recorded after 0.1 M KCl application to the surface of the brain 2 mm anterior to the MPA (Mayevsky et al., 1992). A clear correlation between the oxidation of NADH (CF) and the large increase in relative CBF can be seen during the 4 cycles recorded. The changes in relative CBF are calculated in relation to the control values before the waves were initiated, and in this animal, it was 4–5 times higher. The cycles of CSD were not so clear on the volume trace and we suspect it was a reading problem, since in other studies a clear connection between flow and volume was found using the same laser Doppler flowmeter.

6.1.2 *Effects of Hypoxia and Hyperoxia on the Responses to CSD*

The interaction between CSD and different levels of hypoxia was examined in awake as well as in anesthetized rats. The sensitivity of anesthetized rats to decrease of O_2 in the inspired air was much higher than in the awake one. In most of the anesthetized rats (breathing spontaneously), even 10% of oxygen in the gas mixture causes some respiration problems so most of the experiments were done on awake animals.

Figure 6.3 shows the response of awake rat to CSD in rat exposed to 100% or 10% oxygen (Mayevsky & Chance, 1975). After application of KCI, a few cycles of

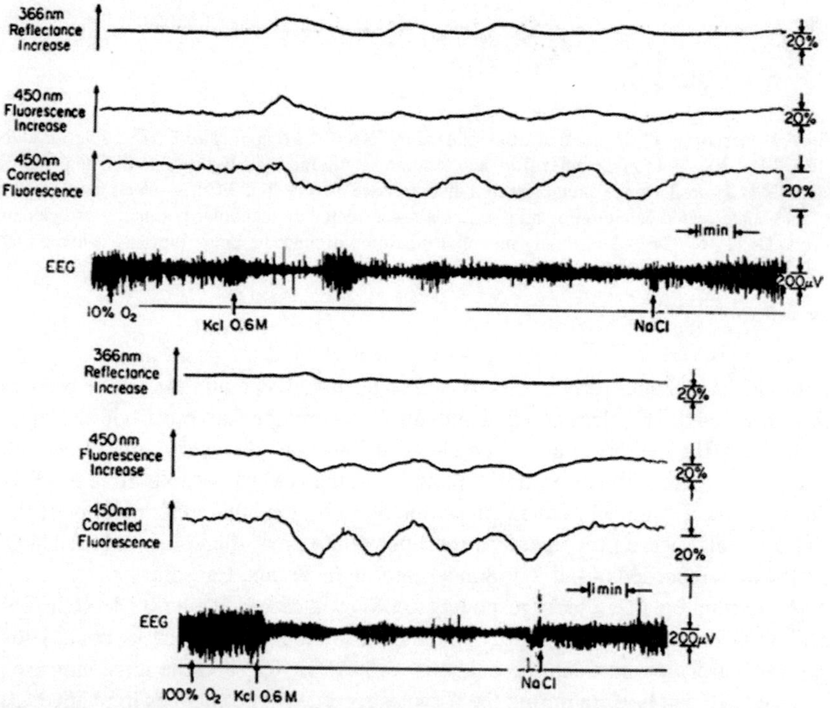

Fig. 6.3 The effects of hypoxia (10% O_2) on the response of NADH fluorescence, 366 nm reflectance, and EEG during CSD (induced by KCl). The lower part was obtained under 100% O_2 atmospheres (Mayevsky & Chance, 1975)

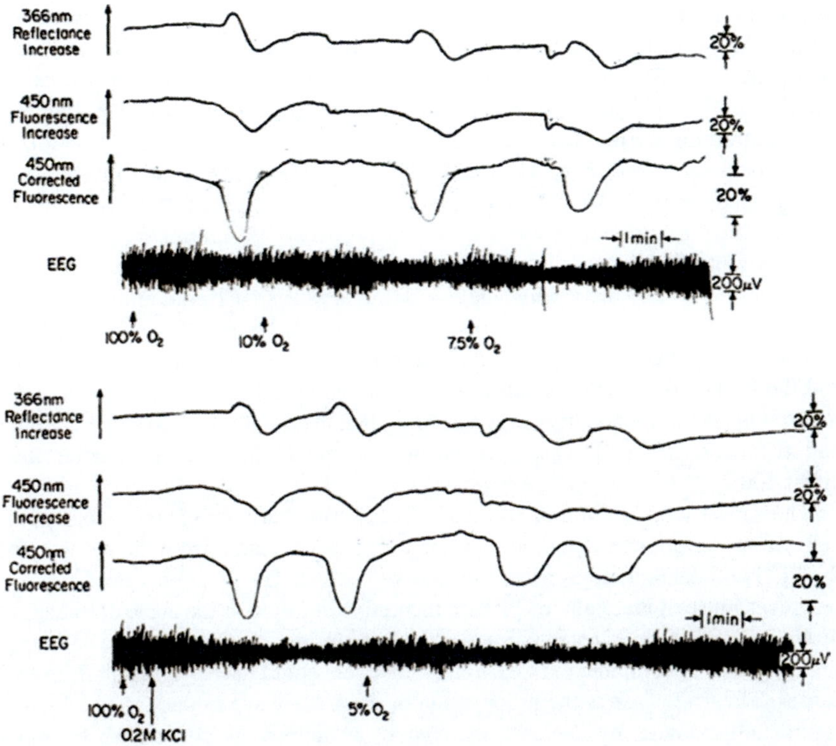

Fig. 6.4 The effects of various levels of oxygen in the inspired air on the response of the awake brain to CSD. The upper part shows the response to 3 levels of O_2 (100%, 10%, and 7.5%) while in the lower part the animal was exposed to 100% and 5% oxygen (Mayevsky & Chance, 1975)

oxidation of NADH were observed. When the gas mixture was changed to 10% O_2 (upper part), the oxidation cycles under these conditions were different in comparison to the normoxic cycles. The amplitude of the minimum NADH level was lower (less oxidized) but the duration of the oxidation cycle was longer. Results obtained in another rat are shown in Fig. 6.4 (Mayevsky & Chance, 1975). In the upper part, the 3 cycles were obtained in 100%, 10%, and 7.5% O_2 while in the lower part, two parts of cycles were obtained under 100% or 5% O_2. When the content of O_2 in the air was reduced, the same effect as in Fig. 6.3 was obtained, namely, that the amplitude of the oxidation cycle was smaller while the duration was longer. The baseline level of NADH was becoming more reduced by giving the animal less O_2. The lower part shows the similarity between two cycles obtained in 100% O_2 as well as at 5% O_2. In the hypoxia experiments, data of 11 animals were used for the calculations. In each animal, two oxidation cycles were used, one under 100% O_2 and the other under 10% O_2 or 7.5% O_2 in some animals. The parameters obtained in the normoxic level were taken as 100% and the effect of hypoxia was calculated in percent change from this level. A t-test on the differences between the means was calculated. Three parameters were measured from the data: the maximal decrease in NADH during the CSD cycle, the duration of the cycle, and finally the integrated area under the CSD oxidation cycle. The results show that hypoxia decreases the amplitude of the NADH oxidation to 70.8 ± 13.9% ($P < 0.01$) of the amplitude of oxidation during normoxia and lengthened the time that the NADH was oxidized below the normoxic level to 130.8 ± 24.8% ($P < 0.001$). The changes in the integrated area were much smaller and found to be 104.8 ± 4.8% ($P < 0.01$) in comparison with the normoxic conditions. In fact, the product of the amplitude and the time is nearly unity (0.93).

In another set of experiments, the blood pressure was measured in the awake rat so that the effect of hypoxia can be tested on this parameter also. The results showed that breathing 10% by an awake animal has a very small effect on the blood pressure and the NADH oxidation–reduction state is more reduced. The main effects of the hypoxia are, as described in Figs. 6.3 and 6.4, that the duration of the cycle is longer and the decrease in fluorescence during the spreading cycle is smaller than under 100%.

The capacity for restoration of the ion distribution between the intra- and extracellular space returned to the normal range as was estimated from the recovery of the EEG and the oxidation–reduction state of NADH. The main effects of hypoxia on the response of the brain to CSD are the longer duration of the cycle of oxidation of NADH and the smaller maximum oxidation during the CSD cycle. The longer time required to complete the CSD oxidation cycle could have originated from two causes. The first cause is the propagation of the CSD wave in the cortex which is significantly slowed by decrease in oxygen availability as was shown by Leão (1947) and Bures (1957). The second possibility is that under hypoxic conditions the availability of oxygen was decreased so that in order to consume the same amount of oxygen a longer period was needed. From the data presented in this study, it seems that the two explanations could be accepted and more detailed experiments are needed in order to distinguish between them. The integrated area below the oxidation cycle was only slightly changed by hypoxia. Thus, the overall energy requirement and utilization during CSD were not altered significantly by hypoxia.

Another aim of this study was to use an awake rat brain model to investigate the effect of hypoxia on the response of the brain to CSD. The question was whether or not brain oxidative metabolism will respond in the same way under normoxic or hypoxic conditions in the awake rat. In anesthetized animals, the effect of hypoxia on the response of brain to CSD is much more pronounced in comparison with the awake animal. In some animals, the hypoxia caused a failure of the brain to recover from a CSD cycle so that the recovery was very slow or even uncompleted during longer periods of time (Mayevsky and Chance, unpublished observations). The main advantage of using the awake model is the elimination of the decrease in blood pressure due to the effect of anesthetics. By using the awake rat, the blood pressure was stable for a few hours and the deterioration of the animal model is negligible.

Figure 6.5 shows the effect of hypoxia on K^+_e and NADH responses to CSD (Mayevsky et al., 1980a). In the first stage, a control response to CSD was measured

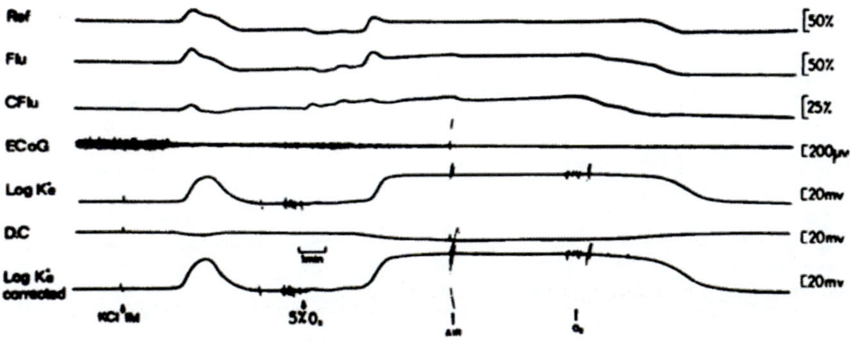

Fig. 6.5 The effects of hypoxia (5% O_2) on K^+_e and metabolic activity in the gerbil. A normal CSD cycle was observed before the hypoxia (Mayevsky et al., 1980a)

in the subject. K^+_e showed a typical response, while NADH showed an oxidation cycle. A small increase in NADH was recorded in the initial phase of this cycle.

When the gerbil was made hypoxic by administration 5% O_2 in 95% N_2, an increased NADH level of about 10% was recorded. At the same time, K^+_e began to increase gradually above it baseline level, finally reaching a point at which a fast, large increase was measured. During this phase, NADH level remained high while the gerbil breathed from the 5% O_2 mixture. When the gerbil was switched back to a room air breathing mixture a small transient oxidation of NADH occurred, but in this animal, only 100% O_2 supplied enough oxygen to oxidize NADH to its original level. K^+_e also started to decrease only after 100% O_2 was introduced.

The effects of hypoxia on the gerbil brain shown in Fig. 6.5 have been described previously for the rat brain (Mayevsky & Chance, 1975). As expected under hypoxic conditions, NADH is reduced, and available energy production decreases. In the present case, it is expected that extracellular levels of K^+_e will accumulate. As is shown in Fig. 6.5 (Mayevsky et al., 1980a), after 5% O_2 respiration, K^+_e went up concomitantly with the increase of NADH. After reaching a critical level, a large increase in K^+_e was recorded under hypoxic conditions. The recovery phase took about 10 min in this case and pure oxygen was required to achieve full recovery.

The recovery of the brain after an episode of CSD is dependent upon energy availability. If energy production is inhibited, the pumping of K^+ is prevented, as shown in Fig. 6.6 (Mayevsky et al., 1986). In this animal, waves of CSD developed spontaneously at a certain stage of the experiment (after a few hypoxic episodes) and the left side of the figure shows the regular response to CSD. The efflux of K^+ reached a level of 14–15 mM and was then pumped back into the cells and the NADH (CF) shows a clear oxidation cycle as shown also in Fig. 6.5. In the second CSD event shown in Fig. 6.6, hypoxia was induced at the beginning of the recovery period and the recovery process was inhibited until 100% O_2 was given to the rat.

In Fig. 6.7, the responses to CSD under normoxic (A) and hypoxic (B) conditions are presented (Mayevsky et al., 1980b). The NADH shows its typical oxidation cycle under both conditions, although during hypoxia the oxidation cycle has a

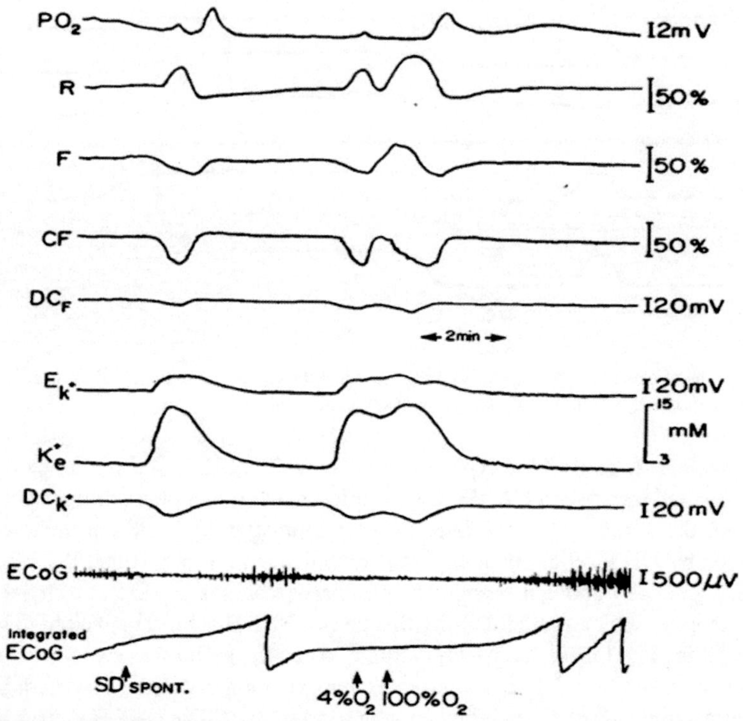

Fig. 6.6 The effects of hypoxia (4% O_2) on the metabolic, ionic, and electrical responses to CSD developed spontaneously (Mayevsky et al., 1986)

lower amplitude with a longer duration. When the gerbil was exposed to 5% O_2, an increase in NADH (10%) was recorded, which was returned to the normoxic level after breathing air. Here again, the correlation between the large increase in pO_2 after the CSD cycle and the reflectance decrease is clear. This overshoot of the pO_2 was blocked almost completely under 5% O_2. When air was given after the 5% O_2, a large overshoot in pO_2 was recorded concomitantly with the oxidation of NADH. After the first application of KCl, the ECoG showed a depression but the recovery was not achieved due to the following CSD cycles.

The effects of graded hypoxia on the responses to CSD are shown in Fig. 6.8 (Mayevsky et al., 1995). A typical CSD response was measured under 95% O_2 showing the oxidation cycle and initial decrease in pO_2, with its large overshoot in parallel to the decrease in reflectance. Under air respiration, the same pattern of changes was recorded, but the overshoot in the pO_2 was smaller. Under 10% O_2, the duration of the oxidation cycle was longer, as well as the interval between the initial decrease in pO_2 and its overshoot. Here again, the pO_2 level during the overshoot was lower. The effects of 5% O_2 were more drastic, as can be seen from the duration of the CSD cycle. By careful analysis of the reflectance trace, one can see there is an interval of time between the initial increase and the dip increase later on. This interval was correlated to the one found in the pO_2 trace.

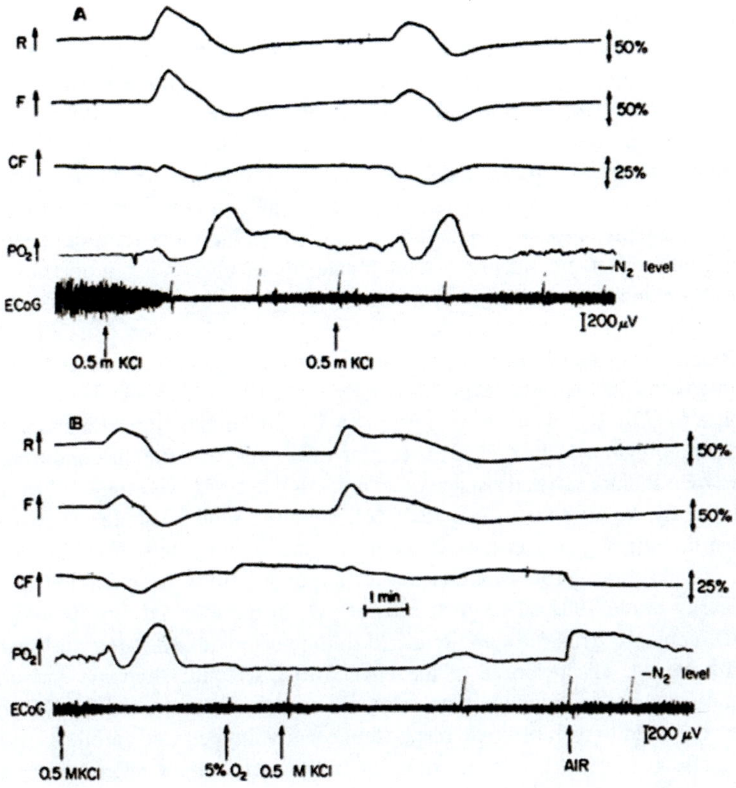

Fig. 6.7 The responses of the normoxic (**A**) and hypoxic (**B**) brain to CSD. (John Wiley and sons, reprinted with permission (Mayevsky et al., 1980b))

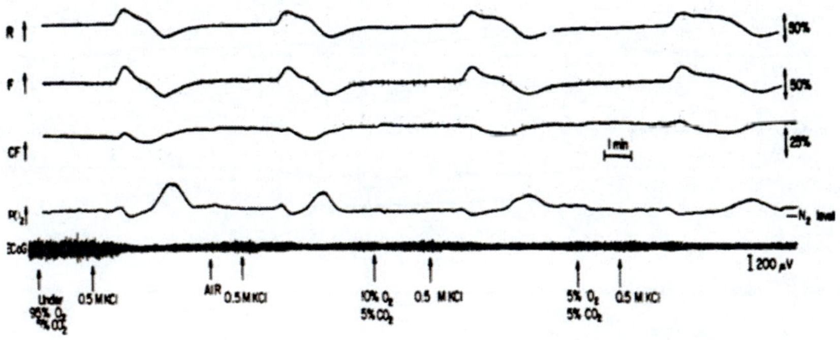

Fig. 6.8 The interrelation between oxygen level in the breathing mixture and the response to CSD in the awake gerbil. (John Wiley and sons, reprinted with permission (Mayevsky et al., 1980b))

The main significance of the pO_2 measurement here is the combination with the reflected and fluorescence signals. From the data presented in our study, it is clear that the reflectance signals include two components in the response to CSD, namely, the initial short increase followed by a long decrease below the baseline level. It seems that the initial increase in reflectance during CSD is not affected by various perturbations such as hypoxia or anesthesia. At the same time, the decrease phase of the reflectance was in a good correlation with the pO_2 measurement under the various perturbations (Figs. 6.7 and 6.8). We may conclude that the initial increase phase of the reflectance may represent changes in the absorbance properties of the brain tissue such as swelling or shrinkage. A support to this conclusion can be obtained from the findings in the blood-free prefused brain exposed to CSD. Under such conditions, the reflected light showed only an increase phase without any decrease below the baseline level (Mayevsky et al., 1981). The NADH fluorescence response to CSD is dependent on oxygen availability in the intracellular space. As we showed previously, the typical "oxidation cycle" under normoxic conditions was altered by ischemia and had a shape of a "reduction cycle" (Mayevsky, 1978). In the present study, we are able to analyze this question in a more accurate and clear way, namely, that the pO_2 can serve as an additional indicator to the state of the tissue.

In Fig. 6.9, the gerbil was exposed to hypoxia before the induction of CSD (Mayevsky et al., 1982). Due to the lower pO_2 in the inspired air (10% O_2), the NADH increased by 5–10% and returned to the normoxic level after air breathing was introduced. The response of the NADH to CSD under hypoxia was a long reduction cycle concomitant with the slower recovery of the K^+_e and the DC potential as compared to the normoxic response measured in the same gerbil.

The effects of repetitive cycles of CSD on the various parameters are shown in Fig. 6.10 (Mayevsky et al., 1992). A special push–pull surface cannula was located

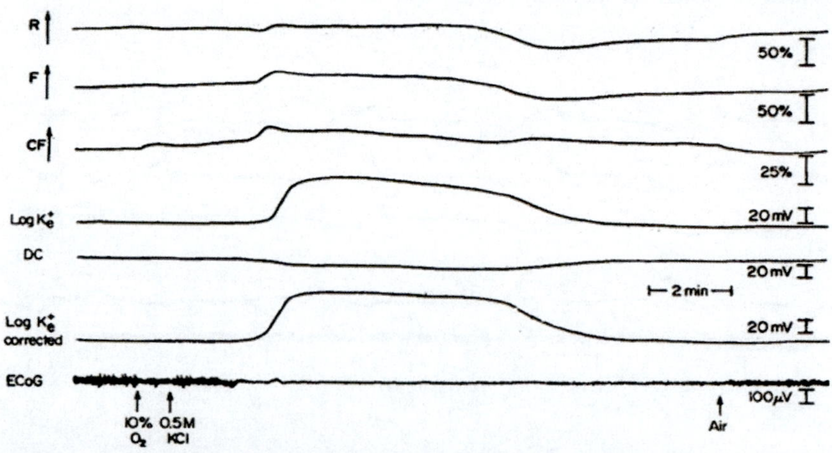

Fig. 6.9 The effects of hypoxia on the metabolic, ionic, and electrical responses to CSD in the gerbil brain. The record is not showing the complete recovery of the ECoG due to its later time (Mayevsky et al., 1982)

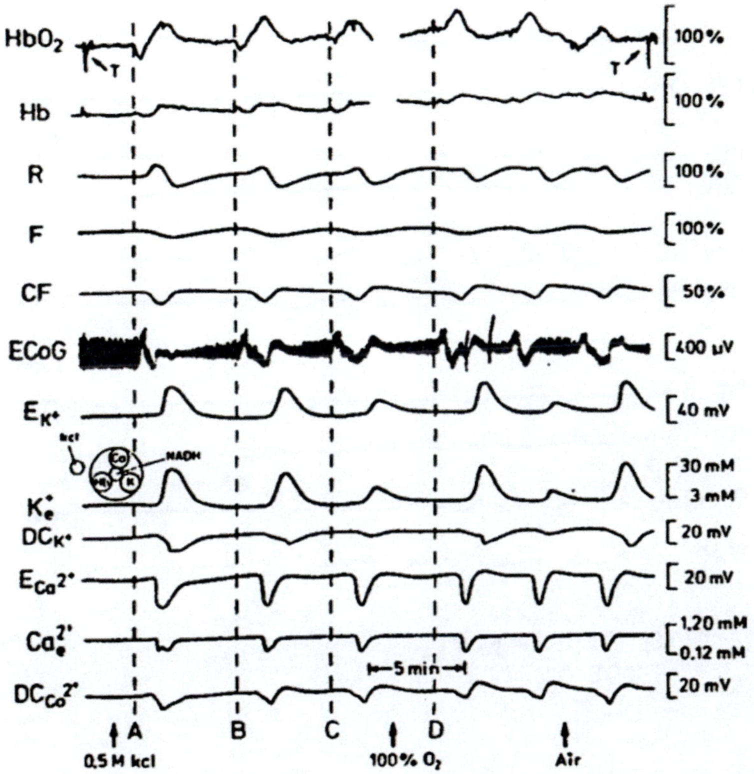

Fig. 6.10 The effects of CSD on the metabolic, ionic, and electrical activities of the gerbil brain. The relationship between the probes' location and the KCI (0.5 M) application site is shown inside the figure (center, left side). The gerbil was breathing air and then was switched to 100% O_2 as well as returned to air during the repeated cycles of CSD. All abbreviations are as in Fig. 6.1 (Mayevsky et al., 1992)

epidurally 2–3 mm anterior to the MPA and was cemented to the skull. The CSD wave was initiated by an 0.5 M KCl epidural application and was stopped by washing the dura with physiological saline. Since the light guide of the EMPHO (measuring the oxygenated hemoglobin) was the closest probe to the KCl application site, it shows the earliest response to the CSD wave. As seen on line A, the HbO_2 trace shows an initial decrease in oxygenation followed by a larger and longer increase. The same response can be seen in the reflectance trace (R). The NADH redox state trace (CF) shows the typical symmetrical oxidation wave, well correlated to the high activity of the Na^+- K^+-ATPase keeping the ion gradient recovery after the large depolarization induced by the CSD. The same responses were recorded when the second and third waves passed below the MPA, as seen in lines B and C (Fig. 6.10). The main difference between the three cycles developed under normoxia was the diminution of the initial decrease of the HbO_2 levels, which could

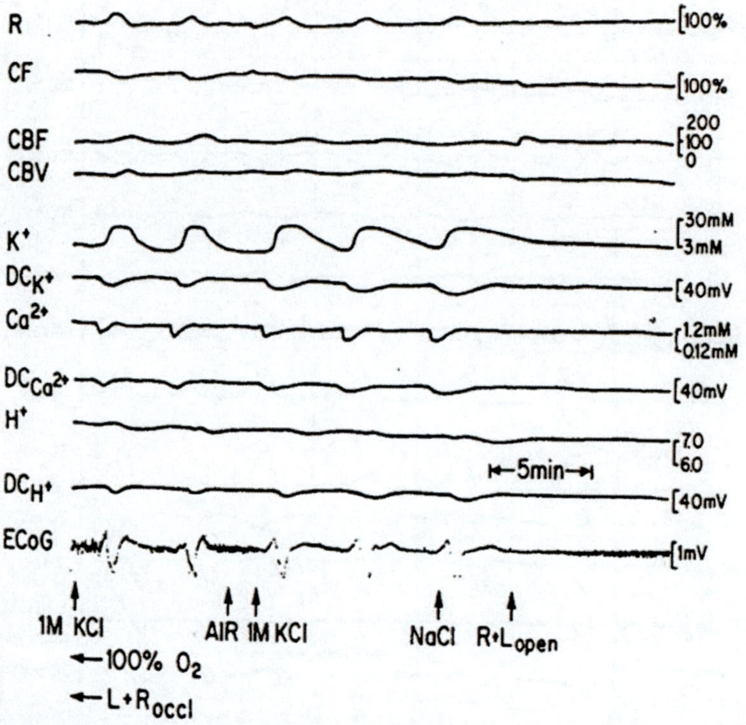

Fig. 6.11 Metabolic ionic and electrical responses to CSD induced in bilateral carotid arteries occluded rat. Prior to the CSD waves, the rat was exposed to 100% O_2 and later on to normal air. All abbreviations are as in Figs. 6.6 and 6.7 (Mayevsky, 1993)

also be seen when the waves were induced under hyperoxic conditions (D). As seen, the potassium electrode did not record a uniform change during the repetitive waves, due to being far away from the KCl application site. Also, it seems that the first cycle led to a slightly different response as compared to the repetitive one.

Figure 6.11 shows two responses recorded under 100% O_2 and bilateral carotid occlusion as well as 3 responses under air breathing situation, while the two carotid arteries were occluded (Mayevsky, 1993). The reopening of the two carotid arteries led to a small increase in CBF and oxidation of NADH. Under bilateral carotid occlusion, the resting CBF is close to normal but under CSD situation the compensation for higher O_2 supply is limited to a certain degree. Under those conditions, the rate of recovery from the CSD event is slower as seen in the kinetics of the K^+ signal. Also, the increase in the CBF signal is small under air breathing compared to the 100% O_2 while the carotid arteries are occluded.

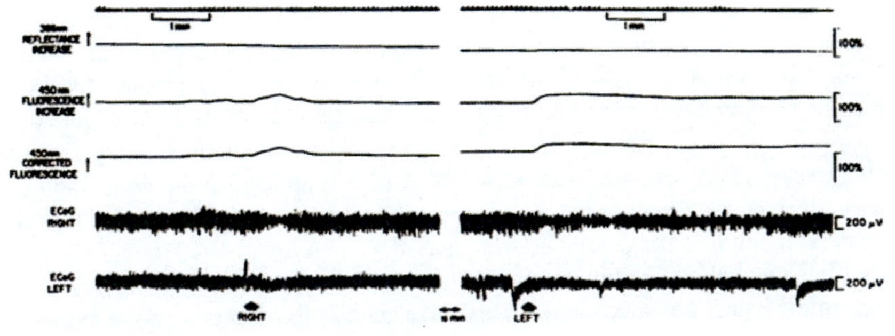

Fig. 6.12 Effects of unilateral and bilateral carotid artery ligation on NADH levels and ECoG. The arrows (right and left) at the bottom show the time of ligation (Mayevsky, 1976)

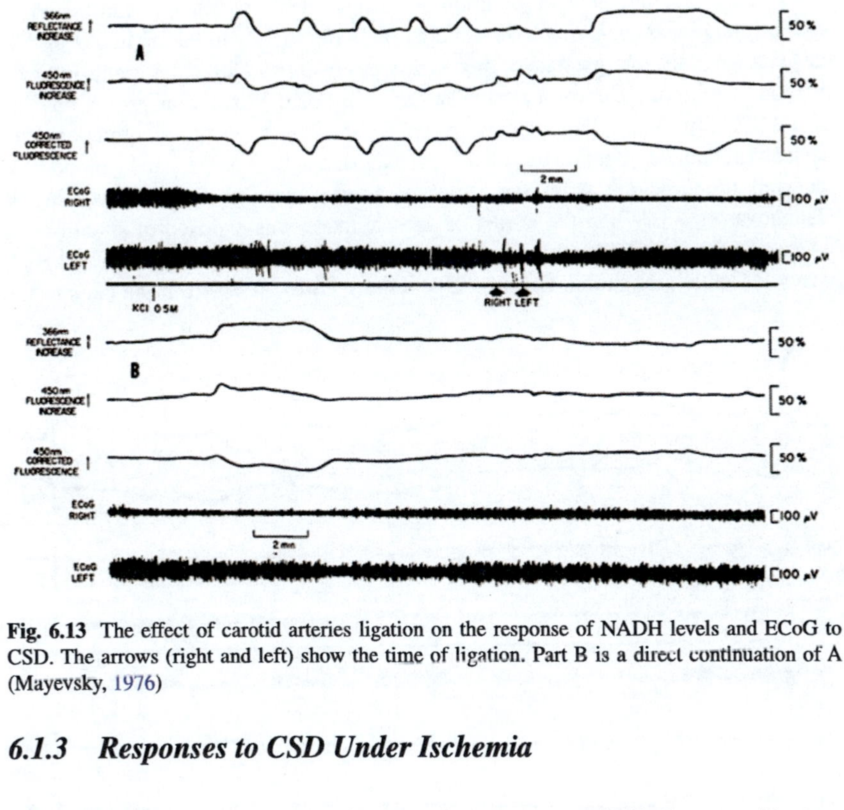

Fig. 6.13 The effect of carotid arteries ligation on the response of NADH levels and ECoG to CSD. The arrows (right and left) show the time of ligation. Part B is a direct continuation of A (Mayevsky, 1976)

6.1.3 Responses to CSD Under Ischemia

Figure 6.12 shows the response of NADH fluorescence and ECoG to carotid artery ligation (Mayevsky, 1976). After ligation of the right artery, an increase of 25% in NADH fluorescence was recorded (second and third traces from the top). The effect on 366 nm reflectance was very small. The ECoG of the right hemisphere showed a short transient depression which recovered very quickly to normal (30 sec). The ligation of the left artery (while the right artery remained closed) resulted in an increase in NADH fluorescence which remained higher compared to that in the preligation period. A gradual increase in NADH fluorescence (about 10%) was recorded during the time interval of ligation of the right and left arteries. Figure 6.13 shows the response of an awake cortex to a bilateral carotid artery ligation (Mayevsky, 1976). Before ligation, 0.5 M KCI was applied to the cortex and the normal metabolic response to CSD was recorded (Mayevsky & Chance, 1974). The bilateral ligation of the carotid arteries caused an increase in NADH, and since the KCl remained on the cortex during this period (about 3 min), the response to CSD was immediately apparent. In this particular animal, the duration of the oxidation cycle became very long after the ligation, as a response to CSD. A similar long oxidation cycle was recorded again later (lower part), but within 2 h the metabolic response to CSD disappeared (not shown in Fig. 6.13).

Eklof and Siesjo (1972) in their studies of the effects of bilateral carotid artery ligation stated that even after a 50% reduction in blood flow to the brain, no significant change in the energy state of the tissue was observed. Our experiments indicate that the ligation of the carotid arteries has an effect on the oxidation–reduction state of NADH (Fig. 6.12). This increase in NADH, however, was not large enough to have an immediate effect on the ECoG, perhaps due to the compensatory blood flow received by the brain from the vertebral arteries through the circle of Willis (Janebova, 1971). When the cortex of such animals was exposed to situations in which extra metabolic work was needed (as in CSD), the cortex became progressively less able to respond. During CSD, a great efflux of K^+ from the cells to the

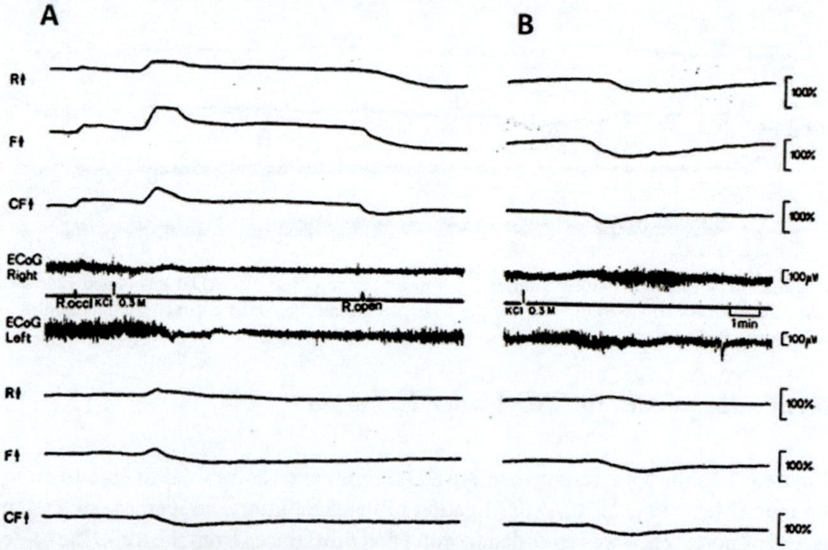

Fig. 6.14 Intramitochondrial metabolic responses to CSD in the partial, unilateral ischemic brain (**A**) and in the normoxic brain after the recovery (**B**). The measurements were performed from the two hemispheres (Mayevsky et al., 1982)

extracellular space occurred (Mayevsky et al., 1974), and the pumping of K^+ back into the cells consumed large amounts of ATP; thus, the NADH became more oxidized (Mayevsky & Chance, 1974; Mayevsky et al., 1974). The effect of the ligation was clear cut within 2-3 h, perhaps due to the cumulative effect of the reduced blood flow and oxygen supply to the cortex. The response of the brain to ligation depends also on the blood pressure of the various animals (Moody et al., 1974), and the same variability in the metabolic response was found in our animals.

Figure 6.14A shows the effects of unilateral carotid occlusion on the metabolic responses to CSD in the awake gerbil (Mayevsky et al., 1982). After the induction of partial ischemia (25–30% increase of NADH in the right hemisphere), the two hemispheres were exposed to CSD and the occluded side showed a reduction cycle as compared to the oxidation cycle measured from the contralateral left "normoxic" hemisphere. After reopening of the right carotid artery, the NADH recovered to the

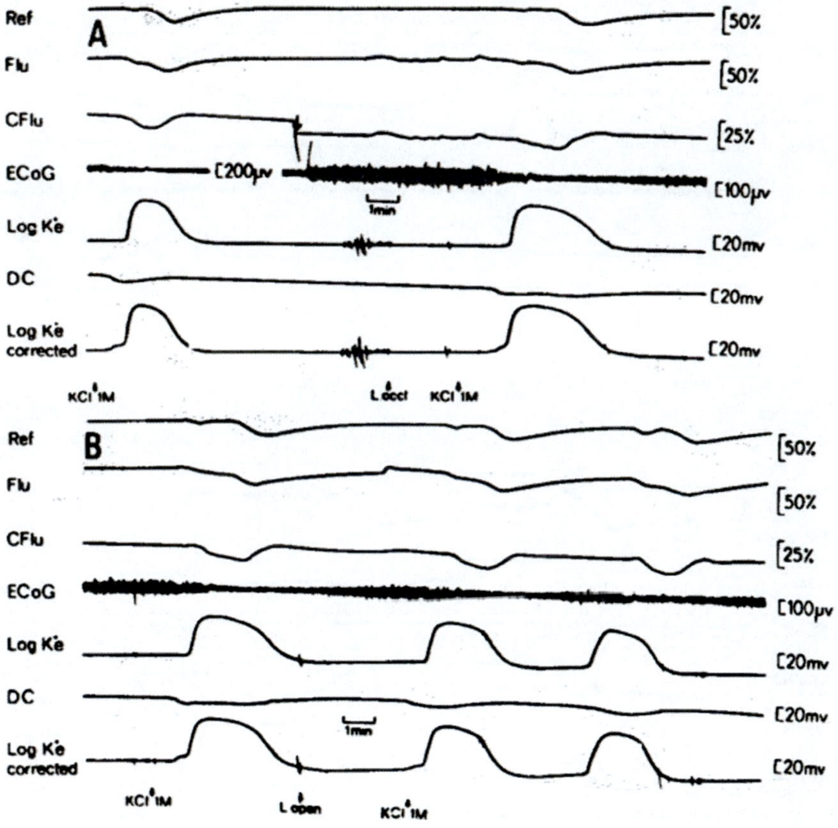

Fig. 6.15 The effects of unilateral carotid occlusion (L_{occl}) on the response of the unanesthetized gerbil brain to CSD. L_{open} indicates reflow in the left carotid artery. B is a continuation of part A (Mayevsky et al., 1980a)

pre-occlusion normoxic level. Figure 6.14B shows the typical responses to CSD induced in the 2 hemispheres, namely 2 oxidation cycles.

In order to evaluate the functional metabolic state of ischemic brain, we used the CSD cycle as a standard stimulus. Fig. 6.15 shows the effects of a contralateral carotid occlusion on metabolic response to CSD (Mayevsky et al., 1980a). The effects of exposure of the un-anesthetized brain to KCl solution are shown in Fig. 6.15A in which a typical response to CSD is recorded. This response was a large increase in K^+_e, which was accompanied by an NADH oxidation cycle. Occlusion of the left carotid artery did not affect the K^+_e level or NADH level, but the response differed from that seen in the control cycle. The time K^+_e returned to its baseline level was over 4 min, as compared to a recovery time of 2.5 min during the control cycle. The same result was found after a second response (Fig. 6.15B).

Following return of flow in the left carotid artery, the response to CSD became similar to the type of response seen in during control cycles. The effect of occlusion

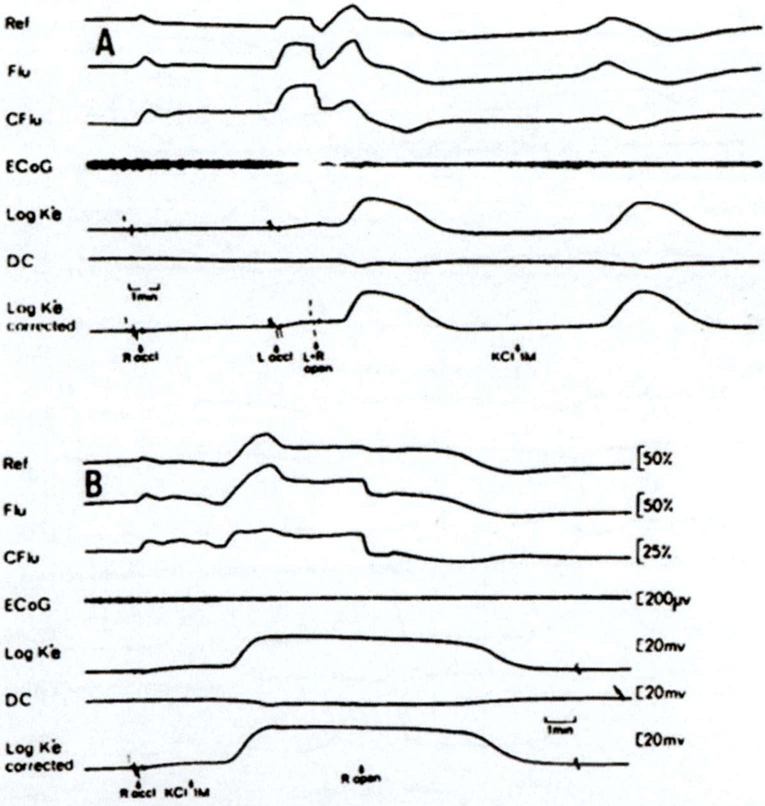

Fig. 6.16 **(A)** Effect of partial and complete ischemia on the metabolic and K^+_e responses of the brain. **(B)** The response to CSD was tested during a unilateral carotid artery occlusion (Mayevsky et al., 1980a)

can be seen also in the shape of the oxidation cycle response to CSD. The duration of the cycle is longer in partially ischemic CSD cycles.

Using another gerbil, the effects of a short bilateral occlusion on the response to CSD cycles were tested (Mayevsky et al., 1980a). When the right carotid was occluded the NADH level increased 10–15%, and K^+_e showed a very slow gradual increase (Fig. 6.16A). When the left carotid was also occluded, a large increase in NADH, which reached a level 40% above the control level, was observed. At the same time, K^+_e continued to increase until the reopening of the two carotids. After recirculation, a CSD cycle developed spontaneously, and typical K^+_e and metabolic responses were shown in the last section of Fig. 6.16.

The effects of partial ischemia-right carotid artery occlusion on the response to CSD are shown in Fig. 6.16B (R_{occl}). In the first stage, occlusion of the right carotid induced metabolic and ionic responses similar to those shown in Fig. 6.16A. Exposing the ischemic brain to CSD (KCl 1 M) caused a typical increase in K^+_e, but the recovery phase was almost completely inhibited. At the same time, NADH increased in a cyclic manner, a finding also seen in the reflected light changes. Full recovery of K^+_e was completed only after recirculation in the right carotid artery and then after a delay of a few minutes. The undershoot of NADH level (CFlu) after recirculation is probably due to the increased pumping activity of K^+ ions. The shape of the NADH trace is that of a small oxidation cycle.

6.1.4 *Responses to CSD Under Partial Ischemia—Statistical Analysis*

In this set of experiments, we ligated the carotid arteries in two ways: (I) ligation of one carotid artery (left or right) and measuring the response of the brain to unilateral ligation for a few hours; (2) simultaneous ligation of the two carotids and measuring NADH and ECoG up to 24 h post-ligation. The protocol of the experiment in this part was the following. The rat was anesthetized and operated on as described in Methods, and 30 min after the operation NADH, as well as ECoG measurements started while the rat was anesthetized. During the control period of the experiment, the brain was exposed to N_2 for one minute in order to check the metabolic state of the brain. Thc response of the brain to CSD was tested by washing the surface of the dura with KCl solution (0.4–0.6 M) through the push-pull cannula. After control values were obtained and when the rat was awake, the carotid arteries were ligated (one or two) while measuring the NADH and ECoG. At various intervals of time after ligation, the brain was exposed to CSD and compared to the control values.

1. *The effect of unilateral ligation.*

Preliminary results of this type of experiment were described by us previously (Mayevsky, 1976). In the present communication, we can summarize results obtained from 8 rats in which unilateral ligation was performed. The main results

Table 6.1 The effects of unilateral carotid artery ligation on metabolic response of the brain (Mayevsky, 1978).

Animal No.	% Reduction under N_2	% Oxidation in CSD	% Oxidation in CSD after unilateral ligation
50	55	28	8
54	52	14	10
57	48	36	12
80	56	22	22
81	60	14	9
83	56	11	9
89	70	10	8
90	60	20	18
Mean ± S.D.	57 ± 6.5	19 ± 9.0	12 ± 5.2

are summarized in Table 6.1. The increase in NADH under N_2 exposure was almost 60% as compared to the normoxic steady-state level. The effect of unilateral occlusion on the response to CSD is in the amplitude of the oxidation cycle. This effect was significant ($P < 0.05$) but, since the cycles after occlusion are longer, we integrate the area below the oxidation cycle and there is no significant difference before and after unilateral occlusion. The steady level of NADH showed a transient response to the ligation as we described earlier (Mayevsky, 1976). The reason for this is the compensation in blood supply from the other hemisphere, so that the increase in NADH after occlusion recovers to normal a few minutes later.

2. *The effects of acute bilateral ligation.*

In this set of experiments, 10 rats were involved. Figure 6.17 shows a typical response of the brain to acute (up to 18 h) bilateral carotid arteries ligation (Mayevsky, 1978). Part A shows the response of the NADH to control CSD and the effect of the ligation on the steady state level of NADH. Parts B, C, and D show the response of NADH to CSD elicited 15 min, 90 min, and 18 h, respectively, after the ligation. In this particular animal, the changes of the reflectance (upper trace in each part) were minimal during the CSD cycles in part A. In part D, one can see very clearly the very long biphasic cycle of the reflectance. The effect of the ligation on the oxidation reduction state of NADH was transient, namely, that a small increase was recorded during the ligation, but within 2 min it recovered to the baseline. The main effect of the ligation is on the response of the brain to CSD which increases oxygen utilization (Mayevsky & Chance, 1974, 1975; Mayevsky et al., 1974; Rosenthal & Somjen, 1973). During the first 60 min after ligation, the oxidation cycle can be detected as a response to CSD, but the amplitude of the oxidation (decrease) was diminished and the duration of the cycle became much longer (almost 4 min in B and only about 2 min in A). The qualitative change in the response of the brain to CSD started several hours after the ligation. In part C, one can see the response to CSD 90 min after the ligation. The uncorrected trace, as well as the corrected one, shows a biphasic response of NADH to CSD. At the beginning of the

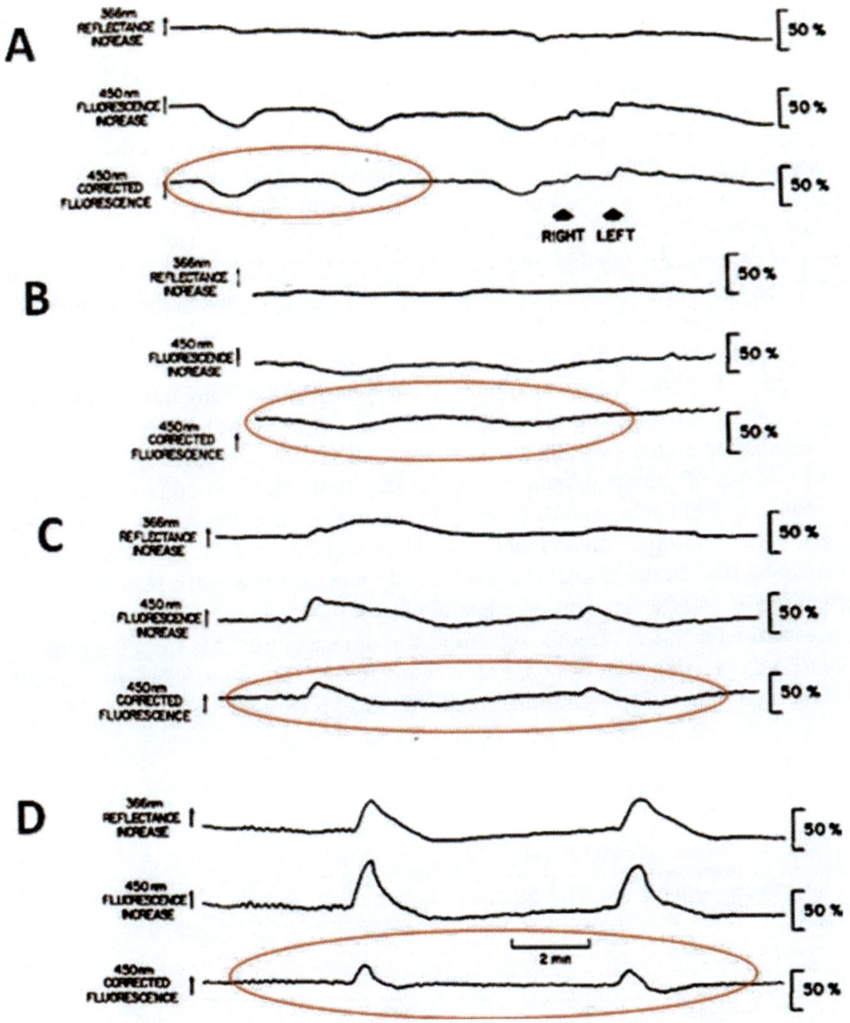

Fig. 6.17 The effect of acute bilateral carotid arteries occlusion on the metabolic response to CSD. Parts B, C, and D were measured 15 min, 90 min, and 18 h after occlusion, respectively (Mayevsky, 1978)

cycle, an increase in NADH was recorded, while afterward, the NADH decreased below the baseline level. In part D (18 h after ligation), the response of the brain was completely opposite to what was measured before the ligation (part A). It seems that instead of an "oxidation cycle" as a response to CSD, we found a "reduction cycle." In this animal, the correction of the NADH for hemodynamic changes is not necessary and the uncorrected trace also shows the same results. The effects of unilateral and bilateral carotid artery occlusion on the ECoG were recorded in all the experiments, but it was difficult to see any change without more sophisticated techniques of analysis.

3. *The effects of chronic bilateral carotid occlusion.*

In this set of experiments, 22 rats were involved. The occlusion of the two carotids was done at different time intervals before the NADH and ECoG were measured. The various animals ranged between 15 h and 14 days of carotid occlusion. After operation and implantation of the light guide holder and ECoG electrodes, the rat was connected to the fluorometer. In order to test the effects of the chronic occlusion, the brain was exposed to anoxia (N_2) and to CSD elicited by washing KCl solution epidurally. According to the results, we divided the 22 rats into three groups: (I) occlusion up to 3 days; (II) occlusion of 4--7 days; (III) occlusion of 8–14 days. The effect of the ligation on the response of the brain to CSD was found to be a reversible effect. In rats of the first group (up to 3 days), the NADH fluorescence showed a "reduction cycle" after application of KCl epidurally as was measured after acute bilateral ligation (Fig. 6.17D). In the third group (8–14 days), the response to CSD was as in the normal brain. In the second group (4–7 days), the results were not clear cut and were seen as dependent upon the physiological and metabolic state of the animal and the brain. Figure 6.18 shows the results obtained from the brain after 10 days of bilateral occlusion (Mayevsky, 1978). Part A was measured a few hours after the operation was done and part B is from the same rat 24 h post-operation. The response of the brain to CSD is the same in parts A and B, which shows that the preparation is stable up to 24 h after the operation.

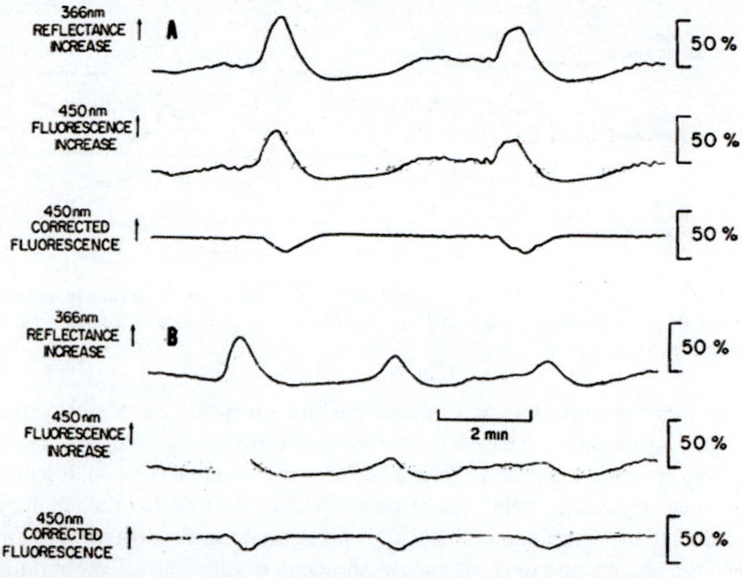

Fig. 6.18 The effects of 10-day bilateral carotid arteries occlusion on the metabolic responses to CSD (Mayevsky, 1978)

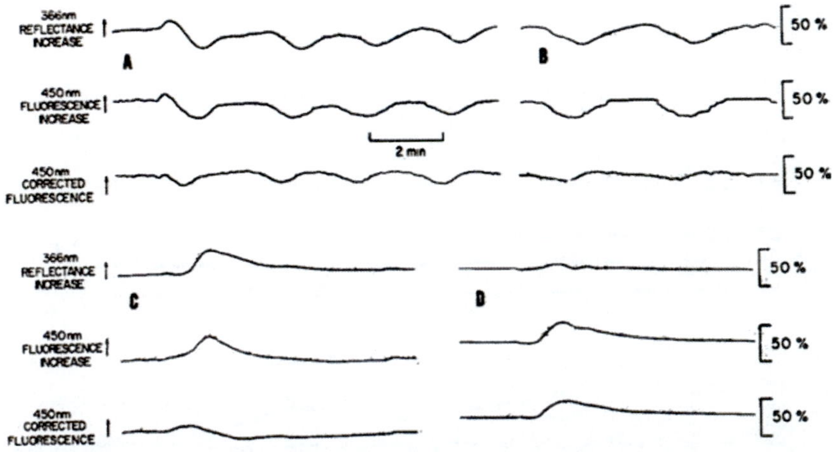

Fig. 6.19 The effects of anesthesia (B, C) on the response of the two brains to CSD after 6 days of bilateral carotid arteries occlusion. Parts A and C were measured during the awake state in the same two rats (Mayevsky et al., 1982)

The same type of transition from oxidation cycle to reduction cycle was found in the partial ischemic adult rat by comparing the awake to the anesthetized brain (4–7 days after occlusion). Fig. 6.19 represents typical results obtained from two rats 6 days after bilateral occlusion (Mayevsky et al., 1982). Figure 6.19A and B was measured from a 6-day occluded rat and Fig. 6.19C and D from another rat. Here again, one can see the variability between two rats, as well as the effect of the anesthesia. Figure 6.19B and D was measured while the rats were anesthetized (Equithesin) and the response to CSD was "reduction cycle" or biphasic. Figure 6.19A, which was measured from an awake rat, shows a clear oxidation cycle, while Fig. 6.19C shows a biphasic response.

6.2 Effect of Anesthesia on the Responses to CSD

We compared the effects of anesthetics in the adult rat (Fig. 6.20) on the NADH response to CSD (Mayevsky et al., 1982). Also, we tested the effects of anesthetics on the metabolic responses to CSD in the very young rat (Fig. 6.21). Figure 6.20 shows a typical metabolic and electrical response to CSD measured from an awake (A) and anesthetized (B, C) adult rat (200 g). CSD was initiated by topical application of KCl solution (0.5–1.0 M) 2–3 mm from the measuring site. In the awake brain (Fig. 6.20A), the reflectance trace (R) shows the typical biphasic response, starting with a short increase of signal followed by a long-lasting decrease phase and a recover to the baseline level. The uncorrected fluorescence trace (F) shows an oxidation of the NADH. The corrected fluorescence trace (CF) response is a symmetrical decrease-increase cycle named "oxidation cycle." The duration of the

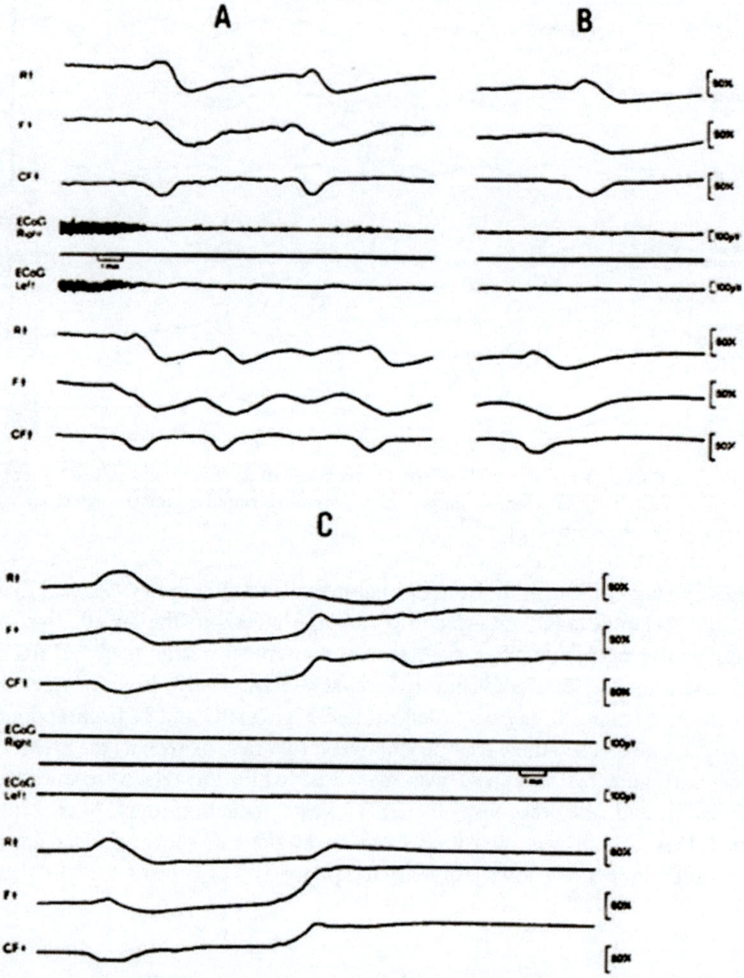

Fig. 6.20 Bilateral measurements of the cortical metabolic response to CSD (CSD) from the awake and anesthetized adult rat (200 g). The upper 4 traces were monitored from the right hemisphere while the other 4 traces from the left one. CSD was induced by topical application of KCI solution (0.5–1.0 M) through a push–pull located 2–3 mm from the measuring site (not shown in the figure). Part A was recorded in the awake state while B and C after injection of anesthetics (Mayevsky et al., 1982)

uncorrected cycle (as shown in the R or F signals) is very long due to the hemodynamic events that have longer duration as compared to the metabolic response of the mitochondria. The same responses were monitored from the two hemispheres of the brain. During the 2 (in the right hemisphere) or the 3 (in the left hemisphere) cycles recorded, the ECoG was depressed and disappeared. After the control awake period, the rat was anesthetized by i.p. injection of Equithesin (ET) and responses to CSD

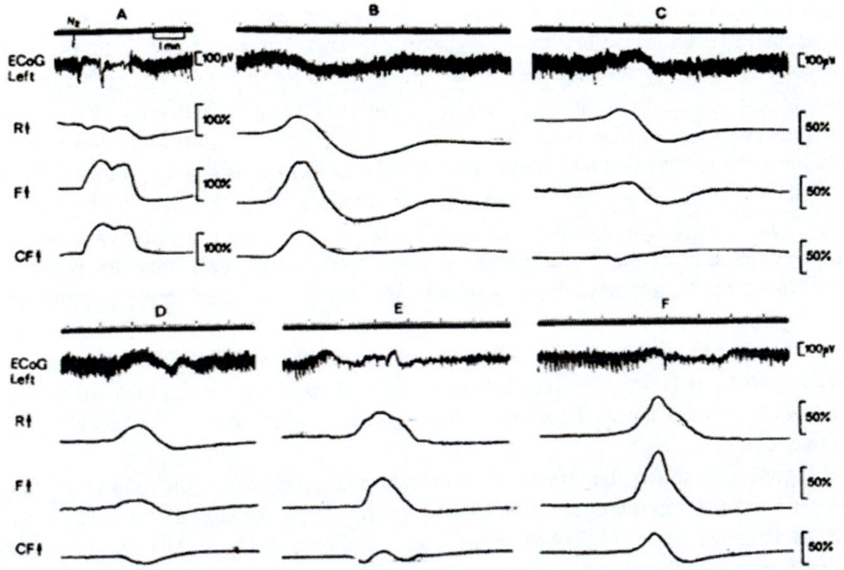

Fig. 6.21 The metabolic responses to anoxia (**A**) and CSD of the awake (**D**) and anesthetized, very young rat (13–14 days old, 20 g). Parts **B** and **C** were recorded when the animal came out from the initial dose of anesthetics, while parts **E** and **F** were obtained after re-anesthetizing the rat by Equithesin. Part **D** was measured in the awake state (Mayevsky et al., 1982)

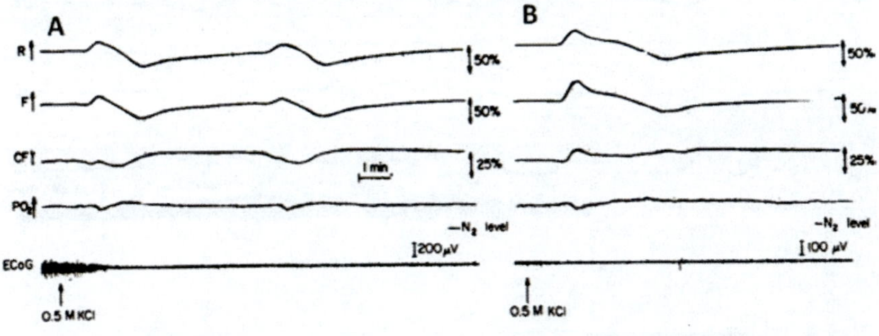

Fig. 6.22 The effects of CSD on NADH, pO_2 in the awake (**A**) and deeply anesthetized (**B**) gerbil brain. (John Wiley and sons, reprinted with permission (Mayevsky et al., 1980b))

were measured every 10 min. The response to CSD shown in Fig. 6.20B was recorded 10 min after injection of ET in the rat. The effect of anesthesia was very small and the typical oxidation cycle was measured bilaterally. Figure 6.20C shows the response to CSD 5 min before cessation of respiration occurred due to an overdose of anesthetics given to the rat (as indicated by an increase of NADH). Here again, the oxidation cycle was recorded although the duration of the cycle was longer.

Figure 6.21 shows responses to CSD of a very young rat (14 days, 20 g) in the awake state as well as under various levels of anesthesia (Mayevsky et al., 1982).

Figure 6.21A shows a typical response to anoxia (increase of F and CF) in which the reflectance response is very small. Figure 6.21B and C was recorded when the very young rat became gradually awake after the initial dose of Equithesin. After monitoring the awake brain (Fig. 6.21D), the rat was reanesthetized and two more responses to CSD are presented (Fig. 6.21E and F). Here again, the awake brain shows a typical oxidation cycle as appeared in Fig. 6.21D. When the young rat was deeply anesthetized, the response to CSD was a reduction cycle as seen in Fig. 6.21B and F. In Fig. 6.21C and E, the response to CSD is a mixture of oxidation and reduction cycles and displays a biphasic-type response. These two opposite types of responses to CSD found in the very young rat disappeared in rats above 15 days old.

Figure 6.22 parts A and B shows the responses to CSD in a gerbil under deep anesthesia (Mayevsky et al., 1980b). The pO_2 response to CSD in the anesthetized state (part B) is similar to that of the awake brain (part A). On the other hand, the NADH corrected trace shows a reduction cycle rather than the typical oxidation cycle.

Figure 6.23 shows the effects of anesthesia on the responses to CSD in a second experiment (Mayevsky et al., 1980b). The control cycle measured before the anesthesia is shown in Fig. 6.23A. The three cycles shown in Fig. 6.23 were measured after induction of anesthesia. The qualitative response of the reflectance is the same in all three cycles, although in part C the interval between the increase and decrease

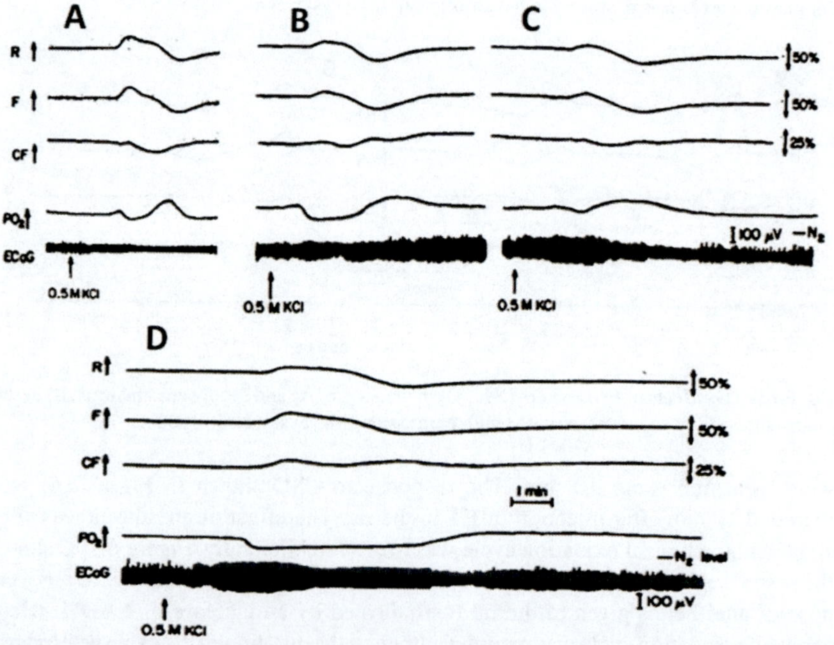

Fig. 6.23 The responses of the brain to CSD under various levels of anesthetics. Part A was measured in the awake state. (John Wiley and sons, reprinted with permission (Mayevsky et al., 1980b))

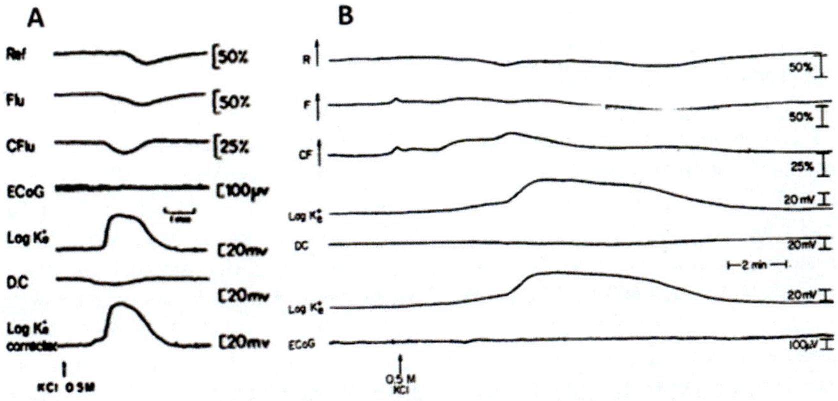

Fig. 6.24 The effects and deep anesthesia (B) on the metabolic, ionic, and electrical responses to CSD in the gerbil brain. The records are not showing the complete recovery of the ECoG due to its later time

was longer. This was in correlation to the longer interval in the pO_2 response. The NADH showed that anesthetics induced a qualitative change, namely, that the oxidation cycle (Fig. 6.23A) became biphasic (Fig. 6.23B, C) and finally showed a long reduction cycle (Fig. 6.23D).

Figure 6.24B represents a response to CSD in the deeply anesthetized gerbil (Mayevsky et al., 1982). Here again, after application of KCI, NADH showed a reduction cycle. The ECoG was depressed during the entire period due to the anesthetics given. The slower recovery of the K^+_e and the DC potential as compared to the normoxic response measured in another typical gerbil (part A) (Mayevsky et al., 1980a). As was shown previously, anesthesia cannot abolish the responses to CSD (Van Harreveld & Stamm, 1953). In the present study, under conditions where NADH showed an increase due to CSD, the rate of K^+ influx into the cells was slower but was returned to the baseline level of extracellular potassium measured before the CSD. This interrelation was found in the gerbil (Fig. 6.24B) as well as in the adult rat, exposed to bilateral carotid occlusion.

In the next 3 figures, the effect of anesthetics will be demonstrated in a rat model using 2 types of monitoring systems. A 4-channel fluorometer reflectometer is described in Chap. 2, Fig. 2.15, and the MPA system is shown in Fig. 2.14.

The MSA System was constructed to monitor reflectance, fluorescence, the mitochondrial NADH redox state, and DC steady potential from four adjacent cortical areas in the same cerebral hemisphere continuously, simultaneously and in real time. The MSA contained four bundles of optical fibers connected to a four-channel fluorometer/reflectometer, which provided information on the mitochondrial NADH redox state (surface fluorometry) and tissue absorption properties (reflected light) from the four adjacent parasagittal locations on the brain cortex (Fig. 2.15). The DC steady potential from four points located parallel to the four light guides was monitored with Ag/AgCl electrodes relative to an Ag/AgCl reference electrode (Mayevsky et al., 1990, 1993; Sonn & Mayevsky, 2001). The special arrangement of the four identical consecutive probes in the MSA allowed us to evaluate the mechanism of the propagation properties of the CSD wave.

The MPA System contained probes to monitor the mitochondrial NADH redox state, extracellular K^+ concentration, DC steady potential, CBF and CBV, and bipolar ECoG during the CSD wave propagation continuously, simultaneously and in real time (Sonn & Mayevsky, 2000). The MPA contained (i) an optical fiber to monitor the mitochondrial NADH redox state (surface fluorometry) and tissue absorption properties (reflected light), (ii) surface mini-electrodes for measuring the extracellular K^+ concentration (K^+_e) and DC steady potential, (iii) a laser Doppler flowmeter to evaluate the relative CBF and CBV, and (iv) two platinum electrodes to detect spontaneous bipolar ECoG. The MPA measured the changes in the various detected parameters from the front line until the recovery period of CSD and helped to identify which parameters start the chain responses and which of them are responding later on during the CSD event.

The Effect of Anesthesia on the Measured Parameters The injection of Equithesin as an anesthetic (0.2–0.25 ml/100 g) caused a significant decrease ($p < 0.05$) in CBF (from 100% to 91.8 ± 4.0%) and in K^+_e (from 3.5 ± 0.52 to

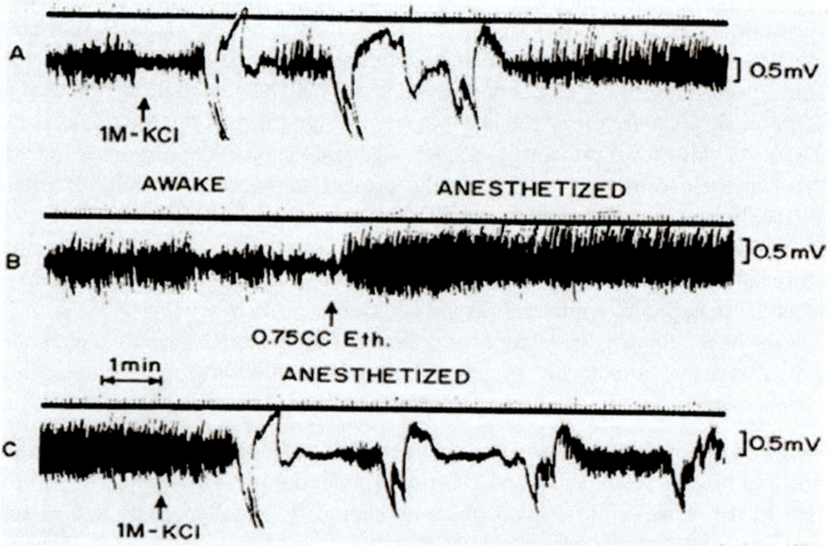

Fig. 6.25 Analog tracings of a bipolar ECoG showing three consecutive CSD waves during wakefulness (A and deep anesthesia (C). Note the characteristic changes in ECoG after anesthesia (Equithesin) as compared with the awake state (B). The arrows in tracings (A) and C) depict the timing of flushing the brain cortex with 1 M KCl solution to initiate CSD waves. The arrow in tracing (B) depicts the timing of an i.p. injection of Equithesin to induce anesthesia (Sonn & Mayevsky, 2006)

2.9 ± 0.38 mM). The Equithesin injection also induced a non-significant decrease in mitochondrial NADH fluorescence and DC potential. Typical analog tracings of bipolar ECoG are shown in Fig. 6.25 (Sonn & Mayevsky, 2006). Figure 6.25B demonstrates a typical augmentation in amplitude and a general increase in the homogeneity of the wave's amplitude after anesthesia. These changes in the ECoG waveform are typical of those recorded when the dominant beta-rhythm (awake state) is replaced by the dominancy of the alpha-rhythm due to a decrease in sensory input. Furthermore, a characteristic depression or silence in ECoG can be seen during CSD wave propagation during both wakefulness (Fig. 6.25A) and under deep anesthesia (Fig. 6.25C).

Amplitude Changes During CSD Wave Propagation CSD waves were induced in brains in the awake state (Figs. 6.26A and 6.27A) and after deep anesthesia (Figs. 6.26B and 6.27B) (Sonn & Mayevsky, 2006). Tracings in Fig. 6.26A and B contain three CSD waves monitored in four adjacent areas in the same cerebral hemisphere using the MSA system. Figure 6.27A and B presents tracings representing reflected light, mitochondrial NADH fluorescence, CBF, CBV, K^{+}_{e}, and the DC potential monitored from a separate experimental group using the MPA system. CSD waves were typically biphasic as measured with reflected light (increase–

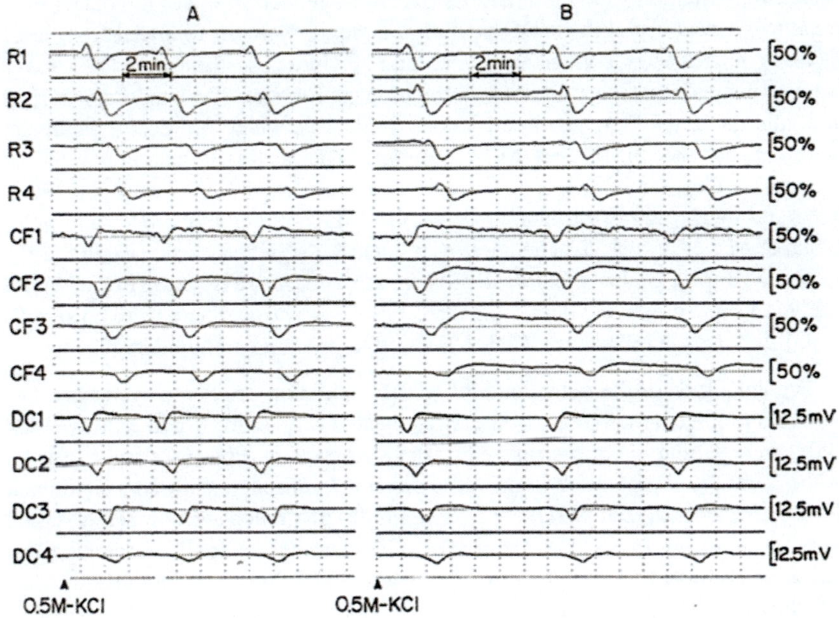

Fig. 6.26 Analog tracings recorded by the MSA system from four adjacent areas in the same cerebral hemisphere showing differences in measured parameters between awake (**A**) and anesthetized (**B**) rats for three consecutive CSD waves, which were induced by a 0.5 M KCl solution. R1-R4, reflectance; CF1-CF4, corrected fluorescence (NADH); DC1-DC4, DC steady potential (Sonn & Mayevsky, 2006)

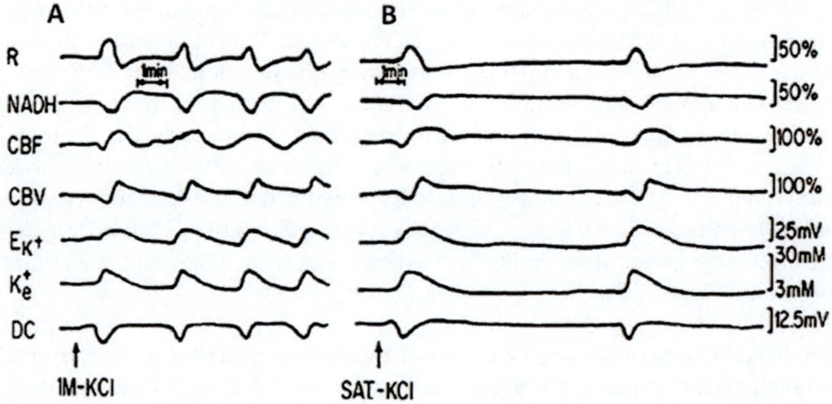

Fig. 6.27 Analog tracings of the metabolic, hemodynamic, ionic, and electrical responses during consecutive CSD waves induced in the awake state (**A**) and after anesthesia (**B**) monitored from a separate experimental group with the MPA system. The tracings represent reflected light (R), mitochondrial NADH (NADH), CBF, CBV, cerebral blood flow and volume, voltage measured by the surface K^+-selective mini-electrode (EK^+), extracellular K^+ ion concentration (K^+_e) (equal to EK^+—DC steady potential), and DC steady potential (DC) (Sonn & Mayevsky, 2006)

decrease), oxidation waves (sometimes oxidation–reduction wave) in NADH fluorescence, a negative shift in DC potential, a small decrease followed by a large increase in CBF and CBV and an increase in K^+_e. Under anesthesia, the CSD waves were prolonged, causing a decrease in the propagation rate. The maximum (VT_{max}) and minimum (VT_{min}) amplitude values were calculated for every parameter recorded by the MSA and MPA systems. The averaged amplitude values (increase = VT_{max} and decrease = VT_{min}) of three successive CSD waves in anesthetized and awake rats were monitored by the MPA system (Figs. 6.28 and 6.29) (Sonn & Mayevsky, 2006). The anesthesia caused a significant decrease in the K^+_e amplitude during the first CSD wave propagation as well as a decrease in the NADH oxidation cycle amplitude during the three following waves. There were no significant differences in reflectance, CBF, CBV, or DC potential amplitudes.

The amplitude of the mitochondrial NADH oxidation–reduction cycles and DC potential that were monitored by the MSA system (Fig. 6.30) (Sonn & Mayevsky, 2006) were similar to those measured by the MPA system (Figs. 6.28 and 6.29). All of the four consecutive areas in the same cerebral hemisphere showed a significant decrease in the oxidized NADH cycle amplitudes and no changes in DC potential amplitudes during two successive CSD cycles. During the third successive CSD wave, the DC potential amplitudes of channels 2 and 4 decreased significantly after anesthesia.

CSD Wave Frequency Anesthesia caused a significant decrease ($p > 0.05$) in CSD wave frequency measured from reflectance, NADH, CBV, and DC using the MPA system. The results were similar for reflectance, NADH fluorescence, and DC potential monitored by the MSA system in the four adjacent areas of the cortex in

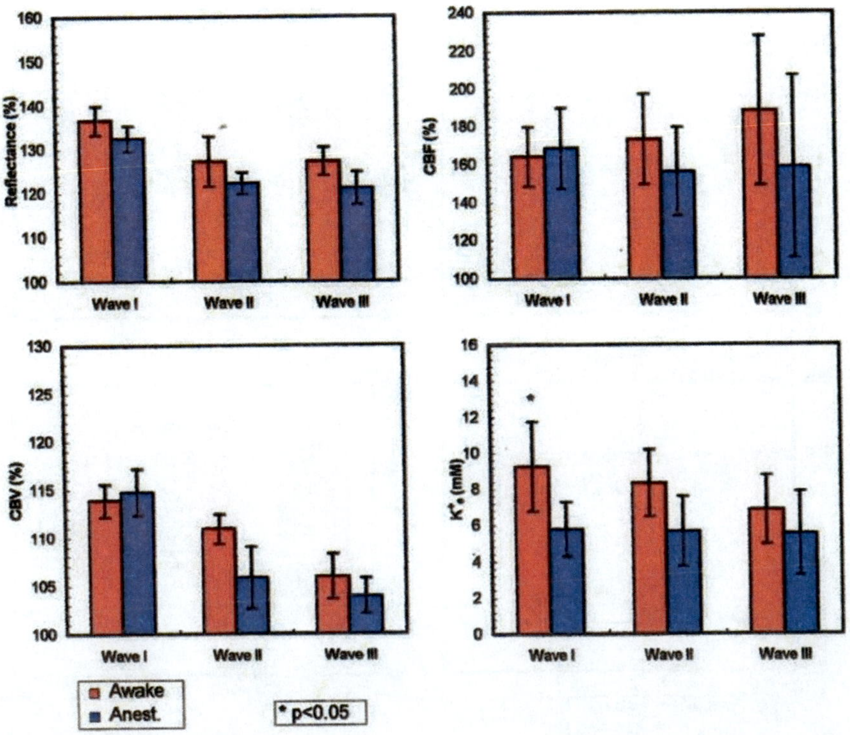

Fig. 6.28 The effect of three consecutive CSD waves (wave I, wave II, and wave III) on maximum (VT_{max})) amplitudes of reflected light (reflectance), CBF, CBV, K^+_e, recorded by the MPA system in awake and Equithesin-anesthetized rats. Anest.—anesthetized. Data are presented as the mean ± SE. *$p < 0.05$ versus awake rats (Sonn & Mayevsky, 2006)

three consecutive waves. There were no significant differences in the wave frequency in any of the measured parameters between the MSA and MPA experimental groups (Table 6.2).

CSD Propagation Rate VTP was calculated at the minimum point (VTP_{min}) of the amplitude during CSD propagation in reflectance, NADH fluorescence, and DC potential for the four channels of the MSA system. There were no significant differences in the VTP_{min} between the awake and anesthetized states (awake state: 2.7 ± 0.18 to 4.6 ± 0.1 mm/minute and anesthesia: 2.39 ± 0.33 to 3.81 ± 0.29 mm/minute). The results were similar for the mean propagation rate at the minimum points ($VTPM_{min}$). The $VTPM_{min}$ for reflectance during the first wave was significantly lower in anesthetized compared with awake rats (Table 6.3). This difference was not significant for reflectance in waves II or III, nor was the $VTPM_{min}$ measured by the NADH redox state and DC potential significantly different for any wave in anesthetized versus awake rats.

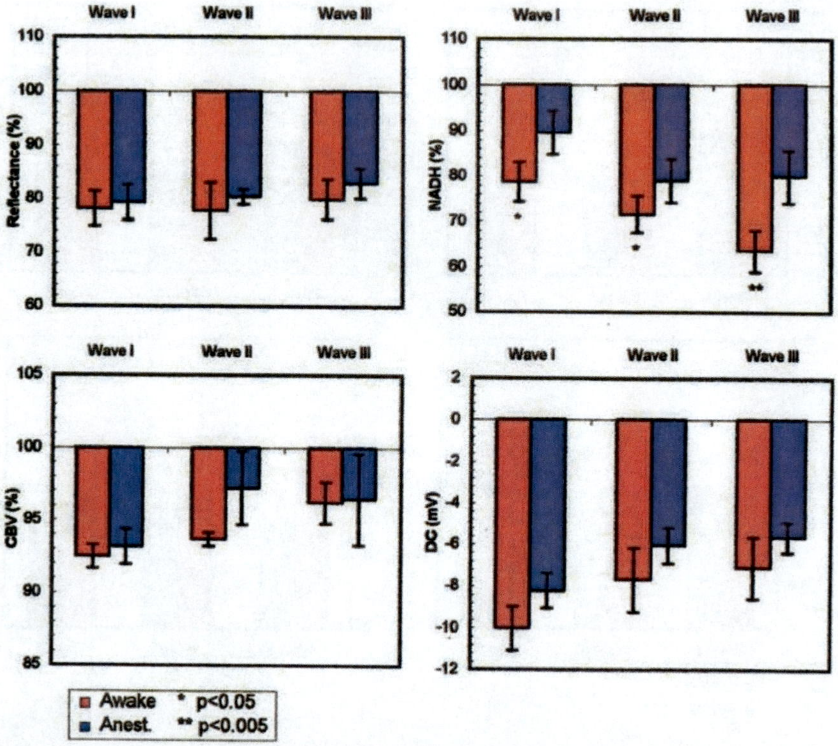

Fig. 6.29 The effect of three consecutive CSD waves (wave I, wave II, and wave III) on minimum (VT_{min}) amplitudes of reflected light (reflectance), CBV, mitochondrial NADH redox state (NADH), and DC recorded by the MPA system in awake and Equithesin-anesthetized rats. Anest.—anesthetized. Data are presented as the mean ± SE. *$p < 0.05$, **$p < 0.005$ versus awake rats (Sonn & Mayevsky, 2006)

CSD Initiation Kinetics There were no significant changes in the initiation kinetics of the first of three CSD waves measured by reflectance, NADH fluorescence, and DC steady potential in any of the four channels of the MSA system after anesthesia (Table 6.4). There were also no differences between wakefulness and anesthesia in the initiation wave kinetics during the other two consecutive waves (data not shown). It is noteworthy that the initiation kinetics of NADH fluorescence and DC potential are negative because there is a decrease in the amplitudes of these parameters during the CSD wave propagation.

CSD Recovery Kinetics The CSD recovery kinetics were calculated by dividing the amplitude value from the maximum or minimum reached during CSD propagation by the time from the maximum or minimum value until full recovery from CSD. The recovery kinetics were calculated for reflectance, NADH fluorescence, and DC steady potential recorded from the four channels of the MSA system for

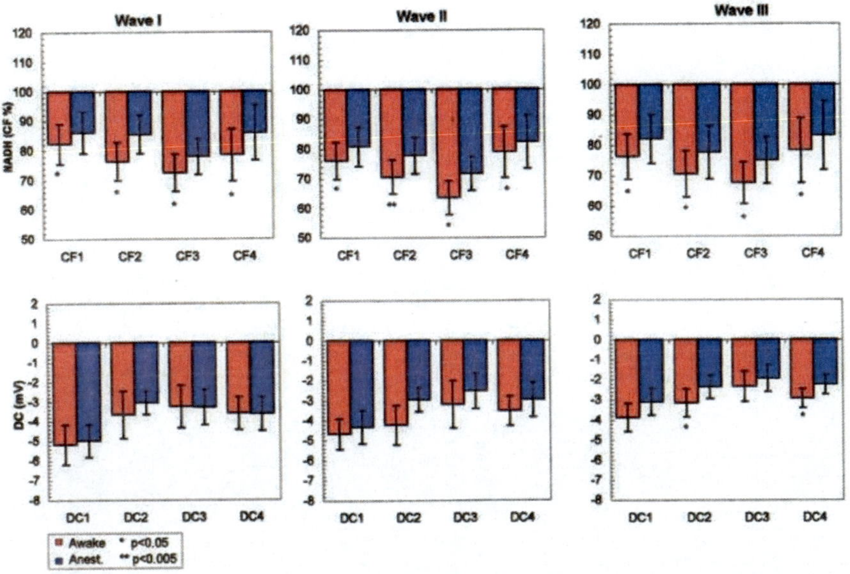

Fig. 6.30 The effect of three consecutive CSD waves (wave I, wave II, and wave III) on the minimum amplitudes of mitochondrial NADH redox state (NADH) and DC recorded by the MSA system in awake and Equithesin-anesthetized rats. CF1-CF4, NADH redox state; DC1-DC4, DC potential; Anest, anesthetized. Data are presented as the mean ± SE (Sonn & Mayevsky, 2006)

each of the three consecutive CSD waves. Anesthesia significantly prolonged the recovery kinetics of the first of three CSD waves as measured by reflectance, NADH fluorescence, and DC steady potential from each of the four channels (Table 6.5). Further, the results were similar for the two subsequent waves (data not shown).

Discussion The combined usage of the MSA and MPA systems allowed us to study the CSD propagation properties as well as the metabolic, hemodynamic, ionic, and electrical characteristics of the brain tissue during CSD wave propagation. The results of this study clarified the responses of the ionic, metabolic, hemodynamic, and electrical changes during CSD wave propagation and their influence on the propagation properties of the CSD wave in the awake state and after anesthesia.

The Effect of Anesthesia on Brain Tissue

Equithesin injection, as a typical anesthetic agent, caused a significant decrease in the basal levels of CBF and K^+_e, but no significant change in mitochondrial NADH levels. The decrease in CBF may indicate a decline in brain function (Drummond, 2018; Drummond & Shapiro, 1990; Siesjo, 1978), which reduces oxygen consumption and can precipitate a decrease in tissue blood flow (Drummond & Shapiro, 1990; Kuroda et al., 1996; Nilsson & Siesjo, 1975; Sakabe & Nakakimura, 1994; Siesjo, 1978). The concurrent decline in oxygen supply (CBF) and consumption due to the anesthesia may maintain the energy balance and thus explain the

Table 6.2 CSD Wave Frequency (wave/minute) during the Awake State and after Equithesin Anesthesia (Anest.) Calculated from the Data Recorded by the MPA and MSA Systems (Sonn & Mayevsky, 2006)

Frequency (FRQ wave/minute) Mean ± SE										
MPA system (n = 7)			MSA system (n = 8)							
			Channel 1		Channel 2		Channel 3		Channel 4	
Parameter	Awake	Anesthetized	Awake	Anesthetized	Awake	Anesthetized	Awake	Anesthetized	Awake	Anesthetized
Reflectance	0.40 ± 0.07	0.27 ± 0.07*	0.38 ± 0.04	0.22 ± 0.04@	0.37 ± 0.04	0.23 ± 0.04#	0.37 ± 0.04	0.22 ± 0.03@	0.35 ± 0.03	0.22 ± 0.03#
NADH	0.42 ± 0.07	0.28 ± 0.07*	0.37 ± 0.04	0.23 ± 0.04#	0.36 ± 0.04	0.23 ± 0.04*	0.35 ± 0.04	0.23 ± 0.04*	0.38 ± 0.04	0.20 ± 0.03*
DC	0.43 ± 0.08	0.24 ± 0.06*	0.37 ± 0.05	0.24 ± 0.04*	0.35 ± 0.05	0.23 ± 0.04#	0.34 ± 0.05	0.21 ± 0.04*	0.35 ± 0.05	0.20 ± 0.03*
CBF	0.40 ± 0.07	0.25 ± 0.05*								
CBV	0.43 ± 0.08	0.25 ± 0.06*								
K^+_e	0.41 ± 0.07	0.22 ± 0.05*								

*$p < 0.05$; # $p < 0.01$ @ $p < 0.005$ or $p < 0.001$

Table 6.3 CSD Mean Propagation Rate (mm/minute) of the Four Consecutive Channels (1–4) During the Awake State and after Equithesin Anesthesia Calculated from the Data Recorded by the MSA System

Mean Propagation Rate, ($VTPM_{min}$ mm/minute) Mean ± SE						
	Wave I		Wave II		Wave III	
Ch.	Awake	Anest.	Awake	Anest.	Awake	Anest.
R (1–4)	3.72 ± 0.23	3.3 ± 0.2**	3.16 ± 0.20	2.79 ± 0.25	3.28 ± 0.23	3.26 ± 0.33
CF (1–4)	4.10 ± 0.34	3.86 ± 0.56	3.32 ± 0.26	2.89 ± 0.3	3.19 ± 0.28	2.59 ± 0.37
DC (1–4)	3.52 ± 0.12	3.0 ± 0.22	2.7 ± 0.18	2.71 ± 0.26	3.23 ± 0.12	3.0 ± 0.22

R, reflectance; CF, NADH redox state; DC, DC potential; Ch., channel. Number of rats, n = 10. **$p < 0.05$ versus awake rats (Sonn & Mayevsky, 2006)

Table 6.4 The initiation kinetics (% min or mV/minute) of the first of three CSD waves during the awake state and after Equithesin anesthesia (Anest.) calculated from the data recorded by the MSA system

Initiation kinetics (to maximum or minimum pick)—Wave I Mean ± SE						
	Reflectance (%/minute)		NADH (%/minute)		DC potential (%/minute)	
Ch.	Awake	Anest.	Awake	Anest.	Awake	Anest.
1	44.3 ± 4.0	30.9 ± 2.9	−33.7 ± 7.7	−30.3 ± 3.0	−8.4 ± 2.2	−10.4 ± 1.8
2	67.4 ± 20.7	43.2 ± 8.3	−36.9 ± 11.3	−32.7 ± 3.8	−5.4 ± 2.6	−5.5 ± 1.0
3	77.9 ± 17.2	50.1 ± 8.0*	−50.3 ± 10.2	−46.1 ± 4.0	−3.2 ± 1.8	−5.2 ± 1.2
4	61.8 ± 13.6	45.9 ± 9.9	−41.2 ± 11.3	−37.6 ± 7.3	−7.0 ± 22.1	−5.7 ± 1.2

Ch., channel; number of rats, n = 13, *$p < 0.05$ versus awake rats (Sonn & Mayevsky, 2006)

Table 6.5 The Recovery Kinetics (%/minute or mV/minute) of the First of Three CSD Waves During the Awake State and after Equithesin Anesthesia (Anest.) Calculated from the Data Recorded by the MSA System

Recovery kinetics—Wave I Mean ± SE						
	Reflectance (%/minute)		NADH (%/minute)		DC potential (%/minute)	
Ch	Awake	Anest.	Awake	Anest.	Awake	Anest.
1	18.2 ± 1.6	13.4 ± 1.4*	31.7 ± 3.0	25.2 ± 3.6*	11.5 ± 1.4	9.3 ± 1.3*
2	27.1 ± 3.7	17.1 ± 1.9*	36.6 ± 3.3	−27.5 ± 3.3*	8.4 ± 1.8	6.0 ± 1.2*
3	32.2 ± 3.6	21.3 ± 2.7*	50.1 ± 5.3	39.2 ± 2.0*	7.3 ± 1.6	4.7 ± 0.9*
4	30.7 ± 0.03	15.2 ± 1.6*	46.2 ± 6.7	35.9 ± 4.9*	10.1 ± 2.4	7.1 ± 1.2*

CH., channel; number of rats, n=13. *$p < 0.05$ versus awake rats (Sonn & Mayevsky, 2006)

unchanged mitochondrial NADH levels. The anesthetic agents can inhibit oxidative phosphorylation (Siesjo, 1978) which may cause a direct decrease in energy production. Furthermore, these authors suggest that inhibition of oxidative metabolism and the depression of functional activity appear to be at least partly independent. On the other hand, Siesjo (Siesjo, 1978) demonstrated that there was an increase in phosphocreatine and ATP after anesthesia, seemingly indicating a "high-energy state." Therefore, the reduced neuronal firing and the relatively high ATP levels under anesthesia (Sha, 1970; Siesjo, 1978) may explain the decrease in K^+_e levels after Equithesin injection. Similar to barbiturates, Equithesin causes a decline in oxygen consumption, although it is not known whether it is due to the chloral hydrate, pentobarbital, or other components of the Equithesin.

The Effect of Anesthesia on CSD Waves
In the present study, Equithesin anesthesia did not block but disturbed CSD elicitation. This was reflected by the need for higher concentrations of the KCl solution to elicit a single or repetitive CSD cycles in the anesthetized phase. Previous studies showed that anesthetic agents such as ketamine and ether block CSD (Amemori & Bures, 1990; Gorelova et al., 1987b; Van Harreveld & Stamm, 1953). The reason for the discrepancy may be due to differences between the experimental procedures in the studies. For example, in the present study, we first induced the anesthetic state by injecting the anesthetic agent and then elicited the CSD wave. On the other hand, Gorelova (1987b) first elicited CSD waves and then injected (i.p.) ketamine and observed a block of the propagation of CSD waves.

Amplitude Changes
There were similar amplitude values for reflectance and NADH fluorescence using either the MPA or MSA systems in both awake and anesthetized rats. Although the electrodes in both systems were identical, the value of the DC potential amplitudes measured by the MSA system was lower than that measured by the MPA system. The values measured by the MPA system may be higher than those measured by the MSA system, because the dura mater was removed during implantation of the MPA system to enable the monitoring of K^+_e. It seems that the removal of the dura mater allowed more sensitive measurements. Anesthesia significantly decreased the amplitudes of the oxidation cycles in mitochondrial NADH during CSD propagation in the MSA and MPA experimental groups (Figs. 6.28, 6.29, and 6.30) and during all three CSD waves. This reduction indicates a decline in energy production in the anesthetized brain during CSD. These results can be explained by the significant reduction in K^+_e during the first CSD after anesthesia (Figs. 6.28 and 6.29) that supports the idea of a reduction in energy production in the anesthetized brain, showing that less energy is needed for the Na^+-K^+-ATPase activity to reach the membrane steady potential. However, during the two subsequent CSD waves, extracellular K^+ also decreased, but not significantly. The finding that anesthesia inhibits oxidative phosphorylation and that this inhibition is only partially related to changes in functional activity and oxygen consumption (Sha, 1970; Siesjo, 1978) supports this alternative explanation (this still needs further study). The significant decrease in NADH oxidation cycles after anesthesia does not correlate with the changes in CBF, which remained stable. It seems that after anesthesia, there is no coupling between the mitochondrial NADH redox state and the CBF or oxygen supply during CSD (Figs. 6.28 and 6.29). Equithesin anesthesia did not affect the DC potential amplitudes. Similarly, Van Harreveld and Stamm (1953) showed that ether, but not pentobarbital, decreased the DC potential amplitude. Thus, the pentobarbital, chloral hydrate, and the other components of the Equithesin anesthesia may account for the lack of effect on DC potential amplitudes in the present study.

Wave Frequency

The wave frequency of all of the parameters that were measured by the MSA and MPA systems was significantly decreased after anesthesia (Table 6.2). The results monitored by the MSA system show that (i) anesthesia causes a decrease in CSD wave frequency (as compared with the awake state) in four consecutive areas along the whole hemisphere, indicating that the whole brain cortex tissue along the hemisphere is homogeneous and behaves similarly with respect to CSD. (ii) This similarity was pronounced in all of the parameters measured by the MSA and MPA systems, demonstrating that anesthesia induces a reduction in CSD wave frequency as measured by metabolic, hemodynamic, ionic, and electrical properties. (iii) These results emphasize the utility of using the MSA system for measuring four adjacent areas for studying brain cortex responses to CSD stimulation. (iv) The prolongation of the recovery wave kinetics (Table 6.5) during anesthesia contributes to the decrease in wave frequency. Anesthesia inhibits mitochondrial respiration and oxidative phosphorylation (Sha, 1970; Siesjo, 1978). As a result, it causes a reduction in energy production that may explain the delay in the recovery kinetics of reflected light (blood volume), mitochondrial NADH fluorescence, and DC steady potential during CSD propagation. It should be emphasized that anesthesia did not affect the initiation kinetics of these parameters during the CSD waves (Table 6.4). These results may indicate that the initiation of CSD waves during Equithesin anesthesia led to a typical "all or none" reaction of the nervous system.

Propagation Rate

Data collected from the MSA system showed that anesthesia by Equithesin did not significantly affect the CSD propagation rate. In addition, the propagation rate of metabolic, hemodynamic, and electrical parameters was similar during the wave propagation from the first area toward the second as well as toward the third and the fourth area measured along the hemisphere by the MSA system. Thus, these results indicate that throughout the entire hemisphere, the rates of propagation of CSD waves were similar during wakefulness and after Equithesin anesthesia. However, other studies demonstrated that i.p. urethane + chloralose or thionembutal administration decreased CSD propagation velocity (Guedes & Barreto, 1992) and ketamine partly or completely blocked CSD propagation (Gorelova et al., 1987a). The differences between the findings of those studies and the present study may be due to the different anesthetic agents that were used.

Conclusions

1. Anesthesia with Equithesin reduced the oxygen demand and supply and kept the oxygen balance stable, which may indicate the protective effect of Equithesin.
2. The propagation properties as well as the metabolic, hemodynamic, ionic, and electrical properties of the CSD wave were elucidated in awake and anesthetized rats using both the MSA and MPA systems.
3. Using the MSA system, we found that the cerebral cortex is homogeneous in terms of the propagation properties and the metabolic, hemodynamic, ionic, and electrical properties of the CSD wave.

4. Anesthesia with Equithesin decreased CSD wave frequency by prolonging the recovery kinetics of the CSD wave, which may indicate a reduction in energy production. In contrast, anesthesia with Equithesin did not block the CSD initiation properties of the brain cortex.
5. Anesthetized brain activation during CSD waves reduced energy production, although oxygen delivery was not limited.
6. The MSA and MPA systems may be used to measure CSD waves that are induced under a variety of oxygen balance conditions to clarify the extent of the tissue metabolic and hemodynamic reserves.

6.3 Effects of Hypothermia on the Responses to CSD

6.3.1 Introduction

The effects of hypothermia on the responses to CSD were tested in animal models since the late 1950s (Bures et al. 1957). The aims of thier study were as follows:

> It was therefore assumed that when studying the effects of temperature on the function of the cerebral cortex, it would be more suitable to use an electrophysiological phenomenon of a purely cortical character, such as Leão's CSD, which has been studied in our laboratory for several years.
>
> An analysis of the effect of temperature changes on the rate of spreading EEG depression might help to throw light on the nature of the processes on which it is based.

They found that with decreasing body temperature the slow potential change—SPC became more prolonged and its conduction slower. On the other hand, the amplitude of the slow negative wave remained unchanged, even when body temperature dropped below 20 °C, that is, at a time when the spontaneous EEG was nearly absent and the SPC remained the only manifestation of CSD.

Very few new studies were published on the connection between hypothermia and CSD repeating the design of Bures et al. (1957) experiments. Those studies dealt with other models of CSD or ischemic depolarization models (Erecinska et al., 2003; Lim et al., 1995; Schiefecker et al., 2018; Yenari et al., 2000).

6.3.2 Our Studies

The effects of hypothermia on the responses to CSD were tested in our 2 types of monitoring systems, namely the MSA and the MPA. We performed 10 experiments using the MSA and three studies using the MPA system. Figure 6.31 presents the ECoG responses to hypothermia induced in un-anesthetized rats. As can be seen, there was a decrease in the amplitude and frequency of the ECoG signal when body temperature decreased from 36 °C to 23 °C.

Figure 6.32 shows typical responses to CSD recorded when the animal was exposed to 3 levels of body temperatures. Part A shows typical responses to CSD in the normothermic animal (37.5 °C). Two typical CSD cycles were developed as seen in all three parameters monitored from 4 different cortical areas. The reflectance trace shows the typical biphasic response, the NADH showed the clear oxidation cycles, and the DC trace shows the negative shift during the CSD event. In part B, the body temperature dropped to 32 °C and the duration of the CSD cycle was clearly prolonged. The reflectance signal shows only the very long increase phase and the NADH fluorescence trace was flipped into an increase (reduction) instead of the oxidation cycle seen in part A. The DC steady potential shows an unclear response. In the third level of temperature tested (27 °C) seen in part C, all the responses were very unclear.

Figure 6.33 presents typical recording showing responses to CSD recorded using the MPA monitoring system at different body temperatures. Part A was measure at the control body temperature of 35 °C. The reflectance shows the typical "increase–decrease" cycle while the NADH signal shows a very small unclear change. The CBF and CBV signals show a clear increase response. The extracellular potassium shows the typical increase and the DC steady potential signal presents the classical negative wave. In the next initial hypothermia seen in part B (30 °C), the reflectance response was changed to the increase phase without the long decrease change that is typical in the normothermic body temperature. The CBF response was "flipped" into a decrease change. The extracellular potassium and the DC steady potential show the same response as in the control response but the duration was much longer. The ECoG was depressed by the low body temperature, but was further depressed by the CSD event. Very similar responses to CSD were recorded in part C (26 °C) and in part D (22 °C). The only significant difference in part C and D was the longer duration of the responses.

In summary, body hypothermia led to

1. A clear decrease in the amplitude and frequency of the ECoG.
2. The duration of the CSD response was significantly higher.
3. Mitochondrial NADH was flipped from an "oxidation cycle" to "reduction cycle."
4. The propagation rate of the responses to CSD under hypothermia was variable.
5. The CBF shows a clear decrease response in all 3 hypothermic stages.

6.4 Responses to CSD in the Aging Rat

Previous results from our laboratory indicate that the nature of the response to CSD is dependent on the ability of the brain to compensate for extra oxygen need by increasing blood flow to the activated site (Mayevsky & Weiss, 1991; Mayevsky

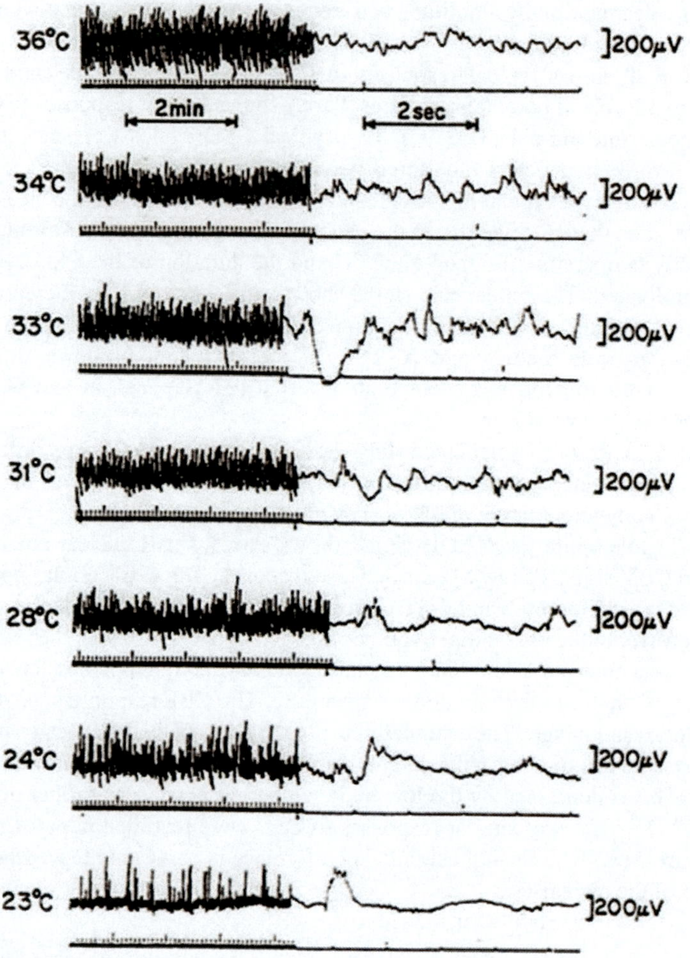

Fig. 6.31 The effect of hypothermia on the ECoG activity in the rat brain. The range of the temperature was 36 °C–23 °C

et al., 1982). It was also found that this ability of the brain is age-dependent, namely, adult brain showed "oxidation cycles" of the NADH in most conditions, while the young rat and the anesthetized gerbil brains exhibited "reduction cycle" of the NADH. On the other hand, other studies demonstrated that aging acts intracellularly to limit the brain's capacity to increase oxidative phosphorylation to meet heightened ATP damage (Benzi et al., 1992). Furthermore, it is also already known that the metabolic functions of the brain as well as its blood supply inevitably decline with age (Meyer et al., 1993).

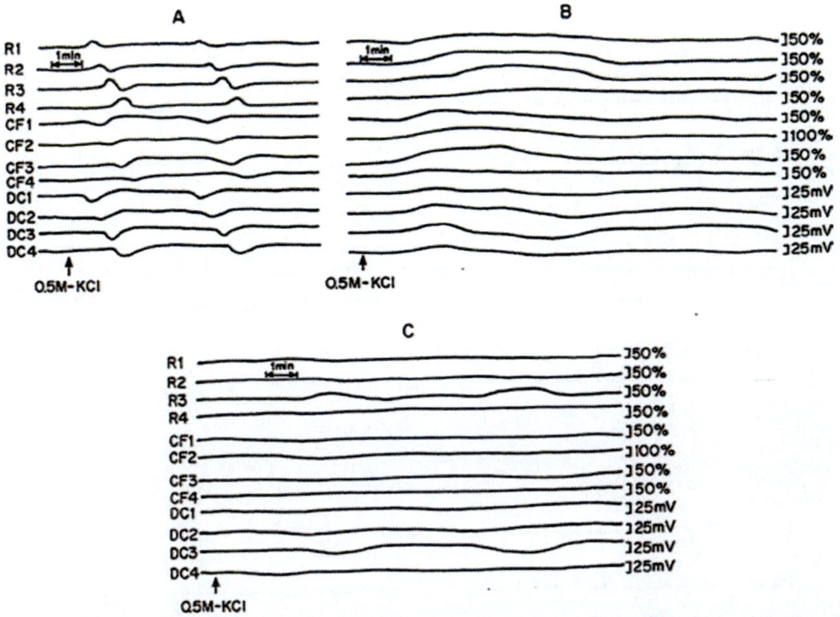

Fig. 6.32 The effect of hypothermia on the multisite responses to CSD induced in the rat brain. The temperatures used were (**A**) 37.5 °C, (**B**) 32 °C, and (**C**) 27 °C. The R1-R4 are traces of the reflectance signal, CF1-CF4 are the corrected fluorescence signal, and DC1-DC4 are the steady current signals. CSD was elicited by 0.5 M KCl

The aim of this study was to evaluate the function during CSD of the aging brain using the multiparametric monitoring system (Mayevsky, 1992a; Mayevsky et al., 1995).

We used the multiparametric assembly (MPA) for the continuous simultaneous online measurement of hemodynamic, metabolic, ionic, and electrical activities of the rat cerebral cortex (Mayevsky et al., 1995). The measurement included NADH redox state (using surface fluorometry), CBF (laser Doppler flowmetry), extracellular K^+, Ca^{2+}, and H^+ concentrations, and DC steady potential as well as ECoG (Fig. 2.12).

CSD was induced in 7 adults and 10 old rats. A typical response to CSD is demonstrated in Fig. 6.34 (Zarchin et al., 1999). This figure is a polygraph tracing taken from an adult animal after induction of CSD by KCl solution. The typical CSD is characterized by a transient increase in CBF followed by a decrease of NADH level (oxidation), electrical changes namely ECoG suppression and DC-negative shift and ion leak from the cells to the extracellular space (elevation in extracellular K^+ concentration, a decrease in extracellular Ca^{2+} concentration, and bi-phasic change in pH).

Figures 6.35, 6.36, and 6.37 are polygraph tracings of old rats' cortex demonstrating the various responses to CSD induction (Zarchin et al., 1999). The general response to CSD induction demonstrated in Fig. 6.35 is similar to the responses

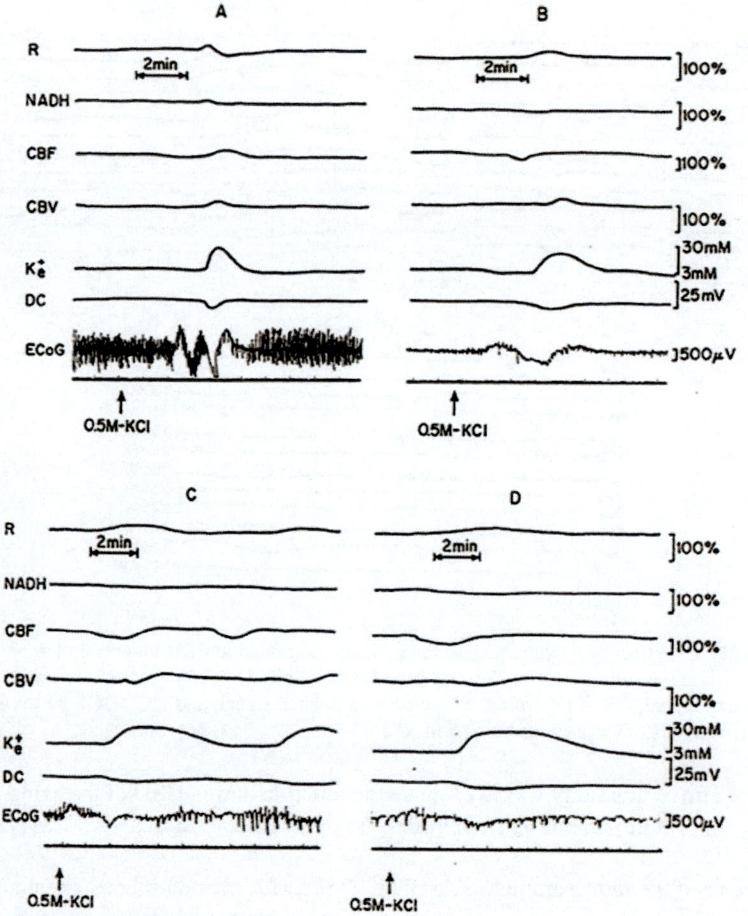

Fig. 6.33 The effects of hypothermia (**A**—35 °C, **B**—30 °C, **C**—26 °C, and **D**—22 °C) on the various physiological monitored parameters by the multiparametric monitoring system. R—reflectance, NADH—mitochondrial redox state, CBF and CBV—cerebral blood flow and volume, K^+_e—extracellular potassium level, DC—DC steady potential, ECoG—electrocortical activity

obtained in adult animal cortex (as was demonstrated previously in Fig. 6.34). Although CBF and NADH responses are smaller than those monitored from the adult rat's cortex and the CSD wave is longer, this CSD response can be categorized as typical normoxic response.

In one old animal, we observed "flipped CSD" namely, as a result of flushing the cortex with KCl the CBF increased but immediately decreased and NADH increased. These changes were followed by a negative shift of the DC steady potential and by changes in the extracellular ions (elevation in K^+ level, reduction in Ca^{2+} level, and biphasic change in pH) (Fig. 6.36). Unfortunately, the ECoG was not monitored in this experiment due to technical problems.

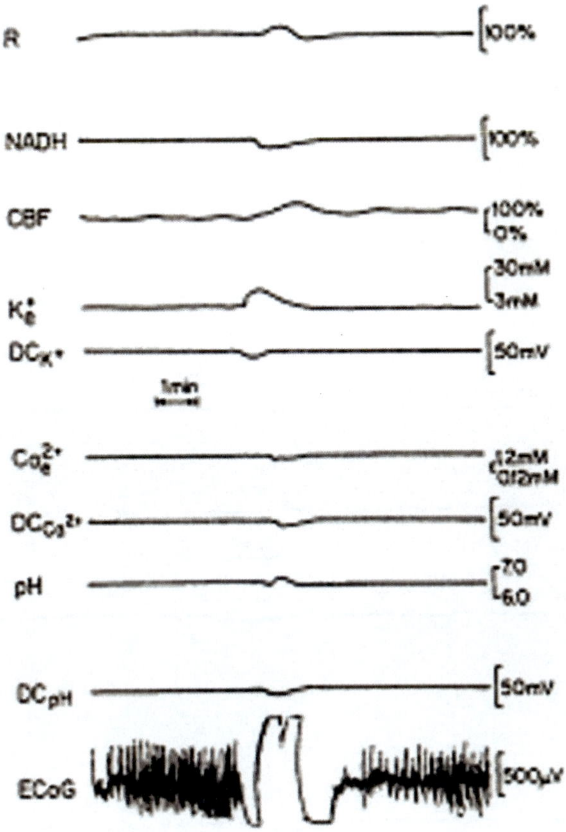

Fig. 6.34 Polygraph tracing of the responses of the adult rat to cortical stimulation by KCl. R—reflectance; NADH—intramitochondrial NADH levels; CBF—cerebral blood flow; K^+e, $Ca^{+2}e$—extracellular potassium and calcium ions; pH—extracellular pH; DC_{K+}, DC_{Ca}^{+2} ; DCpH—direct current concentric the potassium ion, calcium ion, and pH electrodes, respectively; ECoG—electrocorticogram (Zarchin et al., 1999)

Another response of the old rat's cortex to KCl flush is demonstrated in Fig. 6.37 (Zarchin et al., 1999). In this animal, the CBF increased but no metabolic change was observed as was recorded from the NADH signal. All other parameters responded normally, namely, ions leak, DC-negative shift, and ECoG suppression (extracellular Ca^{2+} concentration was not monitored in this experiment due to technical difficulties).

In this study, we observed various responses of the aged rat cortex during CSD. Typical CSD is characterized by a large efflux of potassium to the extracellular space leading to elevation in Na^+/K^+ ATPase pump activity. In order to supply the elevation in ATP demand, the mitochondria are more activated and NADH oxidizes. As a result of these events, CBF increases and this facilitates oxygen supply.

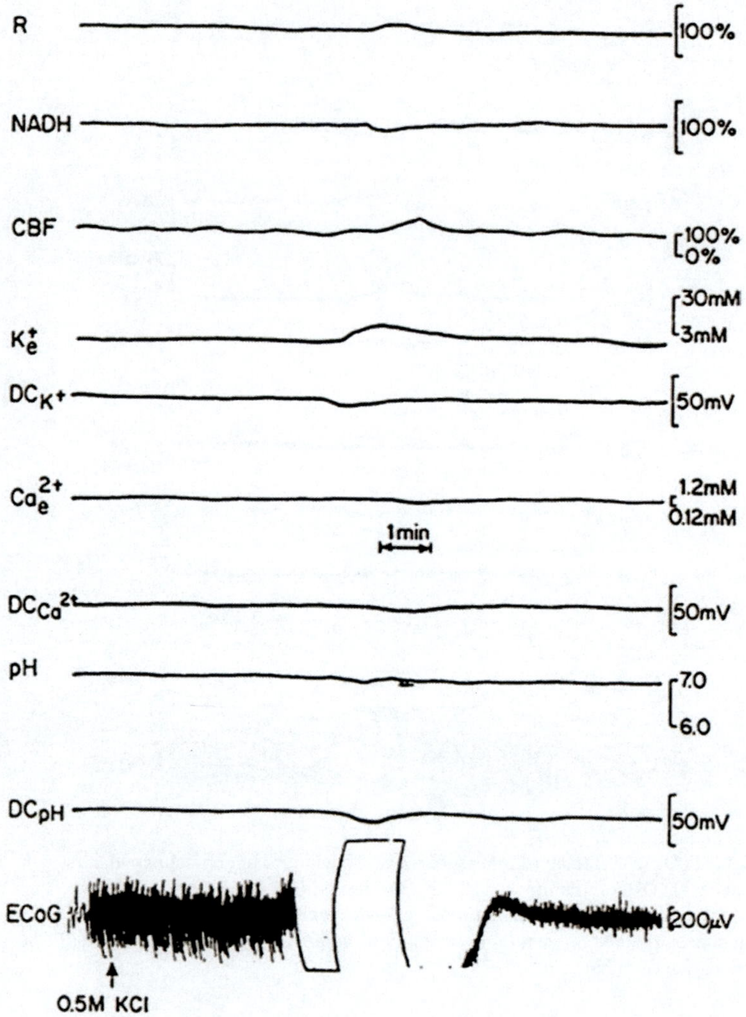

Fig. 6.35 Polygraph tracing of an old rat cortex demonstrating typical normoxic response to CSD induced by KCl application (Zarchin et al., 1999)

Negative shift of DC steady potential and a depression in ECoG occur (Mayevsky, 1992a).

One out of 10 old rats exhibited such a typical response, suggesting that this animal's brain cortex was not influenced by the age in terms of tissue adaptation to an increase of energy demand.

However, the remaining 9 animals exhibited different levels of brain tissue dysfunction. The first non-typical response to KCl flush was "flipped CSD" namely, reduction of NADH and a decrease in CBF. This kind of response is well known as

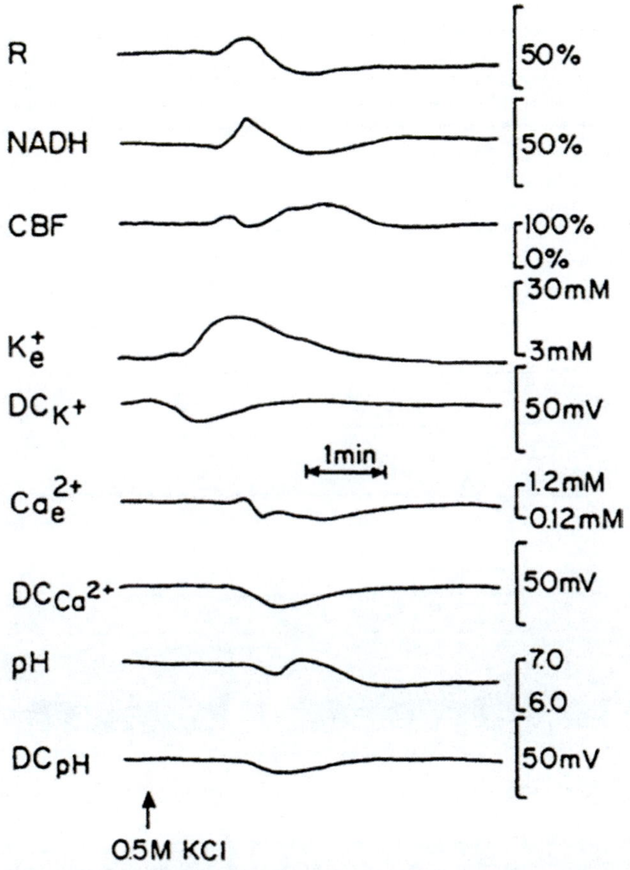

Fig. 6.36 Polygraph tracing of an old rat cortex demonstrating "flipped" response to CSD stimulation by KCl (Zarchin et al., 1999)

typical response of CSD induced under limitation of oxygen delivery (such as ischemia) in adult rats (Mayevsky, 1993; Mayevsky et al., 1996; Mayevsky et al., 1982). All the remaining animals exhibited different levels of brain tissue dysfunction under brain stimulation by CSD. Only the monitoring of the ECoG was the same. These results are similar to a large number of aging studies describing a decline in mitochondrial number and function (Ames et al., 1995; Fattoretti et al., 1996; Takai et al., 1995). A possible explanation for reduction in mitochondrial activity is an appreciable loss of cytochrome oxidase with age, such as reported in heart, skeletal, and liver tissue (Buerk & Nair, 1993; Wallace, 1992).

Furthermore, documentation of morphological changes in the cerebral vasculature of the aging rats included thinning of the endothelium, increase in collagen, and impaired reactivity to vasodilators stimuli which is thought to result from a decline in endothelial function (Davis et al., 1995). In addition, NO is thought to have a vasodilatation function during CSD (Meilin et al., 1995). However, later studies

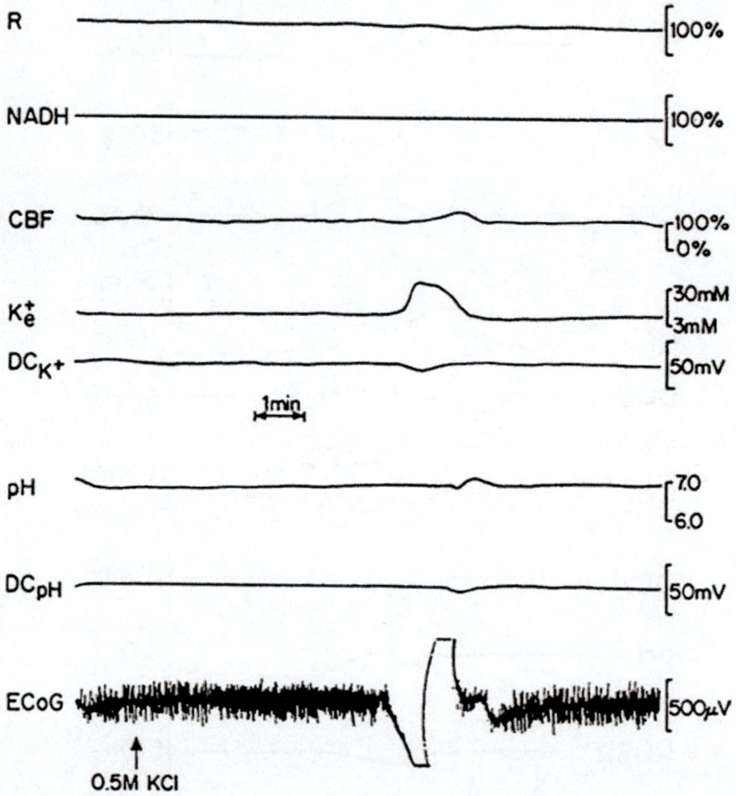

Fig. 6.37 Polygraph tracing of an old rat cortex, demonstrating cortical response to KCl stimulation. No changes were recorded in the NADH due to a problem with the device (Zarchin et al., 1999)

measuring NO synthesis found an elevation in NO synthesis probably to compensate for the reduced sensitivity of the NO system in the aging process (Dorheim et al., 1994; Inada et al., 1996).

In 5 out of 10 old animals, no ion (potassium, calcium, and hydrogen) flux was observed. Roberts et al. found age-related alteration K^+ homeostasis in anoxic rats (Roberts et al., 1997). Early studies carried out by us demonstrated different kinds of age-dependent response to CSD in young animals (Mayevsky et al., 1982). This study underlines the findings suggesting that CSD is an age-dependent phenomenon.

In conclusion, our findings show that in old rats' brain the response to CSD is in one of three conditions: 1. Normal brain-exhibiting typical response for CSD stimulation. 2. Pathological condition type a—"flipped CSD" as occurred in ischemic adult brain. 3. Pathological condition type b—metabolic, hemodynamic, and ionic stability—suggesting dysfunction of the mitochondria with inability of blood vessels responding to the increasing oxygen demand.

References

Amemori, T., & Bures, J. (1990). Ketamine blockade ofCSD: Rapid development of tolerance. *Brain Research, 519*, 351–354.

Ames, B. N., Shigenaga, M. K., & Hagen, T. M. (1995). Mitochondrial decay in aging. *Biochimica et Biophysica Acta (BBA) - Molecular Basis of Disease, 1271*, 165.

Benzi, G., Pastoris, O., Marzatico, F., Villa, R. F., Dagani, F., & Curti, D. (1992). The mitochondrial electron transfer alteration as a factor involved in the brain aging. *Neurobiology of Aging, 13*, 361–368.

Buerk, D. G., & Nair, P. (1993). $PtiO_2$ and $CMRO_2$ changes in cortex and hippocampus of aging gerbil brain. *Journal of Applied Physiology, 74*, 1723–1728.

Bures, J. (1957). The effect of anoxia and asphyxia on spreading EEG depression. *Physiologia Bohemoslovenica, 6*, 447–453.

Bures, J., Buresova, O., & Zacharova, D. (1957). The effect of changes in body temperature on spreading EEG depression. *Physiologia Bohemoslovenica, 6*, 454–461.

Davis, M., Mendelow, A. D., Perry, R. H., Chambers, I. R., & James, O. F. (1995). Experimental stroke and neuroprotection in the aging rat brain. *Stroke, 26*, 1072–1078.

Dorheim, M. A., Tracey, W. R., Pollock, J. S., & Grammas, P. (1994). Nitric oxide synthase activity is elevated in brain microvessels in Alzheimer's disease. *Biochemical and Biophysical Research Communications, 205*, 659–665.

Drummond, J. C. (2018). Baseline cerebral metabolic rate is a critical determinant of the cerebral vasodilating potency of volatile anesthetic agents. *Anesthesiology, 129*, 187–189.

Drummond, J. C., & Shapiro, H. M. (1990). Cerebral physiology. In R. D. Miller (Ed.), *Anesthesia* (pp. 621–658). Churchill Livingstone.

Eklof, B., & Siesjo, B. K. (1972). The effect of bilateral carotid artery ligation upon the blood flow and the energy state of the rat brain. *Acta Physiologica Scandinavica, 86*, 155–165.

Erecinska, M., Thoresen, M., & Silver, I. A. (2003). Effects of hypothermia on energy metabolism in mammalian central nervous system. *Journal of Cerebral Blood Flow and Metabolism, 23*, 513–530.

Fattoretti, P., Bertoni-Freddari, C., Caselli, U., Paoloni, R., & Meier-Ruge, W. (1996). Morphologic changes in cerebellar mitochondria during aging. *Analytical and Quantitative Cytology and Histology, 18*, 205–208.

Gorelova, N. A., Koroleva, V. I., Amemori, T., Pavlik, V., & Bures, J. (1987a). Ketamine blockade ofCSD in rats. *Electroencephalography and Clinical Neurophysiology, 66*, 440–447.

Gorelova, N. A., Koroleva, V. I., Amemori, T., Pavlik, V., & Bures, J. (1987b). Ketamine blockade of corticalCSD in rats. *Electroencephalography and Clinical Neurophysiology, 66*, 440–447.

Guedes, R. C., & Barreto, J. M. (1992). Effect of anesthesia on the propagation of corticalCSD in rats. *Brazilian Journal of Medical and Biological Research, 25*, 393–397.

Inada, K., Yokoi, I., Kabuto, H., Habu, H., Mori, A., & Ogawa, N. (1996). Age-related increase in nitric oxide synthase activity in senescence accelerated mouse brain and the effect of long-term administration of superoxide radical scavenger. *Mechanisms of Ageing and Development, 89*, 95–102.

Janebova, M. (1971). Excitation of Leão's corticalCSD by threshold amounts of electrophoretically applied K^+ ions. *Physiologia Bohemoslovaca, 20*, 447–451.

Kuroda, Y., Murakami, M., Tsuruta, J., Murakawa, T., & Sakabe, T. (1996). Preservation of the ration of cerebral blood flow/metabolic rate for oxygen during prolonged anesthesia with isoflurane, sevoflurane, and halothane in humans. *Anesthesiology, 84*, 555–561.

Leão, A. A. P. (1944). Pial circulation andCSD of activity in the cerebral cortex. *Journal of Neurophysiology, 7*, 391–396.

Leão, A. A. P. (1947). Further observations on theCSD of activity in the cerebral cortex. *Journal of Neurophysiology, 10*, 409–414.

Lim, D. J., Park, J. Y., Park, Y. K., Chung, H. S., Chu, J. H., Lee, K. C., & Lee, H. K. (1995). The changes of corticalCSD during hypothermia and rewarming in rat. *Journal of Korean Neurosurgical Association, 24*, 733–742.

Mayevsky, A. (1976). Brain energy metabolism of the conscious rat exposed to various physiological and pathological situations. *Brain Research, 113*, 327–338.

Mayevsky, A. (1978). Ischemia in the brain: The effects of carotid artery ligation and decapitation on the energy state of the awake and anesthetized rat. *Brain Research, 140*, 217–230.

Mayevsky, A. (1992a). Cerebral blood flow and brain mitochondrial redox state responses to various perturbations in gerbils. *Advances in Experimental Medicine and Biology, 317*, 707–716.

Mayevsky, A. (1992b). Interrelation between intracellular redox state and ion homeostasis in the brain *in vivo*. In M. Kessler (Ed.), *Frank, K* (pp. 155–168). Quantitative Spectroscopy in Tissues.

Mayevsky, A. (1993). Biochemical and physiological activities of the brain as *in vivo* markers of brain pathology. In E. F. Bernstein, A. D. Callow, A. N. Nicolaides, & E. G. Shifrin (Eds.), *Cerebral Revascularization* (pp. 51–69). Med-Orion Pub.

Mayevsky, A., & Chance, B. (1973). A new long-term method for the measurement of NADH fluorescence in intact rat brain with implanted cannula. *Advances in Experimental Medicine and Biology, 37A*, 239–244.

Mayevsky, A., & Chance, B. (1974). Repetitive patterns of metabolic changes during corticalCSD of the awake rat. *Brain Research, 65*, 529–533.

Mayevsky, A., & Chance, B. (1975). Metabolic responses of the awake cerebral cortex to anoxia hypoxia CSD and epileptiform activity. *Brain Research, 98*, 149–165.

Mayevsky, A., & Weiss, H. R. (1991). Cerebral blood flow and oxygen consumption in corticalCSD. *Journal of Cerebral Blood Flow and Metabolism, 11*, 829–836.

Mayevsky, A., Zeuthen, T., & Chance, B. (1974). Measurements of extracellular potassium, ECoG and pyridine nucleotide levels during corticalCSD in rats. *Brain Research, 76*, 347–349.

Mayevsky, A., Crowe, W., & Mela, L. (1980a). The interrelation between brain oxidative metabolism and extracellular potassium in the unanesthetized gerbil. *Neurological Research, 1*, 213–226.

Mayevsky, A., Lebourdais, S., & Chance, B. (1980b). The interrelation between brain PO_2 and NADH oxidation- reduction state in the gerbil. *Journal of Neuroscience Research, 5*, 173–182.

Mayevsky, A., Mizawa, I., & Sloviter, H. A. (1981). Surface fluorometry and electrical activity of the isolated rat brain perfused with artificial blood. *Neurological Research, 3*, 307–316.

Mayevsky, A., Zarchin, N., & Friedli, C. M. (1982). Factors affecting the oxygen balance in the awake cerebral cortex exposed to CSD. *Brain Research, 236*, 93–105.

Mayevsky, A., Zarchin, N., Yoles, E., & Tannenbaum, B. (1986). Oxygen supply to the brain in hypoxic and hyperoxic conditions. In C. Nicolau (Ed.), *Advances in the biosciences. "O_2-Transport in Red Blood Cells"* (pp. 119–132). Pergamon Press.

Mayevsky, A., Duckrow, R. B., Yoles, E., Zarchin, N., & Kanshansky, D. (1990). Brain mitochondrial redox state, tissue hemodynamic and extracellular ion responses to four-vessel occlusion andCSD in the rat. *Neurological Research, 12*, 243–248.

Mayevsky, A., Frank, K., Muck, M., Nioka, S., Kessler, M., & Chance, B. (1992). Multiparametric evaluation of brain functions in the Mongolian gerbil *in vivo*. *Journal of Basic and Clinical Physiology and Pharmacology, 3*, 323–342.

Mayevsky, A., Zarchin, N., & Sonn, J. (1993). Brain redox state and O_2 balance in experimentalCSD and ischemia. In A. Lehmenkuhler, K.-H. Grotemeyer, & F. Tegtmeier (Eds.), *Migraine – Basic Mechanisms and Treatment Urban & Schwarzenberg* (pp. 379–393). Munchen-Wier.

Mayevsky, A., Meilin, S., Rogatsky, G. G., Zarchin, N., & Sonn, J. (1995). Multiparametric monitoring of the awake brain exposed to carbon monoxide. *Journal of Applied Physiology, 78*, 1188–1196.

Mayevsky, A., Doron, A., Manor, T., Meilin, S., Zarchin, N., & Ouaknine, G. E. (1996). CorticalCSD recorded from the human brain using a multiparametric monitoring system. *Brain Research, 740*, 268–274.

Meilin, S., Zarchin, N., Mayevsky, A., & Shapira, S. (1995). Multiparametric responses to corticalCSD under nitric oxide synthesis inhibition. In B. A. Weissman, N. Alon, & S. Shapira (Eds.), *Biochemical pharmacological and clinical aspects of nitric oxide* (pp. 195–204). Plenum Press.

Meyer, J. S., Terayama, Y., & Takashima, S. (1993). Cerebral circulation in the elderly. *Cerebrovascular and Brain Metabolism Reviews, 5*, 122–146.

Moody, W. J., Futamachi, K. J., & Prince, D. A. (1974). Extracellular potassium activity during epileptogenesis. *Experimental Neurology, 42*, 248–263.

Nilsson, L., & Siesjo, B. K. (1975). The effect of phenobarbitone anaesthesia on blood flow and oxygen consumption in the rat brain. *Acta Anaesthesiologica Scandinavica. Supplementum, 57*, 18–24.

Roberts, E. L., Jr., Chih, C. P., & Rosenthal, M. (1997). Age-related changes in brain metabolism and vulnerability to anoxia. *Advances in Experimental Medicine and Biology, 411*, 83–89.

Rosenthal, M., & Somjen, G. (1973). CSD, sustained potential shifts, and metabolic activity of cerebral cortex of cats. *Journal of Neurophysiology, 36*, 739–749.

Sakabe, T., & Nakakimura, K. (1994). Effects of anesthetic agents and other drugs on cerebral blood flow, metabolism, and intracranial pressure. In J. E. Cottrell & D. S. Smith (Eds.), *Anesthesia and neurosurgery* (pp. 149–174). Mosby.

Schiefecker, A. J., Kofler, M., Gaasch, M., Beer, R., Unterberger, I., Pfausler, B., Broessner, G., Lackner, P., Rhomberg, P., Gizewski, E., Hackl, W. O., Mulino, M., Ortler, M., Thome, C., Schmutzhard, E., & Helbok, R. (2018). Brain temperature but not core temperature increases during spreading depolarizations in patients with spontaneous intracerebral hemorrhage. *Journal of Cerebral Blood Flow and Metabolism, 38*, 549–558.

Sha, S. K. (1970). Hypnotics and sedatives. In L. S. Goodman & A. Gilman (Eds.), *The pharmacological basis of therapeutics* (pp. 98–105). MacMillan Company, Collier MacMillan Limited.

Siesjo, B. K. (1978). Anaesthesia, analgesia and sedation. In B. K. Siesjo (Ed.), *Brain energy metabolism* (pp. 233–255). Wiley.

Sonn, J., & Mayevsky, A. (2000). Effects of brain oxygenation on metabolic, hemodynamic, ionic and electrical responses to CSD in the rat. *Brain Research, 882*, 212–216.

Sonn, J., & Mayevsky, A. (2001). The effect of ethanol on metabolic, hemodynamic and electrical responses to corticalCSD. *Brain Research, 908*, 174–186.

Sonn, J., & Mayevsky, A. (2006). Effects of anesthesia on the responses to corticalCSD in the rat brain *in vivo*. *Neurological Research, 28*, 206–219.

Takai, D., Inoue, K., Shisa, H., Kagawa, Y., & Hayashi, J. (1995). Age-associated changes of mitochondrial translation and respiratory function in mouse brain. *Biochemical and Biophysical Research Communications, 217*, 668–674.

Van Harreveld, A., & Stamm, J. S. (1953). Effect of pentobarbital and ether on the spreading cortical depression. *The American Journal of Physiology, 173*, 164–170.

Wallace, D. C. (1992). Mitochondrial genetics: A paradigm for aging and degenerative diseases? *Science, 256*, 628–632.

Yenari, M. A., Onley, D., Hedehus, M., deCrespigny, A., Sun, G. H., Moseley, M. E., & Steinberg, G. K. (2000). Diffusion- and perfusion-weighted magnetic resonance imaging of focal cerebral ischemia and corticalCSD under conditions of mild hypothermia. *Brain Research, 885*, 208–219.

Zarchin, N., Meilin, S., Rifkind, A. J., & Mayevsky, A. (1999). Hemodynamic, metabolic, ionic, and electrical responses to corticalCSD in aging rats. *Advances in Experimental Medicine and Biology, 471*, 223–230.

Chapter 7
Effects of Pharmacological Agents on CSD

Abstract The involvement of various drugs and pharmacological agents in the event named CSD is the subject of the current chapter. The advantage of using the multiparametric monitoring system as well as the multisite recoding of mitochondrial function and other brain activities during CSD is very clear. We are presenting the effects of three pharmacological agents, namely, carbon monoxide, nitric oxide, and alcohol on the CSD event.

Keywords Carbon monoxide (CO) · Nitric oxide (NO) · Alcohol effects · L-NAME inhibition · Multisite monitoring · Multiparametric monitoring

7.1 Effects of Carbon Monoxide

7.1.1 First Study

Understanding the mechanisms behind carbon monoxide (CO) intoxication and brain functions in vivo has a significant value in optimizing the therapeutic approach in patients (Thom & Keim, 1989). A multiparametric monitoring assembly (MPA) connected to the brain of unanesthetized rats was used to assess the toxic effects of CO on various brain functions (Mayevsky, 1992). Energy metabolism was evaluated by optical monitoring of relative CBF (using laser Doppler flowmeter) and intramitochondrial redox state (using NADH surface fluorometry). Ionic homeostasis was assessed by measurements of K^+, Ca^{2+}, and H^+ levels in the extracellular space (using surface minielectrodes). The electrical parameters were ECoG and DC steady potential. Exposing the awake rat to 1000 ppm CO (40 min) followed by 3000 ppm CO (20 min) led to an increase in CBF followed by the development of brain depolarization characterized by changes in ionic homeostasis similar to those recorded under CSD. Exposing the CO intoxicated brain to external stimulation (KCl solution) led to metabolic and hemodynamic responses typical to an ischemic

A. Mayevsky, J. Sonn, *Cortical Spreading Depression of Leao*,
https://doi.org/10.1007/978-3-031-08068-5_7

brain, namely, that due to limited O_2 delivery to the cells, the increase in energy demand led to a decrease in CBF and an increase in NADH redox state. Also in most animals tested, oscillations in CBF and NADH redox state were recorded after exposure to 3000 ppm CO as described in the ischemic brain (Mayevsky & Ziv, 1991). Therefore, we tried to improve brain oxygenation and interfere with the development of CO intoxication by exposing rats to 1–3 ATA 24 h before (Thom, 1990a) as well as immediately after CO exposure.

7.1.1.1 Introduction

The main treatment approach used following acute CO intoxication has been exposure to 100% O_2 at 1–2 ATA or even at increased pressure of O_2 (Penney, 1990; Thom & Keim, 1989). Antagonism of CO mediated lipid peroxidation in rat brain by hyperbaric O_2 was proposed by Thom (1990a). A large improvement in clinical outcome was achieved when HBO treatment was applied as close as possible to the CO exposure. One of the areas of CO intoxication research that suffers from the lack of adequate in vivo techniques is brain hemodynamics, metabolism, and ionic homeostasis in unanesthetized animals. Since CO intoxication occurs in awake humans, it is more appropriate to use awake animals in such studies. During the last decade our laboratory was involved in the development of new methods by which brain functions was evaluated under in vivo conditions and in real-time mode (Friedli et al., 1982; Mayevsky, 1983, 1990; Mayevsky & Zarchin, 1987).

In the present study we used two different in vivo monitoring systems in order to evaluate the effects of CO-poisoning. In the first approach we adopted the multiparametric system for monitoring various parameters in order to examine mechanisms of pathological development in the brain (Mayevsky & Zarchin, 1987). In the second approach we monitored NADH redox state bilaterally in awake rats concurrent with brain activation induced by CSD (Mayevsky, 1984; Mayevsky & Chance, 1982) before and after CO exposure. These measurements were taken inside a hyperbaric chamber so that HBO treatment after CO exposure was possible.

7.1.1.2 Methods

Multiparametric Monitoring System

We developed this approach several years ago (Friedli et al., 1982; Mayevsky, 1983, 1990; Mayevsky & Zarchin, 1987), and it has been used successfully in rats and gerbils. The multiprobe assembly (MPA) used in the present study had the following features:

(a) All parameters have a similar noninvasive surface contact with the brain.
(b) All signals monitored by the MPA show only very slight sensitivity to movements of the animal so that awake animals can be used.
(c) The laser Doppler flowmeter probe was combined with the NADH optical probe.

All details regarding the preparation of electrodes have been described previously (Friedli et al., 1982; Mayevsky, 1983, 1990; Mayevsky & Zarchin, 1987).

For the present studies, a modified assembly with the following differences was developed:

1. Ca^{2+} and H^+ electrodes were added to the original MPA (Crowe et al., 1977; Friedli et al., 1982).
2. The DC steady potential was measured by a highly stable Ag/AgCl electrode located around the K^+, H^+, and the Ca^{2+} electrodes.
3. The reference electrode was AG/AgCl electrode (via a salt bridge), located below the skin of the animal's neck.

CO Exposure Procedure

The rat was located in an air tight plexiglass cage. The CO at different concentrations in air (analyzed and certified by the manufacturer) was connected and flushed through the chamber.

7.1.1.3 Results and Discussion

Ten adult rats were included using multiparametric approach. The typical protocol includes a control period in which the resting levels of all parameters were monitored for 2–3 h. In order to test the effect of brain activation on the various parameters, a wave of CSD was induced by epidural application of KCl solution (Mayevsky & Chance, 1975). The awake rat (at least 3 h after the operation) was exposed to 1000 ppm CO in air for 40 min followed by 20 min of 3000 ppm CO (Thom, 1990b). The most significant effect of 1000 ppm CO was a large increase in CBF as measured by the laser Doppler flowmetry technique (Stern et al., 1977). During this period, the NADH redox state (CF), extracellular K^+ and Ca^{2+}, as well as the DC steady potential were not affected. A slight acidification of brain surface was recorded. This increase in CBF was not recorded in control awake rats monitored during the same time period.

A few minutes after changing to 3000 ppm CO, a significant increase in NADH redox state was recorded, and in the majority of studies a spontaneous CSD-like response occurred. We presume that the CSD-like response is due to disturbances in energy supply (increased CF signal) as a depolarization wave developed. The same pattern of responses as recorded when the rat was exposed to 3000 ppm CO (without the 40 min of 1000 ppm CO exposure). In this case the NADH did show an increase very soon after the exposure to CO although the CBF also increased at the same time. During the depolarization episodes which were very similar to CSD, extracellular K^+ level increased concomitant with a shift in the DC steady potential. The metabolic and hemodynamic responses to this depolarization event were different from the typical response to CSD (Crowe et al., 1977; Friedli et al., 1982) as shown in Fig. 7.1A (Mayevsky et al., 1993a). The response recorded during the

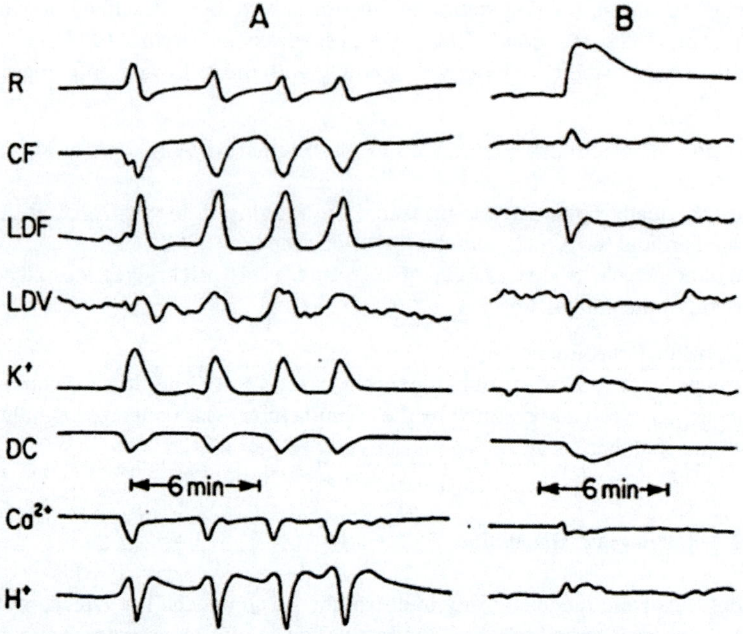

Fig. 7.1 Two responses to CSD recorded during control (**A**) and 23 min after 1000 ppm CO inhalation. (**B**) R, CF: reflectance and corrected fluorescence. LDF, LDV: cerebral blood flow and volume monitored by the laser Doppler flowmeter. K^+, Ca^{2+}, H^+: extracellular levels of the three Ions. DC: steady potential (Mayevsky et al., 1993a)

control period demonstrates the coupling between the maintenance of ionic homeostasis and the increase in CBF due to the activation of mitochondrial activity. The changes in extracellular K^+, Ca^{2+}, and H^+ (Hansen, 1985; Mayevsky et al., 1990b) led to an increase in CBF and the oxidation of NADH (decrease CF). The record shown in Fig. 7.1B (Mayevsky et al., 1993a) was measured in the same rat used for Fig. 7.1A and illustrates depolarization episode which occurred after 23 min of 1000 ppm CO.

In this animal CO intoxication led to a reversed change in NADH fluorescence during the depolarization event, namely, an increase-reduction of NADH (Mayevsky, 1978) instead of a decrease (oxidation). Similar observations were made in a number of other animals. The changes in the extracellular ions were much smaller than those occurring during control CSD although the DC steady potential changes during spontaneous events due to CO poisoning were similar as shown in Fig. 7.1A and B.

7.1.2 *Second Study*

7.1.2.1 Introduction

The goal of the 2nd investigation (Mayevsky et al., 1995) was to evaluate the responses of brain tissue to CO and to attempt to elucidate differences between the physiological responses and toxic effects. The affinity of CO to hemoglobin is 200 times greater than that of O_2; hence, an obvious aspect of exposure to CO is the possibility of impaired O_2 delivery and cellular hypoxia, depending on the concentration of CO. Because >50% of the energy consumed by the brain is used for the normal functions of active ion transport mechanisms, it is important to evaluate O_2 or energy balance under CO exposure. Oxygen balance can be evaluated by monitoring the level of O_2 supply compared with the O_2 demand. To monitor brain O_2 balance in vivo and in a continuous mode, we adopted a multiparametric monitoring approach (Friedli et al., 1982; Mayevsky, 1983; Mayevsky et al., 1990b) by which the hemodynamic, metabolic, ionic, and electrical activities could be examined. Our approach was to monitor the brain on the "tissue level" rather than the "cellular level." We believe that the various component elements of the brain, namely, neurons, glia cells, and capillaries, are acting as an integrated system. Therefore, we monitored the functions of the brain by using "miniprobes" rather than "microprobes" that are normally used to monitor ionic homeostasis (Hansen, 1985). Also, the evaluation of other parameters such as cerebral blood flow (CBF) and NADH redox state were adapted to the tissue level rather than to the cellular level. Hence, our strategy was to develop a multiparametric monitoring assembly in which all the probes had the same type of contact with a small area of the brain. None of the probes penetrated the brain itself, thus avoiding severe damage to the brain or the formation of an artificial environment around the sensor that can be created around a penetrating microelectrode.

7.1.2.2 Methods

All technical details of the MPA used in the CO exposure described in the following few figures were published by our group (Mayevsky et al., 1995).

7.1.2.2.1 Data Collection and Analysis

Relative CBF The LDF provided arbitrary units (blood flow and volume) that were calibrated in relative terms compared with the normoxic brain that served as the control (O-100% range).

NADH Redox State Measurements of NADH provided a method for evaluating the mitochondrial redox state. The normoxic NADH level is considered to be 100%, and the changes due to the treatment given were calculated in comparison with the normoxic level. Mayevsky (1984) has shown in the past that the NADH redox state correlates well with the brain's functional activity. The origin of the NADH signal in our system is mainly mitochondrial (for details see Ref. (Mayevsky, 1984)), whereas the cytoplasmic NADH contribution to the signal is negligible. Nevertheless, because the NADH is not measured in a purified form, we prefer to label the trace in the records as corrected fluorescence (CF).

Multiparameter Calculation The MPA provides analog signals of electrical activities and NADH redox states as well as extracellular concentrations of K^+, H^+, and Ca^{2+}. The mV values measured by the electrodes were transformed to mM values by using the standard calibration technique (Hansen, 1985) and the principles of the Nernst equation. All calculations were performed via special software programs.

Real-Time Data Acquisition Because of the large number of parameters to be measured and the need for logarithmic transformation of the data from the K^+, Ca^{2+}, H^+ and Na^+ electrodes, it was necessary to use a computerized system for data acquisition and storage. Furthermore, to record all the analog signals (at least 16 channels), it was more appropriate to use a computer display instead of the paper capacity for continuously collecting information.

7.1.2.3 Results

Figure 7.2 shows the effect of exposure of the rat to 3000 ppm of CO and not to a preceding interval of 1000 ppm of CO used in other rats (Mayevsky et al., 1995). The CBF was doubled and stayed at the high level during the entire exposure. Here again, the CBF showed a transient decrease during the depolarization event together with a short increase in NADH (CF). The lower NADH level after the wave suggests a degree of NADH oxidation. The ionic responses in this animal were similar to those shown in Fig. 7.3A. The fact that a clear decrease in the extracellular potassium level trace was recorded suggests an activation of the Na^+-K^+-ATPase due to the CSD or changes in K^+ permeability. Differences in the responses to spontaneously developed CSD in air or different CO concentrations can be appreciated in Fig. 7.3 (Mayevsky et al., 1995) where the patterns in one rat during different CO exposures are displayed. In Fig. 7.3A, the rat was in an air-breathing environment, and the changes in the ionic and metabolic responses were typical of CSD. The responses shown in Fig. 7.3B were recorded under 1000 ppm of CO while those in Fig. 7.3C were under 3000 ppm of CO. The main changes among these conditions were in the hemodynamic responses, namely, that under CO the large hyperemic response to CSD did not occur. Also, the duration of the wave was slower under CO, as seen by the various parameters. In another rat exposed to 1000 ppm followed by

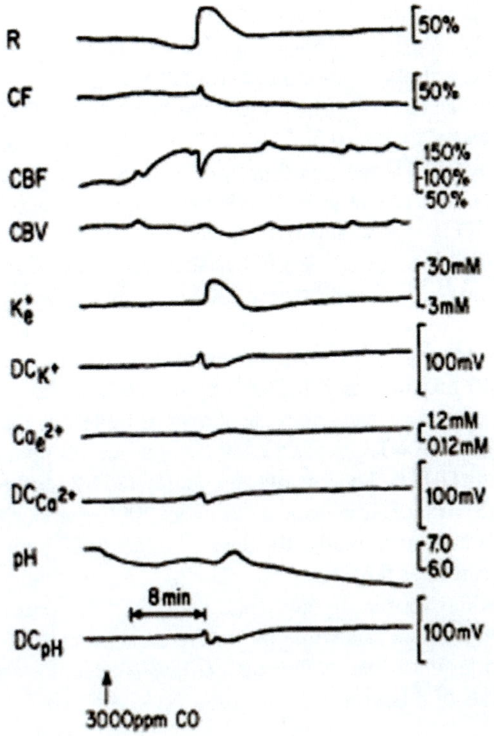

Fig. 7.2 Exposure of an awake rat to 3000 ppm of CO without pretreatment with 1000 ppm of CO. Spontaneous CSD-like wave was developed under CO (Mayevsky et al., 1995)

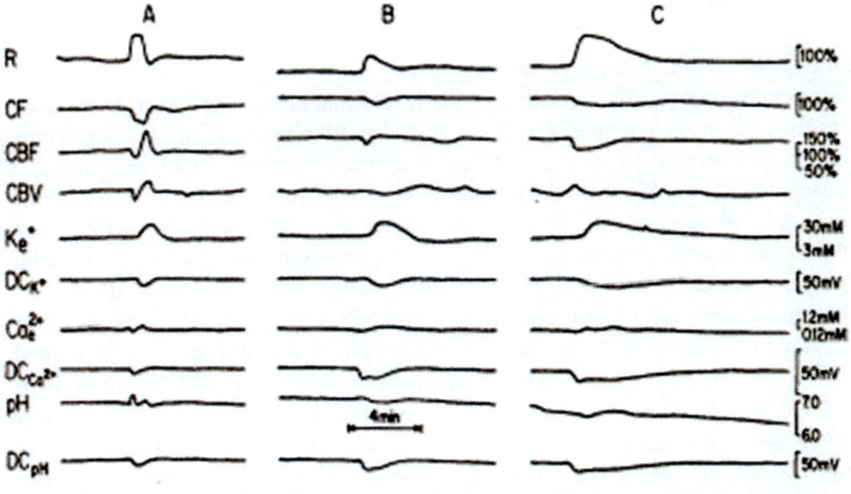

Fig. 7.3 Responses to spontaneously developed cortical CSD under air (**A**), 18 min of 1000 ppm of CO (**B**), and 2.5 min of 3000 ppm of CO (**C**). All scales in C are identical to A and B. Initiation of CSD was spontaneous and was similar to other rats in which CSD was induced by epidural washing of KCl solution 2–5 mm anterior to MPA shown in Fig. 2.12A (Chap. 2) (Mayevsky et al., 1995)

3000 ppm of CO, spontaneous responses to CSD waves were recorded. As seen in Fig. 7.4, under 1000 ppm of CO the responses to CSD (Mayevsky et al., 1995) were very similar to normal control responses (Fig. 7.3A). In this animal, the extracellular calcium did show clear responses to the CSD. The appearance of the two CSD waves under 1000 ppm of CO did not affect the general trend of the clear increase in CBF. Under 3000 ppm of CO, the CBF response was reversed, namely, that a large decrease was recorded instead of the large increase found under 1000 ppm of CO. The R was also different under 3000 ppm of CO, and the CFs were lower. The duration of the wave under 3000 ppm of CO was 3–4 times longer as measured by the DC potential trace.

To show the quantitative effects of CO, we analyzed the results obtained from six rats exposed to the defined protocol. The rats were exposed to 1000 ppm of CO for 40 min and then to 3000 ppm of CO for 20 min. Parameters that demonstrated reproducible changes are shown in Table 7.1. CBF increased significantly in all rats under 1000 ppm of CO ($P < 0.05$) as well as under 3000 ppm of CO ($P < 0.005$). The NADH level was stable and unaffected by 1000 ppm of CO, whereas an $11 \pm 10.6\%$ (SE) increase was recorded under 3000 ppm of CO ($P < 0.05$ by t-test). Extracellular potassium was slightly elevated under 3000 ppm of CO, but this change was significant ($P < 0.05$ by t-test). The pH clearly changed under 3000 ppm of CO but not with exposure to only 1000 ppm of CO. Spontaneous CSD waves were recorded in four of six rats while they were exposed to 1000 ppm of CO. Spontaneous oscillations described in the paper (Mayevsky et al., 1995) were recorded in five of the six rats of this group.

7.1.2.4 Discussion

Spontaneous CSD-like responses were often recorded, suggesting that a process of depolarization was initiated either by an insufficient energy supply or by the accumulated K^+ in the extracellular space. An alternative mechanism may involve the effects of increased intracellular calcium that may occur due to an increased level of guanosine 3′,5′-cyclic monophosphate induced by either the high CO level or the increased biosynthesis of NO (Crutchfield et al., 1990). Depolarization waves developed in all rats exposed to 3000 ppm of CO and in four of six rats exposed to 1000 ppm of CO (Table 7.1). The mechanism of the initiation of the CSD or CSD-like responses may be the same under 1000 and 3000 ppm of CO. However, as shown in Figs. 7.3 and 7.4, the CSD responses in the absence of CO, or with 1000 ppm, were different from those found under 3000 ppm of CO. The response to CSD in the absence of CO and the usual response in the presence of 1000 ppm of CO was a large transient hyperemia, whereas with 3000 ppm of CO a short period of hypoperfusion occurred. This suggests that the ability of the brain to autoregulate CBF was impaired by CO when the microvasculature was nearly maximally vasodilated. In a few cases, the NADH response was also different, namely, an increase in NADH was recorded instead of a decrease (oxidation).

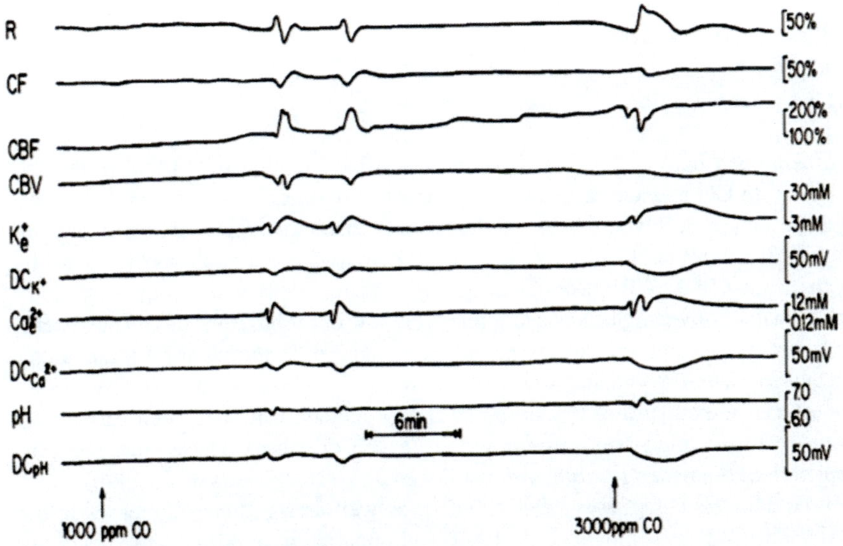

Fig. 7.4 Spontaneous CSD waves developed under 1000 and 3000 ppm of CO (Mayevsky et al., 1995)

Table 7.1 Effects of 1000 ppm CO followed by 3000 ppm on various parameters in 6 awake rats (Mayevsky et al., 1995)

	Rat No.						
	1	2	3	4	5	6	Mean ± SE
Flow %							
Control	100	100	100	100	100	100	133 ± 5.2
CO-1	132	123	127	150			
CO-2	150	123	148	150	172	120	144 ± 7.2
NADH %							
Control	100	100	100	100	100	100	
CO-1	100	100	100	100			
CO-2	110	110	110	113	113	112	111 ± 0.6
K^+, mM							
Control	3.4	2.7	3.1		4.6		3.45 ± 0.35
CO-1	3.4	2.5	3.2				
CO-2	3.6	4.6	6		5.4		4.9 ± 0.45
pH (acidic)							
CO-1	–	–	–	–			
CO-2	+	+	+	+	+	+	
Spontaneous CSD oscillations							
CO-1	+	+	–	–	+	+	
CO-2	+	–	+	+	+	+	

CO-1, end of 1000 ppm CO; CO-2, end of 3000 ppm CO; CSD, spreading depression. Plus, change in pH and recording of spontaneous CSD oscillations; minus, no change in pH and no recording of spontaneous CSD oscillations

7.1.3 *Third Study*

7.1.3.1 Introduction

As mentioned before, Thom and Keim (1989) have demonstrated that in the rat, exposure to CO may cause tissue hypoxia. Since brain energy stores are very limited (Erecinska & Silver, 1989) and the brain is absolutely dependent upon oxygen supply via cerebral blood flow (CBF), exposure to high concentrations of CO results in an increase in the CBF (Somjen et al., 1993; Sonn, 1996), which may be partially due to lack of oxygen. Studies using relatively low concentrations of CO have demonstrated that the autoregulatory increase in blood flow is sufficient for maintaining adequate cerebral oxygen supply (Rosenblum et al., 1987). Most of the studies on the effects of exposure to CO on brain energy metabolism have been carried out using relatively high concentrations of CO (2% or more). These studies report depletion of brain energy stores (Mayevsky et al., 2021; Wiebers et al., 1990), intracellular acidosis (Mayevsky et al., 2020a), as well as significant increases in the $NADH/NAD^+$ ratio (Somjen et al., 1993) and impairment of mitochondrial activity (Mayevsky, 2020; Mayevsky et al., 2020b). Little information is available on brain energy metabolism under low CO concentrations in general and on the effect of CO on the activated brain in particular. Inducing CSD is one way of causing brain activation and increasing oxygen demand (Mayevsky & Weiss, 1991). CSD is a phenomenon that has been described in the literature as a depression of spontaneous electrical activity (Leão, 1944b). It is accompanied by increased metabolic activity (Chance et al., 1964; Mayevsky & Chance, 1974; Mies & Paschen, 1984; Rosenthal & Somjen, 1973; Shinohara et al., 1979), changes in cerebral blood flow (Piper et al., 1991; Shibata et al., 1990; Wahl et al., 1987), and fluctuations in ionic homeostasis (Mayevsky & Meirovithz, 2020; Mayevsky et al., 1982, 1990a). We have reported the recording of CSD in a patient suffering from severe head injury (Mayevsky et al., 1996). The aim of this study was to examine the effect of 1000 ppm CO on the metabolic and electrical activity of the rat brain in vivo and its interrelation with brain activity following induced CSD waves.

7.1.3.2 Methods

This multisite assembly (MSA) contains four bundles of optical fibers for the simultaneous measurement of NADH from four consecutive points on the cerebral cortex and parallel to them four electrodes for measuring the DC potential near each bundle. The bundles were assembled in a rectangular Delrin cannula (6.8 × 3.8 mm). The principles for NADH surface fluorometry have been described previously (Mayevsky, 1984; Mayevsky & Chance, 1983). A schematic representation of the MSA located on the rat brain is presented in Fig. 7.5 (Meilin et al., 1998).

All experimental protocols were approved by the institutional Animal Care and Use Committee of the Bar-Ilan University. Male Sprague-Dawley rats weighing

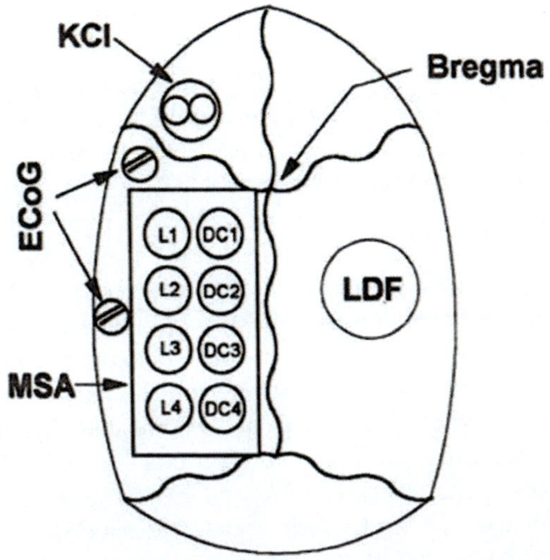

Fig. 7.5 Schematic representation of the location of the MSA on the cortex. *KCl* push-pull cannula for KCl, *DC* position of direct current steady potential probe, *L* light guide, *LDF* laser Doppler flowmeter fiber, *ECoG* electrocorticogram measurement (Meilin et al., 1998)

180–250 g were anesthetized with a 0.3 ml/100 g, IP injection of Equithesin (each ml contains 42.51 mg chloral hydrate, 9.72 mg pentobarbital, 21.25 mg magnesium sulfate, 44.43% w/v propylene glycol, 11.5% alcohol, and distilled water). The cranium was exposed, a rectangular 6.8 x 3.8-mm hole was drilled in the parietal bone for the MSA, and a small hole was drilled in the contralateral bone for laser Doppler measurement of cerebral blood flow (1.5 mm diameter probe, Perimed, Sweden, PF315). The MSA was placed on the brain using a micromanipulator such that no undue pressure was placed on the brain. A 2 mm hole was drilled in the frontal bone ipsilateral to the MSA for a push-pull cannula enabling flushing of the brain with 0.5 M KCl in order to initiate CSD. Holes were also drilled for the screws used for measuring ECoG and also serving to hold the system securely to the cranium (Fig. 7.5). All components were fixed to the brain using dental acrylic cement. The animals were allowed to recover from surgery for at least 90 min. They were then placed in a sealed cage for the various experimental protocols.

The animals were divided into six groups (three control + three experimental groups) of 6–7 animals each. The experiments were carried out in awake rats exposed to 1000 ppm CO for 40 min. In group A, CSD was induced by flushing the brain with KCl during exposure of the animals to CO, approximately 20 min after the beginning of exposure. In group B, CSD was induced immediately upon cessation of exposure to CO. In group C, CSD was induced 90 min after cessation of exposure, while the rat was already breathing air. Control animals underwent the same treatment of inducing CSD, but were not exposed to CO.

7.1.3.3 Results

Rats were exposed to 1000 ppm CO, and CSD was induced at different times during and after exposure. In all animals, blood flow began increasing already after the first 10 min of exposure to CO. This increase was sustained until the end of exposure to CO and reached an average maximal level of 144% (n = 21, 144 ± 6%, $P < 0.05$). Ten minutes after reventilating the cage with air, CBF remained elevated and was not significantly different from the maximal level (139 ± 5%). Two hours after reventilating the cage with air, almost complete recovery was observed and blood flow neared its initial value (108 ± 3%). Control rats that breathed air for the entire experimental period did not exhibit any change in blood flow, and at the end of this time period blood flow was very similar to control levels (n = 8, 100 ± 2%). The changes in blood flow in this study were not accompanied by changes in NADH levels. Figure 7.6 presents a tracing from a rat brain in which CSD waves were induced before exposure to CO (Fig. 7.6A) and 90 min after cessation of exposure (Fig. 7.6B) (Meilin et al., 1998). It can be seen that when CSD was induced 90 min after cessation of exposure to CO, the frequency of the CSD waves increased. Rats in which CSD was induced during exposure to CO did not exhibit any significant changes in the CSD wave frequency as calculated from the NADH in all four

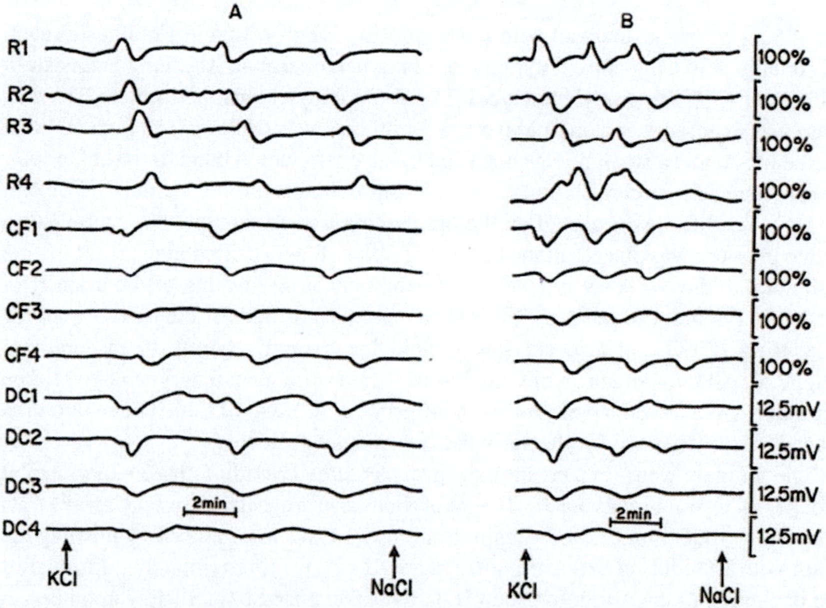

Fig. 7.6 Tracing of the responses to cortical CSD induced by flushing the brain with KCl in control animals (**A**), and 90 min after cessation of exposure to CO (**B**). R1, R2, R3, R4 reflectance; CF1, CF2, CF3, CF4, corrected NADH fluorescence; DC1, DC2, DC3, DC4, direct current potential measured in parallel to the NADH recorded from channels 1, 2, 3, 4, respectively (Meilin et al., 1998)

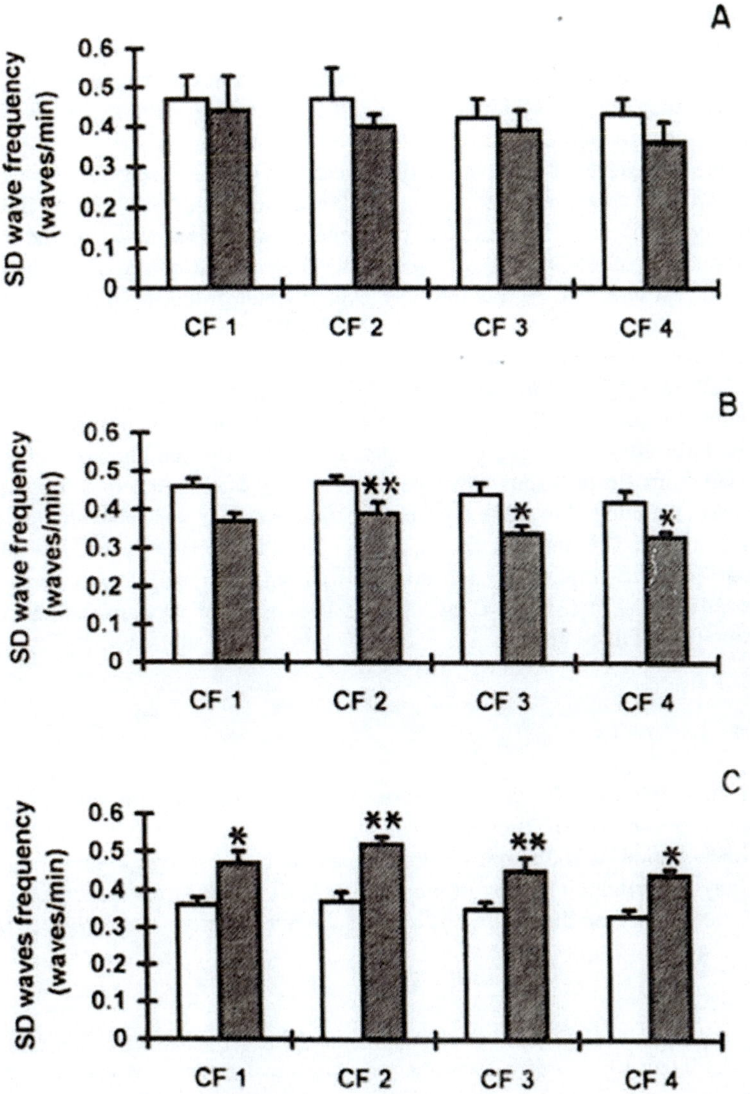

Fig. 7.7 Histogram presenting the frequency of the CSD waves (wave/min). CF1, CF2, CF3, CF4, corrected NADH fluorescence recorded from channels 1,2,3,4, respectively. (**A**): CSD was induced during exposure of the animals to CO. (**B**): CSD was induced immediately upon cessation of exposure to CO. (**C**): CSD was induced 90 min after cessation of the exposure. Light bars, control; dark bars, experimental animals. *$p < 0.05$; **$p < 0.01$ (Meilin et al., 1998)

channels (Fig. 7.7A) (Meilin et al., 1998). A significant decrease in the frequency of the NADH oxidation waves (CF waves) was observed in the rats of this group as compared to the waves induced before exposure to CO (mean of all four channels = 0.45 ± 0.02 waves/min vs. 0.36 ± 0.04; $P < 0.02$ (Fig. 7.7B). These changes did not appear in the control rats, in which CSD was induced at an equivalent time but were not exposed to CO. In contradistinction, rats in which CSD was induced 90 min after cessation of exposure to CO exhibited a significant increase in the CSD wave frequency mean of all four channels = 0.35 ± 0.02 waves/min vs. 0.47 ± 0.02, $P < 0.02$ (Fig. 7.7C) (Meilin et al., 1998).

Similar changes were observed in the reflectance waves (R waves). The average changes in the R waves recorded during the CSD episodes are presented in Table 7.2. The frequency of the DC waves in all experimental groups remained similar to that of the control groups. Group B, in which CSD was induced immediately upon cessation of exposure to CO, was the only experimental group in which significant changes were observed in the amplitude of the NADH waves. A significant decrease was recorded in the amplitude of the CF (NADH) waves of the second CSD wave (29 ± 2.3% average of four channels before exposure to CO vs. 18 ± 2.4% after exposure, $P < 0.05$, and of the third wave 28 ± 0.9% before exposure to CO vs. 20.7 ± 0.6% after exposure, $P < 0.05$) (Table 7.3). The duration of the waves increased from 1.6 to 3 min. (average of three first waves. Due to difficulties in calculating the integral of these NADH oxidation waves, it was calculated only for waves 2 and 3. This integral decreased significantly, apparently because of the decrease in wave amplitude, such that the total energy required was apparently decreased (Fig. 7.8) (Meilin et al., 1998). The speed of propagation remained unchanged in all three groups.

7.1.3.4 Discussion

One of the prominent effects of exposure of rats to a low concentration of CO is a transient increase in CBF (Brown & Piantadosi, 1990). In this study exposure to CO at a concentration of 1000 ppm for 40 min caused an increase in CBF but did not cause any measurable changes in the basal NADH redox levels, i.e., the normal metabolic activity of the brain was apparently not affected, since NADH is an

Table 7.2 Frequency of CSD waves (waves/min) during (group A), immediately after exposure to 1000 ppm CO (group B), and 90 min after cessation of exposure (group C) (Meilin et al., 1998)

		Group A	Group B	Group C
R1	Control	0.47 ± 0.06	0.45 ± 0.02	0.37 ± 0.02
	After CO	0.44 ± 0.09	0.36 ± 0.02[b]	0.46 ± 0.03*
R2	Control	0.47 ± 0.08	0.44 ± 0.02	0.35 ± 0.02
	After CO	0.44 ± 0.09	0.36 ± 0.02*	0.48 ± 0.03[b]
R3	Control	0.42 ± 0.05	0.40 ± 0.02	0.33 ± 0.02
	After CO	0.39 ± 0.05	0.33 ± 0.01[b]	0.44 ± 0.01[b]
R4	Control	0.43 ± 0.04	0.40 ± 0.02	0.33 ± 0.01
	After CO	0.36 ± 0.05	0.33 ± 0.028	0.40 ± 0.02[b]

R1, R2, R3, R4, reflectance of channels 1, 2, 3, 4, respectively. Data are represented as mean ± S.E.M.

Table 7.3 Changes in NADH amplitude during CSD before (Control) and immediately after (CO) exposure to CO (Meilin et al., 1998)

	Channels	Control	CO
Wave 1	CF1	17.42 ± 2.1	15.5 ± 4.25
	CF2	15 ± 3.5	11 ± 0.25
	CF3	23.75 ± 6.6	21.24 ± 4.9*
	CF4	18.7 ± 17	17 ± 2.7
Wave 2	CF1	29.4 ± 2.85	19.4 ± 0.95[b]
	CF2	31.4 ± 7.6	19.3 ± 0.75[b]
	CF3	32.47 ± 4.3	23.55 ± 6.65*
	CF4	28.6 ± 8.8	20.6 ± 4.85*
Wave 3	CF1	27.82 ± 3.6	20.42 ± 2.7*
	CF2	27.9 ± 7.95	18.8 ± 3.95*
	CF3	28.89 ± 1.3	22.77 ± 0.8*
	CF4	27 ± 6.05	21.05 ± 9.05[b]

CF1, CF2, CF3, CF4, corrected NADH fluorescence recorded from channels 1, 2, 3, 4, respectively. All values are expressed as means ± S.E.M. of percent change. $^*P < 0.05$; $^bP < 0.01$

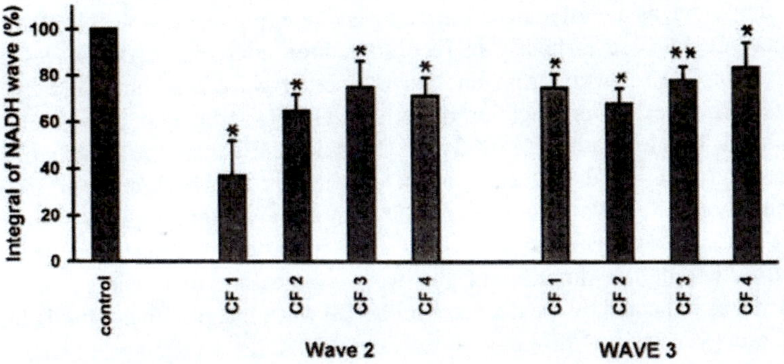

Fig. 7.8 Histogram exhibiting the integral of the NADH waves before control and immediately after exposure to CO (waves 2 and 3). $^*p < 0.05$, $^{**}p < 0.01$ (Meilin et al., 1998)

inverse indicator of oxygen consumption and therefore energy metabolism. These results are in agreement with previous studies carried out in this laboratory (Mayevsky et al., 1995; Meilin et al., 1996). According to the literature, the carboxyhemoglobin level under these conditions is approximately 46% (Hansen & Zeuthen, 1981) and the increase in CBF may be part of an autoregulatory reaction to the decrease in oxygen supply. Mayevsky et al. (Mayevsky et al., 1992) have shown that under other conditions in which the oxygen supply is restricted, such as during hypoxia, a correlation exists between oxyhemoglobin levels, tissue oxygen levels, and NADH levels. The increased blood flow may have afforded sufficient compensation for the decrease in blood oxygen such that tissue hypoxia did not occur and therefore NADH levels remained stable. Increased oxygen extraction may also have increased the oxygen supply to the brain (Mayevsky & Weiss, 1991). The fact that CBF began decreasing only a few minutes after the end of exposure to CO, although carboxyhemoglobin levels do not change so fast, suggests the participation of a dilatory agent such as NO with a direct effect on cerebral vessels (Kanten et al., 1983). However, CO-hemoglobin levels or blood oxygen content were not measured in this study.

In this study exposure to 1000 ppm CO for 40 min caused a decrease in the frequency of the induced CSD waves as expressed by NADH waves, when the CSD was induced immediately after cessation of exposure, whereas when CSD was induced 1.5 h after cessation of the exposure to CO the frequency of the waves increased. Exposure to CO may result in an impairment of the sodium–potassium pump which is expressed under conditions of increased ionic and metabolic activities, such as those existing during CSD. The increased frequency of the waves observed 1.5 h after reventilating the animals with air may be the result of the formation of free radicals which affect the ion permeability of the membrane (Bures et al., 1974). In this study DC waves remained unchanged, negating the possibility that a decrease in the threshold potential of the tissue could increase CSD wave frequency. This suggests that CO did not affect CSD initiation. On the other hand, CO apparently interfered with recovery from CSD repolarization, which is an ATP-dependent state, as recorded from the changes in the frequency of the NADH waves and may explain the difficulty in recording three consecutive DC waves, since before complete recovery from one wave, another wave may have already been initiated. The speed of propagation of both the NADH and the DC waves remained unchanged. This indicates that 1000 ppm CO did not affect the mechanism of propagation and that it did not result in an impairment of membrane permeability. Activation of brain cells by inducing CSD immediately after cessation of exposure to CO caused a smaller change in NADH oxidation than during the control CSD waves. Although the duration of the waves increased, total energy expended decreased as indicated by the decrease in the CF wave integral. In contradistinction to hypoxia (Mayevsky, 1984) in this study basal NADH levels remained stable, and the integral of the CF waves decreased, indicating another possible source for ATP other than the respiratory chain, such as increased glycolysis. Another possibility for the increased energy supply is increased coupling between the respiratory chain and sodium potassium pump activity, i.e., a more efficient reaction between energy formation and utilization. Furthermore, it must be pointed out that CSD was induced after cessation of exposure to CO, at a time when no significant changes were recorded in the CBF, and the possibility that some of the above mentioned changes may be due to reventilation which cannot be ruled out. In order to clarify this point, studies are now being performed with the MPA system, which enables simultaneous measurement of ionic fluctuations during CSD concomitantly to measuring CBF and NADH (Mayevsky et al., 1992). In conclusion, exposure to 1000 ppm CO for 40 min does not have a deleterious effect on the resting brain, but apparently interferes with metabolic activity during brain activation CSD arousal. The results of this study indicate that 1000 ppm CO does not affect the CSD initiation mechanism nor the CSD propagation mechanism. However, it apparently has a delayed effect, which is expressed in a change in repolarization and energy metabolism during recovery from CSD.

7.1.4 *Fourth Study*

7.1.4.1 Introduction

In the 4th study we tested the effects of hyperbaric oxygenation on brain functions after exposure to carbon monoxide (Rogatsky et al., 2002). Studies on animals have suggested that these neurological disturbances may be related to both a hypoxic insult, such as the result of COH_2 formation (Brown & Piantadosi, 1990), as well as an ischemic insult due to carboxymyoglobin-induced cardiovascular dysfunction (Kanten et al., 1983; Penney et al., 1989). At the same time, in the experiments it is known that intracellular uptake of CO can directly impair cerebral energy metabolism, inhibition of mitochondrial electron transport in the brain (Brown & Piantadosi, 1992; Piantadosi et al., 1988). Besides, toxic exposure to CO also produces significant oxidative stress in the brain, as indicated by increased lipid peroxidation (Thom, 1992, 1993b), increased hydrogen peroxide production, intramitochondrial glutathione depletion, and nonenzymatic hydroxylation of salicylate (Zhang & Piantadosi, 1992).

Hypoxia and also this direct adverse effect of CO on cerebral energy metabolism was ameliorated by treatment of animals with HBO_2 (Brown & Piantadosi, 1992; Thom, 1990a, 1993a). However, the mechanisms of HBO_2 action on brain mitochondrial function under the circumstances are not yet understood completely.

We have shown that the nature of the response to a brain activation event induced by CSD could serve as an indicator to the integrity of the cerebral cortex under various conditions (Mayevsky & Chance, 1982; Mayevsky et al., 1982). During the CSD event, a large increase in extracellular K^+ occurs, leading to the activation of the Na^+K^+ATPase. In order to restore the ionic homeostasis, the energy metabolism is stimulated, and the turnover of ATP is much higher as compared to the baseline activity.

In the present study, in order to test the effects of HBO_2 treatment on cerebral energy metabolism after CO exposure, we analyzed the capacity of the brain to respond to brain activation induced by CSD.

7.1.4.2 Methods

In the present study, we used an approach for bilaterally monitoring NADH redox state in awake rats under brain activation induced by CSD (Mayevsky, 1984; Mayevsky & Chance, 1982) before and after CO exposure.

In previous studies we showed that under CSD, the mitochondrial NADH was oxidized, meaning that production of ATP was stimulated. If the CSD was induced in partial ischemic brain (Mayevsky et al., 1982), the NADH response was completely different, namely, the amplitude of the NADH oxidation (decrease) was smaller or even the direction of the change was reversed. Under these conditions, the NADH will increase during the CSD cycle. Therefore, the CSD cycle can be used as a test for the integrity of the CBF autoregulation mechanism in the brain. The correlation between oxygen consumption and CBF under CSD was described in detail (Mayevsky & Weiss, 1991). It was shown that the pumping of K^+ into the cells was due to the stimulation of energy metabolism. The response of the brain to anoxia can also serve as an indicator of the integrity of the respiratory chain in the mitochondria. The amplitude of the NADH increase during anoxia will decrease when the mitochondria is damaged and the NADH was elevated before exposure to anoxia. Since the total increase in the NADH level is constant, the more it increases during CO exposure, the less it will increase during the anoxic test.

The measurements were made inside a hyperbaric chamber so that HBO_2 treatment after CO exposure was possible (Mayevsky et al., 1993a). In the monitoring approach used, a light guide was implanted above the two hemispheres of the brain together with 2 pairs of ECoG (electrocorticogram) electrodes (Fig. 7.9) (Rogatsky et al., 2002).

A push-pull cannula for KCl application (induction of CSD) was implanted 2 mm from the light guide holders (Mayevsky & Chance, 1974, 1982; Mayevsky et al., 1974). Rats were exposed to 3000 ppm CO for 30 min (group A, n = 8) or for 60 min (group C, n = 8). Two other groups of rats were exposed to 3000 ppm CO for 30 or 60 min followed by 30 min HBO_2 at 3 atmospheres absolute (atm abs) in

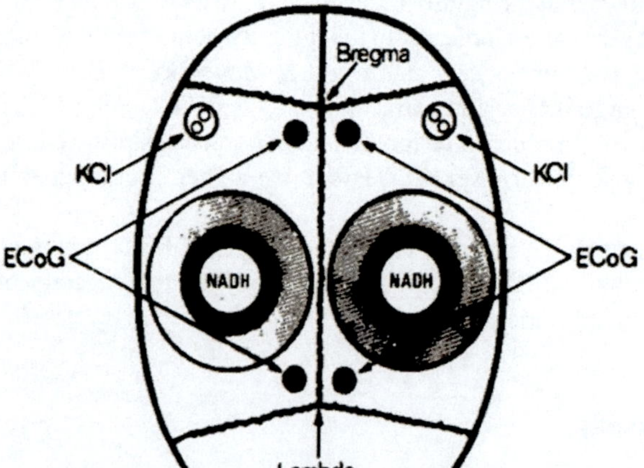

Fig. 7.9 Schematic representations of the location of various cannulas and electrodes on the skull of a rat. ECoG, electrocorticogram (Undersea Medical Society, reprinted with permission (Rogatsky et al., 2002))

a Bethlehem Steel Corp. FM-21-A chamber (group B, n = 6 and group D, n = 8, respectively). CSD was induced before and 2, 3, and 4 h after CO exposure as well as after HBO_2 treatment. The amplitude of the NADH response to CSD was found to be affected by the treatments. At the beginning and end of each experiment, an anoxic test (spontaneously breathing 100% nitrogen) was done in order to measure the maximal deviation of CF (increase of NADH). The purpose of this test was to evaluate the mitochondrial energy state in the brain (Mayevsky, 1984; Mayevsky & Chance, 1975; Mayevsky et al., 1993a).

7.1.4.3 Results

Figure 7.10 shows typical NADH fluorometry and ECoG responses to anoxic test and KCl application leading to CSD cycles (Rogatsky et al., 2002).

Figure 7.11 shows the various protocols used in the current study (Rogatsky et al., 2002).

Figure 7.12 represents an example of typical changes of NADH waves, induced by KCl application after CO exposure (Rogatsky et al., 2002).

These changes can be described as a trend toward decrease in the amplitude of the NADH response and an increased number of cortical CSD waves upon a single passage of KCl solution. In this study, we have found that following CO exposure of groups A and C (control groups), an obvious and significant decrease in the amplitude of NADH waves of cortical CSD was recorded (Fig. 7.13, groups A, C) (Rogatsky et al., 2002).

As one can see from the histogram, the decrease of the amplitude in group C is more pronounced than in group A. Use of HBO_2 after CO exposure accompanied a

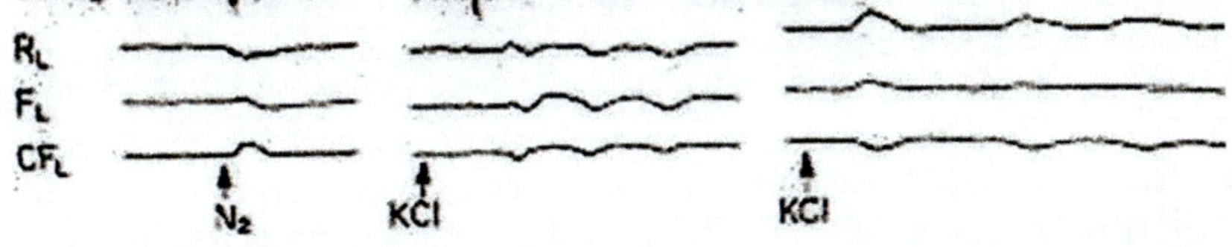

Fig. 7.10 Metabolic and electrical responses to anoxic test (**A**), CSD during control (**B**) and after 3000 ppm CO exposure (**C**). R, F, CF, reflectance, fluorescence, and corrected NADH fluorescence recorded from right (R) and left (L) hemispheres. ECoG, electrocorticogram (Undersea Medical Society, reprinted with permission (Rogatsky et al., 2002))

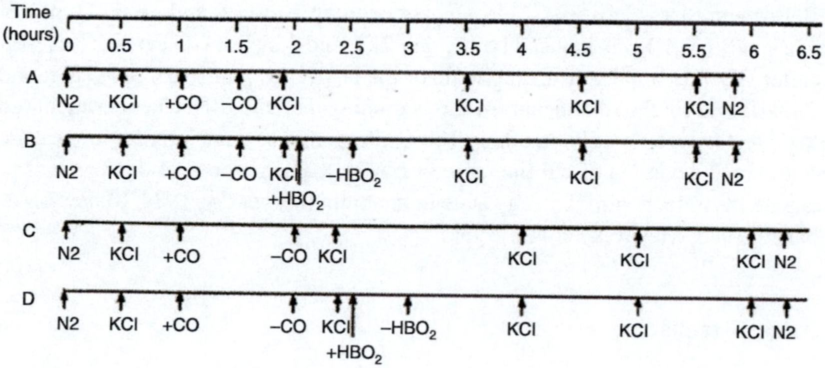

Fig. 7.11 Details of the various protocols in the experimental groups A, B, C, and D. (Undersea Medical Society, reprinted with permission (Rogatsky et al., 2002))

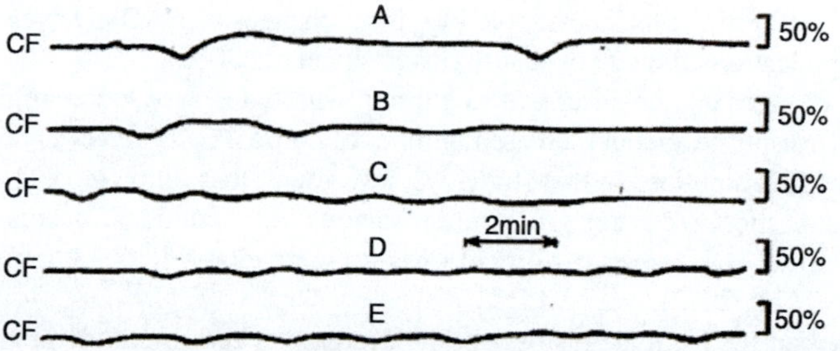

Fig. 7.12 Typical gradual increase of the number of waves CSD (CF) in experimental groups A and C. A, B, C, D, E—the induction CSD before or 20 min, 2, 3, and 4 h after CO exposure, accordingly (Undersea Medical Society, reprinted with permission (Rogatsky et al., 2002))

maintenance of the initial level of NADH amplitude in group B. This effect of preservation of the wave amplitude of CSD was absent in the rats of group D.

The number of CSD waves after CO exposure is presented in Fig. 7.14 (Rogatsky et al., 2002). There is a clear tendency toward an increase in this parameter in the control groups (Fig. 7.14, groups A, C). However, after HBO_2 application, the significant increase of CSD wave number was not observed in group B (Fig. 7.14, group B). In group D, this effect of HBO_2 protection was absent (Fig. 7.14, group D).

The anoxic test in the control groups A and C shows a significant decrease of maximal deviation levels of NADH (reduction) at the end of the observation in comparison to the initial experimental period (Table 7.4, groups A, C). This parameter was not different in group B from the magnitude of NADH deviation at the beginning of the experiments (Table 7.4, group B). The magnitude of NADH deviation by anoxic test at the end of the experiments in group D remained significantly lower as compared to the initial state (Table 7.4, group D). This parameter in the untreated group was invariable (Table 7.4, "Untreated").

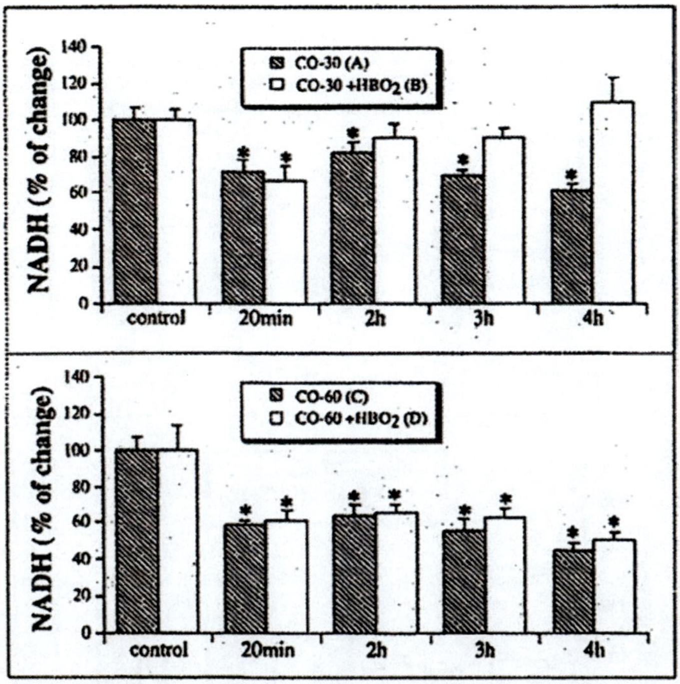

Fig. 7.13 Histogram presenting the amplitude of NADH waves of CSD induced by passage of 1% KCl solution before and after CO exposure in control groups A and C (n = 8 and 8, respectively), as well as after HBO_2 treatment groups B and D (n = 6 and 8, respectively). Abscissa is the time (hours) after CO exposure and ordinate is amplitude of CF (NADH in % of control). *$p < 0.05$ compared with control (Undersea Medical Society, reprinted with permission (Rogatsky et al., 2002))

7.1.4.4 Discussion

In the present study, NADH redox state was used in order to evaluate the responses to CO exposure. The interpretation of the results is somewhat limited since cerebral blood flow (CBF) and cerebral blood volume (CBV) were not monitored.

Progressive decrease in the amplitude and an increased number of NADH waves upon a single passage of KCl solution in control groups (A and C), combined with the decrease in the maximal deviation (reduction) of NADH level upon the anoxic test, have been interpreted as manifestation of mitochondrial dysfunction. Probably one explanation for the character's changing of the NADH waves may be in the compensationary increase of the number of CSD required for elimination of local ionic disbalance and reactivation of transmembrane ionic transport, particularly after CO induced decrease in energetic status of mitochnondria.

The results obtained in group B show that the development of disbalance in energy metabolism in the cerebral cortex can be delayed or prevented by HBO_2 treatment. However, whereas a complete restoration of studied parameters was

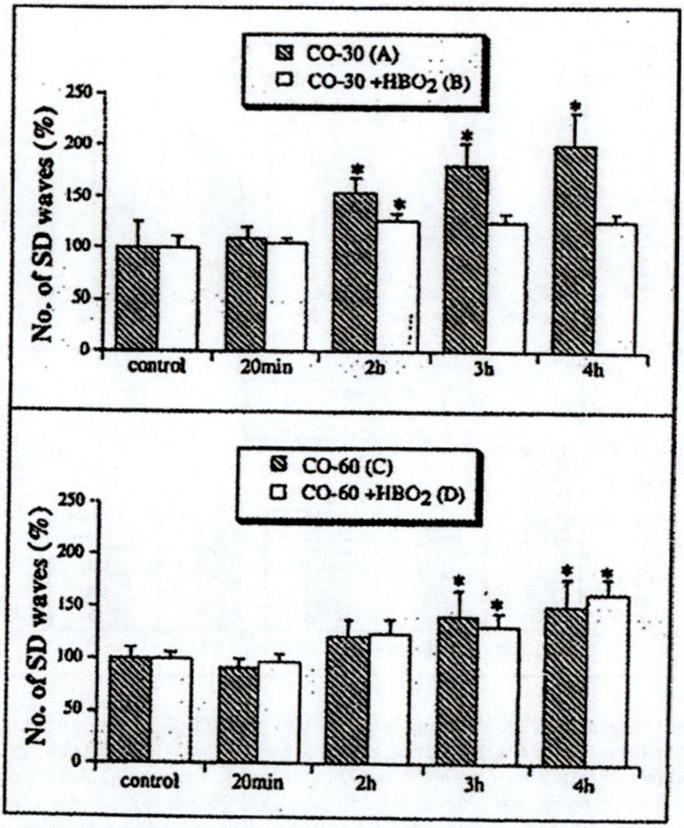

Fig. 7.14 Histogram presenting the number of NADH waves due to cortical CSD, induced after single passage of KCl solution. The ordinate presents the number of NADH waves in % of control. Other designations are as in Fig. 7.13 (Undersea Medical Society, reprinted with permission (Rogatsky et al., 2002))

Table 7.4 Effects of 3000 ppm CO on the magnitude of NADH deviation during anoxic test in the various experimental groups (% of initial state) (Rogatsky et al., 2002)

Group	Initial state		Terminal state	
Untreated	100 ± 4.7	5	99.1 ± 3.35	5
A (30' CO)	100 ± 8.3	14	60.8 ± 3.3*	13
B (30' CO + HBO_2)	100 ± 6.2	13	92.2 ± 3.2	10
C (60' CO)	100 ± 6.8	13	49.59 ± 1.24*	12
D (60' CO + HBO_2)	100 ± 8.6	11	67 ± 4,6*	9

Untreated, A, B, C, D, groups of experiments; initial state, before CO exposure; terminal state animal's death. Mean ± SE; *$P \leq 0.05$ compared with initial state

observed in group B, group D did not exhibit tendency to maintain these parameter levels. These data indicate that the restoration of NADH parameters upon HBO_2 application is dependent on the overall severity of hypoxic and cellular distrubances in the brain, which in group D was more pronounced due to the twice greater duration of CO inhalation, relative to group B.

We propose that the signs of NADH dysfunction found by us can be caused by direct injurious action on the brain CO (Brown & Piantadosi, 1992; Piantadosi et al., 1988) that probably inhibits mitochondrial electron transport. Displacement of CO via HBO_2 exposure from the proteins it has bound to (Brown & Piantadosi, 1989; Chance et al., 1970), also provides the preconditions for the restoration of normal levels of electron transport along mitochondrial respiratory chain, because HBO_2 has some intracellular protective effect on the brain independently from increased O_2 carrying capacity (Brown & Piantadosi, 1992; Thom, 1990a, 1993a). The restoration of mitochondrial function upon HBO_2 exposure in our experiments was adequately reflected by the dynamics of those NADH parameters examined.

Notably, the anoxic test has proven in our studies to be a sensitive and quantitative indicator of mitochondrial respiration. As an alternative to the oxygen test (Lukyanova & Bures, 1967), it provides an adequate prognostic value in the evaluation of damage as well as a potential for restoration of mitochondrial energy metabolism.

Our data also suggest a role for HBO_2 as a means by which cerebral function may be restored after CO intoxication. Earlier it has been obtained that HBO_2 at 2.8–3 atm abs markedly accelerates protection against injurious sequelae occurring in CO poisoning in rats (Brown & Piantadosi, 1989; Jiang & Tyssebotn, 1997; Thom, 1990a, 1993a). However, comparison betweeen the data in groups B and D suggest that the duration of HBO_2 exposure is probably also an important factor for optimizing the positive effect.

In recent studies it was revealed that HBO_2 at 3 atm abs for 60 min. Significantly increased the survival time and rate in rats poisoned with CO (2700 ppm for 60 min) (Jiang & Tyssebotn, 1997). Moreover, it was also established that exposure to hyperbaric O_2 at 3 atm abs for 1–2 h actually improves outcome in rats after acute cerebral ischemia (Badr et al., 2001; Krakovsky et al., 1998; Sunami et al., 2000). It is important to note that in these experiments, HBO_2 treatment regulated the brain energy metabolites (Badr et al., 2001) and prevented the aggravation of lipid peroxidation in the ischemic periphery (Sunami et al., 2000).

7.2 Nitric Oxide

7.2.1 First Study

7.2.1.1 Introduction

CSD described initially by Leão (1944b) is a multifactorial event affecting the electrical, ionic, metabolic, and hemodynamic activities in the brain (Bures et al., 1974; Mayevsky & Weiss, 1991; Mayevsky et al., 1974; Vyskocil et al., 1972). Due to disturbances in the ion homeostasis, the Na^+K^+-ATPase activity and energy metabolism are stimulated in order to restore the normal extracellular ion levels (Hansen, 1985; Mayevsky & Weiss, 1991; Mayevsky et al., 1974). The hemodynamic response to CSD was a challenge to many investigators since the initial observation of dilation of pial vessels (Leão, 1944a). He concluded that the vascular responses are secondary to the local changes in the activity of neural elements. The changes in cerebral blood flow (CBF) just before, during, and after the depolarization wave of CSD were described by various investigators (Buresova, 1957; Hansen et al., 1980; Lauritzen, 1984; Lauritzen & Diemer, 1986; Lauritzen et al., 1982; Mies & Paschen, 1984; Van Harreveld & Ochs, 1957; Van Harreveld & Stamm, 1952) and have been reviewed by Lauritzen (1987a, b). In all studies, a large increase in cerebral blood flow was recorded during the wave; Lauritzen and collaborators described a post-spreading depression wave hypoperfusion, while a preceding vasoconstriction (immediately before the wave) was not established or proved. The mechanism behind the changes in CBF due to the CSD wave is not clear although recently nitric oxide (NO) was proposed to be involved (Duckrow, 1993; Goadsby et al., 1992). NO was suggested as an important factor in CBF regulation (Iadecola et al., 1994; Irikura et al., 1994; Toda et al., 2009a, b) as well as having direct effects on neuronal elements (Culotta & Koshland, 1992).

7.2.1.2 Methods

In order to clarify the role of NO in this multifactorial event (CSD), we adopted the multiparametric monitoring system (MPA) by which the hemodynamic, metabolic, ionic and electrical activities were measured simultaneously from the surface of the brain (Mayevsky et al., 1992). Energy state and metabolism were evaluated by monitoring the cerebral blood flow and volume (Laser Doppler flowmetry) as well as intramitochondrial redox state (surface NADH fluorometry reflectometry). Ion homeostasis was evaluated by the extracellular K^+, Ca^{2+}, and H^+ levels (surface minielectrodes). Electrical activities were assessed by the spontaneous electrocortical activity (EcoG) as well as by DC steady potential.

7.2.1.3 Results

Figures 7.15 and 7.16 show the typical responses to CSD (induced by KCl application) before (A) and after injection of L-NAME (B) (Meilin et al., 1995). As seen, extracellular K^+ level increased while Ca^{2+} decreased. The pH response was toward acidification. The metabolic responses are seen in the oxidation of NADH (decrease in CF) and the hyperfusion of the tissue (increase in CBF). Also, the reflectance trace (R) showed a biphasic response correlated to the changes in blood volume (CBV).

As seen in Figs. 7.15 and 7.16, inhibition of NO synthesis led to changes only in the hemodynamic and metabolic responses. Figure 7.17 demonstrates the changes

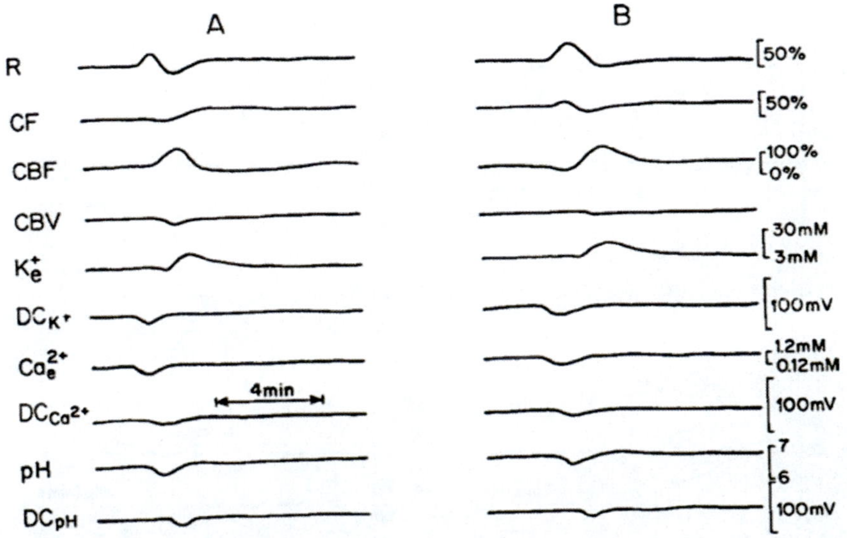

Fig. 7.15 Hemodynamic, metabolic, ionic, and electrical responses to CSD before (**A**) and after (**B**) IP injection of L-NAME (NO synthetase inhibitor). R, CF, reflectance and corrected NADH fluorescence; CBF, CBV, cerebral blood flow and volume; K^+_e. Ca^{2+}_e. pH, extracellular levels of potassium, calcium, and hydrogen; DCK^+ DCa^{2+}. DC_{pH}, direct current steady potential measured concentric to each of the ion-specific electrodes (Meilin et al., 1995)

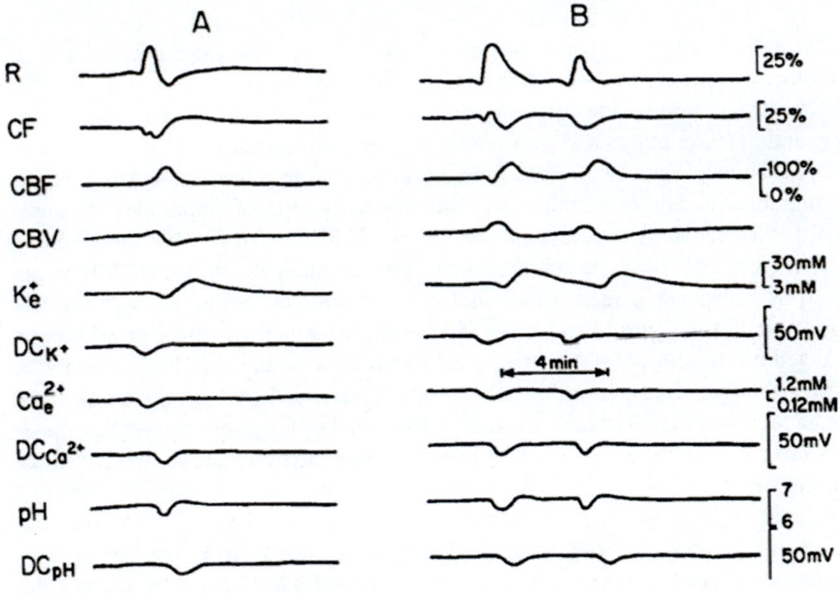

Fig. 7.16 Responses to CSD wave before (**A**) and after (**B**) L-NAME injection measured in a second rat shown in Fig. 7.15. All abbreviations are the same. In this rat two waves were recorded after L-NAME injection (Meilin et al., 1995)

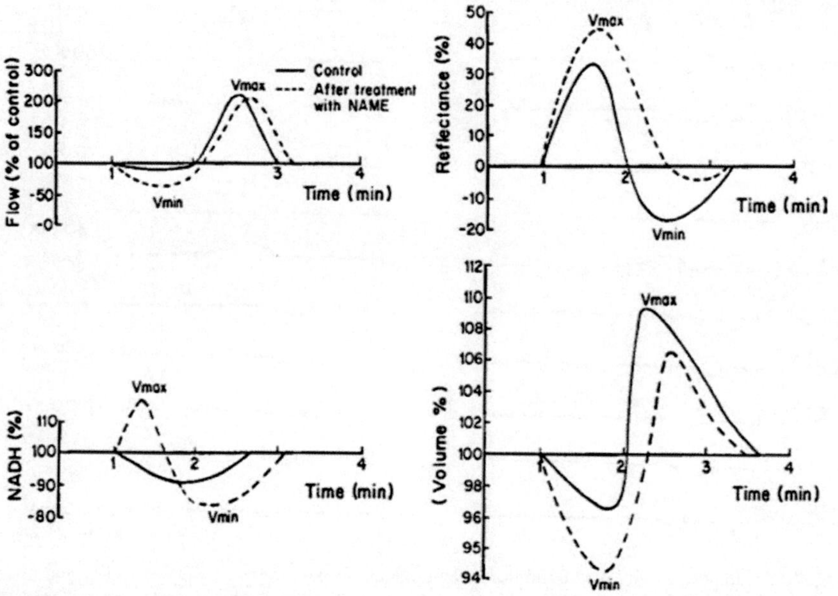

Fig. 7.17 Schematic demonstration of the changes in the responses to CSD induced by L- NAME injection. The changes presented in the figure are the average of all the rats (n = 7) used in each group. V_{min}, V_{max}, minimal and maximal values of average amplitudes calculated for the various parameters, in the control and L-NAME treated rats (Meilin et al., 1995)

in the responses to CSD before and after L-NAME injection (Meilin et al., 1995). All other parameters that were analyzed did not respond to treatment by L-NAME.

Due to variations in the control responses to CSD between rats, it was necessary to quantitate the changes and statistically analyze the responses to NO inhibition. In each of the responses, two values were calculated, namely, the V_{min} and V_{max} representing minimal and maximal levels of the signal in terms of amplitude. The duration of the responses was also affected by L-NAME treatment, but the statitical analysis was done using the amplitude change. The main effect of L-NAME on the blood flow response is seen as an initial hypoperfusion phase (V_{min}) which was not detectable in the control rats. In the NADH trace the typical oxidation wave was changed into a biphasic wave, namely, an initial increase in NADH followed by an oxidation wave. Changes in blood volume characterized by the reflectance trace (using fluorometry) and the blood volume calculated by the laser Doppler flowmeter were also affected by L-NAME injection. Statistical analyses of the two groups of rats are shown in Figs. 7.18, 7.19, 7.20, and 7.21 (Meilin et al., 1995). The effect of L-NAME on the response in CBF to CSD wave is shown in Fig. 7.18. As seen the values of V_{min} after L-NAME treatment (labeled as 2 in the figure) were significantly lower, namely, that a phase of hyperfusion was created. The V_{max} values, namely, the large hyperfusion, was not affected by the injection of L-NAME. The V_{max} values, namely, the large hyperperfusion, were not affected by the injection of L-NAME. The

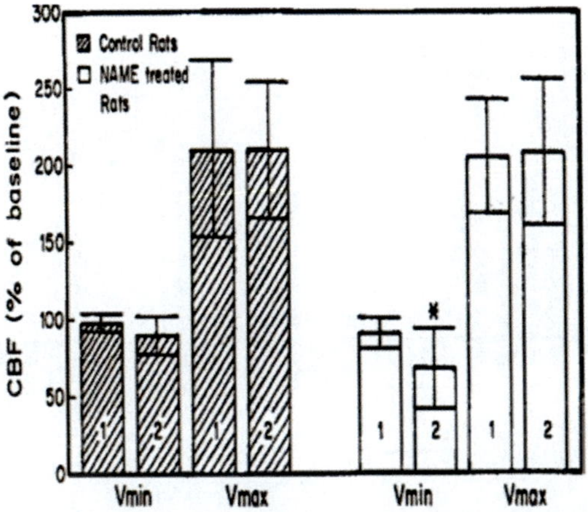

Fig. 7.18 Effects of L-NAME treatment on the cerebral blood flow (CBF) responses to CSD. The numbers 1 and 2 represent the wave before and after treatment on the same interval of time in the control animals (Meilin et al., 1995)

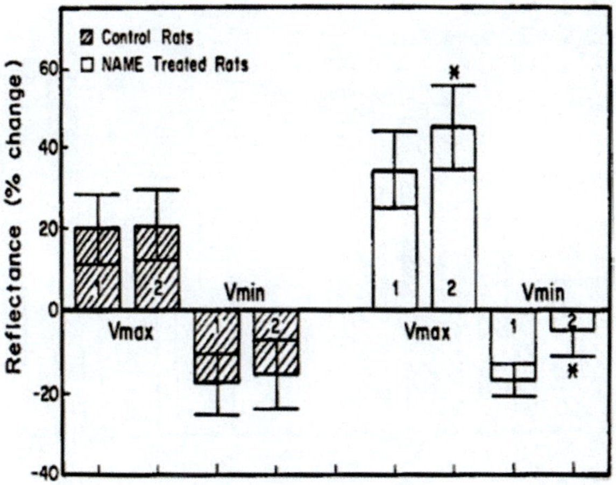

Fig. 7.19 Effects of L-NAME treatment on the NADH responses to CSD. The numbers 1 and 2 represent the wave before and after treatment on the same interval of time in the control animals (Meilin et al., 1995)

typical response of the NADH signal to CSD, a wave of oxidation (decrease in NADH), was changed into a biphasic wave (Fig. 7.19). The changes in the reflectance and blood volume (CBV) are shown in Figs. 7.20 and 7.21. The initial increase phase of the reflectance was significantly higher (V_{max}) while the decrease phase (V_{min}) almost disappeared (Fig. 7.20). As seen in the control animals, the 2 waves

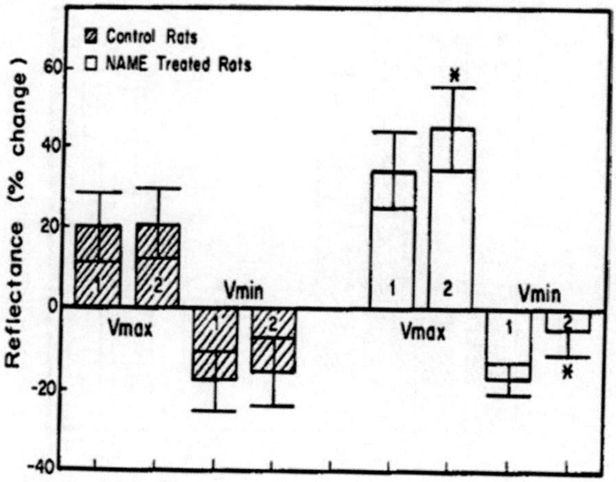

Fig. 7.20 Effects of L-NAME treatment on the reflectance responses to CSD. The numbers 1 and 2 represent the wave before and after treatment on the same interval of time in the control animals (Meilin et al., 1995)

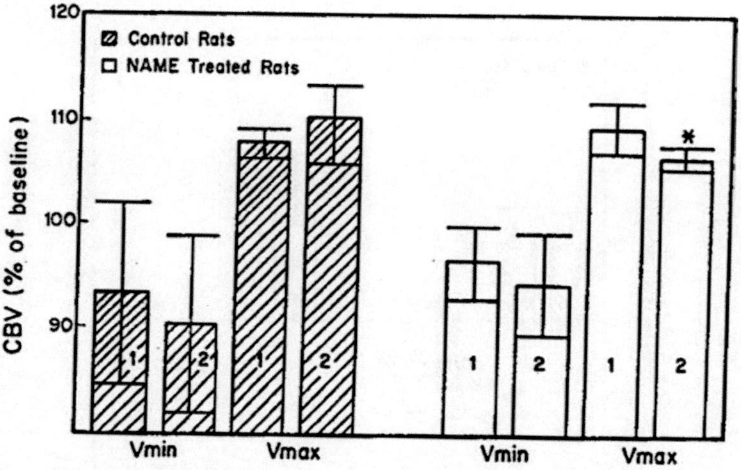

Fig. 7.21 Effects of L-NAME treatment on the cerebral blood volume (CBV) responses to CSD. The numbers 1 and 2 represent the wave before and after treatment on the same interval of time in the control animals (Meilin et al., 1995)

led to the same changes in V_{max} and V_{min}. The same type of changes were recorded by the cerebral blood volume signal (CBV) shown in Fig. 7.21. The V_{min} showed a larger decrease in wave 2 but was not statistically significant.

The V_{max} was significantly smaller after the L-NAME treatment. In order to show the effect of L-NAME on the baseline levels of CBF during the duration of the experiment, Fig. 7.22 is presented (Meilin et al., 1995). The traces are averages of 7 rats in each group. As seen the CBF was very stable in the control rats (open circles) while a significant decrease in CBF was recorded after L-NAME injection. The two L-NAME treated groups were very similar in the responses although one of them was exposed to CSD waves.

7.2.1.4 Discussion

The involvement of NO in CBF regulation and especially under metabolic activated state is not clear. In the present study we adopted the multiparametric monitoring approach to test the involvement of NO in various brain activities under CSD. According to our results, we could conclude that inhibition of NO synthesis by L-NAME injection did affect significantly the resting CBF level without any detectable change in mitochondrial redox state, extracellular levels of ions, or electrical activities. As shown in Fig. 7.22 injection of L-NAME led to 20–30% decrease in CBF although NADH levels were unaffected (not shown in the figure). One possible explanation to the uncoupling between CBF and NADH is that O_2 extracction was increased gradually to compensate for the decrease in CBF. The fact that extracellular K^+ levels were unaffected suggests that energy availability was unchanged as indicated by the stable NADH redox state. We have shown in the past that ischemia induced decrease in CBF was accompanied by an increase in NADH and an increase in extracellular K^+ (Mayevsky, 1990).

When the brain was exposed to a CSD wave, changes were recorded in all parameters monitored (Figs. 7.15 and 7.16) using the system. It was hard to analyze the responses to CSD in terms of sequence of events. In other studies performed in our laboratory (Sonn and Mayevsky, unpublished results), we were able to analyze the

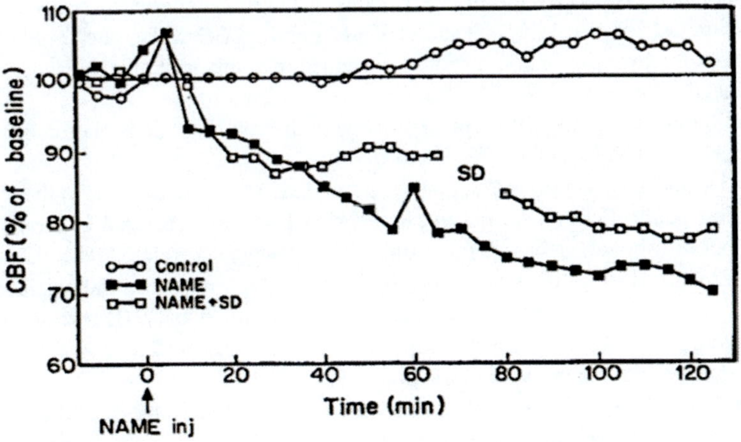

Fig. 7.22 Changes in resting levels of CBF in controls or due to L-NAME injection with and without CSD wave. Each line represents an average of 7 rats (Meilin et al., 1995)

sequence of CBF, NADH, K^+, and DC potential responses to CSD wave. According to those studies, we conclude that the initial changes were in extracellular K^+ and DC potential. The responses of NADH and CBF were recorded later on. This result, for the first time, approved the coupling between the stimulated Na^+-K^+-ATPase activity and the metabolic and hemodynamic responses. Therefore, it is possible to conclude that most of the events recorded under CSD wave are coupled to each other.

There are three main phases of the CSD wave that required discussion in terms of CBF response and other changes:

I. Initial depolarization phase.
II. Metabolic and hemodynamic compensation phase.
III. Post wave period.

According to the results of the present study, the leakage of K^+ and changes in DC potential were not affected by inhibition of the CSD wave. The fact that an initial hypoperfusion was recorded during the initial phase due to NO-inhibition (Figs. 7.15, 7.16, and 7.18) suggests that NO may play a role during this period. It is documented that during depolarization (extracellular increase in K^+ and decrease in Ca^{2+}) blood vessels constrict. The suggestion of Duckrow (1993) that this constriction is balanced by dilation due to NO release is confirmed in the present study. During the second phase, NO inhibition was not effective, and the large hyperperfusion due to the increased metabolic demand was recorded and was similar to the control wave. We were unable to confirm the findings of Goadsby et al. (1992) who were able to block the hyperfusion response induced by the CSD wave. This may suggest that NO is not a major factor in the regulation of CBF during the recovery phase from the CSD wave. This finding cannot confirm the findings of Duckrow (1993) who showed that NO inhibition by L-NAME led to a smaller hyperperfusion (although he claimed that it was not statistically significant). Therefore, we are suggesting that the hyperperfusion phase of the CSD wave is tightly coupled to the increased energetic demand created by the high K^+ levels in the extracellular space. Since NO inhibition did not affect the leakage of K^+ during the CSD wave, one could expect that the same hyperperfusion response is needed to compensate for the same metabolic need.

An important finding of the present study is the significant correlation between the changes in CBF and the NADH redox state.

Under normal conditions CSD led to an oxidation cycle correlated to the hyperperfusion phase. During the initial phase NADH was not changed in the control cycle but did show a clear increased phase (less energy) correlated with the initial hypoperfusion seen in the CBF trace. We were unable to confirm the existence of any signicant post wave hypoperfusion described previously (Duckrow, 1993; Lauritzen, 1984). In the present study the resting CBF as well as NADH was the same after the wave compared to the pre-wave CBF.

We conclude that NO may play a role during the initial phase of CSD, namely, during the depolarization phase. The NO dilation effect is needed to balance the vasoconstriction effect created by the high extracellar K^+ levels. Later on in the CSD event NO may not play any significant role.

7.2.2 *Second Study: The Effect of CSD Under Nitric Oxide (NO) Synthase Inhibition*

7.2.2.1 Introduction

The involvement of NO during CSD wave propagation was tested by NO synthase inhibition, by L-NAME administration. Three CSD waves were recorded: the first CSD wave was a control wave (Control), and the two other CSD waves were recorded after L-NAME administration—1 h after L-NAME administration (L-NAME1) and 2 h after L-NAME administration (L-NAME2). These results were compared to the results of two control groups: (1) A control group, where no CSD waves were induced as well as no inhibition of NO synthase (no administration of L-NAME). Recordings of the measured parameters were taken in parallel time to the CSD experiments: at the beginning of the second hour of the experiment, a control sample was taken to which all the data was compared during 3 h of additional recording. (2) The changes during CSD waves before and after L-NAME administration were compared to control experiments where four CSD waves were induced in a difference of an hour between each CSD wave (SD1, SD3, SD4).

7.2.2.2 Methods

Using the MPA system (Fig. 4.9), we performed a control study in rats under normoxia as seen in Fig. 7.23A. One CSD wave was induced every hour (total 4 waves). This group of 11 rats served as a control group to the other experimental groups that will be presented in this chapter.

The second group of rats were involved in a study that tested the effect of L-NAME on the responses to CSD waves. This set of experiments included 10 animals that were exposed to a control response to CSD before the injection of L-NAME and another two waves of CSD induced 1 and 2 h after the L-NAME IP injection. We used the L-NAME distributed by Sigma 50 mg/kg dissolved in saline. The protocol of this study is presented in Fig. 7.23B.

7.2.2.3 Results

7.2.2.3.1 The Variation of the Measured Parameters in the Control Study

In this protocol we compared the changes in the amplitude and time measured in the first CSD (SD1) wave to those measured in the other three CSD cycles (SD2, SD3, SD4). The changes in the amplitude of the various parameters during the four CSD cycles are presented in Fig. 7.24.

In this paragraph we are presenting the summary of the results.

A. No significant and consistent differences were found between the amplitudes of the parameters measured in the first CSD cycle to the other three.

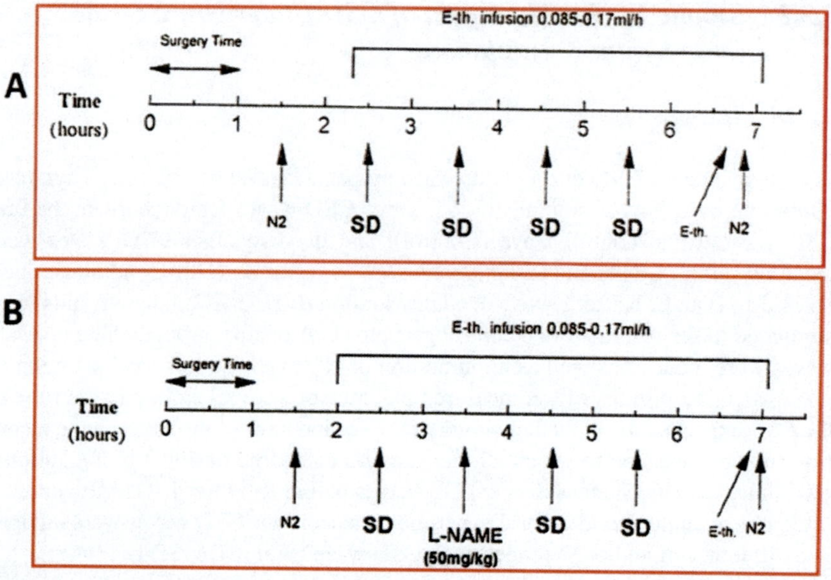

Fig. 7.23 (**A**) Schematic presentation of the experimental protocol in which the responses to four induced CSD waves are presented. The interval between the waves was 1 h. (**B**) Presentation of the protocol used in the experiments that performed in order to test the effects of L-NAME on the responses to CSD. The 1st wave served as a control and the other two CSD waves were induced 1 or 2 h after the injection of L-NAME

B. The oxidation of NADH during the CSD wave appeared before the increase in CBF. The rate of potassium leakage during the four CSD waves was very similar. Some variability was recorded in the rate of the recovery of potassium between the four cycles.

7.2.2.3.2 Results: The Effect of L-NAME Injection on the Measured Parameters

The effect of L-NAME on the measured parameters was tested 1 and 2 h after L-NAME injection and before exposing the brain to CSD event. The samples taken 2 h after L-NAME injection included 7 rats from 10 (in 3 rats the experiment was stopped after 2 h because of continuous spontaneous of CSDs). The histograms in Fig. 7.25 summarize the effect of L-NAME on the measured parameters as compared to rats where the inhibitor was not induced and CSD waves were not initiated. The differences between these groups of experiments were tested by the student t-test.

Mitochondrial NADH Mitochondrial NADH increased after L-NAME injection: 1 h after injection, NADH increased by 8.6 ± 5%, and after 2 h NADH increased by

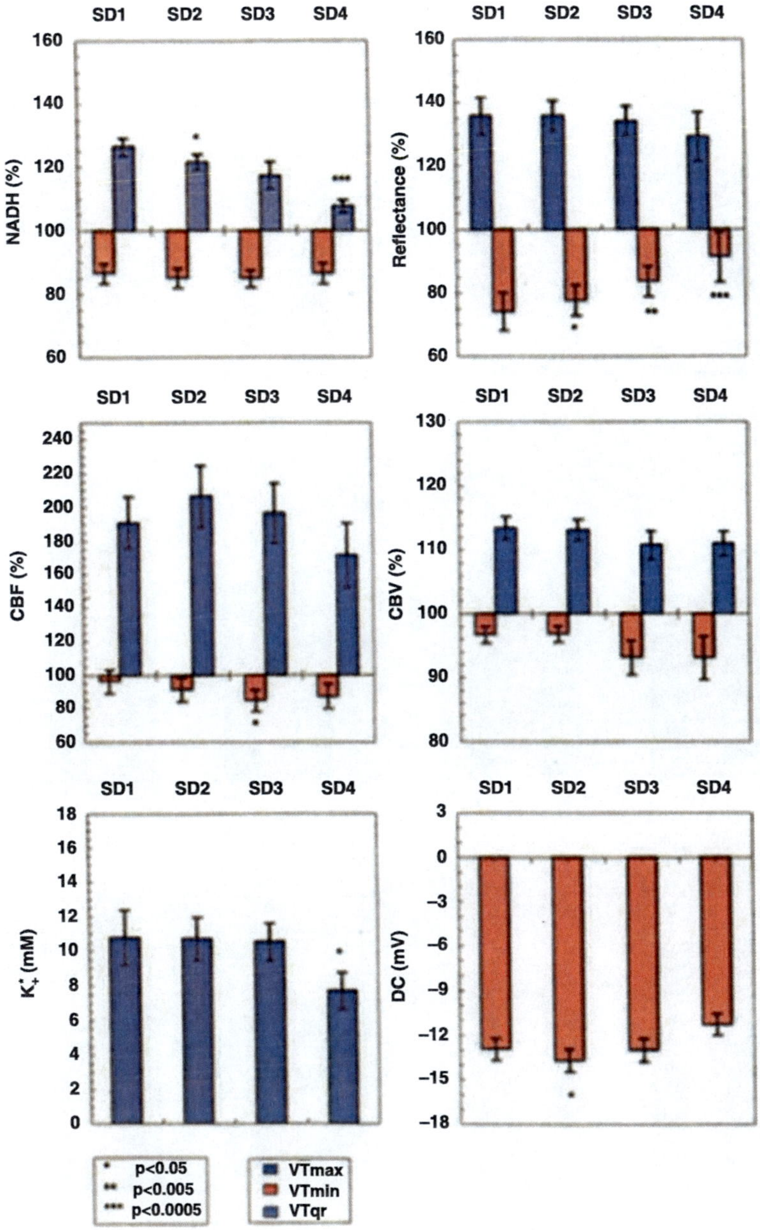

Fig. 7.24 The effects of four repeated CSD cycles (SD1, SD2, SD3, SD4) induced in the control animal, on the various parameters monitored according to the experiment presented in Fig. 7.23A. The values are the averages ± the SEM. The extracellular potassium and the DC potential values of the amplitude during the spread of the wave in Delta mM and Delta mV, respectively. VT_{min}: the decrease in the amplitude—red histogram. VT_{max}: increase in amplitude—blue histogram. VTqr: reduction cycle of NADH—blue crossed. The * represents the level of significant differences between the four CSD cycles

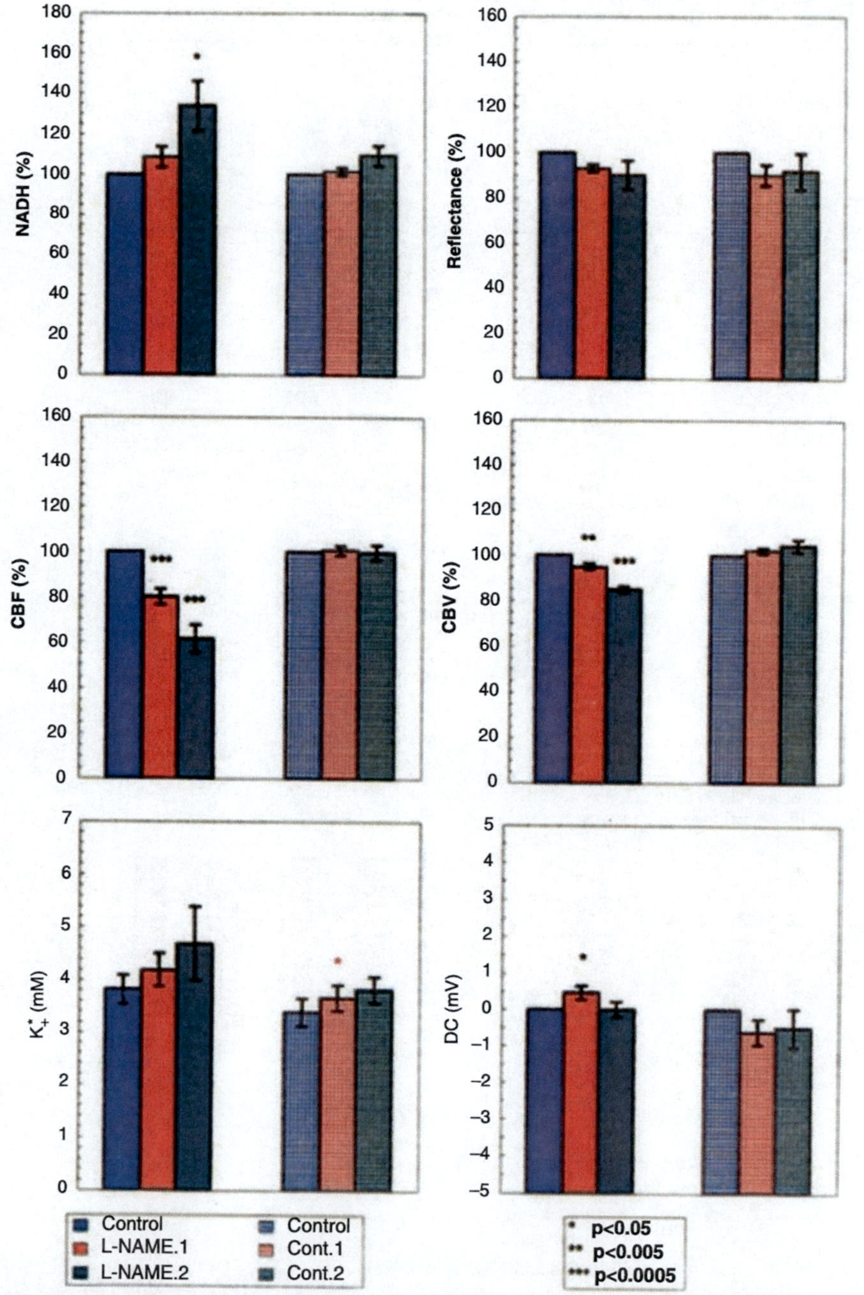

Fig. 7.25 The effects of L-NAME injection on the measured parameters. In each measured parameter, the three left histograms were collected from the experimental animals and the three right ones from the control animals. The blue histograms are from the control time while the red and green were collected after 1 h and 2 h after the injection of the L-NAME or the saline in the control animals

33.9 ± 12.4%. The increase in NADH 2 h after L-NAME injection was significant ($p < 0.05$).

Reflectance No significant changes were found in reflected light after L-NAME injection.

CBF One hour after L-NAME injection, CBF decreased to −19.9 ± 3.9%, and after 2 h CBF decreased to −38.4 ± 6.2%. The decrease in CBF, 1 and 2 h after L-NAME injection, was found significant ($p < 0.0005$).

CBV One hour after L-NAME injection, a decrease of −5.05 ± 1.5% was found ($p < 0.005$), and after 2 h CBV decreased to −15.1 ± 1.3% ($p < 0,0005$).

Extracellular K⁺ Concentration (K_e^+) K_e^+ increased after L-NAME injection. One hour after the injection K_e^+ increased by 0.4 ± 0.2 mM and after 2 h increased by 0.9 ± 0.7 mM. These changes were found significant as compared to K_e^+ values before L-NAME induction ($p < 0.005$ and $p < 0.05$, respectively). No significant changes in K_e^+ were found between control and experimental groups.

DC Potential One hour after L-NAME injection DC potential increased by 0.5 ± 0.2 mV which was found significant ($p < 0.05$) as compared to the control value before L-NAME administration as well as compared to the control group. No significant changes in DC potential were found 2 h after L-NAME injection.

7.2.2.3.3 Results: The Effect of CSD Waves on the Measured Parameters

Typical responses to CSD before and after L-NAME injection are shown in Fig. 7.26.

Mitochondrial NADH During control CSD (Fig. 7.26A) NADH showed an oxidation-reduction wave, whereas 1 h after L-NAME administration (Fig. 7.26B)

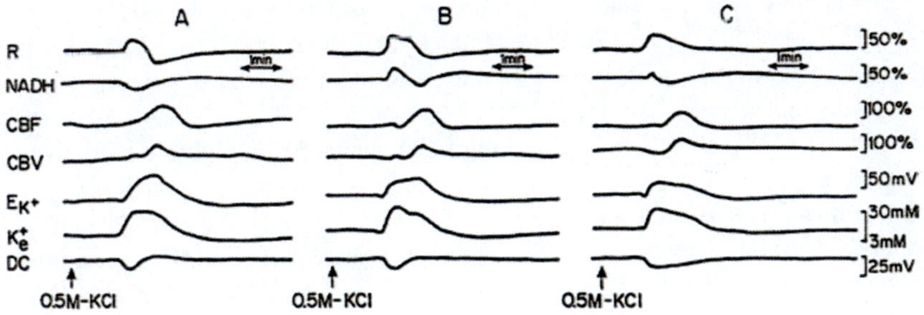

Fig. 7.26 Metabolic hemodynamic and ionic responses to a wave of CSD. (**A**), control; (**B**), 1 h; and (**C**), 2 h after the injection of L-NAME

NADH showed a reduction-oxidation wave. Two hours after L-NAME was induced (Fig. 7.26C), the initial reduction wave in mitochondrial NADH decreased as compared to 1 h after L-NAME injection (Fig. 7.26B).

CBF During CSD, 1 h after L-NAME injection (Fig. 7.26B), CBF showed an initial decrease before the typical increase in CBF during control (Fig. 7.26A).

Reflectance
Reflectance changed in a biphasic pattern, an increase which was followed by a decrease. The increased phase was found higher 1 h after L-NAME injection and the decline was smaller (Fig. 7.26B). Two hours after L-NAME injection (Fig. 7.26C), the decreased wave in reflectance, in some experiments, disappeared.

7.2.2.3.4 Results: CSD Amplitudes

Averaged data ± SE of the experimental group is shown in Fig. 7.27. The amplitudes of CSD waves initiated after L-NAME administration (of the measured parameters) were compared to the parallel amplitudes during the control CSD (SD1) in the experimental group (the significant values are marked by *) and to CSD waves in the control group SD1, SD3, and SD4 (Fig. 7.24). The significant changes in the measured parameters during CSD between the experimental and control groups are marked by #. The CSD waves, L-NAME1 and L-NAME2, were induced during the third and fourth hour of experiment and therefore were compared to SD3 and SD4, respectively.

NADH Waves The average amplitudes of NADH during control CSDs were biphasic, oxidation-reduction (VTmin-VTqr). After L-NAME injection we found a reduction wave (increase) that preceded the oxidation-reduction wave, creating a three-phase wave (VT_{max}-VT_{min}-VTqr). The increase in VT_{max} amplitude in L-NAME1 and L-NAME2 (as compared to control before CSD initiation) was by 19 ± 4.8% and 12 ± 4%, respectively, seen between the oxidation cycles during control CSD and L-NAME1 and between control CSD and L-NAME2 CSDs (−13.25 ± 2.1%, −15.64 ± 4.8%, −18.6 ± 6.5%, respectively). A decrease was seen in the reduction wave amplitude (VTqr) during L-NAME1 and L-NAME2 as compared to control CSD (about 17.6% ($p < 0.0005$) and about 26.9% ($p < 0.05$). These values were not found significantly different as compared to control experiments.

CBF In the L-NAME group, CBF during CSD changed in a biphasic pattern—a small decrease which was followed by a large increase. The initial decrease in CBF in the control CSD (before L-NAME injection) was found only in one rat (the group contained 10 rats), whereas 1 h after L-NAME injection (L-NAME1) the initial decline was seen in 5 rats and in L-NAME2 in 7 rats. Two hours after L-NAME injection, the initial decline in CBF during L-NAME2 (Fig. 7.27) was −15.05 ± 6.09% which was found significant ($p < 0.05$) as compared to control CSD in the same group (Fig. 7.27) and not significant as compared to SD4 (the parallel CSD) in the control group (Fig. 7.24).

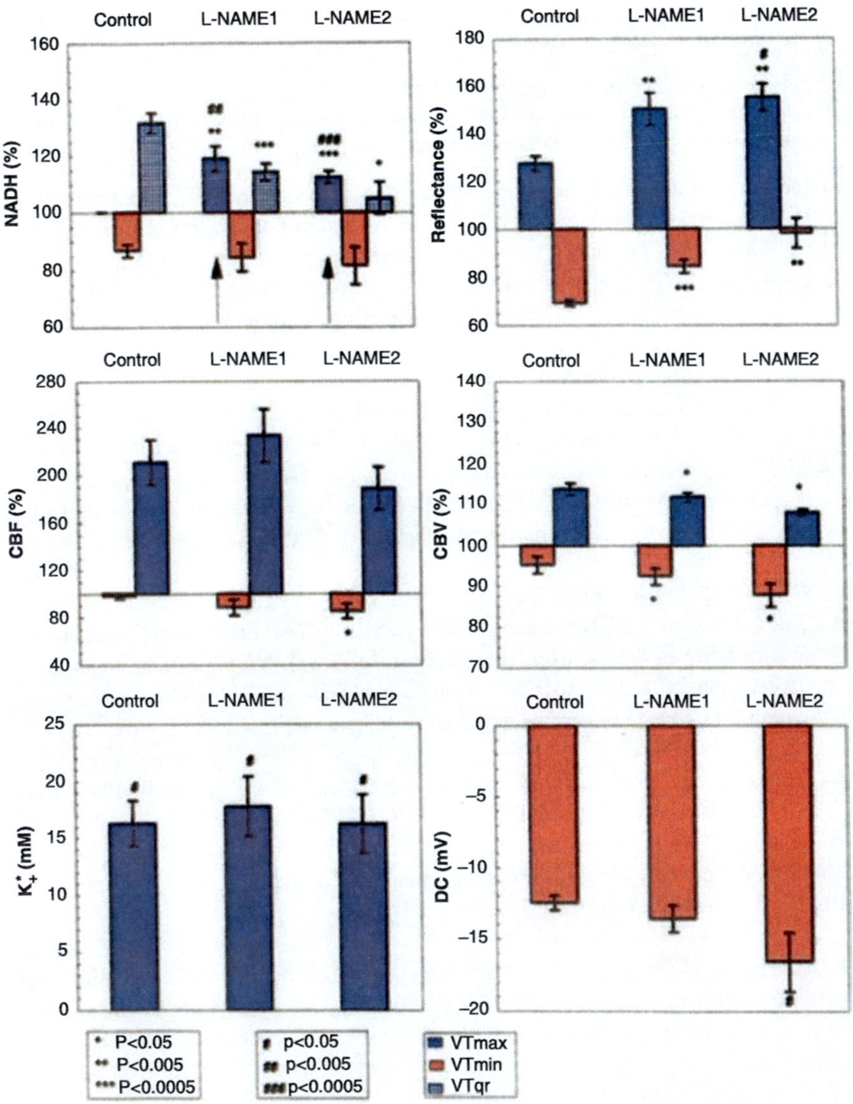

Fig. 7.27 The effect of L-NAME injection on the amplitudes of the CSD waves. The values presented in the histograms averages ±SEM in the three waves: control (Control), 1 h after L-NAME (L-NAME1), and 2 h after L-NAME (L-NAME2). VT_{max}, increased amplitude; VT_{min}, decreased amplitude; VTqr increased amplitude after VT_{min} in NADH. #, level of significance compared to the control animals; *, level of significance compared to the control wave in the experimental group

The increase in CBF during CSD before (control CSD) and after L-NAME injection (L-NAME1 and L-NAME2) was by 110.5 ± 18.6%, 133.2 ± 22.1%, and 89 ± 18%, respectively. No significant changes in VT_{max} were found after L-NAME injection as compared to control CSD in the experimental group (Fig. 7.27) and as compared to the parallel CSD during the control group (Fig. 7.24).

Reflectance VT_{max} amplitude in reflected light increased and VT_{min} decreased after L-NAME injection. The increase in VT_{max} in the experimental group (Fig. 7.27) was 27.5 ± 3%, 50.3 ± 7%, and 55.4 ± 5.8%, respectively, and the increase in VT_{max} in the control group (Figure not shown) was 35.7 ± 6%, 33.98 ± 4.7%, and 29.1 ± 7.8%. The increase in reflectance after L-NAME injection as compared to the control CSD in the experimental group was found significant ($p < 0.005$), whereas when compared to the control group only L-NAME2 showed a significant increase ($p < 0.05$) as compared to SD4 in control group. VT_{min} values after L-NAME injection were 15.6 ± 2.6% and 1.9 ± 6.2%. This decrease was found significant ($p < 0.0005$, $p < 0.005$) as compared to control CSD before L-NAME injection (−30.75 ± 1.3%). No significant differences in VT_{min} in reflectance were found as compared to the parallel CSD in the control experiment (Fig. 7.24).

CBV A significant decrease ($p < 0.05$) in VT_{max} and a significant increase ($p < 0.05$) in VT_{min} were found in CBV during CSDs after L-NAME injection as compared to the control CSD in the experimental group (Fig. 7.27). No significant changes in CBV were found after L-NAME injection (L-NAME1 and L-NAME2) as compared to the parallel CSDs (SD3, SD4) in the control group (Fig. 7.24).

Extracellular K^+ No significant changes were found in VT_{max} amplitudes during CSD before and after L-NAME injection (16.26 ± 2.1 mM, 17.8 ± 2.6 mM, 16.2 ± 2.6 mM) in the experimental group. Significant changes in VT_{max} ($p < 0.05$) were found between the experimental group as compared to the control group (10.75 ± 1.6 mM, 10.5 ± 1.1 mM, 7.7 ± 1.06 mM).

DC Potential No significant changes were found in VT_{min} in DC potential after L-NAME injection as compared to VT_{min} before L-NAME administration. The VT_{min} amplitude during the second CSD after L-NAME injection (L-NAME2) was found significantly higher (−16.6 ± 2 mV, $p < 0.05$) as compared to SD4 (−11.3 ± 0.7 mV) in the control group.

7.2.2.3.5 Results: CSD Wave Duration

Wave duration was calculated as difference between Tbr-T1 or Tr-T1 where T1 is time until beginning of change, Tbr is time to initial recovery, and Tr is time of recovery, steady state. The effect of L-NAME on CSD wave duration from T1 until the quick recovery point (Tqr) and the recovery point (Tr) in the measured parameters are summarized in Tables 7.5 and 7.6. The wave duration until the Tqr point

Table 7.5 The effect of L-NAME injection on the wave duration from the initial stage in the wave (T1) until the initial and fast recovery (Tqr) in the control CSD and the 2 following CSDs after L-NAME injection (L-NAME1 and L-NAME2)

	Control	L-NAME1	L-NAME2
	n = 10	n = 10	n = 7
Reflectance	1.88 ± 0.13	2.59 ± 0.25 *	4.43 ± 1.03 *
NADH	1.90 ± 0.10	1.97 ± 0.11	3.30 ± 1.22
CBF	1.98 ± 0.13	1.80 ± 0.08 *	1.66 ± 0.14
CBV	1.64 ± 0.14	2.22 ± 0.32	2.9 ± 0.56 *
K^+_e	2.06 ± 0.11	2.67 ± 0.16 +	3.23 ± 1.09
DC	1.51 ± 0.12	2.02 ± 0.35	3.37 ± 1.09

* $p < 0.05$; + $p < 0.005$

Table 7.6 The effect of L-NAME injection on the duration of the CSD wave from the initial stage of the wave (T1) until the recovery (Tr)

	Control	L-NAME1	L-NAME2
	n = 10	n = 10	n = 7
Reflectance	3.25 ± 0.22	3.92 ± 0.41	5.76 ± 1.13
NADH	5.84 ± 0.51	3.79 ± 0.41 +	6.10 ± 1.27
CBF	4.46 ± 0.41	3.83 ± 0.35	4.0 ± 0.60
CBV	4.34 ± 0.92	3.62 ± 0.41	4.56 ± 0.92
K^+_e	6.66 ± 0.56	5.81 ± 0.63	6.69 ± 1.13
DC	4.41 ± 0.31	4.11 ± 0.56	5.50 ± 1.25

+ $p < 0.005$

(L-NAME1) as compared to Tqr in the control CSD (Table 7.5) increased in reflectance ($p < 0.05$) and in K^+_e ($p < 0.005$) after L-NAME injection and decreased significantly ($p < 0.05$) in CBF. Wave duration during L-NAME2: reflectance and CBV showed a significant increase in Tqr ($p < 0.05$) after L-NAME administration.

The Speed of Changes in Extracellular K^+ During CSD

The parameter **T1/2** was the time in minutes to 50% of depolarization or repolarization in extracellular K^+ during CSD wave propagation.

No significant changes were found in T1/2 in the depolarization phase of K^+_e after L-NAME injection. Whereas the recovery speed of extracellular K^+ during CSD wave, the repolarization phase, decreased significantly ($p < 0.05$) after L-NAME injection during L-NAME1 and L-NAME2 (Table 7.7). It must be emphasized that T1/2 of the repolarization phases of the three CSD waves (experimental group) were significantly lower ($p < 0.001$) than T1/2 of the depolarization phases.

T1/2 values of the depolarization wave of SD4 (control group) were 0.31 ± 0.04 min, and the T1/2 of the repolarization was 1.01 ± 0.1 min. A significant decline ($p < 0.05$) in T1/2 in both phases was found in L-NAME2 (Table 7.7) as compared to SD4. In addition, T1/2 values of the depolarization phases in all three CSD waves in the experimental group were lower as compared to the control group.

Table 7.7 The effect of L-NAME injection on the rate of potassium leakage (T1/2) in the depolarization phase of the CSD wave and the rate of potassium recovery during the repolarization phase

Treatment	T1/2 Depolarization (min)		T1/2 Repolarization (min)	
	Mean ± SE (n)	Significance	Mean ± SE (n)	Significance
Control	0.23 ± 0.03 (n)		0.74 ± 0.06 (10)	
L-NAME1	0.21 ± 0.02 (10)	n.s.	1.04 ± 0.09 (10)	$p < 0.05$
L-NAME2	0.19 ± 0.04 (7)	n.s.	1.40 ± 0.16 (7)	$p < 0.05$

The Interrelation Between the Measured Parameters During CSD Propagation

The interrelation between CBF and NADH during CSD propagation was tested by comparing the time it took the parameters to get to maximum and minimum peak of the amplitude changes. L-NAME injection caused the biphasic wave in NADH to change to three-phase wave (Fig. 7.26). The time to T_{min} in NADH and T_{max} in CBF and the time to T_{max} in NADH (which preceded the oxidation wave) and T_{min} in CBF were tested. Table 7.8 shows that:

1. During control CSD propagation, the time to the minimum point in NADH (T_{min}) preceded significantly ($p < 0.0001$) the maximum peak (T_{max}) in CBF, whereas, after L-NAME injection, the time to T_{min} in NADH occurred in time of or after T_{max} in CBF.
2. T_{max} in NADH (reduction wave) was not seen in control CSD. After L-NAME injection T_{max} in NADH took place in equal time to T_{min} in NADH during control CSD. The maximum decrease in CBF (T_{min}) preceded significantly the T_{max} in NADH during L-NAME1 ($p < 0.001$), whereas during L-NAME2 no significant changes were found between T_{min} in CBF and T_{max} in NADH.
3. Comparing the changes in time to peaks T_{min} and T_{max} in CBF and NADH during CSD in control experiments: Time to T_{min} in NADH during L-NAME1 and L-NAME2 delayed the T_{min} during SD3 and SD4. T_{min} during L-NAME1 occurred significantly later ($p < 0.0001$) as compared to T_{min} in NADH during SD3 (0.56 ± 0.09 min).

The Interrelation Between CBF and Extracellular K^+

During control CSD T_{max} in extracellular K^+ was 0.59 ± 0.1 min; during L-NAME1 T_{max} was 0.81 ± 0.1 min, and during L-NAME2 wave was 0.76 ± 0.1 min. The time to T_{max} in extracellular K^+ preceded the T_{max} in CBF (Table 7.8). During control CSD T_{max} in K^+_e preceded significantly ($p < 0.0005$) the T_{max} in CBF wave.

A high and significant correlation was found between the amplitudes of the peaks between NADH and CBF during CSD propagation. During control: $r = -0.799$, $N = 20$, $p < 0.01$; during L-NAME1 waves, $r = -0.695$, $N = 22$, $p < 0.01$; during L-NAME2 waves, $r = -0.633$, $N = 19$, $p < 0.01$.

The Interrelation Between CBF and Reflectance

An initial minimum wave T_{min} in CBF, before the T_{max}, was not seen during the control CSD wave, whereas after L-NAME injection it was present during L-NAME1 and in L-NAME2 waves. T_{min} in CBF was not found significantly different as

Table 7.8 The effect of L-NAME injection on the time reaching the peak: minimum (T_{min}), maximum (T_{max}), and before recovery (Tqr) in NADH and CBF parameters. The time values were calculated in minutes (min) from the appearance of the CSD wave and presented as averages ± SEM

Treatment	NADH T_{min} (min)	CBF T_{max} (min)	N	NADH T_{max} (min)	CBF T_{min} (min)	N	NADH Tqr (min)	CBF Tqr (min)	N
Control	0.46 ± 0.07	1.13 ± 0.05 #	10	—	—		1.9 ± 0.1	1.98 ± 0.13	10
L-NAME1	1.26 ± 0.15	1.01 ± 0.09	9	0.44 ± 0.05	0.09 ± 0.04 +	9	1.97 ± 0.17	1.64 ± 0.08	10
L-NAME2	1.98 ± 0.88	0.96 ± 0.08	7	0.53 ± 0.20	0.30 ± 0.09	7	3.59 ± 1.21	1.66 ± 0.14	7

N number of animals

compared to the parallel CSDs during control experiments. The T_{min} in CBF after L-NAME injection preceded the peak of T_{max} in reflectance. A similar phenomenon was seen in control experiments. The peak of T_{max} in reflectance appeared later after L-NAME injection. During L-NAME2 it appeared significantly later ($p < 0.005$) as compared to control CSD and later as compared to SD4 during control experiments (0.46 ± 0.03 min, $p < 0.001$). T_{min} in reflectance preceded significantly the T_{max} in CBF during control CSD and appeared significantly later after L-NAME injection. In addition, T_{min} in reflectance appeared later after L-NAME injection: L-NAME1 $p < 0.005$, L-NAME2 $p < 0.05$. In the control experiments, the time to T_{min} in reflectance did not change significantly during the four waves: SD1, SD2, SD3, SD4. In addition, T_{min} in reflectance appeared significantly later during SD1 as compared to control CSD in the experimental group ($p < 0.05$) and appeared earlier during SD4 as compared to L-NAME2 ($p < 0.005$). L-NAME injection caused the peak T_{min} in reflectance to appear later by about 2.5 min as compared to the parallel CSD (SD4) in control experiment (1.27 ± 0.1 min).

No significant values were found in time to peak T_{max} in CBF during CSD after L-NAME injection as compared to control CSD in the experimental group as well as compared to the parallel CSDs during control experiment.

A high and significant reverse correlation was found between the peak amplitudes in CBF and reflectance during CSD waves propagation. During control CSD: $r = -0.799$, $N = 20$, $p < 0.01$; after L-NAME injection: during waves L-NAME1: $r = -0.825$, $N = 20$, $p < 0.01$; during waves L-NAME2: $r = -0.733$, $N = 12$, $p < 0.01$.

7.2.2.3.6 Results: Recovery from CSD Wave

The recovery time from CSD before and after L-NAME injection is shown in Fig. 7.28. Significant differences were found in the basic values of CBF, CBV, and NADH after L-NAME injection. Therefore, the recovery curves from CSD were tested by statistical comparisons summarized in Tables 7.9 and 7.10:

1. For testing the differences in recovery time from the three CSDs (Control, L-NAME1, L-NAME2), the changes in the measured parameters during 60 min were tested by the paired t-test as compared to the values before each CSD was induced.
2. The statistical ANOVA test with repeated measurements was used for testing the differences between the recovery time after each of the three CDSs during the experimental group (Control, L-NAME1, L-NAME2) to the parallel CSDs, induced during normoxia, in the Control group (SD1, SD3, SD4). The results are summarized in Table 7.9 (left side).
3. The differences between the recovery curve after control CSD (Control) and the recovery curves from CSDs after L-NAME injection (L-NAME1 and L-NAME2) were tested by analysis of variance (ANOVA) with repeated measurements. The results are summarized in Table 7.10 (left side).

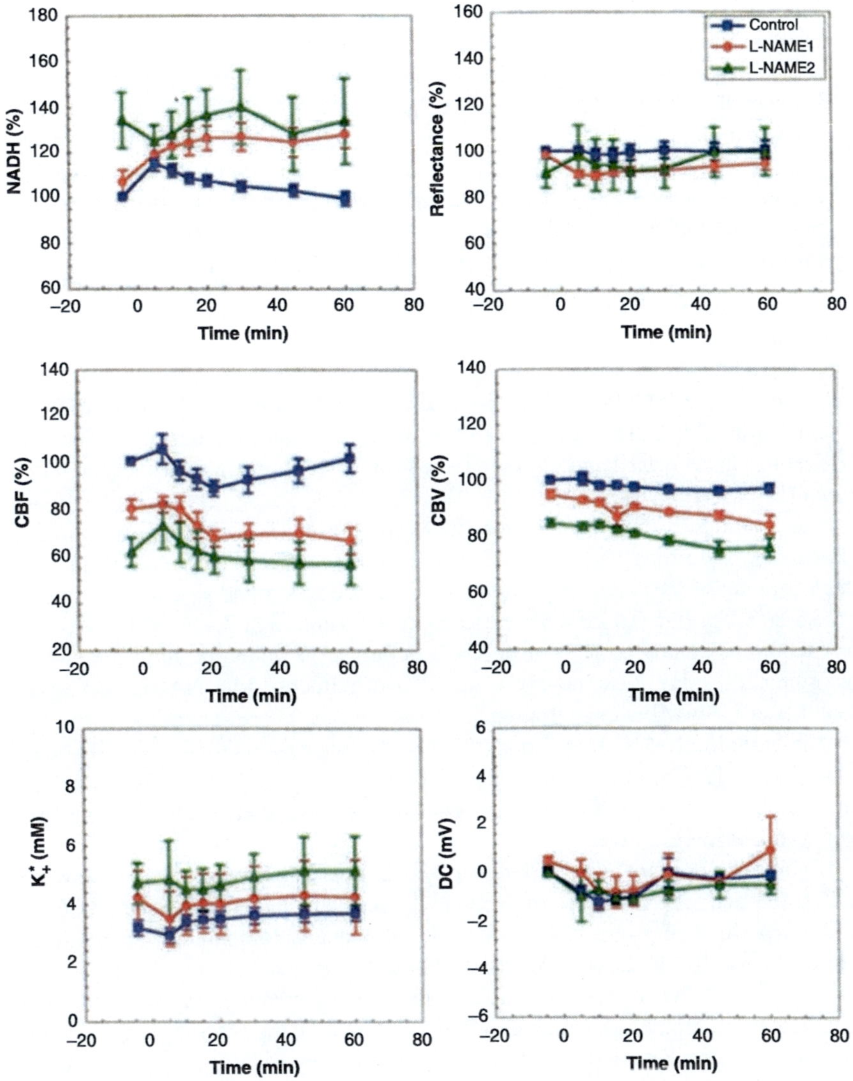

Fig. 7.28 The effects of L-NAME injection on the recovery step from the three CSD waves in the experimental groups of rats

Table 7.9 The effect of L-NAME injection on the relationship between the time to minimum (T_{min}) and maximum (T_{max}), of the peaks in reflectance and CBF

Treatment	Reflectance T_{max} (min)	CBF T_{min} (min)	N	Reflectance T_{max} (min)	CBF T_{min} (min)	N
Control	0.35 ± 0.02	—	10	0.93 ± 0.05	1.13 ± 0.05 +	10
L-NAME1	0.42 ± 0.03	0.09 ± 0.04 #	10	1.59 ± 0.16	1.01 ± 0.09 *	10
L-NAME2	1.10 ± 0.14	0.30 ± 0.09 *	6	3.75 ± 0.87	0.96 ± 0.08 #	7

4. The interaction between the effect of CSD and the effect of L-NAME, during 60 min from the beginning of CSD propagation, was tested by the multiple analysis of variance (MANOVA) test. We assumed that there was not interaction between the effect of CSD and L-NAME. Therefore, not significant results will show that there is no interaction between CSD and L-NAME. These results show that the recovery patterns from CSD during control and after L-NAME induction are parallel, whereas significant differences will indicate that there is an interaction which indicates that the recovery pattern from CSD is not parallel. These results are summarized in the right side of Tables 7.9 and 7.10.

Mitochondrial NADH Redox State Levels

After induction of control CSD, NADH levels were significantly higher during the times 5–20 min (5, 10, 15, and 20) as compared to NADH levels before CSD induction ($p < 0.005$, $p < 0.005$, $p < 0.01$, $p < 0.05$). The maximum increase in NADH was found 5 min after CSD initiation 114.9 ± 3.5% ($p < 0.005$). Later, mitochondrial NADH levels decreased and returned to control levels showing a recovery in mitochondrial NADH levels.

One hour after L-NAME injection CSD was induced (L-NAME1). Five minutes after L-NAME1 propagation, NADH levels increased progressively. At 30 min from the beginning of the wave propagation, the increase was found significantly higher (126.5 ± 5.4%, $p < 0.005$) and remained at the same high levels 127.6 ± 6.1%, $p < 0.005$, as compared to control levels before CSD initiation, until the end of recording (60 min). No recovery from CSD in mitochondrial NADH curve was found after L-NAME administration.

The basic level of NADH, 2 h after L-NAME injection (before CSD initiation), was 133.9 ± 12.4%. Five minutes after L-NAME2 propagation, NADH decreased and increased afterward. No significant changes were found in NADH during recovery from L-NAME2 wave.

Significant changes in the recovery state after CSD were found between L-NAME1 and L-NAME2 as compared to the recovery state from the control CSD in the experimental group (Table 7.10, left side) and as compared to the recovery from SD3 and SD4 in the control group (Table 7.9, left side). The trend of the recovery curve in NADH after the control CSD in the experimental group was not found significantly different as compared to SD1 CSD wave in the control group. In addition, a significant interaction ($p < 0.05$) in the process of the recovery curve in mitochondrial NADH from CSD was found after L-NAME injection (Tables 7.9, 7.10 right side). Therefore, the trend of the recovery curve from CSD in mitochondrial NADH after L-NAME administration was significantly different as compared to the trend of the recovery curve in NADH after CSD, before L-NAME administration.

Cerebral Blood Flow (CBF)

L-NAME injection caused a significant decrease in baseline levels of CBF (Fig. 7.25); therefore, the recovery curve of CBF was seen in three different levels. The recovery curves from CSD propagation after L-NAME injection were found significantly lower as compared to the recovery curves SD3 ($p < 0.001$) and SD4

Table 7.10 Significant differences in the various parameters between the recovery curves in animals injected with L-NAME and control animals (ANOVA test). The level of interaction between the CSD and the effect of L-NAME in the various time recovery changes between the experimental and control animals are presented in the right side of the table

Parameter	Control compared to SD1		L-NAME1 compared to SD3		L-NAME2 compared to SD4		Control compared to SD1		L-NAME1 compared to SD3		L-NAME2 compared to SD4	
	F	p	F	p	F	p	F	p	F	p	F	p
Reflectance	1.26	n.s	0.72	n.s	1.31	n.s	0.48	n.s	1.04	n.s	0.63	n.s
NADH	0.01	n.s	8.40	$p < 0.05$	12.46	$p < 0.005$	1.35	n.s	3.14	$p < 0.05$	4.79	$p < 0.05$
CBF	0.05	n.s	17.8	$p < 0.001$	14.9	$p < 0.005$	0.78	n.s	1.88	n.s	0.51	n.s
CBV	10.26	$p < 0.005$	105.6	$p < 0.001$	16.6	$p < 0.005$	2.75	n.s	3.39	$p < 0.05$	1.54	n.s
K^+_e	3.47	n.s	3.17	n.s	2.78	n.s	0.25	n.s	0.63	n.s	1.48	n.s
DC	0.01	n.s	0.09	n.s	1.43	n.s	0.91	n.s	2.08	n.s	2.17	n.s

n.s, not significant; SD1, SD3, SD4, the CSD waves in the control animals (n = 11)

($p < 0.05$) during control experiments (Table 7.11, left side). No significant differences were found during the recovery from CSD between control CSD in experimental group and SD1 in control group (Table 7.10).

Changes in CBF during the recovery from CSD in experimental group: At the beginning of recovery a slight increase in CBF was seen which was followed by a decrease (hypoperfusion) that continued for 20 min ($-11.5 \pm 3.3\%$, $p < 0.05$ as compared to CBF values before CSD induction). After that, CBF elevated and returned to control values.

Changes in CBF after L-NAME1 (CSD after L-NAME injection): CBF decreased and did not return to control values before CSD induction. The decrease in CBF was found significant 20, 30, and 60 min from the beginning of CSD propagation ($p < 0,005$, $p < 0.05$, $p < 0.05$, respectively). No significant differences were found in CBF during the recovery from L-NAME2 as compared to CBF values before CSD induction. In addition, no significant interaction was found between the recovery curves from CSD in the experimental group. Therefore, it can be concluded that the trend of the three recovery curves, in CBF, from CSD propagation are similar. That is, L-NAME injection caused a reduction in basic CBF levels and no changes in the trend of the recovery curve from CSD propagation. In addition, no significant changes were found between CSDs in the experimental group and the parallel CSDs (SD1, SD3, SD4) in the control group, during the recovery from CSD (Tables 7.10 and 7.11, right side). Hence, in the experimental group we found three parallel recovery curves in CBF. It should be noted that as CBF levels decreased after L-NAME injection, NADH level increased accordingly.

Cerebral Blood Volume (CBV)

The basic levels of CBV decreased after L-NAME injection (Fig. 7.25). A decrease in CBV was seen during the recovery from CSD as compared to control levels before CSD induction (Fig. 7.28).

The recovery from CSDs in the experimental group: during control CSD, a significant decrease ($p < 0.05$) in CBV was found between 10 and 45 min from the beginning of CSD propagation (where after 20 and 30 min $p < 0.01$); after L-NAME1 a significant decrease in CBV ($p < 0.05$) 15–60 min (where after 20 and 30 min $p < 0.005$); after L-NAME2, a significant decrease ($p < 0.05$) in CBV was seen 20 min after the beginning of CSD propagation.

The recovery from CSDs in CBV in the experimental group was significantly different as compared to the parallel CSDs (SD1, SD3, SD4) in the control group (Table 7.10). In addition, the recovery curves from CSDs after L-NAME administration (L-NAME1 and L-NAME2) were significantly different as compared to control CSD in the experimental group (Table 7.11). No interaction was found in CBV between the recovery curve from control CSD to the recovery curves after CSDs after L-NAME injection. Hence, the recovery curves in CBV were found to be parallel, and the course of recovery after injection of L-NAME was similar to the course of recovery in the control wave before the injection of the inhibitor.

Extracellular K^+

The basic levels of extracellular K^+ increased after L-NAME injection (Fig. 7.25). It must be emphasized that extracellular K^+ were found higher after L-NAME

Table 7.11 Significant differences in various parameters (from the ANOVA test) between the recovery after CSD in the control group and the recovery after L-NAME1 and 2 (left side of the table. The interaction level between the CSD and the effect of L-NAME during the recovery phase appears in the right side of the table

Parameter	Control compared to L-NAME1		Control compared to L-NAME2		Control compared to L-NAME1		Control compared to L-NAME2	
					Interaction between time and L-NAME			
	F	p	F	p	F	p	F	p
Reflectance	0.71	n.s	0.63	n.s	0.67	n.s	0.85	n.s
NADH	9.35	$p < 0.01$	9.1	$p < 0.05$	3.44	$p < 0.05$	6.58	$p < 0.05$
CBF	14.38	$p < 0.005$	21.58	$p < 0.001$	0.95	n.s	3.6	n.s
CBV	53.3	$p < 0.001$	186.9	$p < 0.001$	0.51	n.s	0.64	n.s
K^+_e	1.68	n.s	4.45	n.s	1.37	n.s	2.05	n.s
DC	0.26	n.s	0.75	n.s	1.33	n.s	4.04	$p < 0.05$

administration. Significant differences were found in each of the recovery lines (after CSD wave propagation) as compared to control values before CSD initiation (Fig. 7.28): in control CSD a decrease which was followed by moderate increase (~0.28–0.5 mM) and significant between 15 and 60 min (between 15 and 30 min $p < 0.05$ and between 45 and 60 min $p < 0.005$); after wave L-NAME1 a significant decrease of (~0.2–0.7 mM between 5 and 15 min ($p < 0.0005$, $p < 0.005$, $p < 0.05$). As the recovery progressed the values returned to control values; after L-NAME2 wave, no significant changes were found during the recovery curve as compared to control CSD. No significant differences were found between the three recovery curves from CSDs in the experimental group as compared to the parallel CSDs in the control group (Table 7.10). In addition, no interaction was found between the effect of L-NAME and CSD with the changes in time after the wave propagation. Hence, the course of recovery and the shape of the recovery curves of potassium after L-NAME were parallel to the course of recovery after control CSD.

DC Potential

In the recovery curves after CSD, a general trend of decline and then an increase to the control values was observed (Fig. 7.28). After the control CSD, a significant decrease of ~0.83–1.07 mV ($p < 0.05$, $p < 0.005$, $p < 0.001$, $p < 0.005$) was found between 5 and 20 min as compared to the control value before CSD initiation: After L-NAME1 DC potential declined and was found significant at 45 min after the beginning of the wave propagation (0.82 ± 0.2 mV, $p < 0.001$); during the recovery from L-NAME2 propagation, a significant decrease of ~0.5–1.06 mV ($p < 0.05$) was seen between 15 and 30 min and 60 min. No significant differences in DC potential were found between the three recovery curves from CSDs in the experimental group as compared to the parallel CSDs in the control group (Table 7.10).

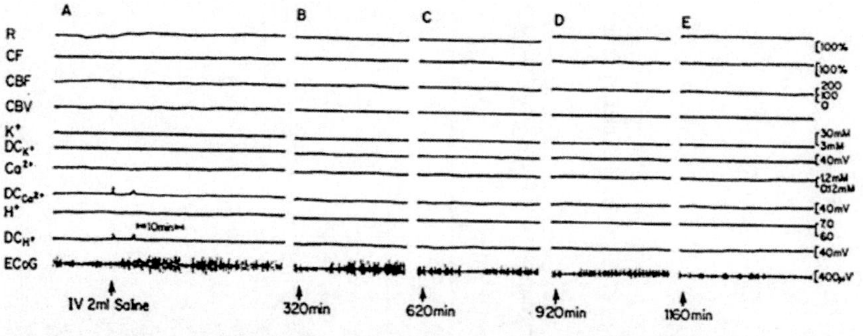

Fig. 7.29 Multiparametric recording of various brain physiologic and biochemical activities measured from unanesthetized rat injected with 2 ml saline. R, CF: reflected light (366 nm) and corrected NADH fluorescence (CF). CBF, CBV: cerebral blood flow and volume. K^+, Ca^{2+}, pH: extracellular levels of potassium, calcium, and H^+. DCK^+, $DCCa^{2+}$, DC_{pH}: DC steady potential around the ion specific electrodes. (John Wiley and sons, reprinted with permission (Mayevsky et al., 2000))

Reflectance

During the recovery from CSD, no significant differences were seen in reflectance at 5–60 min as compared to the control values before each CSD induction. Hence, it is possible that recovery after CSD in reflectance occurs during the first few minutes from the beginning of the wave propagation. In addition, no significant differences were found between the recovery curve trend after CSDs propagation in the experimental group as compared to parallel CSDs in the control group.

7.2.2.4 Summary

A. L-NAME injection caused an increase in mitochondrial NADH, a decline in CBF and in CBV.

B. The effect of L-NAME on CSD amplitudes:

After L-NAME injection, the increase in energy during CSD propagation caused a decline in oxygen delivery which led to a preliminary reduction in NADH. At the same time, in CBF a wave of decrease was observed which preceded the wave of increase in CBF. The number of animals that showed a preliminary drop wave increased after injection of the inhibitor. However, only in the second hour (L-NAME2), the drop wave amplitude was significantly higher compared to the control wave in the same experimental group.

C. An increase in wave duration after L-NAME injection, compared to the control CSD in the same group, or to the parallel waves in the control group was observed in some parameters only.

D. A decrease in the rate of potassium extraction into the cells during the repolarization phase was recorded after the inhibitor injection. In the second wave, after the inhibitor injection (L-NAME2), the decrease was significant compared to the parallel wave in control animals.

E. Following L-NAME injection, significant differences were recorded in time to peak of NADH waves, CBF, and reflectance during CSD wave progression. These differences may indicate a change in the interrelationship between metabolic and hemodynamic parameters during CSD wave propagation.

F. An inverse and significant correlation was recorded between NADH amplitudes and CBF amplitudes and between CBF and reflectance amplitudes.

G. The trend of the recovery curves after CSD waves propagation, which were induced after L-NAME injection, was like those corresponding in the control animals in the parameters: CBF, CBV, reflectance, extracellular K^+, and DC potential. Differences were found during the recovery in the NADH curves after L-NAME injection.

7.3 Alcohol Effects

7.3.1 *First Study*

In order to demonstrate the monitoring capacity and potential use the of MPA seen in Fig. 2.12 (Mayevsky et al., 2000), a typical long-term continuous monitoring of the normal brain is shown in Fig. 7.29 (Mayevsky et al., 2000). The segments A–E were taken at various intervals of time after 2 ml saline IV injection to the awake rat. The figure presents the stability of the monitoring system during more than 15 h. Most of the parameters monitored were very stable while the rat was located in a restraining cage. This kind of model could be used for new compound testing after injection to the body via IP or IV routine. According to the changes recorded in the various parameters, it is possible to hypothesize regarding the mechanism of action of a tested drug. For example, if the CBF was the sole parameter to be affected by the compound while the NADH redox state and all other parameters were stable, one could conclude that the compound has a unique effect on the CBF. The fact that NADH was stable indicated that mitochondrial function was adequate. Another possibility is that a compound may specifically inhibit the Na^+-K^+-ATPase regardless of the energy state of the brain. Under such conditions, it is possible that the extracellular K^+ level will be elevated while all other parameters will be stable. In case the tested compound is a potential epileptic activity inducer, all the parameters will respond to the injection of the compound. It is important to note that due to the changes measured by the various sensors along the time axis, it is necessary always to use a sham control group of animals injected by saline.

A practical option is to use the same type of experimental setup as presented in Fig. 2.12 and to use a known perturbation in order to stimulate the brain while monitoring all the parameters. One of the possible perturbations to be used is the CSD a well-reversible known pathophysiological stimulus described over 50 years ago (Leão, 1944a, b; Leão and Morison, 1945). During the passage of an CSD wave, a massive depolarization propagates from the stimulation site to the rest of the hemisphere (Bures et al., 1974). Due to the disturbance in ionic homeostasis (Mayevsky & Chance, 1974, 1975, 1982), the energy metabolism is stimulated as indicated also by an increase in O_2 consumption and CBF (Mayevsky & Weiss, 1991). Figure 7.30 shows a typical response of the brain to CSD after IV infusion of alcohol to the awake rat (Mayevsky et al., 2000). As can be seen in A, the control CSD, a large increase in extracellular K^+ was correlated to the activated energy metabolism as seen by the oxidation of NADH (CF) and the increase in cerebral blood flow (RCBF). In this rat, the typical "oxidation cycle" was recorded 240 min after the end of alcohol infusion (Lothman et al., 1975). It appears that the autoregulation of CBF was affected. This may also be seen in the RCBF trace. There was a correlation between the responses of the CBF and NADH (CF) during the various stages, suggesting that alcohol induced abnormal microcirculation and energy metabolism. The situation appears quite complicated since the absolute CBF level at stage C is higher than at stages A + B, yet the mitochondria did not exhibit a typical oxidation cycle.

When CSD is induced under relatively normal conditions, complete coupling occurs between the increase in O_2 consumption and the large increase in CBF. On the other hand, in under limited O_2 supply (ischemia, hypoxia), the responses of the brain to CSD are altered, as shown also in Fig. 7.30 when alcohol was infused to the rat. The combination of the NADH probe with the CBF probe used in the present study indicated that the reversed NADH response (an increase instead of a decrease) is due to the limited CBF compensation, as indicated by the initial decreased CBF response (Fig. 7.30).

7.3.2 *Second Study: The Effects of Ethanol on the Responses to CSD*

7.3.2.1 Introduction

Clinical and experimental evidence has demonstrated many toxic effects of acute and chronic alcohol administration on neuronal structure and function (Harper & Kril, 1991; Martin & Compton, 1997; Rawat, 1975; Wallgren, 1971). Ligeti et al. (1991) found that acute alcohol administration caused a decrease in cerebral blood

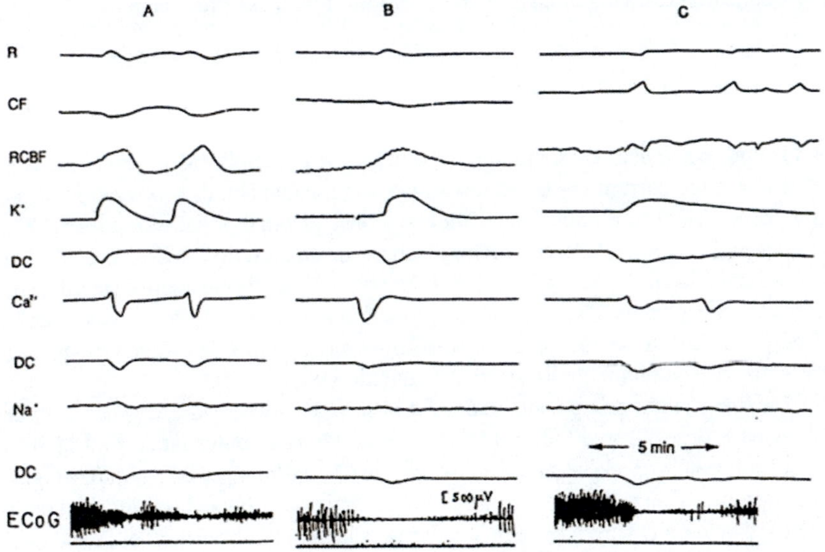

Fig. 7.30 Three responses to CSD recorded during control (**A**), immediately (**B**), as well as 240 min after the end of intravenous infusion of 3 ml 30% alcohol (**C**). In this figure we are presenting the three DC traces recorded around the K^+, Ca^{2+}, and Na^+ electrodes exhibiting similar responses. RCBF, relative CBF; other abbreviations are as in Fig. 7.29. (John Wiley and sons, reprinted with permission (Mayevsky et al., 2000))

flow and in metabolic rate, which was related to blood ethanol levels. In addition, Goldman et al. (1973) showed that cerebral blood flow (CBF) responses to ethanol varied with region and ethanol dose. Furthermore, it was shown that acute ethanol administration caused formation of free oxygen radicals (Bondy, 1992), slowed down mitochondrial oxidation, and inhibited by dose-dependent effect the Na-K-ATPase pump function (Foley & Rhoads, 1992; Novack et al., 1982; Rawat, 1975). We hypothesized that ethanol which affects CBF and inhibits Na-K-ATPase activity will interact with brain responses to CSD. The few previous studies that investigated the effect of ethanol on CSD gave contradictory results. Novack et al. (1982) found that ethanol decreased the CSD amplitude and slowed the oxidation rate of mitochondrial reactants (as compared to normal conditions), suggesting that ethanol inhibits the processes that led to increased oxygen consumption, whereas Guedes and Frade (1993) found that intragastric administration of ethanol for 7 days increased CSD velocities. These investigations indicate that ethanol causes changes in metabolic processes and in the mechanism of CSD propagation but the mechanism of these changes is not clear. Furthermore, since the brain is heterogeneous in function and in cerebral blood flow distribution, increasing oxygen demand, by inducing CSD waves may generate heterogeneity in the propagation of CSD across the entire cortex. This heterogeneity can influence the initiation and/or the recovery phases of CSD wave.

The aim of this study was to investigate the changes in the metabolic, and the electrical activities of CSD wave propagation through the entire ipsilateral cortex in real time, and the effects of ethanol administration on these processes.

7.3.2.2 Methods

To this end, we constructed and applied a multisite assembly (MSA), presented in Fig. 2.15 (in the current book), to investigate the mechanism of propagation of the CSD phenomenon continuously and simultaneously, from four parasagittal arranged adjacent areas through the entire rat hemisphere, during a series of CSD cycles. This approach enabled us to determine whether the brain's metabolic, hemodynamic, and electrical responses to CSD stimulation spread homogeneously from its source to the other cortical areas and helped us determine the rate of CSD wave propagation before and after ethanol administration in the awake rat.

The experimental group contained eight rats. In the awake state, a series of three CSD waves were induced (control CSD waves). Thirty minutes later, 3 ml of 30% ethanol solution in saline was infused for 30 min (0.1 ml/min per 200 g body weight) via the femoral vein catheter. The rate of infusion was normalized for different body weights. Three CSD waves were induced immediately after cessation of the ethanol infusion (time "0") and every 30 min during the following 4 h. The same protocol was applied in the control group (12 rats) after saline administration. CSD waves were induced by flushing the brain with 0.5 M KCl solution. The KCl solution remained until the CSD was stopped by flushing the brain with saline solution immediately after the initiation of the third wave of the series. The concentration of

KCl solution was gradually increased (maximum 2 M or in some cases KCl saturated solution), in cases where it was difficult to induce CSD waves, or when less than three subsequent waves were initiated. In some cases, immediately after ethanol administration 2 M KCl or saturated KCl solution induced only one CSD.

Data Analysis and Calculations

A special software program was prepared and used to analyze and calculate various time and amplitude values during the CSD wave cycle, as well as other calculated values such as propagation rate, frequency, duration, slopes, and integrals for each parameter representing brain tissue responses to CSD. The following calculated values will be shown and discussed:

Maximum (VT_{max}) and minimum (VT_{min}) amplitude values (Fig. 2.16).

Wave frequency (FRQ) – calculated by dividing the number of subsequent waves evoked by a single KCl application, with the time that passed from the beginning of the first CSD until the beginning of the last wave of the series.

FRQ was calculated for every parameter and for each of the four parallel channels.

Propagation rate (VTP) – calculated by dividing the distance between two adjacent monitored areas (1.8 mm) by the time it took the wave to propagate from one area to the next.

Mean propagation rate (VTPM) at the minimum and/or the maximum time points – calculated by dividing the distance between channels 1 and 4, by the time taken for the CSD wave to propagate from the first monitoring site to the fourth.

Initial NADH and reflectance levels were determined when the animals were awake and in steady-state conditions prior to CSD wave initiation. These baseline values were defined as 100%. The initial DC levels were determined as Zero mV under the same conditions.

Values at maximum (the maximum level of the increased amplitude, Fig. 2.16), minimum (the minimum level of the decreased amplitude, Fig. 2.16), and other timing points during a CSD wave were calculated relative to their levels before the CSD episode. All amplitude values, during CSD, were normalized in percent relative to the control CSD wave before ethanol or saline administration.

7.3.2.3 Results

Propagation of CSD waves on the cerebral cortex (of a control untreated animal) showed typical biphasic waves in the reflected light, an oxidation–reduction wave in NADH fluorescence, a negative shift in DC potential, and a depression in ECoG (Fig. 2.16). It can be seen that the CSD wave propagated from the first channel (closest to initiation area in the frontal lobe) towards the fourth, with a time delay of approximately 20–30 s between two adjacent channels.

The effect of ethanol infusion on the electrical activity (ECoG) is shown in Fig. 7.31 (Sonn & Mayevsky, 2001). Intravenous (IV) ethanol infusion has been initiated in awake conditions. About 1.5 min after the infusion started, a sedative

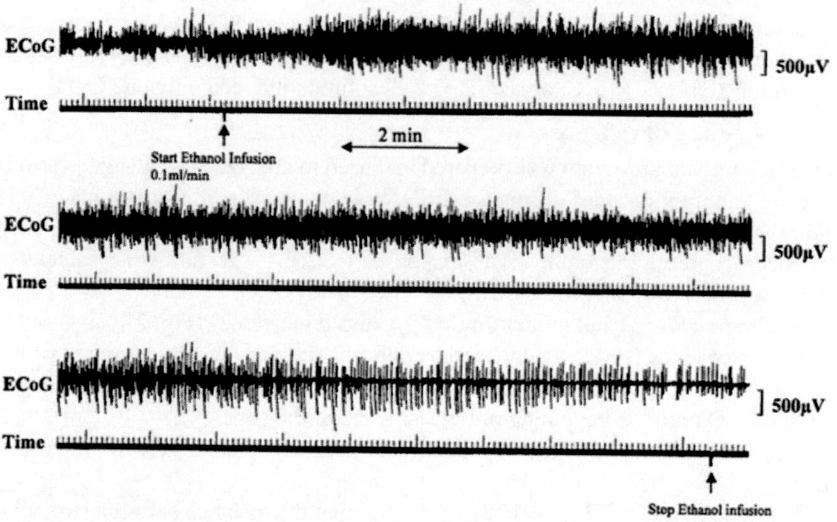

Fig. 7.31 Typical bipolar ECoG (electrocorticography) responses to IV infusion of 30% ethanol. Note the sedative effect of ethanol infusion on ECoG. The rat was in the awake state before the infusion of ethanol was started (first arrow) (Sonn & Mayevsky, 2001)

effect of ethanol was noticed by an increase in ECoG amplitude. Later, (about 20 min of infusion) a decrease in ECoG frequency was seen which was reduced to a few isolated spikes at the end of the infusion period, similar to the burst suppression phenomenon.

Figure 7.32 presents tracings of two CSD waves before (control, A) and after ethanol administration (B) in the same rat (Sonn & Mayevsky, 2001). Ethanol caused a decrease in NADH oxidation cycles (amplitudes) and a decline in the CSD wave frequency (B) as compared to the control CSD wave, induced before ethanol infusion (A). The distorted response in channels 2 and 3 of the DC potential in this specific rat was not the typical response monitored in all rats.

7.3.2.3.1 Amplitude Values

The effect of ethanol on the amplitudes of the measured parameters will be shown only immediately after ethanol (n = 8) or saline infusion (n = 12) at time "0," during the first CSD wave (Fig. 7.33) (Sonn & Mayevsky, 2001). Since ethanol induced significant differences on the amplitudes of mitochondrial NADH oxidation cycles over all monitored time points after ethanol administration, this data will be shown in Fig. 7.34 (Sonn & Mayevsky, 2001). Statistical differences of the repeated measurements will be presented during all the monitored time (4.5 h). Figure 7.34 presents averaged data of the decreased (VT_{min}) in NADH wave amplitudes in ethanol and control groups from the four areas (Channels 1–4).

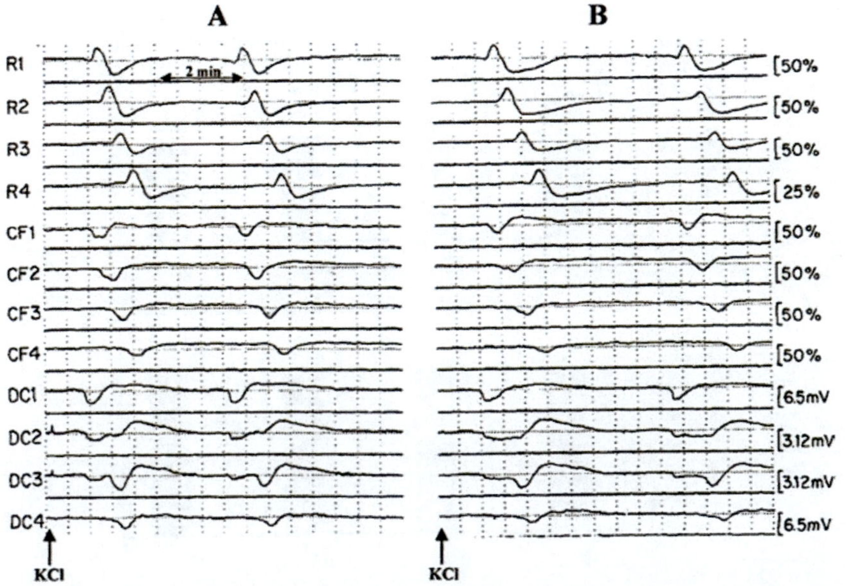

Fig. 7.32 Typical analog responses measured during two CSD cycles following saline (**A**) and 30 min after ethanol infusion (**B**). Note the changes in the amplitude of NADH oxidation cycles and in wave frequency between these two states. Reflectance, RI-R4; mitochondrial NADH redox state, CFI-CF4; direct current steady potential, DCI-DC4. CSD waves were induced by topical application of KCI solution (KCl) (Sonn & Mayevsky, 2001)

Reflectance – VT_{max} During CSD wave propagation, the amplitude of the increased phase in reflected light was in a range of minimum 10% and maximum 26% in the experimental group and between 8 and 41% in the control group. The final statistical model was significant [F = 6.33, (df = 7, 608), P < 0.0001] and included two effecting variables: treatment and number of channels. Ethanol treatment, number of channels, and the interaction between them were significant (P < 0.02, P < 0.002, P < 0.0001, respectively). Time (over 240 min) did not show significant differences. Repeated measurements showed (Fig. 7.33) that immediately after ethanol infusion only channel 3 in VT_{max} decreased significantly (p < 0.05) (Tables 7.12 and 7.13).

Reflectance – VT_{min} Amplitude values of VT_{min} in the ethanol group were found between –12% and – 29%, and in the control group, between –8% and – 32%. The final statistical model of the reflectance decrease (VT_{min}) was significant [F = 16.28, (df = 25, 586), P < 0.0001] and included three effecting variables: treatment, time, and number of channels. The interactions between treatment and time and treatment and number of channels were significant (P < 0.02, P < 0.0001, respectively). These interactions show dependence of time and number of channels as a function of treatment. Repeated measurements showed that ethanol treatment significantly enhanced the decrease in reflectance in channels 2 and 3 (Fig. 7.33) immediately after infusion, as compared to the control group. Significant differences between the two

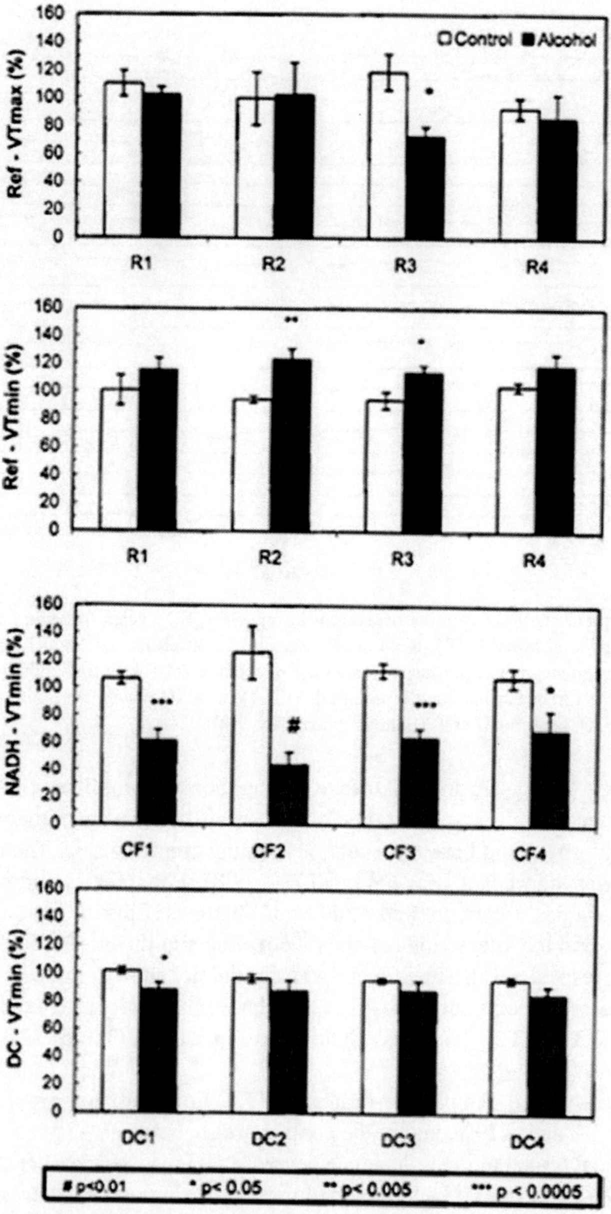

Fig. 7.33 Effect of CSD wave on reflectance, NADH, and DC amplitude values induced immediately after ethanol or saline infusion ended (at time "0") in control and ethanol treated rats. VT_{max}, value at maximum; VT_{min}, value at minimum. Data presented as mean ± S.E (Sonn & Mayevsky, 2001)

treatments were seen over almost the entire recording period, showing differences between treatments in blood volume during CSD. In the control group, VT_{min} in reflectance was significantly reduced during the 4 h after saline administration, whereas in ethanol experiments, immediately after ethanol infusion VT_{min} increased, and 60 min later a decrease tendency was found (data not shown). Significant differences in amplitude levels (VT_{min} in reflectance) between the four channels, over 240 min, in both experimental groups are shown in Table 7.12. Control experiments showed significant changes between channel 4 and channels 1 and 3, whereas, after ethanol infusion, channel 2 showed significant differences relative to the other three channels. These results indicate heterogeneous reactions between the four cortex areas in blood volume, during CSD in the two treatments and this heterogeneity was not specific to a particular channel or cortex area.

NADH Oxidation Cycles – VT_{min} The decreases in NADH amplitude during CSD wave propagation in the ethanol group were between −10% and − 42%, and in the control group between −16% and − 40%. The final statistical model was significant [$F = 9.41$, ($df = 25, 615$), $P < 0.0001$] and included three effecting variables: treatment, time, and number of channels. Significant interactions were found between treatment and time ($P < 0.0001$) and between treatment and number of channels ($P < 0.0001$). These interactions show dependence of time and number of channels as a function of treatment. Figure 7.34 shows that over 240 min of monitoring after ethanol infusion (from time 0), the NADH oxidation cycles declined, whereas at the same time in control experiments, VT_{min} in NADH remained stable or increased. As seen in Fig. 7.33 immediately after ethanol infusion (time "0") the decrease in NADH oxidation cycles was significant in all four channels. Repeated measurements showed that ethanol treatment significantly reduced NADH oxidation cycles ($P < 0.001$ or $P < 0.0001$) at almost all time points (0–180 min). More details are shown in Table 7.12.

Between the four parallel channels, significant differences were found only between channels 1 and 3 in the control group, whereas, after ethanol administration differences were seen between the first and other three channels, and also between channels 3 and 4 (Table 7.13). These results may indicate heterogeneous metabolic reactions (in NADH amplitude values) to CSD over the cerebral cortex.

The negative deflection in DC potential – VT_{min} The negative DC potential wave was between −7.5 mV and − 12 mV in the control group and between −3.9 mV and − 6.7 mV in the ethanol group. The final statistical model was significant [$F = 5.52$, ($df = 25, 737$), $P < 0.0001$] and included three effecting variables: treatment, time, and number of channels. Significant interactions were found between treatment and time ($P < 0.0002$) and between treatment and number of channels ($P < 0.005$). These interactions show dependence of time and number of channels as a function of treatment. Both treatments showed a decrease in VT_{min} in DC potential over time (data not shown). In ethanol experiments, the decrease in VT_{min} started immediately after ethanol infusion whereas in the control experiments the significant decrease started only from 150 min. As seen in Fig. 7.33, immediately after ethanol infusion a

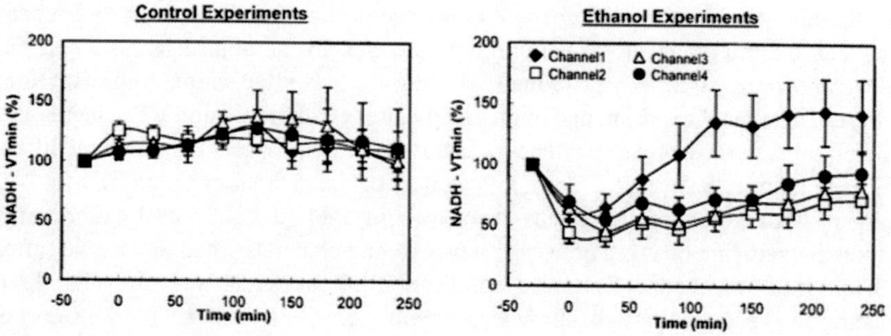

Fig. 7.34 Effect of CSD wave on NADH amplitude values measured every 30 min during 4.5 h in control and ethanol treated rats. Time 0 – CSD wave was induced immediately after ethanol or saline infusion ended. VT_{min}, value at minimum. The data include the control CSD wave (before ethanol or saline infusion, at 30 min) and are presented as mean ± S.E (Sonn & Mayevsky, 2001)

Table 7.12 Statistical difference in VT_{min} between ethanol and control treated groups during 240 min of monitoring after saline or ethanol infusion (Sonn & Mayevsky, 2001)

Time	Reflectance	NADH	DC
0	$P<0.01$	$P<0.0001$	$P<0.03$
30	$P<0.03$	$P<0.0001$	n.s
60	n.s	$P<0.0001$	$P<0.0001$
90	n.s	$P<0.0001$	$P<0.0001$
120	$P<0.02$	$P<0.0001$	$P<0.0001$
150	$P<0.0001$	$P<0.001$	n.s
180	$P<0.0001$	$P<0.001$	n.s
210	$P<0.0001$	n.s	n.s
240	$P<0.0001$	n.s	n.s

The comparison between treatments was done on CSD cycles at each time point detected except the first CSD wave before treatment initiation at time – 30 min (control); n.s, not statistically significant

decrease in VT_{min} was seen in all four channels but only channel 1 showed a significant reduction ($P < 0.05$). Detailed significant results found by repeated measurements in DC potential VT_{min} after ethanol administration are presented in Table 7.12.

Significant differences between the four parallel channels were found: in control experiments between channel 1 and the other three channels; between the two treatments differences were found between channels 2, 3, and 4 (Table 7.13). No significant differences were found between the four channels in the ethanol treated rats, showing that ethanol treatment induced homogeneous electrical changes in the four cortex areas as a result of CSD initiation.

Table 7.13 Statistical differences between the four parallel channels found during and between the two treatments in VT_{max} and VT_{min} (Sonn & Mayevsky, 2001)

Chann	Control Experiments				Ethanol Experiments				Ethanol Control				
	R VT_{max}	R VT_{min}	CF VT_{min}	DC VT_{min}	R VT_{max}	R VT_{min}	CF VT_{min}	DC VT_{min}	Chann	R VT_{max}	R VT_{min}	CF VT_{min}	DC VT_{min}
1–2	n.s	n.s	n.s	$P < 0.02$	$P < 0.02$	$P < 0.0001$	$P < 0.0001$	n.s	1–1	n.s	$P < 0.01$	n.s	n.s
1–3	n.s	n.s	$P < 0.01$	$P < 0.001$	$P < 0.0001$	n.s	$P < 0.0001$	n.s	2–2	n.s	$P < 0.0001$	$P < 0.0001$	$p < 0.002$
1–4	$p < 0.01$	$p < 0.001$	n.s	$p < 0.05$	$p < 0.003$	n.s	$p < 0.004$	n.s	3–3	$p < 0.001$	$p < 0.001$	$p < 0.001$	$p < 0.001$
2–3	$P < 0.04$	n.s	n.s	n.s	$p < 0.01$	$P < 0001$	n.s	n.s	4–4	n.s	n.s	$p < 0.001$	$p < 0.02$
2–4	n.s	n.s	n.s	n.s	n.s	$P < 0.0001$	n.s	n.s					
3–4	$p < 0.002$	$p < 0.001$	n.s	n.s	n.s	n.s	$P < 0.04$	n.s					

Chann channel, *n.s* not statistically significant, *R* reflectance, *CF* corrected fluorescence (NADH), *DC* DC potential

7.3.2.3.2 Wave Frequency (FRQ)

Wave frequency values are shown in Fig. 7.35 for the three parameters and the four parallel channels in every parameter (Sonn & Mayevsky, 2001). Since the control values of the first CSD wave (time − 30 min) were significantly different between treatments ($P < 0.0001$), for the statistic tests (as repeated measurements) we subtracted them from all the other following values. In general, Fig. 7.35 shows that the CSD wave frequency during control experiments remained stable or increased whereas, after ethanol infusion, FRQ decreased in all three parameters and four parallel cortical areas. FRQ in reflectance and DC potential showed a tendency of recovery at about 120 min after ethanol infusion. The final statistical ANOVA model of wave frequency was significant [$F = 79.81$, ($df = 15, 1475$), $P < 0.0001$] and included three effecting variables: treatment, time, and number of channels. For each treatment, FRQ values for all three parameters (reflectance, NADH, and DC.) were similar and the four parallel channels behaved alike.

Reflectance (Fig. 7.35) The final statistical model of FRQ in reflectance was significant [$F = 37.51$, ($df = 10, 428$), $P < 0.0001$] and included two effecting variables: treatment and time. Repeated measurements showed that ethanol significantly decreased wave frequency at all-time points measured ($P < 0.0005$ or $P < 0.0001$). No statistical differences were found in FRQ of the reflected light for a given treatment, showing homogeneity in hemodynamic wave frequency between the four cortical areas.

NADH (Fig. 7.35) The wave frequency of the oxidation cycles in NADH declined after ethanol treatment. The final ANOVA model was significant [$F = 162.5$, ($df = 2, 474$), $P < 0.0001$] and only treatment was found as the effecting variable. These differences between treatments were found at all monitored time points ($P < 0.0005$ or $P < 0.0001$). No differences were found between the four-recorded channels in NADH wave frequency, indicating homogeneity in metabolic wave frequency for a given treatment.

DC Potential (Fig. 7.35) The DC potential FRQ declined after ethanol infusion as compared to control experiments. The final statistical model was significant [$F = 22.13$, ($df = 21, 553$), $P < 0.0001$] and included three effecting variables: treatment, time, and number of channels. No interactions between the effecting variables were found. The decrease in DC wave frequency after ethanol infusion remained significant during the 4 h of recording ($P < 0.001$, or $P < 0.0001$).

7.3.2.3.3 Propagation Rate: VTP (Mm/Min)

The propagation rates were calculated at the maximum and minimum points during CSD wave propagation: VTP_{max} and VTP_{min} for reflectance, VTP_{min} for NADH, and VTP_{min} for DC potential. Significant statistical differences between ethanol and control groups ($P < 0.0001$) were found in VTP_{max} in reflectance, and in VTP_{min} in

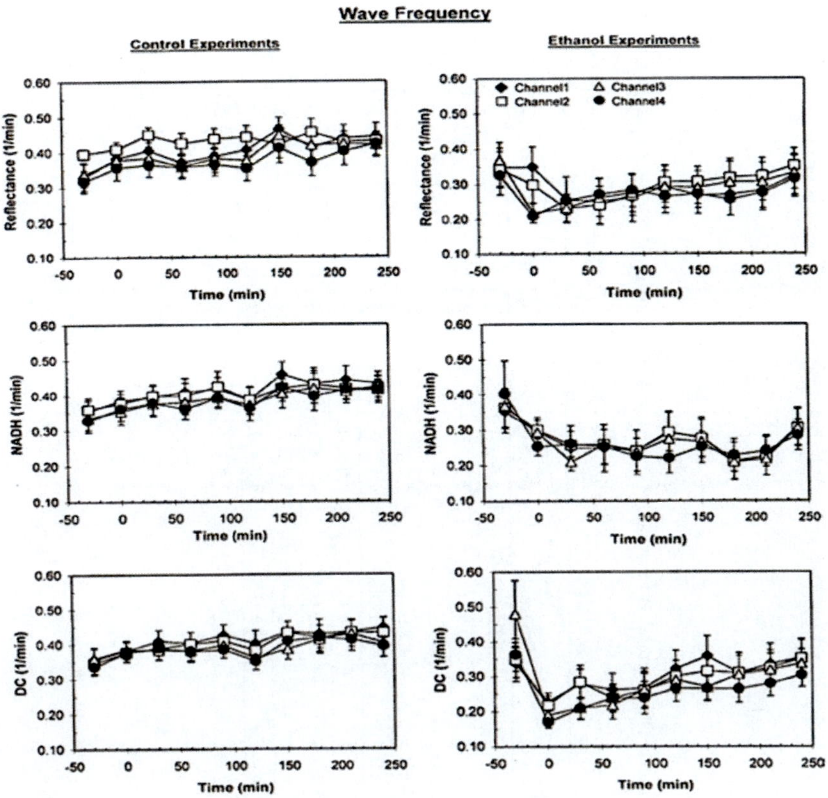

Fig. 7.35 Calculated CSD wave frequencies (mm/min) of reflectance, NADH, and DC potential waves in control and ethanol treated rats. Time 0 – CSD wave was induced immediately after ethanol infusion ended. Data presented as mean ± S.E.M. (Sonn & Mayevsky, 2001)

NADH and DC potential (data not shown). Since no statistical differences were found in propagation rate – VTP (except in VTP_{min} in reflectance) between the four cortical parallel channels and between the mean propagation rate (VTPM), detailed data are shown for VTPM. VTPM was calculated by dividing the distance between channels 1 and 4 with the time taken by a CSD wave to propagate from the first monitoring site to the fourth. Since VTPM values of the control CSD waves (time – 30 min) were significantly different between treatments ($P < 0.0001$), for the statistic tests (repeated measurements) we subtracted them from all the other following values. The values of the mean propagation rates of the CSD waves during the two treatments, and over 4.5 h of monitoring, are shown in Table 7.14. In general, after ethanol infusion VTPM decreased, whereas in control animals VTPM increased over time. The final ANOVA statistical model for VTPM was significant [$F = 2.59$, ($df = 80, 537$), $P < 0.0001$] and included three effecting variables: treatment, time, and parameters ($VTPM_{max}$ and $VTPM_{min}$ for reflectance, $VTPM_{min}$ for

Table 7.14 Mean propagation rate (VTPM) in reflectance, NADH, and DC potential at maximum ($VTPM_{max}$) and minimum ($VTPM_{min}$) points (Sonn & Mayevsky, 2001)

Time (min)	Reflectance				NADH		DC	
	$VTPM_{max}$ (mm/min)		$VTPM_{min}$ (mm/min)		$VTPM_{min}$ (mm/min)		$VTPM_{min}$ (mm/min)	
	Control	Ethanol	Control	Ethanol	Control	Ethanol	Control	Ethanol
Control	4.4 ± 0.1	3.6 ± 0.1	4.3 ± 0.2	4.0 ± 0.6	4.5 ± 0.2	4.5 ± 0.7	4.2 ± 0.2	3.5 ± 0.3
0	4.5 ± 0.2	3.5 ± 0.2	4.5 ± 0.3	3.4 ± 0.2	4.5 ± 0.3	3.7 ± 0.4	4.4 ± 0.2	3.3 ± 0.1
30	4.7 ± 0.2	3.5 ± 0.2	4.6 ± 0.5	3.5 ± 0.3	4.9 ± 0.5	3.9 ± 0.4	4.4 ± 0.2	3.2 ± 0.2
60	4.7 ± 0.2	3.7 ± 0.3	4.6 ± 0.5	3.4 ± 0.3	4.9 ± 0.4	3.7 ± 0.3	5.0 ± 0.6	3.2 ± 0.2
90	4.7 ± 0.3	3.8 ± 0.3	5.5 ± 0.8	3.4 ± 0.3	4.6 ± 0.3	4.1 ± 0.5	6.2 ± 1.8	3.4 ± 0.2
120	4.8 ± 0.3	3.6 ± 0.4	5.3 ± 0.8	3.3 ± 0.3	5.0 ± 0.4	3.5 ± 0.3	5.2 ± 0.5	3.4 ± 0.2
150	5.3 ± 0.5	3.6 ± 0.2	5.9 ± 0.9	3.3 ± 0.2	5.4 ± 0.4	3.6 ± 0.3	4.9 ± 0.2	3.5 ± 0.2
180	5.5 ± 0.5	3.8 ± 0.3	5.5 ± 0.7	3.7 ± 0.2	5.5 ± 0.4	3.8 ± 0.3	5.1 ± 0.2	3.5 ± 0.2
210	5.2 ± 0.7	3.7 ± 0.2	4.7 ± 1	3.5 ± 0.2	5.2 ± 0.5	3.8 ± 0.3	4.9 ± 0.3	3.5 ± 0.2
240	5.7 ± 0.5	3.8 ± 0.2	4.7 ± 1.2	3.8 ± 0.2	6.2 ± 0.7	3.8 ± 0.3	72 ± 2.1	3.6 ± 0.2

Data presented as mean ± S.E

NADH, and $VTPM_{min}$ for DC potential). Significant differences in VTPM ($P < 0.0001$) were found between the two treatments, and significant interactions were found between treatment and parameters found between the four monitored parameters ($P < 0.03$) and between treatment and time ($P = 0.05$). These interactions show dependence of time and parameter as a function of treatment. No significant differences were found between the four monitored parameters.

7.3.2.4 Discussion

This study was done in order to follow the metabolic, hemodynamic, and electrical changes during CSD wave propagation under acute ethanol administration. Due to the experimental procedure used, the rats were restrained in a small plexiglass cage. Therefore, it was impossible to evaluate and assess behavioral parameters of those animals. Nevertheless, the monitored significant physiological parameter used, ECoG, served as a qualitative good indicator to evaluate the animal state.

The anesthetic effect of ethanol infusion on ECoG (Fig. 7.31) was noticed by a decrease in firing rate, which was reduced to a few isolated spikes at the end of the infusion period, similar to the burst suppression phenomenon. In this study, blood alcohol levels were not measured, but a high dose of alcohol infusion (30% ethanol) was used as performed by Ligeti et al. (1991). They found that blood alcohol concentrations varied between 250 and 400 mg %.

7.3.2.4.1 Reflectance

Previous data showed (Mayevsky, 1984) that changes in reflected light represent mainly changes in blood volume. An increase in reflectance during CSD wave propagation can be related to a decrease in blood volume, while a decrease in reflectance indicates an increase in blood volume. This inverse correlation between reflectance and cerebral blood volume was also found in a different study where we measured cerebral blood flow and volume (by laser Doppler flowmetry), parallel to changes in reflectance during CSD propagation cycles (unpublished data). As shown by Lauritzen and others (Lauritzen, 1987a, 1994; Piper et al., 1991), a CSD wave is accompanied by at least three phases in cerebral blood flow: an initial decrease followed by a significant increase and a long-lasting hypoperfusion wave. The cyclical phases the reflected light may reflect these changes in blood volume, and thus in CBF during the first two phases of the CSD wave (the third phase was not seen since we monitored three following waves). In the present study, the tendency in VT_{min} after ethanol infusion increased significantly as compared to the control group, indicating an increase in blood volume in the second phase of the CSD wave after ethanol administration. Lauritzen (1987b) showed that the transient increase in CBF during CSD propagation was higher when the resting level of CBF was lower. Therefore, the augmentation in the hyperperfusion phase can occur as a result of vascular bed changes after ethanol administration, that is, a decrease in the basic levels of CBF, and therefore a higher increase in CBF levels during CSD. A decline in CBF levels after acute alcohol administration was shown by Ligeti et al. (1991) during normocapnia. Furthermore, a variability in maximum and minimum and amplitude values between the measured four channels were found in both experimental groups (Table 7.13), showing a variability in the initial hypoperfusion and hyperperfusion waves.

7.3.2.4.2 NADH Redox State

The responses of the NADH corrected fluorescence signal represent changes in mitochondrial function. As shown in our previous publication (Mayevsky, 1984), the NADH redox state represents the lack of oxygen under ischemia or hypoxia (increase NADH) as well as the increase in O_2 consumption in the activated brain (decrease in NADH). The significant correlation between the oxidation of NADH and the increase in oxygen consumption was shown by Mayevsky and Weiss in (1991). Also, the typical oxidation NADH cycle found in the normoxic brain during CSD propagation was shifted into a "reduction" NADH cycle as described in experimental animals (Mayevsky, 1984; Mayevsky et al., 1993b; Sonn & Mayevsky, 2000) as well as in the human brain (Mayevsky et al., 1996). Therefore, a decrease in the amplitude of the NADH oxidation cycle may represent a decrease in energy production and O_2 consumption.

The decline in the amplitude of the NADH oxidation cycles in ethanol treated animals (Figs. 7.33 and 7.34; Table 7.12) resulted in a decrease in energy production. It is not clear whether ethanol induced a decrease in energy demand during CSD propagation as compared to control conditions, or that this decline resulted from the direct inhibitory effect of ethanol on mitochondrial oxidative phosphorylation (Bondy, 1992). Anesthesia was shown to inhibit oxidative phosphorylation (Sharpless, 1970; Siesjo, 1978). Novack et al. (1982) found that ethanol and acetaldehyde decreased the amplitude and slowed the oxidation rate of mitochondrial reactants as compared to those normally occurring when energy demand was increased, suggesting that these drugs alter energy metabolism through inhibition of Na^+-K^+-ATPase. Nielsen et al. (1975) discussed the sequence of events occurring after acute administration of ethanol: ethanol inhibits the activity of Na^+-K^+-ATPase which leads to a decrease in brain requirements for ATP and as a result, oxygen consumption decreases. This decrease in NADH oxidation cycles after ethanol infusion can also develop due to the suppression of nerve cell activity induced by ethanol on the GABA-chloride ion channels and the NMDA receptors. The significant decrease in oxidation cycles persisted for about 90 min immediately after ethanol infusion, following which, recovery appeared (Fig. 7.34; Table 7.12). This decrease in NADH cycles after ethanol treatment may result from the anesthetic effect of ethanol. The anesthetic effect was attenuated 120 min after ethanol infusion, and then a recovery tendency in the NADH oxidation cycles was observed. The depression in ECoG amplitude found during (Fig. 7.31) and after ethanol administration can be an evidence for the sedative effect of ethanol. The recovery time of the ECoG amplitude was parallel to the recovery time in the NADH oxidation cycle amplitudes (data not shown).

7.3.2.4.3 DC Potential

The decrease in the DC steady potential amplitude (DC-VT_{min}) immediately after ethanol administration (Fig. 7.33; Table 7.12) may result from changes in extracellular ion concentrations caused by the inhibition in Na^+-K^+ ATPase pump activity and in the ionic transport processes (Israel et al., 1965; Nielsen et al., 1975; Rawat, 1975).

7.3.2.4.4 CSD Wave Frequency and Propagation Rate

CSD wave frequency decreased significantly after ethanol administration (Fig. 7.35) in all three parameters measured and the four following cortical areas monitored, as compared to control experiments. Exposure to ethanol may affect GABAergic mechanisms and voltage-dependent calcium and sodium channels inducing alterations in excitability, which can be involved in CSD initiation and propagation (Siesjo & Bengtsson, 1989). The inhibitory effect of ethanol on Na^+-K^+-ATPase pump activity may also contribute to a decrease in recovery rate from CSD, thus prolonging wave frequency. Calculations of NADH oxidation cycle slopes showed that ethanol significantly decreased the initiation and recovery slope velocities (data not shown) which may induce a decrease in CSD waves frequency. Propagation rates were calculated at the following time points during CSD propagation: at the wave initiation point (the wave front), at the maximum and minimum points and at two points on the recovery phase. The time points of the front of the waves and the recovery time points were sometimes difficult to accurately define. Therefore, the maximum and minimum time points were selected for final calculation and presentation of the propagation rates, as these values were the most accurate, although this way of calculation is affected by the propagation velocity and the kinetic process of the wave. Ethanol caused a significant decrease in CSD propagation rate expressed in the hemodynamic, metabolic, and electrical parameters monitored. These changes may also result from the sedative effect of ethanol on tissue excitability and on metabolic and hemodynamic functions. Furthermore, it was found that alcohol prevents opening of gap junctions, which is an essential step to the mechanism of CSD propagation (Largo et al., 1997). Guedes and Frade (1993) reported that rats treated with ethanol displayed higher mean CSD velocities. These differences may be due to the differences in ethanol dose or the experimental procedure (Guedes and Frade used intragastric administration of ethanol for 7 days). Comparing between treatments, the differences between channels became more pronounced, after ethanol infusion in reflectance and NADH amplitudes, whereas in DC potential, no changes were found after ethanol treatment. The reason for this variability is unclear. Ligeti et al. (1991) showed a decrease in cerebral blood flow and metabolic rate under high doses of ethanol. Goldman et al. (1973) found that these high doses of ethanol caused sedation and decreased cerebral blood flow in cortical areas. Since the areas monitored were equal in size, and the monitoring device was identical for the four channels, it seems that this heterogeneity can derive only from changes in brain tissue function. Therefore, ethanol treatment augmented the variability in hemodynamic and metabolic parameters and produced homogenies reactions in electrical responses (DC potential). The differences between the recording sites show no consistent pattern, and therefore, this phenomenon remains to be clarified. In contrast, more homogeneous changes were found in wave frequency or propagation rates between the four equal recorded channels of the hemodynamic (reflectance), metabolic (NADH), and electrical (DC potential) parameters. From these results we can conclude that the mechanisms behind the propagation of the CSD wave of the four cortical monitored areas do not differ during either treatment. In addition, ethanol caused identical mean propagation rates in the metabolic, hemodynamic, and electrical parameters, which may derive from the direct sedative effect of ethanol. However, the role of this mechanism remains to be clarified.

References

Badr, A. E., Yin, W., Mychaskiw, G., & Zhang, J. H. (2001). Effect of hyperbaric oxygen on striatal metabolites: A microdialysis study in awake freely moving rats after MCA occlusion. *Brain Research, 916*, 85–90.

Bondy, S. C. (1992). Ethanol toxicity and oxidative stress. *Toxicology Letters, 63*, 231–241.

Brown, S. D., & Piantadosi, C. A. (1989). Reversal of carbon monoxide-cytochrome c oxidase binding by hyperbaric oxygen *in vivo*. *Advances in Experimental Medicine and Biology, 248*, 747–754.

Brown, S. D., & Piantadosi, C. A. (1990). *In vivo* binding of carbon monoxide to cytochrome c oxidase in rat brain. *Journal of Applied Physiology, 68*, 604–610.

Brown, S. D., & Piantadosi, C. A. (1992). Recovery of energy metabolism in rat brain after carbon monoxide hypoxia. *The Journal of Clinical Investigation, 89*, 666–672.

Bures, J., Buresova, O., & Krivanek, J. (1974). *The mechanism and applications of Leão's spreading depression of electroencephalographic activity*. Academic.

Buresova, O. (1957). Changes in cerebral circulationin rats during spreading EEG depression. *Physiologia Bohemoslovenica, 6*, 1–11.

Chance, B., Estabrook, R. W., & Ghosh, A. (1964). Damped sinusoidal oscillations of cytoplasmic reduced pyridine nucleotinde in yeast cells. *Proceedings of the National Academy of Sciences of the United States of America, 51*, 1244–1251.

Chance, B., Erecinska, M., & Wagner, M. (1970). Mitochondrial responses to carbon monoxide toxicity. Ann. N.Y. *Academy of Sciences, 174*, 193–204.

Crowe, W., Mayevsky, A., & Mela, L. (1977). Application of a solid membrane ion selective electrode to *in vivo* measurements. *The American Journal of Physiology, 233*, C56–C60.

Crutchfield, J. S., Narayan, R. K., Robertson, C. S., & Michael, L. H. (1990). Evaluation of a fiber-optic intracranial pressure monitor. *Journal of Neurosurgery, 72*, 482–487.

Culotta, E., & Koshland, D. E., Jr. (1992). NO news is good news. *Science, 258*, 1862–1865.

Duckrow, R. B. (1993). A brief hypoperfusion precedes spreading depression if nitric oxide synthesis is inhibited. *Brain Research, 618*, 190–195.

Erecinska, M., & Silver, I. A. (1989). ATP and brain function. *Journal of Cerebral Blood Flow and Metabolism, 9*, 2–19.

Foley, T. D., & Rhoads, D. E. (1992). Effects of ethanol on Na^+-dependent amino acid uptake: Dependence on rat age and Na^+, K^+-ATPase activity. *Brain Research, 593*, 39–44.

Friedli, C. M., Sclarsky, D. S., & Mayevsky, A. (1982). Multiprobe monitoring of ionic, metabolic, and electrical activities in the awake brain. *The American Journal of Physiology, 243*, R462–R469.

Goadsby, P. J., Kaube, H., & Hoskin, K. L. (1992). Nitric oxide synthesis couples cerebral blood flow and metabolism. *Brain Research, 595*, 167–170.

Goldman, H., Sapirstein, L. A., Murphy, S., & Moore, J. (1973). Alcohol and regional blood flow in brains of rats. *Proceedings of the Society for Experimental Biology and Medicine, 144*, 983–988.

Guedes, R. C., & Frade, S. F. (1993). Effect of ethanol on cortical spreading depression. *Brazilian Journal of Medical and Biological Research, 26*, 1241–1244.

Hansen, A. J. (1985). Effect of anoxia on ion distribution in the brain. *Physiological Reviews, 65*, 101–148.

Hansen, A. J., & Zeuthen, T. (1981). Extracellular ion concentrations during spreading depression and ischemia in the rat brain cortex. *Acta Physiologica Scandinavica, 113*, 437–445.

Hansen, A. J., Quistorff, B., & Gjedde, A. (1980). Relationship between local changes in cortical blood flow and extracellular K^+ during spreading depression. *Acta Physiologica Scandinavica, 109*, 1–6.

Harper, C., & Kril, J. (1991). If you drink your brain will shrink. Neuropathological considerations. *Alcohol Alcohol, 1*, 375–380.

Iadecola, C., Pelligrino, D. A., Moskowitz, M. A., & Lassen, N. A. (1994). Nitric oxide synthase inhibition and cerebrovascular regulation. *Journal of Cerebral Blood Flow and Metabolism, 14*, 175–192.

Irikura, K., Maynard, K. I., & Moskowitz, M. A. (1994). Importance of nitric oxide synthase inhibition to the attenuated vascular responses induced by topical L-nitroarginine during vibrissal stimulation. *Journal of Cerebral Blood Flow and Metabolism, 14*, 45–48.

Israel, Y., Kalant, H., & Laufer, I. (1965). Effects of ethanol on na, K, mg-stimulated microsomal ATPase activity. *Biochemical Pharmacology, 14*, 1803–1814.

Jiang, J., & Tyssebotn, I. (1997). Normobaric and hyperbaric oxygen treatment of acute carbon monoxide poisoning in rats. *Undersea & Hyperbaric Medicine, 24*, 107–116.

Kanten, W. E., Penney, D. G., Francisco, K., & Thill, J. E. (1983). Hemodynamic responses to acute carboxyhemoglobinemia in the rat. *The American Journal of Physiology, 244*, H320–H327.

Krakovsky, M., Rogatsky, G., Zarchin, N., & Mayevsky, A. (1998). Effect of hyperbaric oxygen therapy on survival after global cerebral ischemia in rats. *Surgical Neurology, 49*, 412–416.

Largo, C., Tombaugh, G. C., Aitken, P. G., Herreras, O., & Somjen, G. G. (1997). Heptanol but not fluoroacetate prevents the propagation of spreading depression in rat hippocampal slices. *Journal of Neurophysiology, 77*, 9–16.

Lauritzen, M. (1984). Long-lasting reduction of cortical blood flow of the brain after spreading depression with preserved autoregulation and impaired CO2 response. *Journal of Cerebral Blood Flow and Metabolism, 4*, 546–554.

Lauritzen, M. (1987a). Cerebral blood flow in migraine and cortical spreading depression. *Acta Neurologica Scandinavica. Supplementum, 113*, 1–40.

Lauritzen, M. (1987b). Regional cerebral blood flow during cortical spreading depression in rat brain: Increased reactive hyperperfusion in low-flow states. *Acta Neurologica Scandinavica, 75*, 1–8.

Lauritzen, M. (1994). Pathophysiology of the migraine aura. *The spreading depression theory. Brain, 117*(Pt 1), 199–210.

Lauritzen, M., & Diemer, N. H. (1986). Uncoupling of cerebral blood flow and metabolism after single episode of cortical spreading depression in the rat brain. *Brain Research, 370*, 405–408.

Lauritzen, M., Jorgensen, M. B., Diemer, N. H., Gjedde, A., & Hansen, A. J. (1982). Persistent oligemia of rat cerebral cortex in the wake of spreading depression. *Annals of Neurology, 12*, 469–474.

Leão, A. A. P. (1944a). Pial circulation and spreading depression of activity in the cerebral cortex. *Journal of Neurophysiology, 7*, 391–396.

Leão, A. A. P. (1944b). Spreading depression of activity in cerebral cortex. *Journal of Neurophysiology, 7*, 359–390.

Leão, A. A. P., & Morison, R. (1945). Propagation of spreading cortical depression. *Journal of Neurophysiology, 8*, 33–45.

Ligeti, L., Hines, K., Dora, E., Sinnwell, T., Huang, M. T., & McLaughlin, A. C. (1991). Cerebral blood flow and metabolic rate in the conscious, freely moving rat: The effects of hypercapnia, and acute ethanol administration. *Alcoholism: Clinical and Experimental Research, 15*, 766–770.

Lothman, E., LaManna, J., Cordingley, G., Rosenthal, M., & Somjen, G. (1975). Responses of electrical potential potassium levels, and oxidative metabolic activity of the cerebral neocortex of cats. *Brain Research, 88*, 15–36.

Lukyanova, L. D., & Bures, J. (1967). Changes in pO_2 due to spreading depression in cortex and nucleus caudatus of rat. *Physiologia Bohemoslovaca, 16*, 449–455.

Martin, B. R., & Compton, D. R. (1997). Contemporary drug abuse. In C. R. Craig & R. E. Stitzel (Eds.), *Modern pharmacology with clinical applications* (pp. 429–451). Little Brown and Company.

Mayevsky, A. (1978). Ischemia in the brain: The effects of carotid artery ligation and decapitation on the energy state of the awake and anesthetized rat. *Brain Research, 140*, 217–230.

Mayevsky, A. (1983). Multiparameter monitoring of the awake brain under hyperbaric oxygenation. *Journal of Applied Physiology, 54*, 740–748.

Mayevsky, A. (1984). Brain NADH redox state monitored *in vivo* by fiber optic surface fluorometry. *Brain Research Reviews, 7*, 49–68.

Mayevsky, A. (1990). Level of ischemia and brain functions in the Mongolian gerbil *in vivo*. *Brain Research, 524*, 1–9.

Mayevsky, A. (1992). Cerebral blood flow and brain mitochondrial redox state responses to various perturbations in gerbils. *Advances in Experimental Medicine and Biology, 317*, 707–716.

Mayevsky, A. (2020). Assessment of brain metabolic score (BMS) *in vivo* based on mitochondrial activity in neuropathology. *American Journal of Biomedical Science and Research, 7*, 141–172.

Mayevsky, A., & Chance, B. (1974). Repetitive patterns of metabolic changes during cortical spreading depression of the awake rat. *Brain Research, 65*, 529–533.

Mayevsky, A., & Chance, B. (1975). Metabolic responses of the awake cerebral cortex to anoxia hypoxia spreading depression and epileptiform activity. *Brain Research, 98*, 149–165.

Mayevsky, A., & Chance, B. (1982). Intracellular oxidation-reduction state measured *in situ* by a multicannel fiber-optic surface fluorometer. *Science, 217*, 537–540.

Mayevsky, A., & Chance, B. (1983). Multisite measurements of NADH redox state from cerebral cortex of the awake animal. *Advances in Experimental Medicine and Biology, 159*, 143–155.

Mayevsky, A., & Meirovithz, E. (2020). Oxygen homeostasis and mitochondrial function in COVID-19 and ARDS patients: The ultimate vital signs. *Journal of Bioscience and Bioengineering, 1*, 1–24.

Mayevsky, A., & Weiss, H. R. (1991). Cerebral blood flow and oxygen consumption in cortical spreading depression. *Journal of Cerebral Blood Flow & Metabolism, 11*, 829–836.

Mayevsky, A., & Zarchin, N. (1987). Metabolic ionic and electrical activities during and after incomplete or complete cerebral ischemia in the Mongolian gerbil. *Advances in Experimental Medicine and Biology, 215*, 265–273.

Mayevsky, A., & Ziv, I. (1991). Oscillations of cortical oxidative metabolism and microcirculation in the ischaemic brain. *Neurological Research, 13*, 39–47.

Mayevsky, A., Zeuthen, T., & Chance, B. (1974). Measurements of extracellular potassium, ECoG and pyridine nucleotide levels during cortical spreading depression in rats. *Brain Research, 76*, 347–349.

Mayevsky, A., Zarchin, N., & Friedli, C. M. (1982). Factors affecting the oxygen balance in the awake cerebral cortex exposed to spreading depression. *Brain Research, 236*, 93–105.

Mayevsky, A., Duckrow, R. B., Yoles, E., Zarchin, N., & Kanshansky, D. (1990a). Brain mitochondrial redox state, tissue hemodynamic and extracellular ion responses to four-vessel occlusion and spreading depression in the rat. *Neurological Research, 12*, 243–248.

Mayevsky, A., Yoles, E., Zarchin, N., & Kaushansky, D. (1990b). Brain vascular ionic and metabolic responses to ischemia in the Mongolian gerbil. *Journal of Basic and Clinical Physiology and Pharmacology, 1*, 207–220.

Mayevsky, A., Frank, K., Muck, M., Nioka, S., Kessler, M., & Chance, B. (1992). Multiparametric evaluation of brain functions in the Mongolian gerbil *in vivo*. *Journal of Basic and Clinical Physiology and Pharmacology, 3*, 323–342.

Mayevsky, A., Rogatsky, G. G., Zarchin, N., & Thom, S. R. (1993a). Interrelation between hyperbaric oxygenation and carbon monoxide intoxication in the rat brain *in vivo*. In P. B. Bennett & R. E. Marquis (Eds.), *Basic and applied high pressure biology* (pp. 409–420). University of Rochester Press.

Mayevsky, A., Zarchin, N., & Sonn, J. (1993b). Brain redox state and O_2 balance in experimental spreading depression and ischemia. In A. Lehmenkuhler, K.-H. Grotemeyer, & F. Tegtmeier (Eds.), *Migraine – Basic mechanisms and treatment urban & Schwarzenberg* (pp. 379–393). Munchen-Wier.

Mayevsky, A., Meilin, S., Rogatsky, G. G., Zarchin, N., & Sonn, J. (1995). Multiparametric monitoring of the awake brain exposed to carbon monoxide. *Journal of Applied Physiology, 78*, 1188–1196.

Mayevsky, A., Doron, A., Manor, T., Meilin, S., Zarchin, N., & Ouaknine, G. E. (1996). Cortical spreading depression recorded from the human brain using a multiparametric monitoring system. *Brain Research, 740*, 268–274.

Mayevsky, A., Rogatsky, G. G., & Sonn, J. (2000). New multiparametric monitoring approach for real-time evaluation of drug tissue interaction *in vivo*. *Drug Development Research, 50*, 457–470.

Mayevsky, A., Kutai-Asis, H., & Tolmasov, M. (2020a). Mitochondrial function and brain metabolic score (BMS) in ischemic stroke: Evaluation of "neuroprotectants" safety and efficacy. *Mitochondrion, 50*, 170–194.

Mayevsky, A., Meirovithz, E., Livnat, A., Tolmasov, M., Kutai-Asis, H., & Mandelbaum, M. (2020b). Urethral tissue metabolic score (TMS) as a surrogate marker of brain oxygen balance in stroke, ARDS and Critical Care. *EC Emergency Medicine and Critical Care, SI.01*, 05–40.

Mayevsky, A., Tolmasov, M., Kutai-Asis, H., & Mandelbaum-Livnat, M. (2021). Tissue oxygen homeostasis and LifenLight score (LLS) of patients in critical care medicine. *Highlights on Medicine and Medical Research*, 18–56.

Meilin, S., Zarchin, N., Mayevsky, A., & Shapira, S. (1995). Multiparametric responses to cortical spreading depression under nitric oxide synthesis inhibition. In B. A. Weissman, N. Alon, & S. Shapira (Eds.), *Biochemical pharmacological and clinical aspects of nitric oxide* (pp. 195–204). Plenum Press.

Meilin, S., Rogatsky, G. G., Thom, S. R., Zarchin, N., Guggenheimer-Furman, E., & Mayevsky, A. (1996). Effects of carbon monoxide exposure on the brain may be mediated by nitric oxide. *Journal of Applied Physiology, 81*, 1078–1083.

Meilin, S., Sonn, J., Zarchin, N., Rogatsky, G., Guggenheimer-Furman, E., & Mayevsky, A. (1998). Responses of rat brain to induced spreading depression following exposure to carbon monoxide. *Brain Research, 780*, 323–328.

Mies, G., & Paschen, W. (1984). Regional changes of blood flow, glucose, and ATP content determined on brain sections during a single passage of spreading depression in rat brain cortex. *Experimental Neurology, 84*, 249–258.

Nielsen, R. H., Hawkins, R. A., & Veech, R. L. (1975). The effects of acute ethanol intoxication on cerebral energy metabolism. *Advances in Experimental Medicine and Biology, 59*, 93–109.

Novack, R. L., LaManna, J. C., & Rosenthal, M. (1982). Ethanol and acetaldehyde alter brain mitochondrial redox responses to direct cortical stimulation *in vivo*. *Neuropharmacology, 21*, 1051–1058.

Penney, D. G. (1990). Acute carbon monoxide poisoning: Animal models: A review. *Toxicology, 62*, 123–160.

Penney, D. G., Verma, K., & Hull, J. A. (1989). Cardiovascular, metabolic and neurologic effects of acute carbon monoxide poisoning in the rat. *Toxicology Letters, 45*, 207–213.

Piantadosi, C. A., Lee, P. A., & Sylvia, A. L. (1988). Direct effects of CO on cerebral energy metabolism in bloodless rats. *Journal of Applied Physiology, 65*, 878–887.

Piper, R. D., Lambert, G. A., & Duckworth, J. W. (1991). Cortical blood flow changes during spreading depression in cats. *The American Journal of Physiology, 261*, 96–102.

Rawat, A. K. (1975). Effects of ethanol on brain metabolism. *Advances in Experimental Medicine and Biology, 56*, 165–177.

Rogatsky, G. G., Meilin, S., Zarchin, N., Thom, S. R., & Mayevsky, A. (2002). Hyperbaric oxygenation affects rat brain function after carbon monoxide exposure. *Undersea & Hyperbaric Medicine, 29*, 50–58.

Rosenblum, B. R., Bonner, R. F., & Oldfield, E. H. (1987). Intraoperative measurement of cortical blood flow adjacent to cerebral AVM using laser Doppler velocimetry. *Journal of Neurosurgery, 66*, 396–399.

Rosenthal, M., & Somjen, G. (1973). Spreading depression, sustained potential shifts, and metabolic activity of cerebral cortex of cats. *Journal of Neurophysiology, 36*, 739–749.

Sharpless, S. K. (1970). Hypnotics and sedatives. In L. S. Goodman & A. Gilman (Eds.), *The pharmacological basis of therapeutics* (pp. 98–105). MacMillan Company.

Shibata, M., Leffler, C. W., & Busija, D. W. (1990). Cerebral hemodynamics during cortical spreading depression in rabbits. *Brain Research, 530*, 267–274.

Shinohara, M., Dollinger, B., Brown, G., Rapoport, S., & Sokoloff, L. (1979). Cerebral glucose utilization: Local changes during and after recovery from spreading cortical depression. *Science, 203*, 188–190.

Siesjo, B. K. (1978). Anaesthesia, analgesia and sedation. In B. K. Siesjo (Ed.), *Brain energy metabolism* (pp. 233–255). Wiley.

Siesjo, B. K., & Bengtsson, F. (1989). Calcium fluxes, calcium antagonists, and calcium-related pathology in brain ischemia, hypoglycemia, and spreading depression: A unifying hypothesis. *Journal of Cerebral Blood Flow and Metabolism, 9*, 127–140.

Somjen, G. G., Aitken, P. G., Czeh, G., Herreras, O., Jing, J., & Young, J. N. (1993). Spreading depression in hippocampus: Membrane currents and ion mechanisms. In A. Lehmenkuhler, K.-H. Grotemeyer, & F. Tegtmeier (Eds.), *Migraine: Basic mechanisms and treatment* (pp. 329–344). Urban and Schwarzenberg.

Sonn, J. (1996). Effect of spreading depression on the interrelation between brain energy metabolism and ion homeostasis *in vivo*, Life Sciences. Bar-Ilan, Ramat-Gan, Israel.

Sonn, J., & Mayevsky, A. (2000). Effects of brain oxygenation on metabolic, hemodynamic, ionic and electrical responses to spreading depression in the rat. *Brain Research, 882*, 212–216.

Sonn, J., & Mayevsky, A. (2001). The effect of ethanol on metabolic, hemodynamic and electrical responses to cortical spreading depression. *Brain Research, 908*, 174–186.

Stern, M. D., Lappe, D. L., Bowen, P. D., Chimosky, J. E., Holloway, G. A., Jr., Keiser, H. R., & Bowman, R. L. (1977). Continuous measurement of tissue blood flow by laser-Doppler spectroscopy. *The American Journal of Physiology, 232*, H441–H448.

Sunami, K., Takeda, Y., Hashimoto, M., & Hirakawa, M. (2000). Hyperbaric oxygen reduces infarct volume in rats by increasing oxygen supply to the ischemic periphery. *Critical Care Medicine, 28*, 2831–2836.

Thom, S. R. (1990a). Antagonism of carbon monoxide-mediated brain lipid peroxidation by hyperbaric oxygen. *Toxicology and Applied Pharmacology, 105*, 340–344.

Thom, S. R. (1990b). Carbon monoxide-mediated brain lipid peroxidation in the rat. *Journal of Applied Physiology, 68*, 997–1003.

Thom, S. R. (1992). Dehydrogenase conversion to oxidase and lipid peroxidation in brain after carbon monoxide poisoning. *Journal of Applied Physiology, 73*, 1584–1589.

Thom, S. R. (1993a). Functional inhibition of leukocyte B2 integrins by hyperbaric oxygen in carbon monoxide-mediated brain injury in rats. *Toxicology and Applied Pharmacology, 123*, 248–256.

Thom, S. R. (1993b). Leukocytes in carbon monoxide-mediated brain oxidative injury. *Toxicology and Applied Pharmacology, 123*, 234–247.

Thom, S. R., & Keim, L. W. (1989). Carbon monoxide poisoning: A review epidemiology, pathophysiology, clinical findings, and treatment options including hyperbaric oxygen therapy. *Journal of Toxicology. Clinical Toxicology, 27*, 141–156.

Toda, N., Ayajiki, K., & Okamura, T. (2009a). Cerebral blood flow regulation by nitric oxide in neurological disorders. *Canadian Journal of Physiology and Pharmacology, 87*, 581–594.

Toda, N., Ayajiki, K., & Okamura, T. (2009b). Cerebral blood flow regulation by nitric oxide: Recent advances. *Pharmacological Reviews, 61*, 62–97.

Van Harreveld, A., & Ochs, S. (1957). Electrical and vascular concomitants of spreading depression. *The American Journal of Physiology, 189*, 159–166.

Van Harreveld, A., & Stamm, J. S. (1952). Vascular concomitants of spreading cortical depression. *Journal of Neurophysiology, 15*, 487–496.

Vyskocil, F., Kritz, N., & Bures, J. (1972). Potassium-selective microelectrodes used for measuring the extracellular brain potassium during spreading depression and anoxic depolarization in rats. *Brain Research, 39*, 255–259.

Wahl, M., Lauritzen, M., & Schilling, L. (1987). Change of cerebrovascular reactivity after cortical spreading depression in cats and rats. *Brain Research, 411*, 72–80.

Wallgren, H. (1971). Effect of ethanol on intracellular respiration and cerebral function. In B. Kissin & H. Begleiter (Eds.), *The biology of Alcholism* (pp. 103–125). Plenum Press.

Wiebers, D. O., Adams, H. P., Jr., & Whisnant, J. P. (1990). Animal models of stroke: Are they relevant to human disease? *Stroke, 21*, 1–3.

Zhang, J., & Piantadosi, C. A. (1992). Mitochondrial oxidative stress after carbon monoxide hypoxia in the rat brain. *The Journal of Clinical Investigation, 90*, 1193–1199.

Chapter 8
Clinical Monitoring of Cortical Spreading Depression (CSD) in the Human Brain

Abstract The technologies for monitoring of NADH and other parameters developed for animal study of CSD was applied to monitor patients that were exposed to various surgical procedures including long-term monitoring in the ICU. The brain was the main organ monitored in neurosurgical patients who underwent vascular procedures or treated in the ICU. Our studies led to the first discovery of the responses to CSD in neurosurgical patient in the ICU. Special devices and optical probes were developed and used in the early stage and were approved for use by the Institutional Review Board of the hospitals. Also, our devices were cleared by the FDA. Typical results of the main reports are presented in this chapter.

Keywords Patient monitoring · Human brain activity · NADH monitoring in patients · Patients in neurosurgical ICU

8.1 Introduction

After the discovery of CSD in animal models by Leão in the early 1940s (Leão, 1943, 1944a, 1944b), Jasper in 1960 mentioned the CSD in human in a discussion of a meeting that "I have seen it occur accidentally while recording the electrical activity from the exposed cortex in man" (Jasper, 1960). It took another 17–18 years to show clearly (by Sramka et al) that CSD could be induced in the human brain caudate nucleus and hippocampus (Sramka et al., 1977). In the introduction to their paper, they claim that "Although CSD was occasionally observed during surgery in man (Jasper, 1960), conditions for eliciting CSD in human brain were not yet systematically explored." More effort was made in the following years (until 1996) to discover and record the CSD event in the human brain but were unsuccessful. Olsen in (1985) studied the effect of migraine on CBF in patients and found that "The changes are similar to those observed in associations with the experimental phenomenon of CSD of cortical electrical activity (spreading depression of Leão)." It's

A. Mayevsky, J. Sonn, *Cortical Spreading Depression of Leao*,
https://doi.org/10.1007/978-3-031-08068-5_8

important to note that he measured only the CBF and not any type of electrical activity typical to CSD of Leão. In 1986 Gloor (Gloor, 1986) criticized Olsen's claim and responded to his paper "In the July issue of Trends in Neurosciences, Olesen reiterates the hypothesis that the pathophysiological substrate of the classical migraine attack is Leão's CSD. He states that 'he has found no evidence against the hypothesis'. I nevertheless feel that this explanation is unconvincing. Over a 30-year period I have personally recorded the electrical activity of the exposed human cerebral cortex in the operating theater in nearly 1000 conscious, locally anesthetized epileptic patients whose cortex was widely exposed prior to surgical excision of their epileptogenic focus. Yet I have never observed anything that remotely resembles CSD during these surgical procedures. This evidence and my lack of ever observing the phenomenon in the exposed human cortex make me doubt that plays any role in human cerebral pathophysiology, including that of migraine."

In 1992, Piper, in his Doctorate of Medicine thesis submitted to the University of New South Wales (Piper, 1992), concluded that "CSD is not easily elicited after focal cortical injury in anesthetized human subjects." Only in 1–2 patients of a group of 14 operated patients, he found sort of irregular increase in CBF measured by laser Doppler flowmeter, but the data was not typical to clear increase cycle of CSD.

In 1994 another study by McLachlan and Girvin was published a paper (McLachlan & Girvin, 1994). They summarized as follows "Spreading depression (CSD) has been implicated in the pathophysiology of a number of neurological conditions. The same methodology was used to elicit CSD in anesthetized rats and in non-anesthetized patients undergoing cortical resections for intractable epilepsy. A slowly spreading DC potential shift (mean −9 mV) occurred in the cortex of 10 of 15 rats in association with attenuation of the electrocorticogram but this could not be reproduced in any of the human cortices ($n = 23$) where the mean potential shift was −0.56 mV ($P < 0.0001$). CSD is more difficult to elicit in human than rodent cortex and may not occur in man."

Recently, Hofmeijer et al. (2018) summarized as follows: "In conclusion, with full band scalp EEG in patients with cortical ischemic stroke and traumatic brain injury, we observed various ISA (Infra Slow Activity), probably modulating cortical excitability. However, we were unable to identify unambiguous characteristics of CSD. We support the recently published consensus statement that invasive recordings are needed for detection of CSD and, so far, scalp EEG alone is not sufficient."

8.2 Monitoring of NADH and Other Parameters in Patients

After more than 15 years (1973–1988) of studies on CSD, in rats, we decided to test the option of monitoring the human brain under different clinical situations that may show the CSD in the human brain.

The initial main question was could we measure reliably mitochondrial function (NADH redox state) and microcirculatory cerebral blood flow in the human cerebral cortex in the operating rooms (OR) or in the neurosurgical ICU.

The first attempt to monitor NADH from the human brain by our group was in 1990. We monitored few neurosurgical patients in the Hospital of the University of Pennsylvania after IRB approval. We monitored in those patients only mitochondrial NADH and microcirculatory blood flow since extra time availability during the surgical procedure is very limited.

Figure 8.1A shows the monitoring probe, of NADH and CBF placed on the brain surface of the patient during neurosurgical procedure. The cross section of the common part of the light guide probe is seen in Fig. 8.1B (Mayevsky et al., 1991). The common part of the light guide (1.8 mm in diameter) contained two groups of fibers. For the laser Doppler flowmetry, one 50 μ input fiber and two 100 μ output fibers were glued in a triangular shape (0.7 mm separation) and connected to the standard commercial plug of the laser Doppler flowmeter (TSI Inc., St. Paul, MN or Perimed Inc. Piscataway, N.J.). For the monitoring of NADH redox state 10 excitation and 10 emissions 200 μ fibers (General Fiber Inc., Cedar Grove, N.Y.) were randomly mixed between and around the fibers used for the laser Doppler flowmetry (Fig. 8.1B). This configuration of the fibers enabled us to monitor CBF and NADH redox state from about the same tissue volume.

The responses of the human brain to ischemia are presented in an event shown in Fig. 8.2 (Mayevsky et al., 1991). In this patient a short brain ischemic episode was induced by the occlusion of the common carotid artery during the preparations for brain aneurysm procedure. As seen in Fig. 8.2 an immediate decrease of CBF was recoded simultaneously with an increase in NADH redox state with only very small changes in the reflectance recorded (R). The recovery of CBF and NADH oxidation were very fast. The time to reach 50% of the change in NADH during the recovery

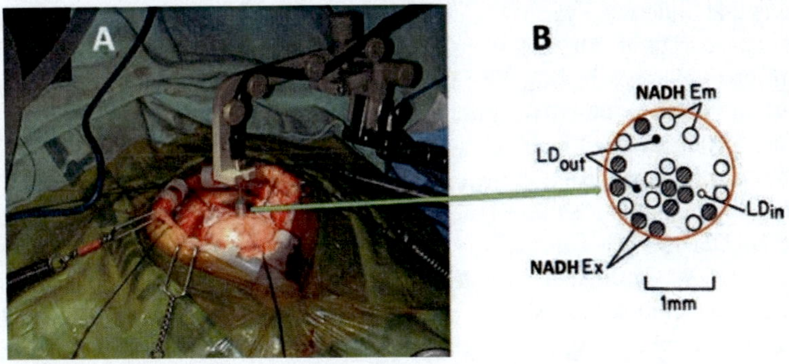

Fig. 8.1 **(A)** The application of the standard optical probe with the floating arm during a neurosurgical operation for the removal of a brain tumor (Mayevsky & Barbiro-Michaely, 2013). **(B)** Cross section of the combined light guide for monitoring of CBF and NADH redox state. A single fiber is used for LDF light input and 2 for the output. Ex and Em represent the two groups of fibers used for NADH excitation (Ex) and emission (Em). (SPIE, reprinted with permission (Mayevsky et al., 1991))

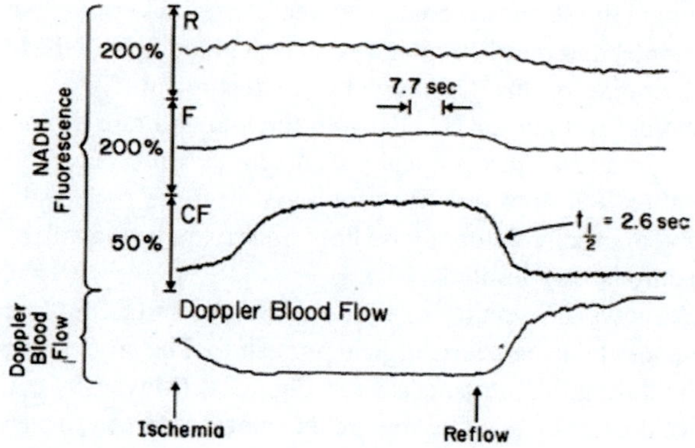

Fig. 8.2 Responses during the operation of a human cerebral cortex to a decrease in blood flow induced by common carotid artery occlusion. R, F, CF—reflectance, fluorescence, and corrected fluorescence. (SPIE, reprinted with permission (Mayevsky et al., 1991))

phase was less than 3 sec. In contrary to the NADH recovery, the CBF signal shows a large hyperemic response (large overshoot in CBF) after the reflow. *This event shown in Fig. 8.2 was the first time that mitochondrial function (NADH redox state together with cerebral blood flow could be monitored in real time, in neurosurgical patient exposed to brain ischemia.*

The effects of changes in CPP induced by a decrease in ICP following the withdrawal of 40 ml cerebrospinal fluid (CSF) from the brain are presented in Fig. 8.3 (Mayevsky et al., 2002). As seen, there is a clear negative correlation between CBF and NADH. As ICP decreased, CBF increased and NADH decreased (more oxidized). The reflectance trace shows a slow gradual decrease during the changes in CPP levels indicating that blood volume was elevated.

In another study, cerebral blood flow and NADH were monitored in a group of patients that underwent various neurosurgical procedures during a short episode of hypercapnia. These patients were divided to three groups according to their responses to changes in CO_2. The first group (n = 7) showed no responses to the change in CO_2. The other two groups responded to this short hypercapnia. In six patients CBF increased significantly while NADH remained stable. In the third group (two patients) the blood stealing phenomenon was recorded (monitoring took place at a region near a removed tumor). As seen in Fig. 8.4A, CBF was decreased, while NADH increased during the exposure to higher levels of CO_2 (Mayevsky et al., 2001). When the data recorded in the responsive group of patients was analyzed, the correlation between the two signals, seen in Fig. 8.4B (Mayevsky et al., 2002), was found to be statistically significant ($R = -0.5$, $p < 0.05$, $n = 15$).

In order to demonstrate the potential use of our technology in clinical neurosurgery, other results are presented. The patient underwent carotid artery aneurysm correction procedure, and Fig. 8.5A (Mayevsky et al., 2002) shows the results. The occlusion of the aneurysm area led to a clear decrease in CBF, while mitochondrial NADH was increased significantly. An opposite response was recorded when

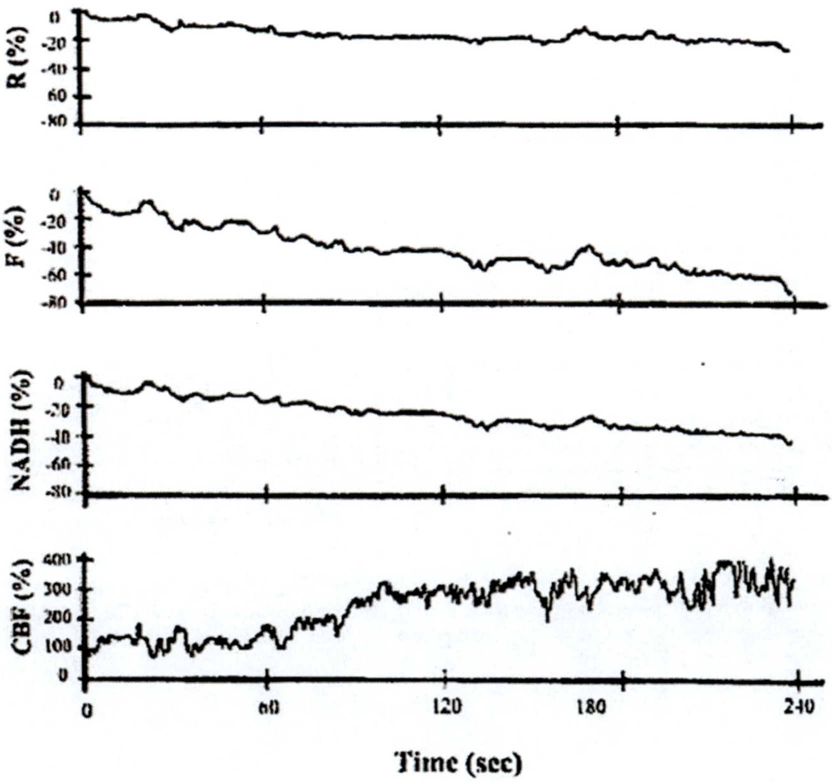

Fig. 8.3 Effects of a change in CPP induced by withdrawal of CSF on brain CBF and NADH. (Mayevsky et al., 2002)

papaverine was injected locally as seen in Fig. 8.5B (Mayevsky et al., 2001) and a clear elevation of CBF was recorded together with the small but clear oxidation of NADH (decreased signal).

After the preliminary results described in Figs. 8.1, 8.2, 8.3, 8.4, and 8.5 in which we proved that it's practical to measure CBF and NADH redox state in the cortex of patients, we decided to move to the next stage.

8.3 The Discovery of CSD in the Human Cerebral Cortex

8.3.1 Introduction

One aim of the study was to demonstrate the feasibility of using the developed patient multiparametric monitoring system in neurosurgical patients. CSD has been implicated in various pathophysiological events including ischemia or stroke (Somjen et al., 1990), seizures (Marshall, 1959), migraine (Akmal et al., 2010), and

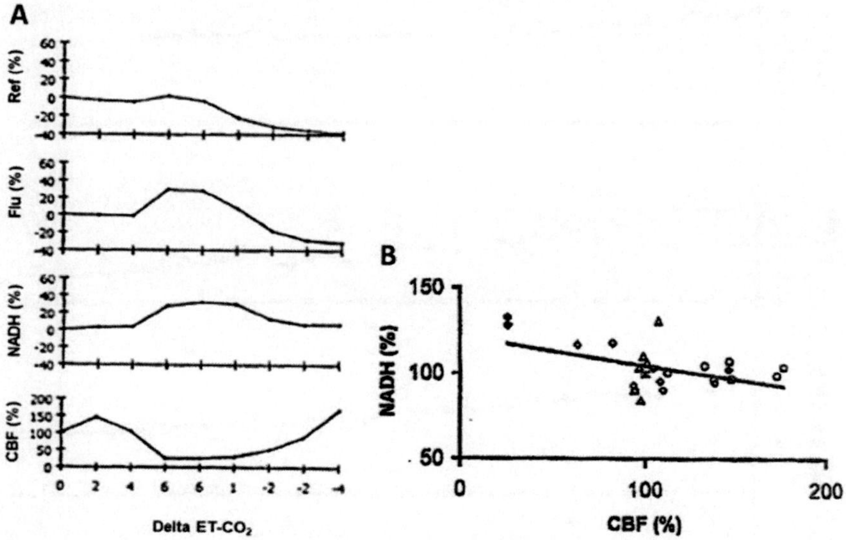

Fig. 8.4 (**A**) The possible stealing effect due to changes in CO_2 recorded in two patients. (SPIE, reprinted with permission (Mayevsky et al., 2001)). (**B**) The correlation between CBF and NADH redox state measured in neurosurgical patients exposed to hypercapnia during the operation. (Mayevsky et al., 2002)

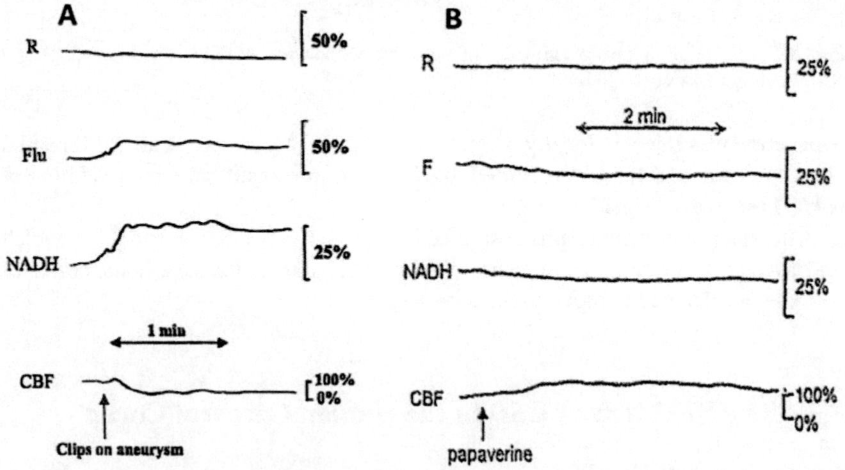

Fig. 8.5 (**A**) Metabolic and hemodynamic responses to occlusion of an aneurysm located in one of the brain arteries (Mayevsky et al., 2002). (**B**) Metabolic and hemodynamic responses to the injection of papaverine. (SPIE, reprinted with permission (Mayevsky et al., 2001))

brain injury (Akmal et al., 2010; Rogatsky et al., 1996). CSD waves cause transient changes in hemodynamic, metabolic, ionic, and electrical activities (Hansen & Zeuthen, 1981; Lauritzen, 1994; Mayevsky et al., 1974, 1985; Rosenthal & Somjen, 1973; Vyskocil et al., 1972), and thus it is preferable to monitor all these parameters simultaneously. A second aim of this study was to present results indicating, for the first time, that CSD occurs also in the human cerebral cortex.

8.3.2 *Methods*

8.3.2.1 The Development of the Monitoring System

The multiparametric assembly (MPA) enables the simultaneous, continuous on-line measurement of relative cerebral blood flow (CBF) and volume (CBV) (Haberl et al., 1989); intramitochondrial NADH redox state (Mayevsky, 1984; Mayevsky & Chance, 1982); extracellular K^+ concentrations (Crowe et al., 1977); spontaneous electrocorticography (ECoG); and DC steady potential and intracranial pressure (ICP) (Crutchfield et al., 1990). The principles of the various sensors and parameters included in the MPA have been described in previous publications (Crowe et al., 1977; Friedli et al., 1982; Mayevsky, 1983; Mayevsky et al., 1991, 1992). Figure 8.6 presents the various hardware parts of the initial monitoring device constructed for monitoring of neurosurgical patients (Mayevsky & Barbiro-Michaely, 2013). Therefore, only a short description of the principles of the various parameters measured by the MPA is presented (Fig. 8.7A). The technical details of the main technology used in the study, namely, laser Doppler flowmetry, NADH surface fluorometry, and extracellular potassium levels were described in our paper (Mayevsky et al., 1996).

8.3.2.2 Construction of the Multiprobe Assembly (MPA)

The solution used for the sensor holder (Fig. 8.7A) was basically a modification of the Lucite cannula described by Crowe et al. (Crowe et al., 1977; Friedli et al., 1982) for the light guide and the potassium-sensitive electrode. The hole accommodating the K^+ electrode was divergent at the top to facilitate handling and sealing of the probes (Friedli et al., 1982). An additional hole was drilled obliquely from the upper surface to merge with the K^+ electrode channel at approximately mid-distance to the lower surface. This hole was used to accommodate a polyethylene tubing for recording the local DC potential concentrically to the K^+ electrode. The long and rigid steel stem of the light guide used in this study occupies a straight vertical hole in the cannula and serves as an axis to hold the cannula and the cable holder at a convenient distance from each other. Spontaneous electrical activity (ECoG) was measured by two polished platinum rods inserted into the multiprobe assembly holder. Parenchymal ICP monitoring was performed using a fiber optic ICP probe (Camino Lab., USA) which was inserted into the MPA so as to have the same

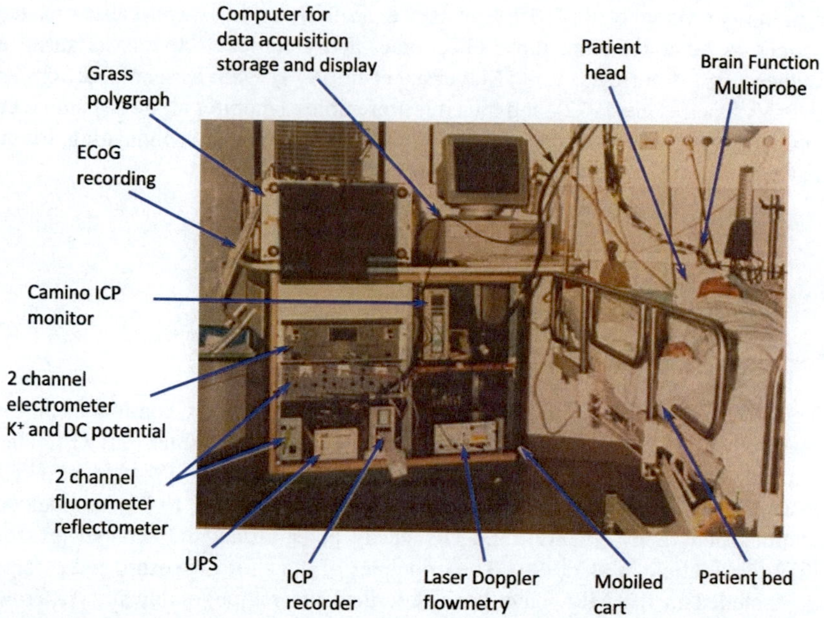

Fig. 8.6 The first clinical monitoring setup ("Brain Monitor") used in the neurosurgical intensive care unit. The multiparametric monitoring system consists of various devices installed in the same cart. The MPA shown in Fig. 8.7A connects the brain of the patient to the brain monitor. (Mayevsky & Barbiro-Michaely, 2013)

contact area with the brain surface as the other probes. The values measured by the optical probe correlated well to the standard ICP monitoring device (Friedli et al., 1982). The complete assembly was protected and shielded by an aluminum sleeve sliding over the cable holder. For stronger construction, the sleeve was permanently screwed into the cannula and cable holder. The MPA was prepared for clinical use by gas sterilization which was found to be harmless to all the sensors.

8.3.2.3 Patient Preparation

Informed consent was obtained from 14 patients in accordance with institutional review board procedures (Tel Aviv Medical Center). The patients monitored by this multiparametric assembly were hospitalized due to a comatose state induced by severe head injury (GCS <8). Preparation of the patient for brain monitoring was performed in the neurosurgical operating theater under general anesthesia. The patients were prepared for monitoring as follows: (1) A 15-mm-diameter hole was carefully drilled in the frontal bone area selected according to the CT scan. The hole was drilled in a "normal" looking section of the brain. (2) After cleaning the exposed area of bone residues, the dura mater was cut to enable direct contact between the MPA probes and the brain. (3) The MPA holder was screwed into the skull to a

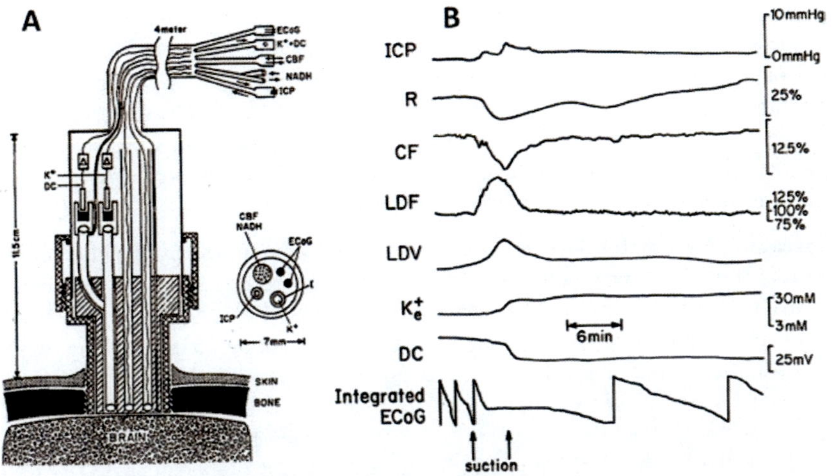

Fig. 8.7 (**A**) The multiprobe assembly (MPA) used for monitoring the brain of head-injured patients. The MPA is connected to the brain via a special holder screwed into the skull. ICP, intracranial pressure probe; CBF, NADH, fiber optic light guide probe to measure local cerebral blood flow, (CBF) and mitochondrial redox state (NADH); K^+, DC, extracellular K^+ mini surface-electrode surrounded by a DC steady potential monitoring space; ECoG, bipolar electrocortical electrodes (Mayevsky et al., 1996). (**B**) The first developed CSD cycle in a head-injured patient. ICP, intracranial pressure measured by the parenchymal probe; R, CF, 366 nm reflectance and 450 nm corrected NADH fluorescence; CBF, CBV, cerebral blood flow (CBF) and volume (CBV); K^+, DC, extracellular potassium and DC steady potential; integrated ECoG-amplitude-integrated electrocortical activity

predetermined depth to avoid any undue pressure on the brain. (4) The operated area was closed by suturing the skin around the MPA holder (covered until the beginning of the monitoring). (5) The patient was transferred to the intensive care unit (either general or neurosurgical), and after stabilization and connection to the routine patient monitoring system, the sterilized MPA was introduced into the holder (for contact with the brain) and fastened to it in order to ensure measurements for up to 48 h (sometimes even longer). Figure 8.6 shows the monitoring system connected to the brain of a comatose patient treated in the neurosurgical ICU. The average monitoring time in the 14 patients was 40 h. Three out of the 14 have died during the monitoring period due to the severity of the trauma upon admission. The other 11 recovered and were discharged from the hospital.

8.3.3 Results

In this study, the brain function of 14 patients suffering from severe head injury was monitored using a multiparametric analyzer. Due to the circumstances of the traumas, it was not possible to measure all the parameters in all patients. In general, it was possible to monitor ECoG, ICP, and CBF in most patients. NADH fluorescence was less easily measured, and the most problematic parameter was the extracellular K^+ concentration. This was probably due to the fact that the K^+ electrode is a high impedance electrode, which is very sensitive to ground noise. The typical response of most patients to treatments aimed at reducing the intracranial pressure (ICP), such as mannitol infusion and hyperventilation, was a measurable decrease in ICP and a resultant increase in CBF and oxidation of NADH.

Spreading Depression Cycles Only one of the 14 monitored patients had developed changes in all parameters similar to the typical responses to CSD recorded in animals (Mayevsky et al., 1992; Mayevsky & Weiss, 1991; Salama et al., 1987). These changes were recorded about 4.5 h after the beginning of monitoring, which was 7 h after admittance to hospital. During the measuring period, this patient was bilaterally irresponsive to pain; his pupils were dilated and non-reactive to light. He was mechanically ventilated, and his brain CT scan showed evidence of severe brain edema in the left hemisphere and right parietal hemorrhagic contusion. The measurements were taken from the right frontal lobe.

The first cycle of CSD was developed in parallel to the standard procedure in comatose mechanically ventilated patient, namely, suction of the respiratory system tubing. This response is presented in Fig. 8.7B. The ICP was elevated slightly and recovered to baseline very fast. The integrated ECoG shows a significant decrease as also very clear from the raw ECoG response seen in Fig. 8.7B.

As seen in Fig. 8.7B, an elevated extracellular K^+ and a negative shift in the DC potential were recorded. We assumed that due to the suction procedure the patient moved and therefore those changes in potassium and DC signal were not recovered to the base line and recalibration of the 2 signals was necessary. Hemodynamic and metabolic compensation for the depolarization was observed in the other parameters. An oxidation cycle was observed in the NADH redox state (decreased CF), while blood flow and volume increased. The calibration of the CBV trace shown in Fig. 8.7B is missing due to an error recorded during this interval of time.

This patient exhibited repetitive CSD cycles every 20–30 min, although at various times during the monitoring period, CSD cycles were recorded at shorter or longer intervals. Five fragments of 1 h analog and integrated ECoG recordings are presented in Fig. 8.8 (Mayevsky et al., 1996). It can be seen that the deep depression lasted between 5 and 10 min while complete recovery was not achieved even after 30–45 min. Part E of the Fig. Fig. 8.8, which corresponds to the CSD cycle shown in Fig. 8.7B, the ECoG demonstrated very slow recovery as compared to the other parameters. The DC potential negative shifts of those 5 depolarization cycles were 4 ± 0.35 mV (S.E.M). The first five cycles recorded in this patient were very similar, i.e., NADH was oxidized while blood flow increased significantly (as in Fig. 8.9A). However, in all the other cycles recorded later in the same patient (starting 4.5 h after the CSD cycle shown in Fig. 8.9A) the metabolic and hemodynamic responses were different (Fig. 8.7B) from the initial cycles (Mayevsky et al., 1998). The leakage of potassium and negative shift of the DC potential were very similar in all CSD cycles, whereas the NADH oxidation cycle was replaced by a biphasic cycle

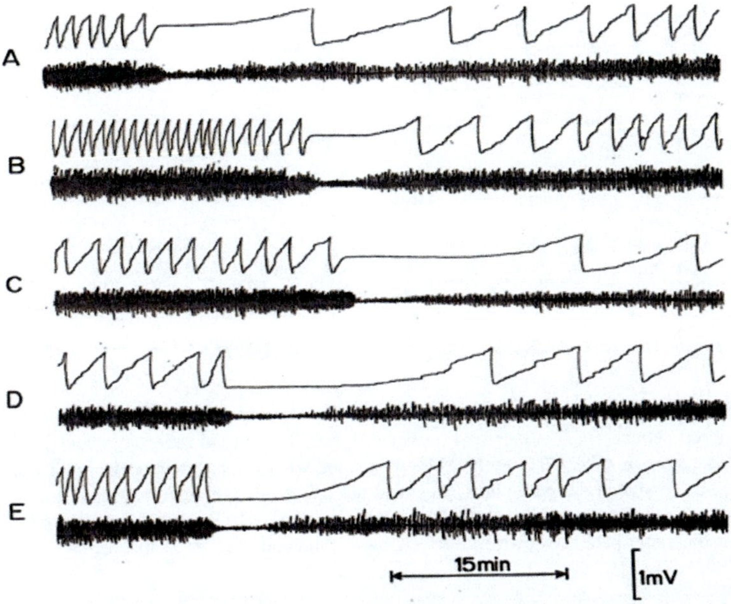

Fig. 8.8 Raw analog and integrated ECoG signals monitored during five CSD cycles. Records (**A–E**) were taken 4.4, 6.6, 7.6, 9.4, and 13.5 h after the beginning of the monitoring. During the initial phase of part A, a suction procedure was used in the patient while the other four CSD cycles occurred spontaneously. (Mayevsky et al., 1996)

comprised mainly of a phase of increased NADH followed by a small oxidation phase. The compensation of blood flow and volume was also reversed at this time. The monophasic increase in CBF and CBV (Figs. 8.7B and 8.9A) was replaced by an initial decrease followed by a smaller increase in flow and volume (Figs. 8.9B and 8.10). Figure 8.11 (Mayevsky et al., 2001) shows the analysis of the six CSD cycles seen in Fig. 8.10 (Mayevsky et al., 1996). The amplitudes of changes in NADH, LDF, LDV, and extracellular potassium were calculated. Significant correlations were found between the amplitude of the changes in the monitored parameters shown in Fig. 8.10: NADH and CBF, $p < 0.05$ (r = 0.903); NADH and K^+, $p < 0.001$ (r = 0.994); CBF and K^+, $p < 0.01$ (r = 0.940).

Figure 8.12 shows eight consecutive CSD cycles recorded from the same patient (Mayevsky et al., 1996). Downward deflections in the DC potential represent a negative shift, and the average value calculated is 5.3 ± 0.3 mV (S.E.M). The comparable values in gerbils were 10 mV (Mayevsky et al., 1985) and in rats 14–16 mV (Sonn, 1996) using the same monitoring approach.

Another set of 4 CSD cycles recorded from the same patient are presented in Fig. 8.13 (Mayevsky et al., 2001). As seen, during the period of the 4 cycles the ICP was elevated, and the metabolic responses were very similar to those seen in Fig. 8.10. During each cycle, the mitochondrial NADH signal shows a clear increase correlated to the large decrease in the CBF and CBV. When the ICP reached a

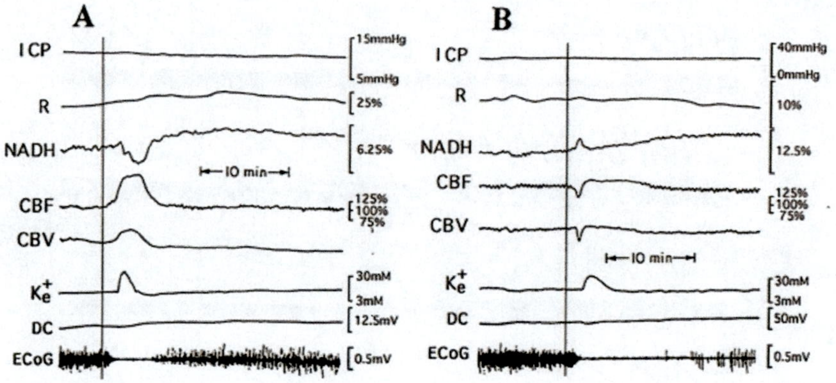

Fig. 8.9 (**A**) The responses of the brain to a wave of CSD developed spontaneously in a severely head-injured patient. (*ICP* intracranial pressure, *R* NADH, 366 nm reflectance and 450 nm corrected fluorescence, *CBF, CBV* cerebral blood flow and volume measured by laser Doppler flowmetry, K^+_e, *DC* extracellular potassium levels and the DC steady potential measured around the K^+ electrode, *ECoG* electrocorticogram (Mayevsky et al., 1998)). (**B**) Repeated response to CSD measured in the same patient shown in Part (**A**). All abbreviations are as in Part (**A**). (Mayevsky et al., 1998)

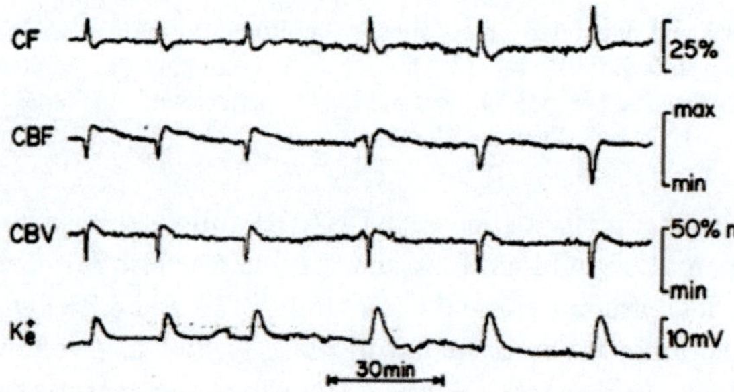

Fig. 8.10 Six repetitive CSD cycles monitored in a head-injured patient 29.5 h after the beginning of monitoring. (*CF* 450 nm corrected NADH fluorescence, *CBF, CBV* cerebral blood flow and volume, K^+_e extracellular potassium (Mayevsky et al., 1996))

certain high level (vertical line A), the base line of the CBF and CBV show a clear decrease (Horizontal line B) correlated to the increase in the NADH(CF) signal (Horizontal line C).

The patient who developed CSD cycles (presented in Figs. 8.7B and 8.9A) showed signs of brain deterioration leading to brain death. The tracing presented in Fig. 2.13B demonstrates that metabolic and ionic changes preceded brain death by several hours. The ECoG became isoelectric 3 h prior to the beginning of the recording in Fig. 2.13B (68 h from beginning of monitoring). A gradual increase in ICP up to the level of 100 mmHg was measured, followed by a sharp decrease in cerebral blood flow (CBF) and volume (CBV) (Fig. 2.13B). During the terminal ischemic phase, a large increase in the extracellular K^+ level was recorded.

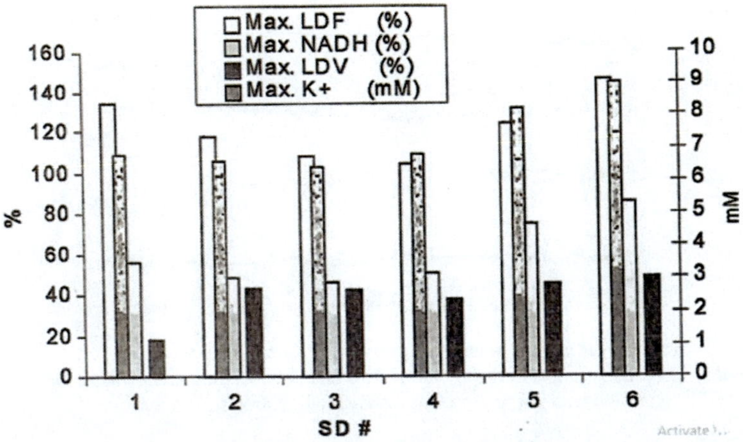

Fig. 8.11 The amplitudes of changes in a series of repetitive CSD waves. NADH, LDF, and LDV are presented as percent of change, when 100% is defined as baseline at the beginning of measurement. (SPIE, reprinted with permission. (Mayevsky et al., 2001))

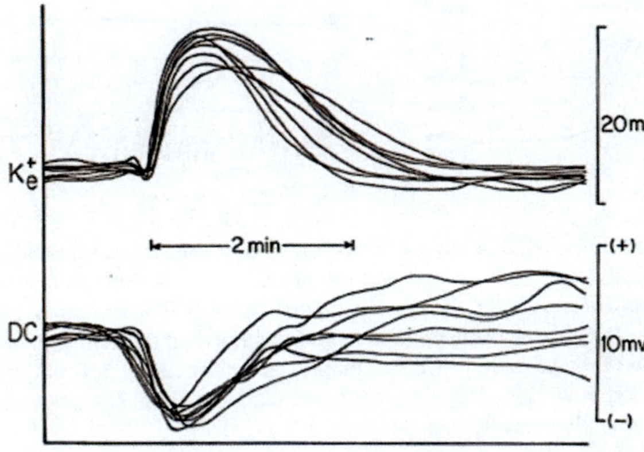

Fig. 8.12 Responses of extracellular K⁺ (K_e^+) and DC steady potential to eight repetitive CSD cycles. The absolute values of K⁺ were estimated to be 3–4 mM, but due to measurements artifacts occurring periodically, absolute values are missing. (Mayevsky et al., 1996)

The typical response of most patients to treatments aimed at reducing the intracranial pressure (ICP), such as mannitol infusion and hyperventilation, was a measurable decrease in ICP and a resultant increase in CBF and oxidation of NADH (Fig. 8.14A) (Mayevsky et al., 1996).

Figure 8.14B shows a recording (about 1 h) obtained from a comatose patient, who was treated with mannitol (Mayevsky et al., 1998). Before treatment ICP gradually increased, and as a result CBF and NADH indicated some signs of ischemia. The mannitol treatment led to hypotension and to an improvement in the energy balance (increased CBF and oxidation of NADH). The extracellular K⁺ level was

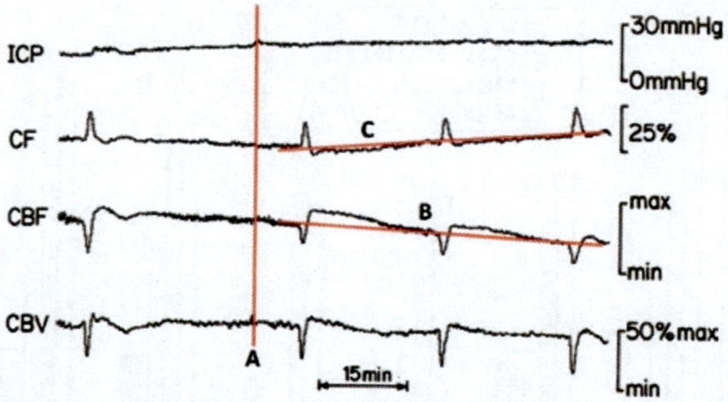

Fig. 8.13 Cycles of "reversed" waves recorded from a head-injured patient. (*ICP* intracranial pressure, *CF, NADH, CBF, CBV* cerebral blood flow and volume, measured by laser Doppler flowmetry. (SPIE, reprinted with permission (Mayevsky et al., 2001))

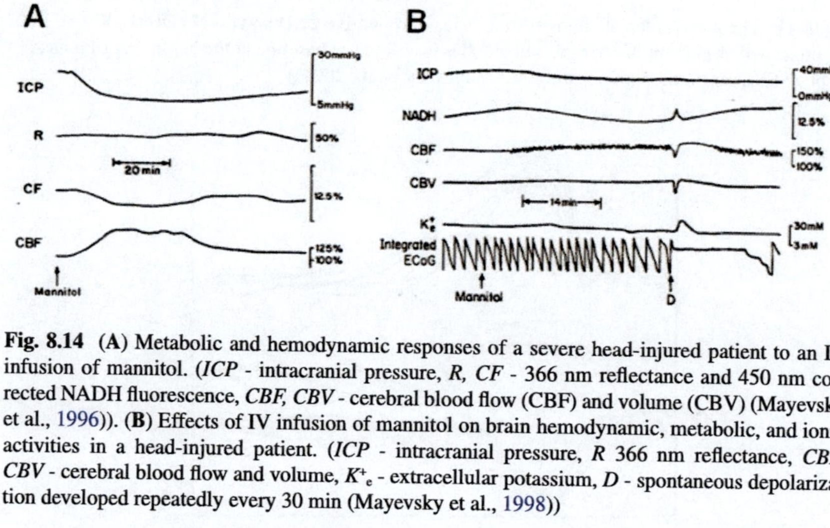

Fig. 8.14 (**A**) Metabolic and hemodynamic responses of a severe head-injured patient to an IV infusion of mannitol. (*ICP* - intracranial pressure, *R, CF* - 366 nm reflectance and 450 nm corrected NADH fluorescence, *CBF, CBV* - cerebral blood flow (CBF) and volume (CBV) (Mayevsky et al., 1996)). (**B**) Effects of IV infusion of mannitol on brain hemodynamic, metabolic, and ionic activities in a head-injured patient. (*ICP* - intracranial pressure, *R* 366 nm reflectance, *CBF, CBV* - cerebral blood flow and volume, K^{+}_{e} - extracellular potassium, *D* - spontaneous depolarization developed repeatedly every 30 min (Mayevsky et al., 1998))

apparently also affected by this treatment (decrease). The spontaneous CSD (CSD) seen in the recording was part of a repetition of more than total 40 cycles developed in this patient.

Figure 8.15 presents the responses to CSD in a rat model at the upper part (Mayevsky et al., 2000) and in patients at the lower part (Mayevsky et al., 1998). When we induced CSD in animal study, we were sure about the oxygen supply to the animal. The blood supply to the brain decreased between part A and B. The shift in response to CSD from the normoxic (A) to the ischemic (B) state is clear as discussed in Chap. 6.1.3. When we move to patients' monitoring, we want to use the NADH trace as an indicator to the physiological state of the patient or the brain in this example. As seen in the lower part, the two responses developed in the neurosurgical patients (C and D) are not the same. There are similarities between A and C

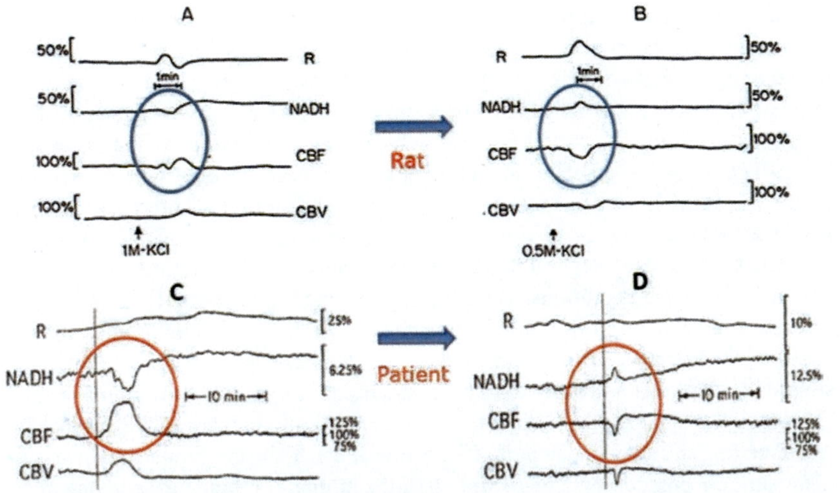

Fig. 8.15 Responses to CSD measured in normoxic (**A**) and ischemic (**B**) brain in the rat and in early "normoxic" (**C**) and later phase of the head-injured patient (**D**). (Mayevsky et al., 2000)

as well as between B and D. Therefore, we concluded that the brain of the patient in state C was probably normoxic while in D the brain was partially ischemic. Indeed, this patient was deteriorating in time, and finally brain death was determined. This example demonstrates that the basic pathophysiological events developed in patients are very similar to those monitored in animal models.

8.3.4 Discussion

Monitoring of the human brain exposed to clinical pathophysiological conditions has long presented a challenge. However, the lack of appropriate measuring techniques has so far prevented progress in this field. Attempts have been made to monitor the human brain in the operating room as well as in the intensive care unit. For example, the standard technique of intraoperative EEG monitoring has been applied to patients exposed to surgical excision of their epileptogenic focus. CSD of the cerebral cortex in human was never recorded. Sramka et al. (1977), using DC steady potential together with EEG recording, were able to elicit and record CSD in the caudate nucleus and hippocampus. Somjen et al. (1993) stated that "The electrical signs of CSD have never been actually demonstrated in any condition of intact human brain," but they were able to do so in hippocampal tissue slices. CBF has been monitored in the operating room (OR) using laser Doppler flowmetry (Rosenblum et al., 1987). Other groups have applied spectrophotometric techniques to the monitoring of mitochondrial redox state in the neurosurgical OR (Jobsis et al., 1971). Nevertheless, until now a standard practical approach for monitoring the brain in the OR or in the intensive care unit has not been adopted. Even in less severe neurological disorders such as stroke, the need exists for a reliable monitoring technique. Wiebers et al. (1990) made the following very lucid statement on the issue of human brain monitoring: "Ultimately, however, the answers to many of our questions regarding the underlying pathophysiology and treatment of stroke do not lie with continued attempts to model the human situation perfectly in animals, but rather with development of techniques to enable the study of more basic metabolism, pathophysiology and anatomical imaging detail in living humans." The results presented in this study suggest that the human brain can be monitored under various pathophysiological conditions. The development and implementation of the surface probes organized in the multiprobe assembly (MPA) for animal studies (Crowe et al., 1977; Friedli et al., 1982; Mayevsky & Chance, 1982) has made it relatively simple to adapt this assembly to clinical situations. Indeed, in a preliminary study (Mayevsky et al., 1991), we demonstrated a correlation between CBF and NADH redox state measured under partial ischemia induced in the human brain during neurosurgical procedures. The recording of the human cerebral cortex in one of the patients, which was found to be qualitatively similar to that observed in rats, gerbils, or cats (Mayevsky et al., 1991; Mayevsky & Weiss, 1991; Rosenthal & Somjen, 1973), suggests that the basic mechanisms of pathophysiological processes may be similar in the animal and human brain. It is well documented that during the CSD wave, which is basically the propagation of depolarization through the entire hemisphere, there is a concomitant large increase in O_2 consumptions due to stimulation of the ion pumps (for review, see Ref. (Mayevsky & Weiss, 1991)). A large increase in CBF (measured by various techniques) compensates for this increased O_2 demand (Chance et al., 1964; Lauritzen, 1994; Van Harreveld & Stamm, 1952). When CSD is induced in animals under normal conditions, complete coupling occurs between the increased O_2 consumptions and the increase in CBF. If the O_2 supply is limited (ischemia, hypoxia), the responses of the brain to CSD are different. We have found that under various pathophysiological conditions, NADH demonstrates a cycle of reduction (increased NADH) instead of the typical normal oxidation cycle (decreased NADH) (Mayevsky et al., 1982). The combined measurement of NADH and CBF used in the present study indicated that the reversed NADH response (an increase instead of a decrease) was due to the limited compensation of CBF as indicated by its initial decrease (Fig. 8.10). The results presented in this report demonstrate that CSD can develop in the human brain. We suspect that with this patient, one of the suction procedures (used to clear the airway system) was the stimulus for initiating the first CSD wave. We do not know what caused the repetitive CSD waves. The fact that the nature of the hemodynamic and metabolic responses changed in this patient as a function of time may be due to the deterioration of CBF autoregulation in the head-injured patient. Indeed, Oka et al. (1977) claim that in head-injured patients, a unique traumatic CSD syndrome could be identified, although they were unable to correlate behavior and physiological parameters. In conclusion, this study demonstrates that the multiparametric analyzer can be used for monitoring the human brain in the neurosurgical as well as the general intensive care unit. This assembly could be used for monitoring other pathophysiological conditions, such as ischemia, epilepsy, and migraine, which are considered inducive of CSD, especially since it has been shown in this study that CSD may occur in vivo in the human cerebral cortex.

References

Akmal, M., Zulkifle, M., & Ansari, A. (2010). Ibn nafis - a forgotten genius in the discovery of pulmonary blood circulation. *Heart Views, 11*, 26–30.

Chance, B., Estabrook, R. W., & Ghosh, A. (1964). Damped sinusoidal oscillations of cytoplasmic reduced pyridine nucleotinde in yeast cells. *Proceedings of the National Academy of Sciences of the United States of America, 51*, 1244–1251.

Crowe, W., Mayevsky, A., & Mela, L. (1977). Application of a solid membrane ion selective electrode to in vivo measurements. *The American Journal of Physiology, 233*, C56–C60.

Crutchfield, J. S., Narayan, R. K., Robertson, C. S., & Michael, L. H. (1990). Evaluation of a fiber-optic intracranial pressure monitor. *Journal of Neurosurgery, 72*, 482–487.

Friedli, C. M., Sclarsky, D. S., & Mayevsky, A. (1982). Multiprobe monitoring of ionic, metabolic, and electrical activities in the awake brain. *The American Journal of Physiology, 243*, R462–R469.

Gloor, P. (1986). Migraine and regional cerebral blood flow. *Trends in Neurosciences, 9*, 21.

Haberl, R. L., Heizer, M. L., Marmarou, A., & Ellis, E. F. (1989). Laser-Doppler assessment of brain microcirculation: Effect of systemic alterations. *American Journal of Physiology. Heart and Circulatory Physiology, 256*, H1247–H1254.

Hansen, A. J., & Zeuthen, T. (1981). Extracellular ion concentrations during spreading depression and ischemia in the rat brain cortex. *Acta Physiologica Scandinavica, 113*, 437–445.

Hofmeijer, J., van Kaam, C. R., van de Werff, B., Vermeer, S. E., Tjepkema-Cloostermans, M. C., & van Putten, M. (2018). Detecting cortical spreading depolarization with full band scalp electroencephalography: An illlusion? *Frontiers in Neurology, 9*, 17.

Jasper, H. H. (1960). Discussion in: The moscow colloquium on electroencephalography of higher nervous activity. In H. H. Jasper & G. D. Smirnov (Eds.), *Electroencephalography Clinical Neurophysiology. Suppl. 13* (p. 375). The Academy of Sciences of the USSR.

Jobsis, F. F., O'Connor, M. J., Rosenthal, M., & Van Buren, J. M. (1971). Fluorometric monitoring of metabolic activity in the intact cerebral cortex. In G. G. Somjen (Ed.), *Excerpta medica international congress series, NO. 253* (Vol. 20–22, pp. 18–26).

Lauritzen, M. (1994). Pathophysiology of the migraine aura. The spreading depression theory. *Brain, 117*(Pt 1), 199–210.

Leão, A. A. P. (1943). *Spreading depression of activity in the cerebral cortex*. Harvard University.

Leão, A. A. P. (1944a). Pial circulation and spreading depression of activity in the cerebral cortex. *Journal of Neurophysiology, 7*, 391–396.

Leão, A. A. P. (1944b). Spreading depression of activity in cerebral cortex. *Journal of Neurophysiology, 7*, 359–390.

Marshall, W. H. (1959). Spreading cortical depression of Leão. *Physiological Reviews, 39*, 239–279.

Mayevsky, A. (1983). Multiparameter monitoring of the awake brain under hyperbaric oxygenation. *Journal of Applied Physiology, 54*, 740–748.

Mayevsky, A. (1984). Brain NADH redox state monitored in vivo by fiber optic surface fluorometry. *Brain Research Reviews, 7*, 49–68.

Mayevsky, A., & Barbiro-Michaely, E. (2013). Shedding light on mitochondrial function by real time monitoring of NADH fluorescence: II: Human studies. *Journal of Clinical Monitoring and Computing, 27*, 125–145.

Mayevsky, A., & Chance, B. (1982). Intracellular oxidation-reduction state measured in situ by a multicannel fiber-optic surface fluorometer. *Science, 217*, 537–540.
Mayevsky, A., & Weiss, H. R. (1991). Cerebral blood flow and oxygen consumption in cortical spreading depression. *Journal of Cerebral Blood Flow & Metabolism, 11*, 829–836.
Mayevsky, A., Zeuthen, T., & Chance, B. (1974). Measurements of extracellular potassium, ECoG and pyridine nucleotide levels during cortical spreading depression in rats. *Brain Research, 76*, 347–349.
Mayevsky, A., Zarchin, N., & Friedli, C. M. (1982). Factors affecting the oxygen balance in the awake cerebral cortex exposed to spreading depression. *Brain Research, 236*, 93–105.
Mayevsky, A., Friedli, C. M., & Reivich, M. (1985). Metabolic, ionic and electrical responses of the gerbil brain to ischemia. *The American Journal of Physiology, 248*, R99–R107.
Mayevsky, A., Flamm, E. S., Pennie, W., & Chance, B. (1991). A fiber optic based multiprobes system for intraoperative monitoring of brain functions. *SPIE, 1431*, 303–313.
Mayevsky, A., Frank, K., Muck, M., Nioka, S., Kessler, M., & Chance, B. (1992). Multiparametric evaluation of brain functions in the Mongolian gerbil in vivo. *Journal of Basic and Clinical Physiology and Pharmacology, 3*, 323–342.
Mayevsky, A., Doron, A., Manor, T., Meilin, S., Zarchin, N., & Ouaknine, G. E. (1996). CSD recorded from the human brain using a multiparametric monitoring system. *Brain Research, 740*, 268–274.
Mayevsky, A., Manor, T., Meilin, S., Doron, A., & Ouaknine, G. E. (1998). Real-time multiparametric monitoring of the injured human cerebral cortex - a new approach. In R. Marmarou & R. Bullock (Eds.), *Acta Neurochir. Suppl (Wien.)* (pp. 78–81). ICPX.
Mayevsky, A., Meilin, S., Manor, T., Ornstein, E., Zarchin, N., & Sonn, J. (1998). Multiparametric monitoring of brain oxygen balance under experimental and clinical conditions. *Neurological Research, 20*, S76–S80.
Mayevsky, A., Sonn, J., Manor, T., Razon, N., & Ouaknine, G. E. (2000). Responses to CSD during normoxia and ischemia: multiparametric monitoring study in animals and the human brain. In Y. Fukuuchi, M. Tomita, & A. Koto (Eds.), *Ischemic blood flow in the brain* (pp. 343–350). Keio University Symposium.
Mayevsky, A., Manor, T., Meilin, S., Razon, N., & Ouaknine, G. E. (2001). Multiparametric monitoring of tissue vitality in clinical situations. *SPIE, 4255*, 33–39.
Mayevsky, A., Ornstein, E., Meilin, S., Razon, N., & Ouaknine, G. E. (2002). The evaluation of brain CBF and mitochondrial function by a fiber optic tissue spectroscope in neurosurgical patients. *Acta Neurochirurgica. Supplement, 81*, 367–371.
McLachlan, R. S., & Girvin, J. P. (1994). Spreading depression of Leão in rodent and human cortex. *Brain Research, 666*, 133–136.
Oka, H., Kako, M., Matsushima, M., & Ando, K. (1977). Traumatic spreading depression syndrome. Review of a particular type of head injury in 37 patients. *Brain, 100*, 287–298.
Olesen, J. (1985). Migraine and regional cerebral blood flow. *Trends in Neurosciences, 8*, 318–321.
Piper, R. D. (1992). *Cortical spreading depression relevance to human pathophysiology and animal experimentation*. University of New South Wales.
Rogatsky, G. G., Mayevsky, A., Zarchin, N., & Doron, A. (1996). Continuous multiparametric monitoring of brain activities following fluid-percussion injury in rats: Preliminary results. *Journal of Basic and Clinical Physiology and Pharmacology, 7*, 23–43.
Rosenblum, B. R., Bonner, R. F., & Oldfield, E. H. (1987). Intraoperative measurement of cortical blood flow adjacent to cerebral AVM using laser Doppler velocimetry. *Journal of Neurosurgery, 66*, 396–399.
Rosenthal, M., & Somjen, G. (1973). Spreading depression, sustained potential shifts, and metabolic activity of cerebral cortex of cats. *Journal of Neurophysiology, 36*, 739–749.
Salama, G., Lombardi, R., & Elson, J. (1987). Maps of optical action potentials and NADH fluorescence in intact working hearts. *The American Journal of Physiology, 252*, H384–H394.

Somjen, G. G., Aitken, P. G., Balestrino, M., Herreras, O., & Kawasaki, K. (1990). Spreading depression-like depolarization and selective vulnerability of neurons. A brief review. *Stroke, 21*, III179–183.

Somjen, G. G., Aitken, P. G., Czeh, G., Herreras, O., Jing, J., & Young, J. N. (1993). Spreading depression in Hippocampus: membrane currents and ion mechanisms. In A. Lehmenkuhler, K.-H. Grotemeyer, & F. Tegtmeier (Eds.), *Migraine: basic mechanisms and treatment* (pp. 329–344). Urban and Schwarzenberg.

Sonn, J. (1996). *Effect of spreading depression on the interrelation between brain energy metabolism and ion homeostasis in vivo*. Life Sciences.

Sramka, M., Brozek, G., Bures, J., & Nadvornik, P. (1977). Functional ablation by spreading depression: possible use in human stereotactic neurosurgery. *Applied Neurophysiology, 40*, 48–61.

Van Harreveld, A., & Stamm, J. S. (1952). Vascular concomitants of spreading cortical depression. *Journal of Neurophysiology, 15*, 487–496.

Vyskocil, F., Kritz, N., & Bures, J. (1972). Potassium-selective microelectrodes used for measuring the extracellular brain potassium during spreading depression and anoxic depolarization in rats. *Brain Research, 39*, 255–259.

Wiebers, D. O., Adams, H. P., Jr., & Whisnant, J. P. (1990). Animal models of stroke: are they relevant to human disease? *Stroke, 21*, 1–3.

Chapter 9
CSD and the LifenLight Score (LLS)

Abstract The shift from single parameter monitoring to multiparametric monitoring approach increased the complexity of data analysis and physiological interpretation. In this chapter we are presenting the evolution of the concept named initially "brain vitality."

To develop the brain metabolic score or LifenLight Score (LLS), we used the four parameter monitoring system including mitochondrial function (NADH fluorescence), cerebral blood flow and volume, as well as the hemoglobin oxygenation or saturation in the microcirculation. The development of the LLS was based on animal experiments including a variety of physiological and pathophysiological perturbations of the brain including hypoxia, ischemia, epileptic activity, and cortical spreading depression—CSD.

Keywords Multiparametric monitoring of the brain · Brain metabolic index · Brain vitality · Brain metabolic score · In vivo monitoring

9.1 Introduction

The two publications of Leão in 1944 (1944a, b) were based on his PhD thesis submitted to Harvard University in 1943 (1943). In most of the studies reported in the thesis, only a single parameter, namely, the "spontaneous" electrical activity (EEG), was measured from the brain during the CSD event. In the last part of his reported thesis, Leão presented the use of a second physiological parameter in a series of experiment. He measured the diameter of the pial blood vessels and concluded that "A wave of marked vasodilation and increased blood flow traveled simultaneously with the wave of depression" (see Fig. 5 page XIII). In the third paper published by Leão in (1947), he added a third parameter measured during the CSD event, namely, the "slow potential variation" seen in Fig. 1.9. Those studies and discoveries made by Leão during the 1940s established the foundation to more than 80 years of

A. Mayevsky, J. Sonn, *Cortical Spreading Depression of Leao*,
https://doi.org/10.1007/978-3-031-08068-5_9

intensive studies covering many aspects of the CSD including clinical monitoring of patients.

In our laboratory we also started in 1973 the studies of CSD using a single parameter measured (mitochondrial NADH) in the rat model up to 8 parameters monitored in rats as well as in head-injured patients in 1996. In order to understand the changes in NADH measured in vivo during CSD, it was necessary to move from a single parameter monitoring to the multiparametric approach. As described in Fig. 1.23 in the front matter, the mitochondrial NADH represents also the balance between oxygen demand and supply. Therefore, the multiparametric approach provides better understanding of the pathophysiological processes developed.

9.1.1 *The Developed Concept*

Figure 9.1 presents the main stages of the development of the LifenLight Score (LLS) to be used in clinical practice in the future. The four monitored parameters seen in A1–A4 are measures by a new medical device seen in Part B. The four analog signals are displayed and presented in Parts C and D. The calculated LifenLight Score (LLS) appears in part E.

Figure 9.2 illustrates the relationship between brain NADH and CBF measured under various pathophysiological states developed in the brain (Mayevsky, 2015). Four types of areas on this sort of map, namely, in two groups of responses, NADH will increase, and in the other two the NADH will decrease. Also, in two groups of responses, CBF will increase, and in the other two CBF will decrease. In Figs. 9.3,

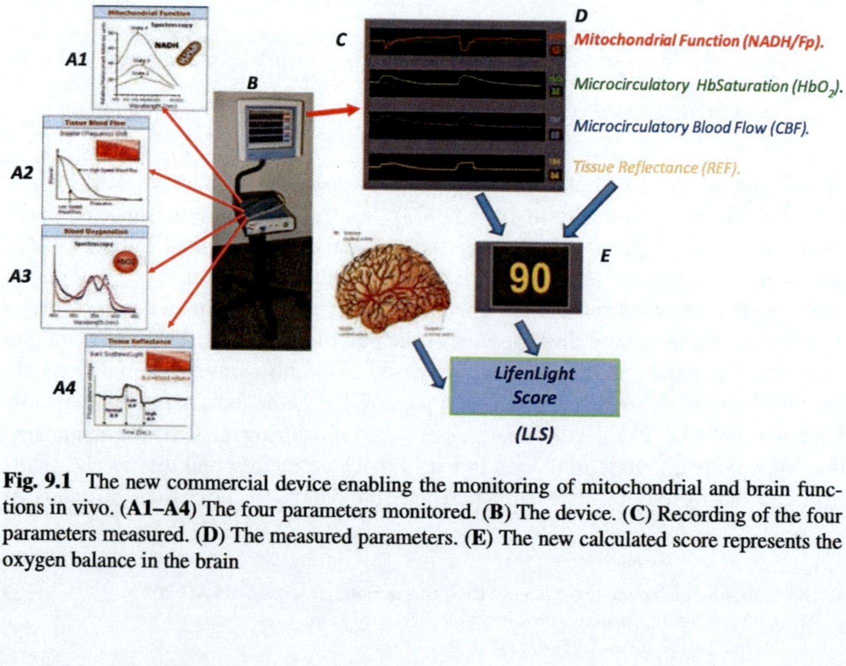

Fig. 9.1 The new commercial device enabling the monitoring of mitochondrial and brain functions in vivo. (**A1–A4**) The four parameters monitored. (**B**) The device. (**C**) Recording of the four parameters measured. (**D**) The measured parameters. (**E**) The new calculated score represents the oxygen balance in the brain

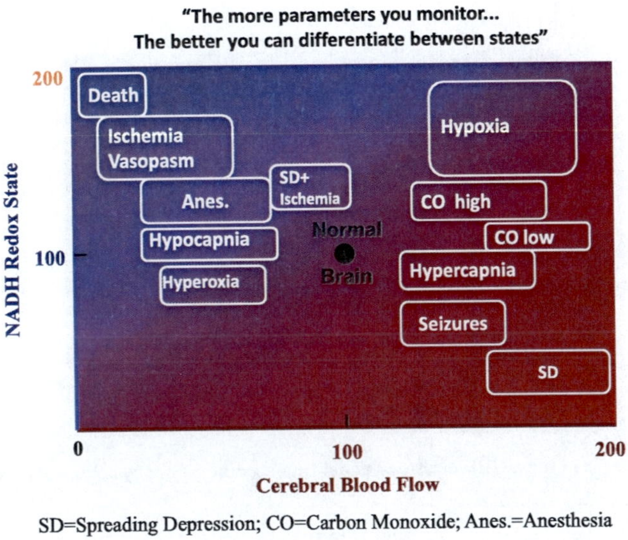

Fig. 9.2 The relationship between brain CBF and brain NADH redox state measured under various pathophysiological conditions. (Mayevsky, 2015)

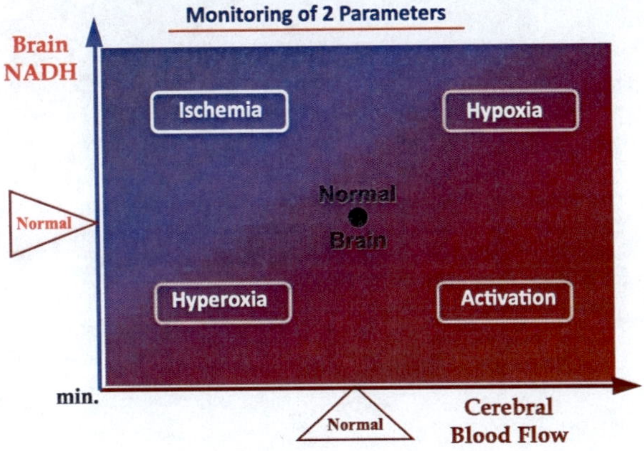

Fig. 9.3 The relationship between brain CBF and brain NADH redox state measured under four specific pathophysiological conditions

9.4, and 9.5 the mentioned problem demonstrates the advantage of moving from monitoring a single parameter to two parameters. In order to simplify the situation, in Fig. 9.3, four pathophysiological conditions are presented, namely, brain ischemia, hypoxia, hyperoxia, and brain activation. Figure 9.4 shows that if only CBF is measured, there is a problem to differentiate between ischemia and hyperoxia and between hypoxia and brain activation. The same is true if only NADH is monitored

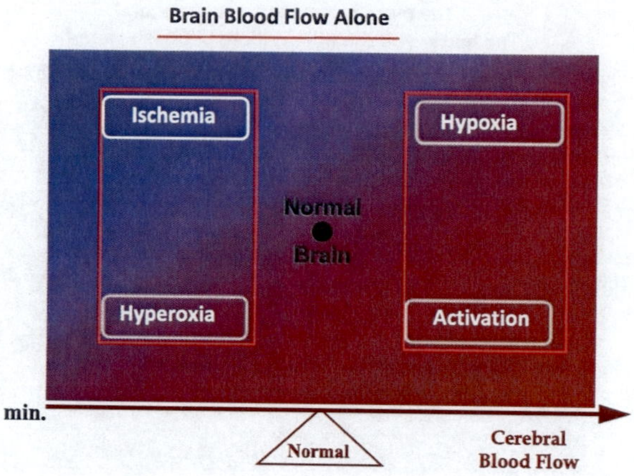

Fig. 9.4 Changes of brain CBF measured under four specific pathophysiological conditions

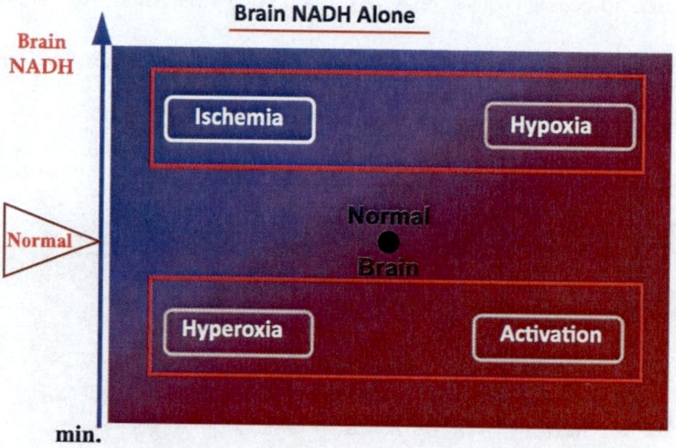

Fig. 9.5 Changes of brain NADH redox state measured under four specific pathophysiological conditions

as seen in Fig. 9.5, and it's impossible to differentiate between ischemia and hypoxia and also between hyperoxia and brain activation. Figure 9.6 shows real data points measured in various studies in animals and also in human (Mayevsky et al., 1991).

The use of other physiological parameters in addition to NADH redox state enabled us to understand better the mechanism behind the development of pathophysiological states in various animal models and especially under CSD used. This approach will enable a clinician to diagnose better the development of pathological states in patients.

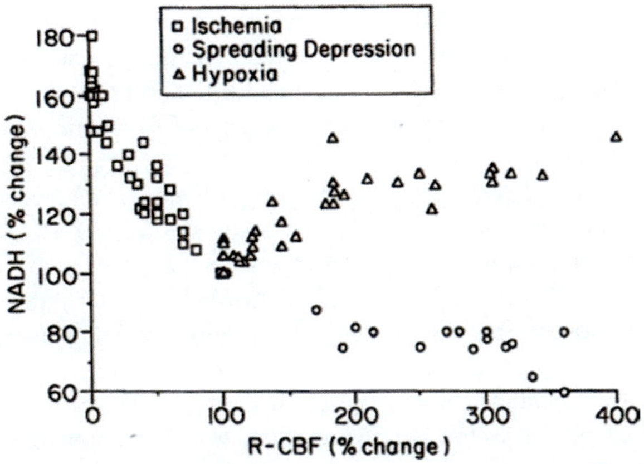

Fig. 9.6 Correlation between relative CBF and NADH redox state in the gerbil brain exposed to hypoxia, ischemia, and spreading depression. The results are expressed as percent change related to the 100% level at the control period. (SPIE, reprinted with permission (Mayevsky et al., 1991))

9.1.2 Brain LifenLight Score (LLS)

The need for new approaches in evaluating the clinical status of patients is very wide. For example, the interpretation of the physiological status of anesthetized patients is very subjective and therefore may be evaluated differentially by various anesthesiologists. The same problem is demonstrated in patients hospitalized in the various intensive care units that need additional monitoring approaches to evaluate objectively their condition. As of today, there is a discrepancy between the large amount of data collected by various techniques and the ability to integrate it into a practical objective tool to be used in clinical practice. Our effort to apply our developed multiparametric system to clinical daily use required the development of a new indexing technique for data interpretation. After the accumulation of huge amount of experimental results, we decided to develop a preliminary tool named *LifenLight Score (LLS)*. In the next figures of this chapter, we are presenting the preliminary developed LLS that could be used in evaluating the oxygen balance of the cerebral cortex in real time. We used the brain as a typical organ for the calculation of the LLS, but the same approach could be used for other organs or tissues as well. The ultimate goal of our study is to develop the LLS for tissues that could be monitored in the intensive care units or in the operating rooms. In those medical environment the need for objective approach decision-making process is very important as discussed by Hanson and Marshall (2001) and by Bates and Young (2003).

The aim of the present study was to develop a new score based on the monitoring of three–four physiological parameters. The multiparametric monitoring system provides simultaneously information on microcirculatory blood flow and hemoglobin oxygenation as well as mitochondrial function and reflectance. Under various pathophysiological conditions, these three or four parameters are changing continuously, and the expert is able to evaluate the vitality level of the tissue. In order to provide an objective evaluation on the state of the tissue, it was important to develop the LifenLight Score—LLS. Few techniques are available in order to calculate and evaluate the changes in various physiological and biochemical parameters related to the metabolic activity of the tissue. In most cases the monitoring device provides information on a single parameter.

Monitoring of tissue blood flow, tissue pO_2, or blood oxygenation (HbO_2) provides information on the integrity of microcirculatory function. Nevertheless, these parameters do not represent the activity of the mitochondria which is the most important functional unit in cellular energy metabolism. Therefore, we decided to calculate the LifenLight Score (LLS) based on various parameters covering the entire range of oxygen gradient, namely, between the microcirculation and the intracellular energy transformation site—the mitochondria.

In order to develop the LLS, we used the multiparametric monitoring system, described in our previous publication (Kutai-Asis et al., 2006).

9.2 Methodology Used

9.2.1 The Multiparametric Monitoring System

The multiparametric monitoring system used in the present study is based on the combination of several independent techniques. Simultaneously monitoring of mitochondrial NADH and HbO_2 from the brain surface was accomplished using the Time Sharing Fluorometer Reflectometer (TSFR). Cerebral blood flow (CBF) is monitored by laser Doppler flowmetry (Wadhwani et al., 1990). All of the optical fibers were joined into one bundle of fibers that was placed on the parietal cortex of the rat's brain (Fig. 9.7).

9.2.2 Time Sharing Fluorometer Reflectometer (TSFR)

As seen in Fig. 9.7 the filter wheel of the TSFR contained eight appropriate filters enabling the measurements of three parameters, namely, tissue reflectance, NADH fluorescence, and blood oxygenation as described previously (Rampil et al., 1992).

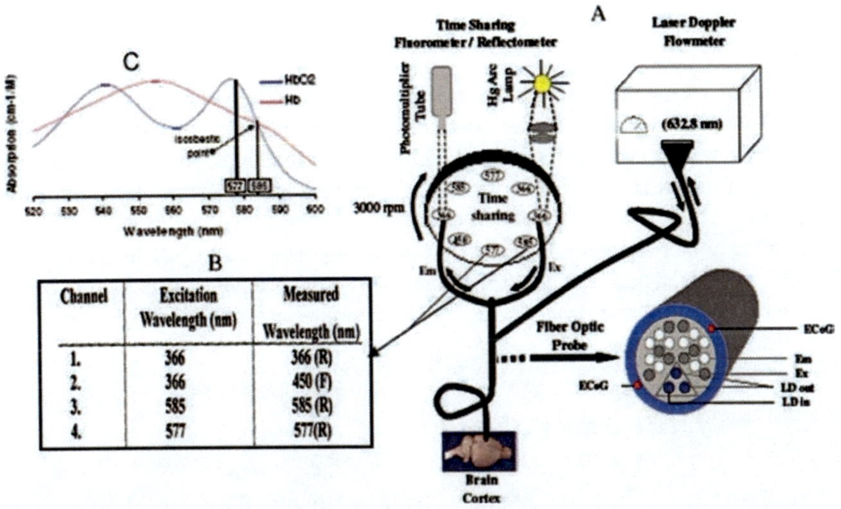

Fig. 9.7 (**A**) Schematic presentation of the Time Sharing Fluorometer Reflectometer (TSFR) combined with the laser Doppler flowmeter for blood flow monitoring (Rampil et al., 1992). The time sharing system includes a wheel that rotates at a speed of 3000 rpm with eight filters: four for the measurements of mitochondrial NADH (366 nm and 450 nm) and four for oxyhemoglobin measurements (585 nm and 577 nm). The source of light is a mercury lamp. The probe includes optical fibers for NADH excitation (Ex) and emission (Em), laser Doppler excitation (LD in), laser Doppler emission (LD out), and needle electrodes for ECoG measurement (ECoG). (**B**) The wavelength of the excitation and emission filters placed inside the time sharing fluorometer/reflectometer (Rampil et al., 1992) that measure the NADH tissue reflectance and oxygenated hemoglobin. (**C**) The absorption spectrum of oxy- and deoxyhemoglobin

9.2.3 Mitochondrial NADH Redox State (NADH)

Mitochondrial NADH Redox State (NADH) is determined by the fluorescence of redox NADH monitored from the surface of the tissue by the fluorometric monitoring technique. Excitation light at the wavelength of 366 nm passes from the fluorometer to the tissue via a bundle of quartz optical fibers. This UV light when absorbed by the NADH, is re-emitted at 450 nm (Fluorescence). The emitted fluorescent light at 450 nm, together with the reflected light at the excitation wavelength (366 nm), is transferred to the fluorometer via another bundle of fibers and through appropriate filters. The changes in the reflected light are correlated to changes in tissue blood volume and are used to correct the fluorescent signal for hemodynamic artifacts, which also affect the fluorescence signal at 450 nm. The changes in the corrected fluorescence signal are normalized relative to calibrated signals for non-treated animals under normoxic conditions (Mayevsky, 1976, 1984; Mayevsky et al., 1990; Mayevsky & Chance, 1982).

9.2.4 Microcirculatory Blood Oxygenation

Tissue blood oxygenation was monitored based on the differences in the absorption properties of HbO_2 and deoxyhemoglobin. The brain surface is illuminated with light at two wavelengths (585 nm and 577 nm). 585 nm is an isosbestic point with the same absorption coefficient for HbO_2 and deoxyhemoglobin. Therefore, the absorption at this wavelength reflects changes in cerebral blood volume. At 577 nm, HbO_2 has a higher absorption coefficient than deoxyhemoglobin, and by subtracting the 585 nm reflectance from the 577 nm reflectance a parameter correlated to blood oxygenation is obtained (Frank et al., 1989; Rampil et al., 1992).

9.2.5 Microcirculatory Blood Flow

Tissue blood flow (TBF) was monitored by laser Doppler flowmetry (LDF), which measures relative changes in microcirculatory blood flow where the basal level is considered as 100% and the level monitored on death is 0% (Wadhwani et al., 1990). The Doppler probe, which uses light of 632.8 nm, is included in the bundle of optical fibers used for the TSFR, and special filters are used to prevent red light from interfering with the fluorometric measurements.

9.2.6 Animal Preparation

All experiments were performed in accordance with the Animal Care Committee of Bar-Ilan University Guidelines. A group of male Wistar rats of 250–300 gr (2.5–3 months old) were used. The rats were anesthetized with Equithesin (0.3 ml/100gr body weight, IP) (each ml contains pentobarbital 9.72 mg, chloral hydrate 42.51 mg, magnesium sulfate 21.25 mg, propylene glycol 44.34% w/v, alcohol 11.5%, and distilled water). During the duration of the experiment, steady anesthesia was maintained by 0.1 ml Equithesin injection every 30 minutes. Additionally, the rats were placed on a warming tray, and body temperature was maintained at 37 °C. The rat was placed on an operation table with a special mouth holder. Two holes were drilled in the parietal bone, one for the monitoring probe and one for the pull-push cannula (4 mm and 2 mm, respectively) in order to induce CSD. Additionally, two screws were attached to the parietal bone, for better fixation of the monitoring probe to the brain surface. The cannula designed for the probe, as well as the KCl cannula, for CSD, was attached to the brain surface and fixated by acrylic dental cement. Then after, the monitoring probe was placed in its cannula, avoiding extra pressure on the cerebral tissue (Mayevsky, 1984; Mayevsky & Chance, 1982). In a group of rats, before craniotomy was induced, the rats were placed on the back, and both carotid arteries were exposed and prepared for later occlusion using a surgical suture 4–0. After a period of stabilization, all rats were exposed to a short anoxia, by pure nitrogen inhalation, until respiratory arrest. This short anoxia is a routine protocol conducted for the evaluation of proper placement of the probe on the brain surface and to ensure reliable monitoring. CSD was induced by washing the brain surface with KCl solution (concentration:1–2 Molar), and stopped with NaCl 0.9%. Bilateral carotid artery occlusion was induced by pulling the sutures that surround the arteries for 30 seconds following with releasing the sutures allowing blood flow through both carotid arteries to recover.

9.2.7 Real-Time Data Acquisition

All parameters were continuously monitored through the entire experimental period, and data was transferred to a multichannel computerized acquisition and recording system (Labview A/D Software, National Instruments Co. USA). The data from the Labview was calculated and transferred to the Excel for further analysis.

9.2.8 LifenLight Score (LLS)

The first step in the development of the LLS was the preparation of special software using MATLAB, enabling various calculations of the data collected from the experiments. Several combinations with various partial importance (in percentage) were tested for each parameter (namely, TBF, NADH, and HbO_2). The ratios used included (1) equal values for each parameter, namely, 33% for TBF, NADH, and HbO_2; (2) 25% CBF, 50% NADH, 25%HbO_2; and (3) 10% CBF, 80% NADH, 10%HbO_2. Each one of this ratio was tested in all three models and was correlated to an expert opinion score.

9.3 Results

The responses of the three monitored parameters (excluding tissue reflectance) to the three types of perturbations are shown in Figs. 9.8, 9.9, and 9.10 (Kutai-Asis et al., 2006). The results of a typical anoxic episode and the calculated LifenLight Score (LLS) are presented in Fig. 9.8. As seen, inhalation of N_2 caused a decrease of 10% in CBF and HbO_2. Concomitantly, NADH increased by 47%. Those changes caused a decrease in the LLS by 38%. Upon recovery, when the rat breathed air, a massive hyperemia in CBF to the level of 185% (as compared to the basal level) was observed and HbO_2 increased to the level of 116%. This massive hyperemia was not affecting the LLS and returned to the basal level. When CBF returned to its basal level HbO_2, NADH and all LLSs fully recovered. Figure 9.9B presents the responses

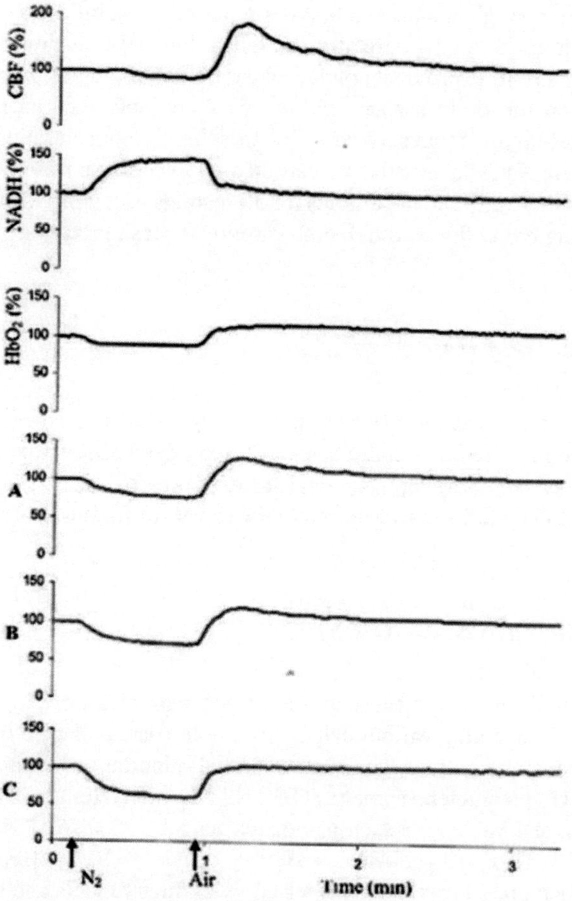

Fig. 9.8 The responses of cerebral blood flow (CBF), mitochondrial NADH level (NADH), and hemoglobin oxygenation level (HbO_2) and three LifenLight Scores [LLS] (**A**, **B**, **C**) calculated to short anoxia (N_2) in rat model. (**A**) A LLS with equal values for each parameter; (**B**) a LLS with 25% CBF, 50% NADH, 25%HbO_2; (**C**) a LLS with 10% CBF, 80% NADH, 10%HbO_2. (SPIE, reprinted with permission (Kutai-Asis et al., 2006))

of the rat brain and LLS to ischemia induced by bilateral occlusion of the carotid arteries. As observed, the occlusion of both arteries caused a decrease in CBF and HbO_2 by 16%, while mitochondrial NADH increased by 23%. At this period the vitality indexes decreased by 27%. When the occlusion was removed, CBF rapidly increased, reaching hyperemic levels of 116%, and HbO_2 increased to the level of 113%. Consequently, NADH was oxidized and decreased by 5% from its basal level. This oxidation was associated with elevation of the vitality index by 5%. Twelve seconds after the oxidation phase, all parameters including the LLS showed full recovery. When CSD was induced (Fig. 9.10), three waves were developed. The first wave was characterized by a biphasic response of HbO_2 starting by HbO_2 decrease of 16% following by a large increase of CBF (138%) and a 22% increase

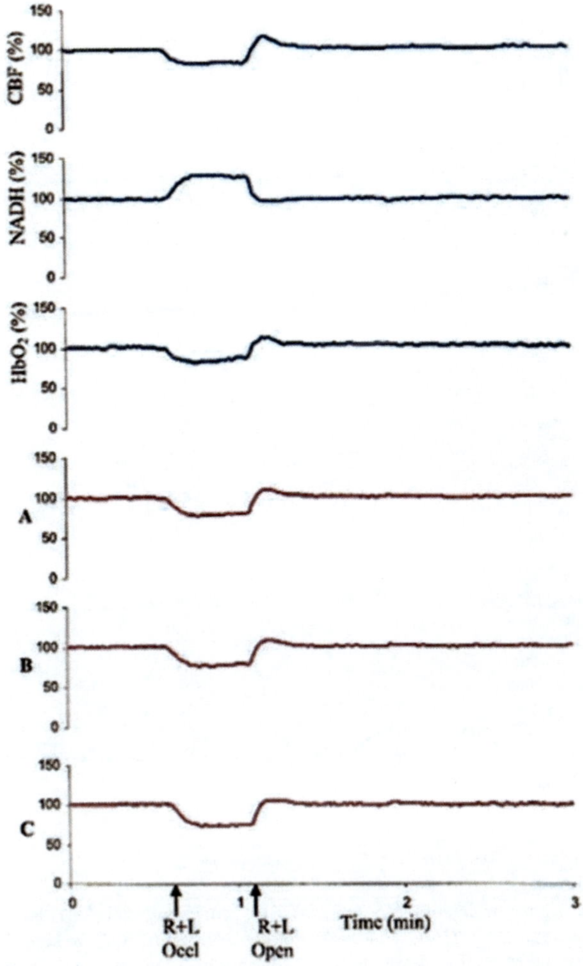

Fig. 9.9 The responses of cerebral blood flow (CBF), mitochondrial NADH level (NADH), and hemoglobin oxygenation level (HbO_2) and three LifenLight Scores [LLS] (**A, B, C**) calculated to bilateral carotid artery occlusion in rat model. (**A**) A LLS with equal values for each parameter; (**B**) a LLS with 25% CBF, 50% NADH, 25%HbO2; (**C**) a LLS with 10% CBF, 80% NADH, 10%HbO_2. (SPIE, reprinted with permission (Kutai-Asis et al., 2006))

in HbO_2. As for the NADH, here also a two-step reaction was recorded. NADH was oxidized rapidly by 21% followed by a further oxidation of 35%. At this stage the LLS showed a biphasic increase of 15% followed by a further increase of 26%. At the stage of recovery from the first wave, not one of the parameters fully recovered. CBF dropped to the level of 110%, HbO_2 decreased to the level of 118%, and NADH was reduced to the level of 107%. As for the vitality index, it returned to its basal level immediately. The second and third waves showed an increase of 34–36% in CBF which was associated with HbO_2 oxygenation to the level of approximately

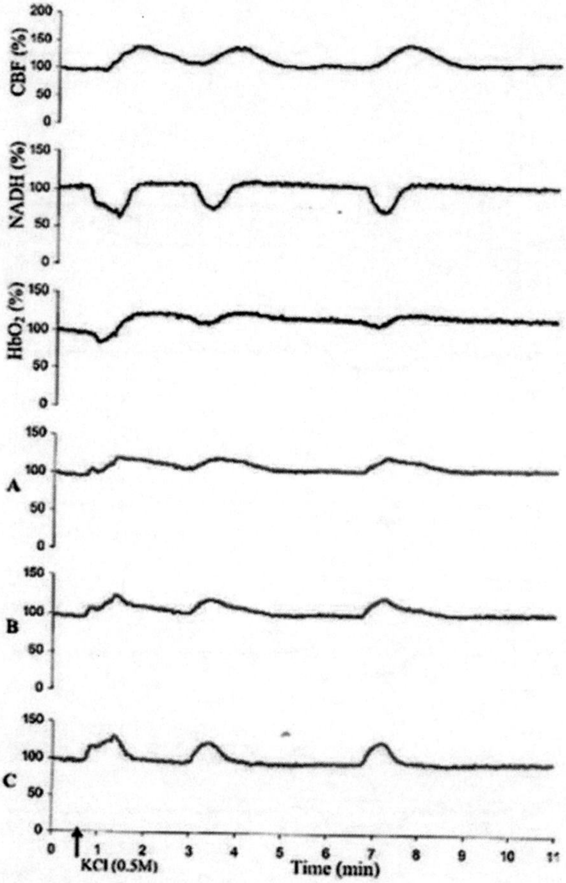

Fig. 9.10 The responses of cerebral blood flow (CBF), mitochondrial NADH level (NADH), and hemoglobin oxygenation level (HbO_2) and three vitality indexes (**A**, **B**, **C**) to spreading depression (CSD) induced by 0.5 M KCl solution applied epidurally in rat model. (**A**) An index with equal values for each parameter; (**B**) an index with 25% CBF, 50% NADH, 25%HbO_2; (**C**) an index with 10% CBF, 80% NADH, 10%HbO_2. (SPIE, reprinted with permission (Kutai-Asis et al., 2006))

110%. As a result, NADH was oxidized by 24–31%. Regarding the changes observed in the LLS, here also the score was more instant and with steeper slopes.

9.4 Discussion

The aim of the present study was to develop a new LifenLight Score (LLS) based on the monitoring of three or four physiological parameters. The multiparametric monitoring system provides simultaneously information on microcirculatory blood flow and hemoglobin oxygenation as well as mitochondrial function. Under various pathophysiological conditions, these three parameters are changing continuously, and the expert is able to evaluate the vitality level of the tissue. In order to provide objective evaluation on the state of the tissue, it was important to develop the LLS. Few techniques are available in order to calculate and evaluate the changes in various physiological and biochemical parameters related to the metabolic activity of the tissue. In most cases the monitoring device provides information on a single parameter. Monitoring of tissue blood flow, tissue oxygen level PO_2, or blood oxygenation (HbO_2) provides information indicating the integrity of microcirculatory function. Nevertheless, these parameters are not representing the activity of the mitochondria which are the most important functional unit in the cellular energy metabolism. Therefore, we decided to calculate the LLS based on various parameters covering the entire range of oxygen gradient, namely, between the microcirculation and the intracellular energy transformation site—the mitochondria. Our results demonstrated that the most accurate LLS to be used is LLS C (Figs. 9.8, 9.9, 9.10, and Table 9.1) in which NADH is represented with the highest percentage, 80%, and CBF and HbO_2 are presented with 10% each. As seen under anoxia, the hyperemia which probably developed in order to compensate for the deficit in oxygen yielded an increase in LLS A and B although this increase in CBF doesn't imply any improvement in the brain vitality state. However, LLS C was not influenced by this hyperemia and returned to basal levels concomitantly with the rat breathing air. Moreover, mitochondrial NADH levels fully recovered as the rat started to breathe air, a few seconds before CBF reached its maximum hyperemia. In other words, the hyperemia seen following anoxia is actually not essential for adequate O_2 supply to the mitochondria in the brain cortex. A similar phenomenon was observed under the ischemic model where following the decrease in CBF all three LLSs were also decreased to the same level. However, the hyperemia observed immediately following both carotid arteries reopening caused a higher increase in LLS A and B, compared with LLS C. It should be clarifying that this hyperemia doesn't improve tissue vitality state but rather is a result of micro-vessels contractility state. During ischemia, cortical arteries are vasodilated in order to improve blood supply to the impact cerebral tissue. In order to restore vasculature tone to its pre-occlusion state, several seconds are needed, meanwhile vessels are relatively vasodilated compared to their normal state. Moreover, here also the hyperemia is needless, since full recovery of tissue energetic state (NADH) was accomplished immediately as arterial obstructions where removed, namely, before full hyperemia

Table 9.1 The correlations (R^2) between calculated vitality indexes and an expert opinion under three pathophysiological conditions

R^2 (CSD)	R^2 (Ischemia)	R^2 (Anoxia)	Vitality index	
0.6468	0.9384	0.6686	A	CBF 33% NADH 33% HbO_2 33%
0.8035	0.9133	0.7960	B	CBF 25% NADH 50% HbO_2 25%
0.7755	0.9406	0.9187	C	CBF 10% NADH 80% HbO_2 10%

CSD spreading depression, ischemia, bilateral carotid artery occlusion; and anoxia, pure N_2 inhalation (Kutai-Asis et al., 2006)

was achieved. The use of the LLS under a condition in which the brain is activated (the CSD model) also indicated that the NADH is a crucial parameter to be taken under consideration for LLS calculations. In other words, although all three scores yielded similar changes following CSD, the waves observed in score C were the clearest, with sharp borders between increased vitality versus decreased vitality. Furthermore, the advantage of LLS C is pronounced in the high correlation coefficients values presented in Table 9.1 under all three models used, where the calculated values were compared with an expert opinion. Since the main purpose of a vitality index is to give indication for the state in which a tissue or an organ is in, and since oxygen is a major factor in tissue viability, it is obvious that this LLS should include the combination between tissue oxygen supply and demand. The mitochondrion is the cell energy factory (Waltemath, 1970), and mitochondrial NADH is the most sensitive indicator for oxygen balance (Mayevsky et al., 1980; Mayevsky et al., 1992; Mayevsky et al., 1998); thus we believe NADH should get the highest importance in a formula for LifenLight Score.

In order to test the correctness of the LLS calculation, we applied the technology to calculate (offline mode) the LLS in few results obtained in animal studies and also in one patient.

Figure 9.11 presents the responses of the rat brain to ischemia induced by unilateral and bilateral occlusion of the carotid arteries (left side) and anoxia (right side) measured by the CritiView (Mayevsky & Barbiro-Michaely, 2013). Actually, the CritiView calculates data on five measured signals that are translated into the four physiological parameters, namely, TBF, R_{375}, NADH, and HbO_2. The raw fluorescence signal is translated into the net NADH signal by subtracting the R_{375} from the fluorescence signal. The HbO_2 signal is obtained by subtracting the R_{530} (isosbestic reflectance) from the R_{470} (the non-isosbestic reflectance). The net reflectance change between the R_{470} and R_{530} is reversed in its polarity in order to demonstrate

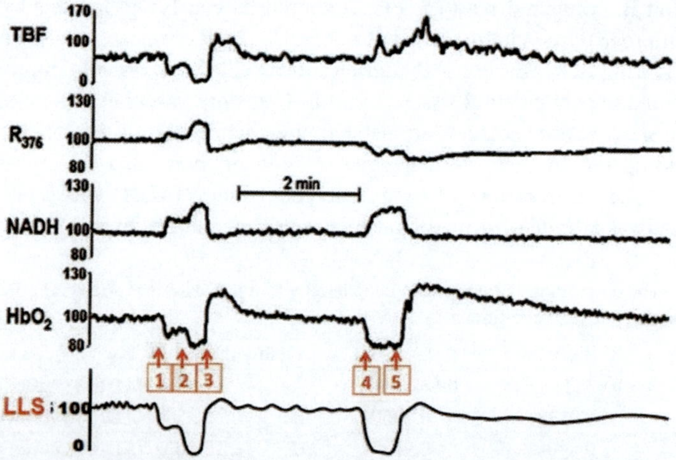

Fig. 9.11 The effects of partial ischemia (1), complete ischemia (2) and recovery (3), induced in a gerbil brain. The effects of anoxia (100% N_2) are shown in 4 and 5. The calculated Lifenlight (LLS) is presented in the lower graph. 1 – right carotid artery occlusion; 2 – left carotid artery occlusion; 3 – open the two blood vessels; 4 – anoxia; 5 – Air. (Mayevsky et al., 2020)

the HbO_2 in values of increase or decrease in HbO_2 and not the reflectance change. The responses to partial and complete ischemia induced by unilateral and bilateral occlusion of the carotid arteries are shown in the left side of Fig. 9.11 (Mayevsky et al., 2020). Due to the decrease in blood flow and volume to the brain (TBF signal), all measured signals are increasing. The net changes are a large decrease in HbO_2 which correlated very well to the NADH. A small hyperemia was noticed in the TBF and HbO_2 traces after the reperfusion. Under anoxia (100% N_2) the initial response was the decrease in HbO_2 followed by an increase in NADH. The TBF showed a later increase which turned into a hyperemic response that was also recorded in the HbO_2 signal after the recovery from anoxia. The time needed for recovery to the base line value is much shorter in the NADH signal as compared to the TBF and other reflectance signals (R_{375}, R_{470}, R_{530}).

In the present study the developed new LifenLight Score (LLS) based on the monitoring of three physiological parameters is presented. The multiparametric monitoring system provides simultaneously information on microcirculatory blood flow and hemoglobin oxygenation as well as mitochondrial function. Under various pathophysiological conditions, these three parameters change continuously, and the expert is able to evaluate the vitality level of the tissue. In order to provide an objective evaluation on the state of the tissue, it was important to develop the LLS. Few techniques are available in order to calculate and evaluate the changes in various physiological and biochemical parameters related to the metabolic activity of the tissue. In most cases the monitoring device provides information on a single parameter.

Our results demonstrated that the most accurate LLS is calculated when NADH is represented with the highest percentage, 80%, and CBF and HbO_2 and reflectance are presented together with about 20% (Kutai-Asis et al., 2006). Under ischemia (Fig. 9.11) the calculated LLS decreases in two steps (1 and 2 in the lowest trace) and recover to pre-ischemic level at 3 (Mayevsky et al., 2020). As seen under anoxia (right side), the hyperemia which probably developed in order to compensate for the deficit in oxygen is not affecting the LLS. Moreover, mitochondrial NADH levels fully recovered as the rat started to breathe air, a few seconds before CBF reached its maximum hyperemia. In other words, the hyperemia seen following anoxia is actually not essential for adequate O_2 supply to the mitochondria in the brain cortex.

It should be clarified that this hyperemia does not improve tissue vitality state but rather is a result of micro-vessels contractility state. During ischemia, cortical arteries are vasodilated in order to improve blood supply to the impact cerebral tissue. In order to restore vasculature tone to its pre-occlusion state, several seconds are needed. Meanwhile vessels are relatively vasodilated compared to their normal state. Moreover, here also the hyperemia is needless, since full recovery of tissue energetic state (NADH) was accomplished immediately as arterial obstructions were removed, namely, before full hyperemia was achieved. Since the main purpose of LLS is to give an indication for the state in which a tissue or an organ is in, and since oxygen is a major factor in tissue viability, it is obvious that this Score should include the combination between tissue oxygen supply and demand. The mitochondrion is the cell energy factory (Waltemath, 1970), and mitochondrial NADH is the

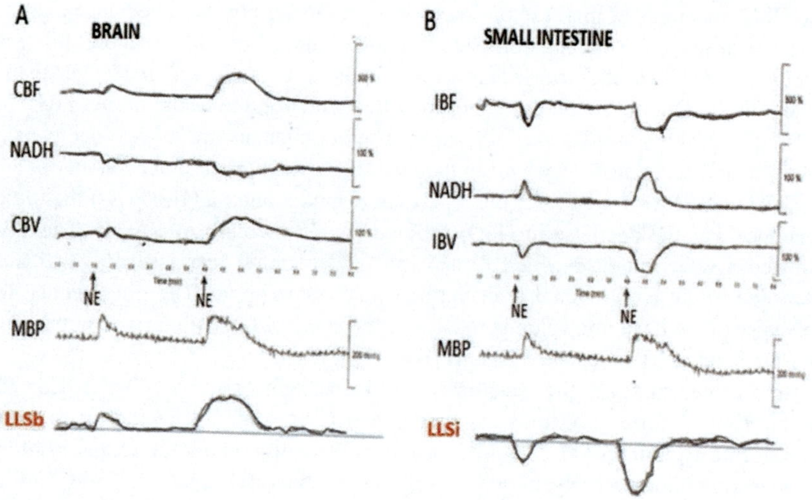

Fig. 9.12 The responses of the brain (**A**) and small intestine (**B**) to two levels of norepinephrine injected to the anesthetized rat. C: The correlation between the tissue blood flow (TBF) and NADH in the organs monitored. (*CBF* and *CBV* – cerebral blood flow and volume, measured by a laser Doppler flowmeter, *MBP* – mean blood pressure, *IBF* and *IBV* – intestinal blood flow and volume, *LLSb* and *LLSi* – *Lifenlight* score calculated for the brain and small intestine (Mayevsky et al., 2021))

most sensitive indicator for oxygen balance (Mayevsky et al., 1980; Mayevsky et al., 1992; Mayevsky et al., 1998). Thus, we believe NADH should get the highest importance in a formula for calculating the LLS.

The utility of the calculated LifenLight Score (LLS) is presented in Fig. 9.12 (Mayevsky et al., 2021). In the rat used in the study, the brain and small intestine were monitored under the injection of two levels of norepinephrine (NE). The effect of the NE led to a clear increase in blood pressure depending of the dose injected. The responses of the three parameters measured in the two organs were in the opposite directions. In the brain (Fig. 9.12A), the clear increase in CBF and CBV was correlated to the oxidation of NADH redox state. In the small intestine (Fig. 9.12B) the NE led to opposite responses compared to the brain. The tissue metabolic score that was calculated, as seen in the lower records (LLS), shows clearly that the less vital organ is exposed to ischemia, while the brain is preserved and well oxygenated under the NE injection.

Figure 9.13 is an analog tracing that illustrates the effect of CSD on the brain cortex during normoxia (A) and hypoxia (B) of the same animal (Mayevsky et al., 2020). CSD during normoxia (A) caused a biphasic response in reflectance, an increase that was followed by a decrease. Mitochondrial NADH decreased showing an oxidation cycle, CBF and $[K^+]e$ increased, and DC potential decreased. During hypoxia (B), the increase in reflectance augmented, and the decrease phase reduced. NADH and CBF under hypoxic conditions changed their responses to CSD: in NADH an initial reduction cycle was followed by a small oxidation one; CBF declined parallel to the increase in NADH, and later on, a small increase was seen,

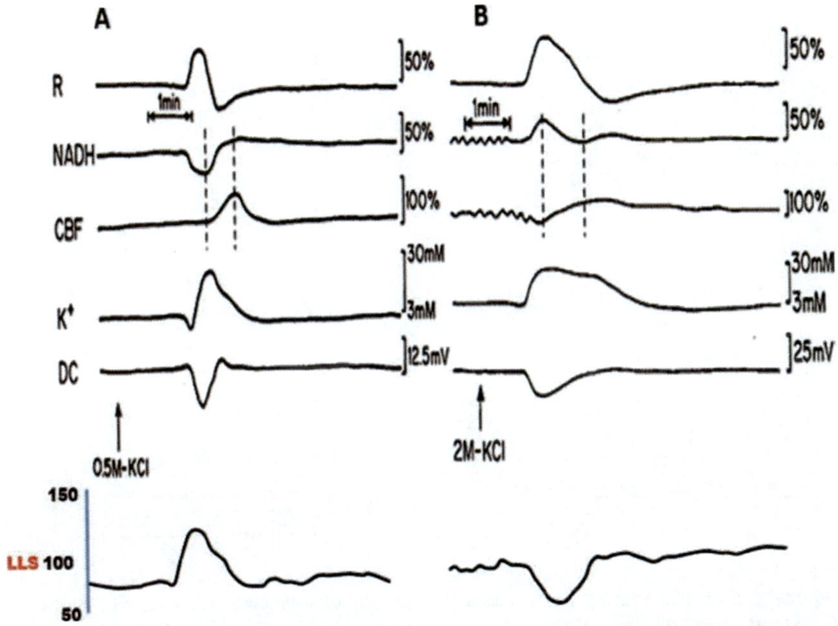

Fig. 9.13 Responses of the normoxic (**A**) and hypoxic or partial ischemic (**B**) rat brain to CSD induced by applying KCl solution to the cerebral cortex

which was accompanied by the small oxidation cycle in NADH. CSD during control conditions showed that the maximum (peak) in [K+]e occurred parallel to the minimum (peak) in NADH (maximum oxidation) whereas the CBF increase, started afterwards reaching the maximum level about 45 seconds later. In contradiction, the maximum (peak) in $[K^+]_e$ occurred parallel to the reduction cycle in NADH and the decrease in CBF (Fig. 9.13B dotted lines). In addition, CSD wave duration prolonged during hypoxia as compared to normoxia.

It was found that the responses to CSD in normoxic and ischemic brain are different (Mayevsky, 1978). Rosenthal and Martel (1979) showed that ischemia affected the response to CSD, namely, instead of NADH oxidation, an increase in NADH was recorded. About 25 years ago we reported, for the first time, on the development of CSD in the brain of a comatose patient (Mayevsky et al., 1996). In this patient, the typical "oxidation cycle" of NADH was replaced by a "reduction cycle" due to a decrease in O_2 compensation capabilities when the patient shows signs of brain deterioration due to high intracranial pressure (ICP) levels. Figure 9.14 presents the responses to CSD in a rat model (A and B) and in patients at the lower part (C and D). When we induced CSD in the animal study, we were sure about the oxygen supply to the animal. The blood supply to the brain decreased between part A and B. The shift in response to CSD from the normoxic (A) to the ischemic (B) state is clear as discussed before (Mayevsky, 1978). When we move to patients' monitoring, we want to use the NADH trace as an indicator to the pathophysiological state of the patient or the brain in this example. As seen in the lower part, the two

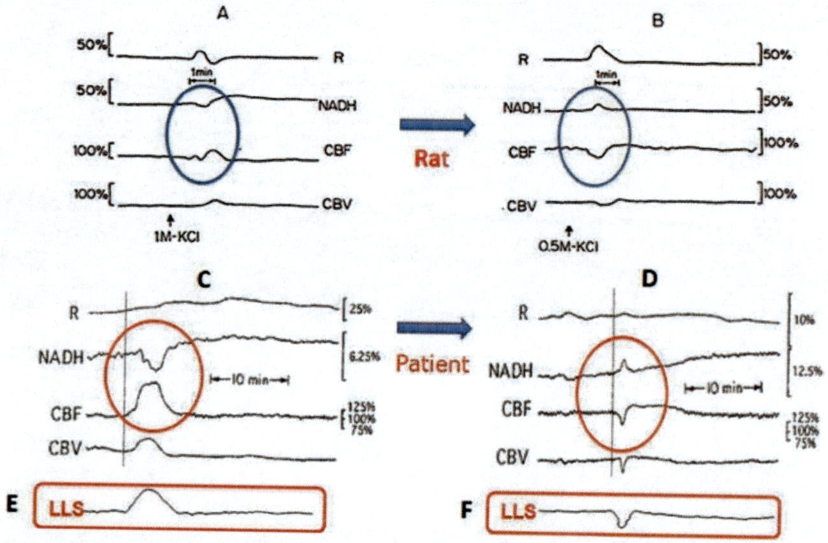

Fig. 9.14 The responses to CSD (CSD) measured in normoxic (**A**) and ischemic brain (**B**) in the rat model. Parts (**C**) and (**D**) were measured in head-injured patients; (**E, F**) LifenLight Score (LLS) calculated for the normoxic (**E**) and partial ischemic (**F**) human brain responses to CSD. (*TBF* cerebral blood flow, R_{376} reflectance, *NADH* mitochondrial NADH redox state, HBO_2 oxygenation of hemoglobin)

responses developed in the neurosurgical patients (C and D) are not the same. There are similarities between A and C as well as between B and D. Therefore, we concluded that the brain of the patient in state C was probably normoxic while in D the brain was partially ischemic. Indeed, this patient was deteriorating in time and finally brain death was determined (Mayevsky et al., 1996). This example demonstrates that the basic pathophysiological events developed in patients present very similar results as compared to those monitored in animal models. The calculated LLS traces, measured from the patient shown in Part C and D, are presented in Part E and F, and the difference in the responses to CSD is very obvious. The initial response measured in the patient (part C) is very similar to the responses measured in rat shown in Part A and in Fig. 9.13A. The same similarity is seen in the LLS responses shown in Part F and Part B in Fig. 9.14.

References

Bates, J. H., & Young, M. P. (2003). Applying fuzzy logic to medical decision making in the intensive care unit. *American Journal of Respiratory and Critical Care Medicine, 167*, 948–952.

Frank, K. H., Kessler, M., Appelbaum, K., & Dummler, W. (1989). The Erlangen micro-lightguide spectrophotometer EMPHO I. *Physics in Medicine and Biology, 34*, 1883–1900.

Hanson, C. W., & Marshall, B. E. (2001). Artificial intelligence applications in the intensive care unit. *Critical Care Medicine, 29*, 427–435.

Kutai-Asis, H., Barbiro-Michaely, E., Deutsch, A., & Mayevsky, A. (2006). Fiber optic based multiparametric spectroscopy in vivo: Toward a new quantitative tissue vitality index. *SPIE Proceedings, 6083*, 10-10–10-11.

Leão, A. A. P. (1943). *Spreading depression of activity in the cerebral cortex*. Harvard University.

Leão, A. A. P. (1944a). Pial circulation and spreading depression of activity in the cerebral cortex. *Journal of Neurophysiology, 7*, 391–396.

Leão, A. A. P. (1944b). Spreading depression of activity in cerebral cortex. *Journal of Neurophysiology, 7*, 359–390.

Leão, A. A. P. (1947). Further observations on the spreading depression of activity in the cerebral cortex. *Journal of Neurophysiology, 10*, 409–414.

Mayevsky, A. (1976). Brain energy metabolism of the conscious rat exposed to various physiological and pathological situations. *Brain Research, 113*, 327–338.

Mayevsky, A. (1978). Ischemia in the brain: The effects of carotid artery ligation and decapitation on the energy state of the awake and anesthetized rat. *Brain Research, 140*, 217–230.

Mayevsky, A. (1984). Brain NADH redox state monitored in vivo by fiber optic surface fluorometry. *Brain Research Reviews, 7*, 49–68.

Mayevsky, A. (2015). *Mitochondrial function in vivo evaluated by NADH fluorescence*. Springer.

Mayevsky, A., & Barbiro-Michaely, E. (2013). Shedding light on mitochondrial function by real time monitoring of NADH fluorescence: I. Basic methodology and animal studies. *Journal of Clinical Monitoring and Computing, 27*, 1–34.

Mayevsky, A., & Chance, B. (1982). Intracellular oxidation-reduction state measured in situ by a multicannel fiber-optic surface fluorometer. *Science, 217*, 537–540.

Mayevsky, A., Lebourdais, S., & Chance, B. (1980). The interrelation between brain PO_2 and NADH oxidation- reduction state in the gerbil. *Journal of Neuroscience Research, 5*, 173–182.

Mayevsky, A., Duckrow, R. B., Yoles, E., Zarchin, N., & Kanshansky, D. (1990). Brain mitochondrial redox state, tissue hemodynamic and extracellular ion responses to four-vessel occlusion and spreading depression in the rat. *Neurological Research, 12*, 243–248.

Mayevsky, A., Flamm, E. S., Pennie, W., & Chance, B. (1991). A fiber optic based multiprobes system for intraoperative monitoring of brain functions. *SPIE, 1431*, 303–313.

Mayevsky, A., Frank, K., Muck, M., Nioka, S., Kessler, M., & Chance, B. (1992). Multiparametric evaluation of brain functions in the Mongolian gerbil in vivo. *Journal of Basic and Clinical Physiology and Pharmacology, 3*, 323–342.

Mayevsky, A., Doron, A., Manor, T., Meilin, S., Zarchin, N., & Ouaknine, G. E. (1996). Cortical spreading depression recorded from the human brain using a multiparametric monitoring system. *Brain Research, 740*, 268–274.

Mayevsky, A., Meilin, S., Manor, T., Ornstein, E., Zarchin, N., & Sonn, J. (1998). Multiparametric monitoring of brain oxygen balance under experimental and clinical conditions. *Neurological Research, 20*, S76–S80.

Mayevsky, A., Kutai-Asis, H., & Tolmasov, M. (2020). Mitochondrial function and brain Metabolic Score (BMS) in ischemic Stroke: Evaluation of "neuroprotectants" safety and efficacy. *Mitochondrion, 50*, 170–194.

Mayevsky, A., Tolmasov, M., Kutai-Asis, H., & Mandelbaum-Livnat, M. (2021). Tissue Oxygen Homeostasis and LifenLight Score (LLS) of Patients in Critical Care Medicine. *Highlights on Medicine and Medical Research*, 18–56.

Rampil, I. J., Litt, L., & Mayevsky, A. (1992). Correlated, simultaneous, multiple-wavelength optical monitoring in vivo of localized cerebrocortical NADH and brain microvessel hemoglobin oxygen saturation. *Journal of Clinical Monitoring, 8*, 216–225.

Rosenthal, M., & Martel, D. L. (1979). Ischemia-induced alterations in oxidative "recovery" metabolism after spreading cortical depression in situ. *Experimental Neurology, 63*, 367–378.

Wadhwani, K. C., Rapoport, S. I., Shepherd, A. P., & Oberg, P. A. (1990). Blood flow in the central and peripheral nervous systems. In A. P. Shephrd & P. A. Oberg (Eds.), *Laser doppler blood flowmetry* (pp. 265–304). Kluwer Academic Pub.

Waltemath, C. L. (1970). Oxygen, uptake, transport, and tissue utilization. *Anesthesia and Analgesia, 49*, 184–203.

Chapter 10
Discussion and Conclusions

Abstract A short summary of the main subjects described in this book is presented. This book is devoted to the metabolic aspects of the phenomenon cortical spreading depression (CSD) discovered in the early 1940s. The main differences between in vitro and in vivo conditions methodology adopted in both situations are discussed. The mitochondrial metabolic states described by Chance and Williams 1955 are compared to the in vivo conditions. Also, the use of NADH fluorescence together with other physiological parameters in patients is discussed.

Keywords In vivo monitoring · Cortical spreading depression · Metabolic state · Patients monitoring · Brain mitochondrial function

10.1 Introduction

In the very last and very short part of his PhD thesis, submitted to Harvard University in 1943, Leão described the effect of CSD on pial circulation. In the discussion part of the thesis, he concluded that "It is clear therefore that the change in the activity of the cortical neurons, when depression is starting in any given region causes local dilatation and increased flow." "An understanding of the interrelations between the two phenomena awaits further investigation."

Most of his thesis content and findings were described in two papers published by him in 1944 (Leão, 1944a, b) after his graduation.

The findings of Leão of a vasodilatation response accompanied the CSD were not clearly proven by other investigators (Marshall, 1959; Van Harreveld & Stamm, 1952). Changes in other types of metabolites in the brain were responded to CSD (Krivanek, 1961). All those metabolic changes related to CSD were summarized in Chap. 6 of the book by Bures et al. published in 1974 (Bures et al., 1974).

A. Mayevsky, J. Sonn, *Cortical Spreading Depression of Leao*,
https://doi.org/10.1007/978-3-031-08068-5_10

It was very obvious in the early 1970s that the involvement of mitochondrial function in CSD was necessary to be investigated. Indeed, in 1973 Rosenthal and Somjen (Rosenthal & Somjen, 1973) and Mayevsky and Chance (Mayevsky & Chance, 1973) showed that brain energy metabolism was activated and increased under CSD using the measurement of mitochondrial NADH redox state. In the following years our research team published more than 50 papers in which the changes of mitochondrial NADH redox state were investigated.

10.2 The Discovery of CSD in the Human Cerebral Cortex

Our most important discovery during our 50 years of studies in the field of CSD was to monitor, for the first time, brain mitochondrial NADH redox state in the cerebral cortex of human during CSD developed in head-injured patient (Mayevsky et al., 1996; Penney et al., 1989). Not only that we measured the mitochondrial NADH, in addition, we measured in the same patient ICP, ECoG, extracellular potassium, DC steady potential, and cerebral blood flow (CBF) and volume.

One year after our publication, Warner D.S. published in 1997 (Warner, 1997), in a Journal club report, the following:

> However, despite efforts of multiple investigators, spreading depression waves have not been detected in the human brain. The current report by Mayevsky et al. makes a strong case for presence of spreading depression waves in the injured human brain.

> The use of the multiparametric probe was pivotal because it allowed multi-modality confirmation of the presence of spreading depression in the human brain.

Six years after our report on the CSD event developed in head-injured patient, Anthony J. Strong and a group of leading scientists (Strong et al., 2002) described changes in ECoG similar to CSD in most of the 14 patients analyzed. They concluded the following:

"These results indicate that CSD or similar events occur in the injured human brain and are more frequent than previously suggested. On the basis of these observations, we suggest that the related phenomenon, peri-infarct depolarization, is indeed likely to occur in boundary zones in the ischemic human cerebral cortex". In their review, Strong et al. mentioned the following:

> "The only evidence suggesting the occurrence of CSD-like events in the human brain after trauma is from Mayevsky et al. (1996) who located their recording system at the standard (single) neurosurgical monitoring site in the right frontal convexity cortex. There were positive findings in only 1 of 14 patients. In light of our new findings from perilesional cortex in at least 11 of our 14 patients, we attribute the minimal findings from Mayevsky and colleagues to their use of a standard recording site, which was most probably not in the perilesional region in most patients".

Four years later in 2006 Strong and Dardis, in the very long review (Strong & Dardis, 2005), described details of our original paper published in 1995 (Penney et al., 1989) and 1996 (Mayevsky et al., 1996) as follows: "Is there any direct evidence for the occurrence of depolarization's in the human brain?" In the course of stereotaxic neurosurgical procedures, Sramka and colleagues were able to demonstrate CSD in deep grey matter (Sramka et al., 1977). *Mayevsky and colleaggues used a multimodal monitoring system located on the right frontal convexity in 14 patients* (Mayevsky et al., 1996). In only one did they find evidence for CSD, but the findings in this individual were striking. Recurrent ECoG suppressions were seen, associated with transient increases in CBF (laser Doppler), oxidation of NADH, and elevations of K_e, a combination of features closely compatible with CSD, although – despite the authors' claims – proper verification of propagation of the events was not possible at the single monitoring point used. As brain swelling progressed, the NAD/H transients became reduction rather than oxidation events, and the CBF transients became negative – the features now of PIDs. Although this group saw CSD/PID events in only one patient, it is important to note that their regular use of a right frontal monitoring site irrespective of the site of any contusion or hematomas, although standard practice at the time of the study, will have precluded detection of events confined to the margins or 'traumatic penumbra' of a lesion elsewhere in the brain. There now appears to be wide recognition of the value of locating detection devices, at the very least for research purposes, near the edge of focal lesions."

Another paper published in 2006 (Fabricius et al., 2006), by Fabricius et al., also mentioned our discovery made in 1995–1996 as follows:

> "The third line of evidence comes from the work of Mayevsky who observed repetitive episodes of CSDs in a head injured patient in vivo (i.e. in only 1 out of the 14 studied) starting at 30 h after start of the monitoring (Mayevsky et al., 1996). The recordings showed characteristic transients of potassium, cerebral blood flow, cerebral blood volume, oxidation of the cortex and ECoG suppression consistent with CSD. Measurements were performed at a single point; therefore, the putative spreading nature of the events could not be demonstrated. The recording equipment was positioned over the hemisphere contralateral to the lesion for practical reasons (Mayevsky et al., 1996), which may explain the low incidence of CSD/PID episodes reported."

In their long review published in 2015 Ayata and Lauritzen mentioned only in 2–3 sentences the discovery of CSD in the human brain as follows (Ayata and Lauritzen, 2015):

> "In human brain, CSD was first detected electrophysiologically upon stereotactic injection of concentrated KCl in caudate and hippocampus in neurosurgical patients (Sramka et al., 1977) and then spontancously in cortex in a brain-injured patient (Mayevsky et al., 1996)".

A few years ago, Lauritzen and Strong (2017) summarized the history of the detection of CSD in human and mentioned our discovery:

> "In 1996, a low incidence of CSD events following TBI (in humans) was reported by Mayevsky et al. (1996) who sought SDs using a single multiparametric monitoring bolt placed at the neurosurgical standard right frontal convexity location in their 14 patients with TBI; SDs were found in only one, and at an ipsilateral site remote from the lesion".

10.3 NADH Responses to CSD In Vivo vs. Isolated Mitochondria In Vitro

In the front matter section, the issue of mitochondrial metabolic states and activity according to Chance and Williams (1955) was presented. In the current chapter, the dilemma of comparison of the results obtained in vivo under CSD, according to the in vitro data, will be presented. Figure 10.1 presents the main transitions between the various metabolic states of the isolated mitochondria and are illustrated in part A (Chance & Williams, 1955). In rectangle A1 the various metabolic states are presented, and in A2, the rate limiting components are presented. It's very hard or impossible to mimic most of the states listed in A1 under in vivo conditions. Only one very definite state could be achieved in vivo, namely, state 5 could be induced by exposing the animal in vivo to lack of oxygen by eliminating the oxygen from the breathing mixture (100% nitrogen) or by complete ischemia. This condition was achieved in vitro as well as in vivo as seen in Fig. 10.1 B, B1–B4 (Chance &

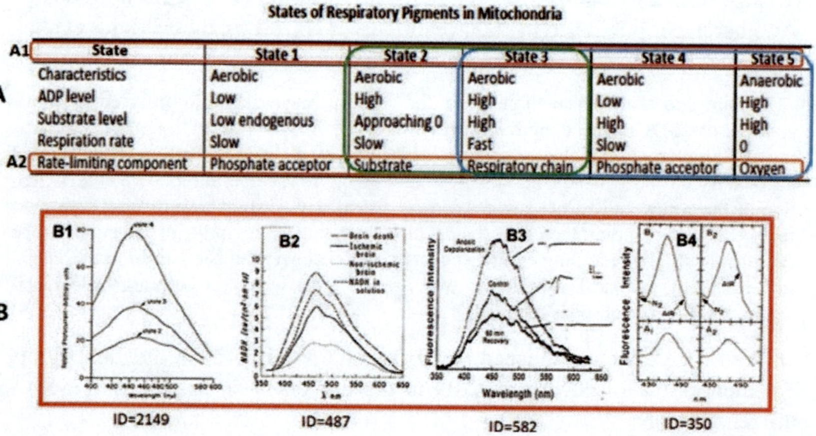
States of Respiratory Pigments in Mitochondria

State	State 1	State 2	State 3	State 4	State 5
Characteristics	Aerobic	Aerobic	Aerobic	Aerobic	Anaerobic
ADP level	Low	High	High	Low	High
Substrate level	Low endogenous	Approaching 0	High	High	High
Respiration rate	Slow	Slow	Fast	Slow	0
Rate-limiting component	Phosphate acceptor	Substrate	Respiratory chain	Phosphate acceptor	Oxygen

Fig. 10.1 Upper Table (**A**) The definition of in vitro mitochondrial metabolic states described in 1955 (Chance and Williams, 1955, 1956). The green and the blue rectangles and the two red rectangles present the more important parts of the table. The lower red rectangle (**B**) presents four examples of NADH fluorescence spectra measured under various metabolic conditions. **B1** – Fluorescence spectrum of mitochondria in three metabolic states. Mitochondria diluted tenfold in phosphorylating medium. State 4 treated with succinate, State 3 treated with succinate and ADP (4 mM), State 2 treated with ADP only (1 mM). Excitation wave length 330 mμ. (Chance and Baltscheffsky, 1958). **B2** – Emission spectral fluorescence curves for NADH in solution, in normal brain, in ischemic cortex, and in cortex 5 min after death (Sundt and Anderson, 1975). **B3** – Examples of NADH fluorescence emission spectra and hippocampal field potentials recorded during control conditions, 2 min after anoxic depolarization, and 60 min following reoxygenation. Failure of evoked potential recovery was related to hyperoxidation of NADH/NAD$^+$. Reprinted with permission from Elsevier (Perez-Pinzon et al., 1997). **B4** – Emission spectra of the brain under excitation of 366 nm light (A_1, A_2, B_1, B_2,) (Mayevsky, 1984)

Baltscheffsky, 1958; Mayevsky, 1984; Perez-Pinzon et al., 1997; Sundt & Anderson, 1975).

In order to create the other four metabolic states of the isolated mitochondria, under in vivo conditions, its critical to eliminate the ADP, substrate, or the creation of maximal activity of the respiratory chain. These conditions are not practical to create in the brain in vivo. In contrast to the defined resting state of the isolated mitochondria-state 4, under in vitro conditions, it's not possible to define the resting state under in vivo conditions. This issue was discussed in Chaps. 2 and 3 of the current book as well as in few of our publications (Mayevsky, 1976, 2015; Mayevsky & Rogatsky, 2007).

Figure 10.2 presents the comparison between the mitochondrial metabolic states in vitro to the states in the brain in vivo. We create this in vivo responses of the brain during real-time measurements of NADH and CBF in most of the perturbation induced. We concluded from many of the experiments performed in our laboratory that the resting state of the brain in vivo is around 3.6 and not state 4 as determined by Chance in his team for the mitochondria in vitro (Chance & Williams, 1955). As seen in the figure, cortical spreading depression in normoxia shifts the NADH redox state toward state 3 (more oxidized state) and an increase of the CBF. As was

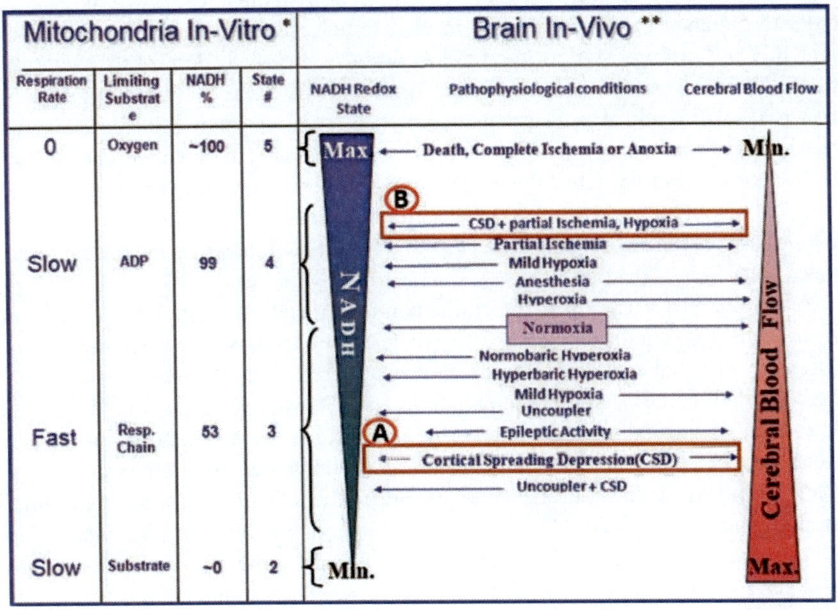

Fig. 10.2 Comparison between the classical definition of Chance and Williams regarding the metabolic state of isolated mitochondria and the metabolic state of the in vivo brain (Mayevsky et al., 1999) (*according to Chance & Williams 1955 (1955); ** according to Mayevsky 1984 (1984); HBO, Hyperbaric oxygenation; CSD, cortical spreading depression. The lower red rectangle A shows the response to CSD in normoxia and the upper one B shows the response to CSD under partial ischemia or mild hypoxia

presented and discussed in Chap. 3 (Figs. 3.7 and 3.8), when an uncoupler was applied locally to the cortex, NADH was more oxidized and the addition of CSD after the uncoupler shifted the NADH to even more oxidized state as seen also in Fig. 10.2.

10.4 Cortical Spreading Depression (CSD) under Pathophysiological Conditions

During our long-term period in studying CSD, it was clear that the development of one or more waves of this unique event depends on the appropriate stimulation given to the cerebral cortex. Once the stimulus intensity is appropriate, the wave is developed and propagates to the other areas of the stimulated hemisphere. Nevertheless, the recovery of the brain tissue is dependent upon the availability of energy (ATP) to restore the ion homeostasis to normal. The entire event was discussed in Chap. 4, Figs. 4.3 and 4.14.

Mitochondrial function and its metabolic states under in vivo condition are completely different from the definition made for isolated mitochondria by Chance and Williams in 1955 (Chance & Williams, 1955) as seen in the left side of Fig. 10.2. In order to compare the in vitro to the in vivo states, only four states (2–5) are presented. The "resting" state of the mitochondria in vitro was defined as "State 4," where NADH was 99% in the reduced form, and ADP was the rate limiting substance. If ADP is added to a suspension, oxygen consumption will increase and the rate limit will be determined by the activity of the respiratory chain. During this State 3, or the "active state," the NADH redox state will decrease or become more oxidized (about 50%). When the "resting" mitochondria are deprived of oxygen, the activity of the respiratory chain will stop, and NADH will reach its maximum redox state (State 5).

A definitive description of the mitochondrial metabolic state has never been given for in vivo conditions. Therefore, we described the in vivo mitochondria conditions as recorded by NADH fluorescence in a representative tissue or organ—e.g., the brain. While the range between minimal NADH (~0) and its maximal level was determined in vitro, it is almost impossible to determine it in the intact brain or other organs in vivo. For example, State 2, with a substrate free medium, could not be achieved in vivo since the tissue would die due to the lack of substrate. On the other hand, the maximal level of NADH (State 5) could be monitored in vivo under complete deprivation of oxygen by anoxia or complete ischemia. We used the changes in NADH levels monitored in vivo to create a new scale ranging from a maximal definite point to the minimal level as shown in the right side of Fig. 10.2. Details of this approach have been published (Mayevsky, 1984).

As seen in Fig. 10.2, the maximal NADH level is achieved under complete O_2 deprivation that can be induced both under in vitro and in vivo conditions. This signifies that this definitive point can be used to determine State 5 in vivo as well.

The problem is to determine the metabolic state of a tissue in an in vivo situation. If we adopt the in vitro value of a resting state (State 4), this would signify that the increase in NADH during state 5, induced by anoxia (0% O_2), would be only 1%. According to all in vivo studies, this is not the case, and during anoxia the increase in NADH is larger than the decrease. Therefore, we concluded that, under in vivo conditions, the "resting" metabolic state of the brain is found between States 4 and 3 rather than in State 4 as defined in vitro (Mayevsky, 1984). In order to determine the maximal levels of NADH in vivo, it is almost impossible to use cyanide; therefore, we used 100% nitrogen until the cessation of respiration and the recovery to air was monitored. It is almost impossible to achieve the minimal level of NADH (maximum oxidation) in vivo as compared to the level determined in state 2 when substrate availability is minimal. We were able to inject a non-fluorescing uncoupler (pentachlorophenol) into the brain, and NADH became more oxidized (Mayevsky, 1976). Using fiber optic fluorometry, we measured the brain in anesthetized as well as awake rats exposed to many perturbation in vivo and measured the changes in NADH redox state together with other physiological parameters shown also in Fig. 10.2 (Mayevsky, 1984; Mayevsky & Rogatsky, 2007).

Between 1973 and 1980 my lab was equipped and able to monitor, under CSD, only mitochondrial NADH regarding the direct main parameter that represents the cellular energy metabolism process. In 1980 we added the measurement of tissue pO_2 using oxygen electrode to the NADH measurement, but it was clear that this is not the ultimate best parameter to measure in the brain due to its dependence on cerebral blood flow. Therefore, after seeking a better parameter to add to the NADH, we adopted the measurement of CBF using the newly developed technology based on the laser Doppler flowmetry in 1990–1991 (Mayevsky et al., 1991, 1990; Mayevsky & Ziv, 1991).

The main issue that came up was, could we interpret the in vivo data according to the concept developed for the in vitro conditions? A good example to demonstrate the dilemma is presented in Fig. 10.2. The lower red rectangle (A) shows that when CSD developed in the normoxic brain, the NADH was oxidized (decrease fluorescence) and a large increase in CBF was recorded. But when the CSD developed under partial ischemia, for example, as seen in the upper red rectangle (B), NADH was elevated, namely, more reduced due to the decrease in CBF and a limitation in oxygen due to the partial ischemic condition. The same anomaly was seen in other conditions in addition to partial ischemia as seen in Figs. 10.3 (animal studies) and 10.4. (head-injured patient). According to Chance and Williams (1955, 1956), and Fig. 10.2 (left side), ATP synthesis is not possible if the tissue is in state 4 due to the unavailable ADP needed. This is one of the main reasons that we concluded that the "resting" brain in vivo is in state 3.5–3.6 and not in state 4. The fact that the brain tissue is recovered from a wave of CSD even under partial ischemia (Fig. 10.3F), exposure to carbon monoxide (CO) or to alcohol infusion (Fig. 10.3B and D) support our conclusions. The same conclusion is true for the human cerebral cortex as seen in Fig. 10.4C.

The differences in the responses to CSD in the normoxic and ischemic brain are demonstrated schematically in Fig. 10.5.

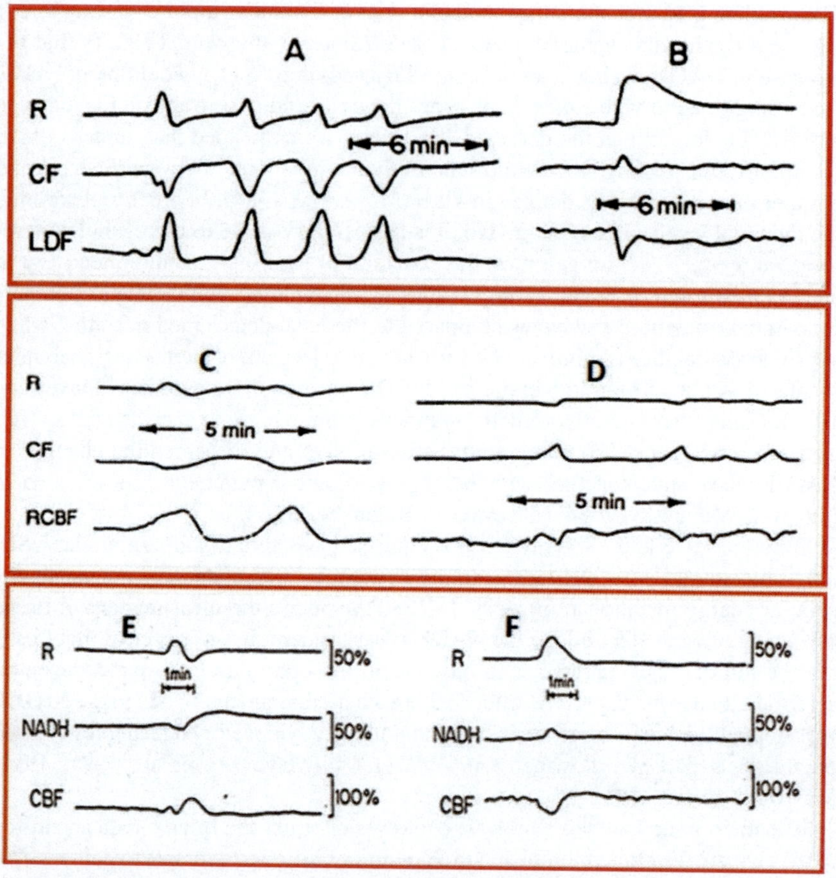

Fig. 10.3 Responses to CSD induced under various pathophysiological situations. **A** + **B** were recorded during control rat (**A**) and under exposure to 1000 ppm carbon monoxide (CO) in **B** (Mayevsky et al., 1993). **C** + **D** were recorded in a control rat (**C**) and 240 min after IV infusion of 3 ml of 30% alcohol (**D**) (Mayevsky et al., 2000a). **E** + **F** were recorded CSD induced in a control rat (**E**) and partial ischemic rat (**F**) (Mayevsky et al., 2000b)

The metabolic responses to the Leão spreading depression (Leão, 1944b) phenomenon (CSD) have been studied by various investigators in different animal models. Few investigators showed that CSD caused an increase in oxygen utilization as evaluated by measuring pO_2 change in the cortex (Mayevsky et al., 1980b). In order to supply the need for extra O_2, hemodynamic changes take place, and indeed, some researchers showed blood flow changes during cortical CSD (Buresova, 1957; Leão, 1944a; Van Harreveld & Ochs, 1957). Using fluorometric/reflectometric techniques for monitoring mitochondrial signals, the same conclusion was drawn, namely, that CSD caused an increase in energy utilization. In 1973 Rosenthal and Somjen (1973) showed an oxidation of NADH due to CSD in the cerveau isole cat preparation. Using the awake rat model, we also showed oxidation of NADH due to CSD

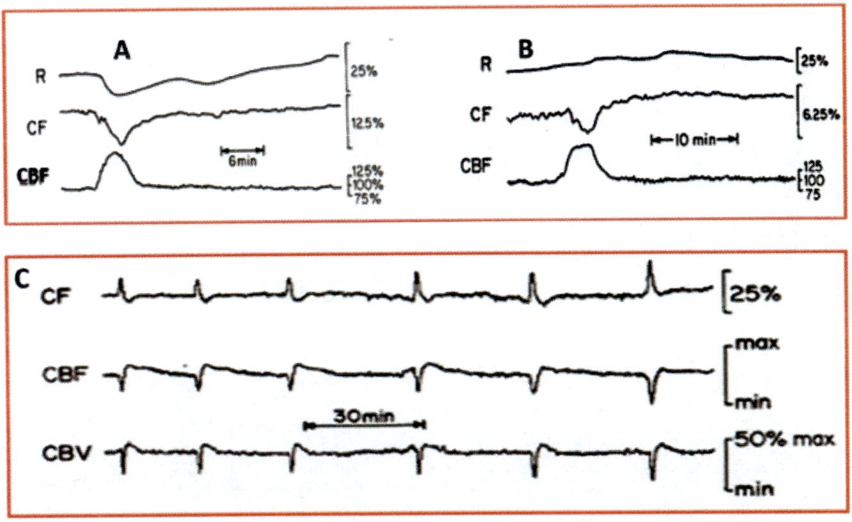

Fig. 10.4 **A + B**, two responses identified as responses to CSD recorded in a head-injured patient during the early stage of the recording. **(C)** Six repetitive responses identified as responses to CSD in the later stage in the recording of the same patient (Mayevsky et al., 1996)

(Mayevsky & Chance, 1973, 1974). We were able to correlate the NADH oxidation cycle with the pumping activity of the extracellular K^+ in the recovery phase of CSD in rats (Mayevsky et al., 1974). The effects of pentobarbital and Quabain on the metabolic responses to CSD in the cat brain were described by LaManna and Rosenthal (1975). We showed, in the awake rat, the effects of hypoxia and partial ischemia (Mayevsky, 1978) on the responses to CSD. The results presented here augment the knowledge gained up to this time on the various types of responses to CSD obtained under various pathological situations in the awake and anesthetized animal. Three animal models were used: the adult rat, the very young rat (13–14 days old), and the adult gerbil. Table 10.1 summarizes the results obtained in the various models exposed to three physiological and pathological conditions, namely, hypoxia, partial ischemia, and anesthesia.

As shown in the past, during CSD, there is a large, passive efflux of K^+ into the extracellular space (Mayevsky et al., 1974), regardless of energy availability. The pumping back of the K^+ by the ATPase into the cells is an energy-dependent process and increases the energy demand of the tissue (Crowe et al., 1979; Mayevsky et al., 1974, 1980a; Rosenthal & Somjen, 1973). As a result, the mitochondria are activated (toward State 3) in order to produce the extra ATP needed and oxygen utilization is increased (Lothman et al., 1975; Somjen et al., 1976). In the normoxic brain, the increased blood flow will compensate for the extra O_2 needed (Mayevsky and Weiss 1991) and will keep the balance of oxygen in the normal range so that oxidation of NADH is observed. When oxygen delivery to the tissue is limited, an increase of the energy demand during CSD will lead to an imbalance between supply and

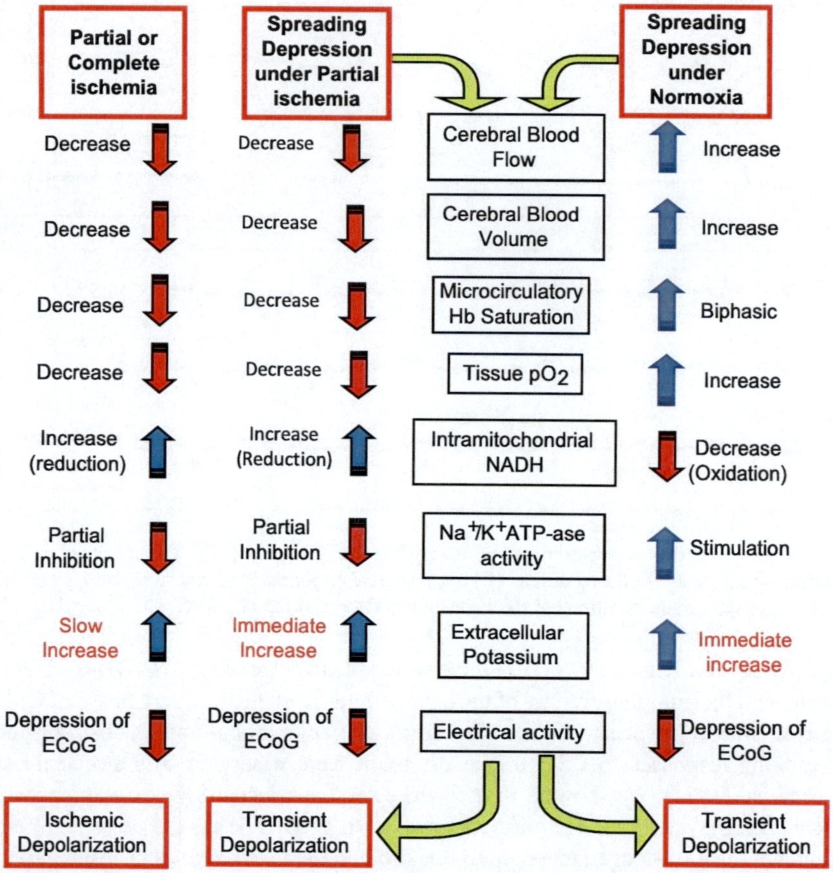

Fig. 10.5 Schematic comparison between two responses to CSD induced in the normoxic brain (right side) and in the partial ischemic brain (left side). Also, the response to partial or complete ischemia without CSD is presented in the left side

Table 10.1 NADH responses to CSD in rats and gerbils under various physiological and pathological conditions. In rats the reduction response to CSD obtained after bilateral carotid occlusion was reversible 8 days after occlusion (Mayevsky, 1978). After 4–7 days, the awake rats showed oxidation cycles while the anesthetized rats showed a reduction cycle

	Normoxia	Hypoxia	Partial ischemia	Anesthesia
Adult rat	Oxidation	Oxidation (Lim et al., 1995)	Reduction (Sharpless, 1970)	Oxidation
Young rat	Oxidation	Not tested	Reduction	Reduction
Adult gerbil	Oxidation	Reduction	Reduction	Reduction

demand, resulting in a reduction of NADH. In severe oxygen imbalance, the ATP production will be so slow (anoxia or complete ischemia) that the K^+ will not return into the cells during the CSD episode and NADH will be fully reduced. It seems from the results shown in Table 10.1 that various animals respond differently to pathological conditions. Partial ischemia had a uniform effect on the metabolic response to CSD in all three models tested, namely, a reduction of NADH observed as compared to the normoxic state, due to the large demand for extra energy needed. Deep anesthesia and hypoxia have different effects on the three models used. The young rat and the adult gerbil showed the same change in NADH due to CSD (increase) and the adult rat was basically unaffected by the two treatments. We can conclude from Table 10.1 that the brain compensation mechanisms activated in the young rat and the adult gerbil by CSD under hypoxia and deep anesthesia cannot deliver the amount of O_2 needed for the complete recovery of the CSD so that a reduction (increase) of NADH was recorded. As was shown previously, anesthesia cannot abolish the responses to CSD (Van Harreveld & Stamm, 1953). Under conditions where NADH showed an increase due to CSD, the rate of K^+ influx into the cells was slower but was returned to the baseline level of extracellular potassium measured before the CSD. This interrelation was found in the gerbil as well as in the adult rat, exposed to bilateral carotid occlusion.

The effects of the functional state on the response to CSD were reported by others (LaManna & Rosenthal, 1975) showing the slower kinetics of the re-reduction after CSD stimulation or even during the direct cortical stimulation (Rosenthal & Martel, 1979; Rosenthal et al., 1976). The same type of interrelation between K^+ pumping and NADH redox state was described during hippocampal stimulation (Lewis & Schuette, 1975) or pentylenetetrazol seizures in the cat brain exposed to ischemia (Vern et al., 1979). The results presented in our study may explain the opposite types of responses to direct cortical stimulation described by various investigators (Dora & Kovach, 1979; Rosenthal & Jobsis, 1971; Rosenthal et al., 1976). We believe that the nature of the NADH response to direct cortical stimulation or to CSD will depend upon the ability of the brain to respond for the extra oxygen needed under such conditions. We can conclude from the results presented that the type of response to CSD is a good indicator of the state of the oxygen balance in the brain. Also, in terms of the metabolic states of the brain, we believe that in vivo the brain can produce extra ATP either when NADH shows an oxidation (toward State 3) or reduction (toward State 4). The main difference between the two conditions is the rate of K^+ pumping activity from the extracellular space into the cells.

10.5 Modeling Brain Energy Metabolism and Function

10.5.1 Introduction

In order to test the validity of the two types of NADH responses to CSD induced in the normoxic and ischemic brain, for example, we used another technique. We constructed and tested a mathematical model (Vatov et al., 2006), capable of simulating changes in brain energy metabolism that develop in real time under various pathophysiological conditions.

The goal of our study was to develop and test a mathematical model that permits the exploration of brain tissue metabolism in real time, based on several observable and hidden parameters computed via the model. Such a model has the potential to serve as a tool in simulating the effects of new pathophysiological conditions on the brain responses. A sufficiently detailed model of this kind, when properly tuned, can be used to assess the brain tissue functioning even when not all the parameters are monitored in an actual experiment. As mentioned above, in the framework of our research, we refer to the tissue as a whole, i.e., as a self-contained system with inputs and outputs. This simplification enables us to utilize the so-called integral- or point-model approach, when the physiological variables are regarded as functions of time only (their interdependence apart), rather than as space-distributed variables. Hence, the "diffusion" terms and the spatial derivatives that appear in the space-distributed models (Revett et al., 1998; Ruppin et al., 1999) are omitted, and we arrive at a classical dynamic system, whose states are described by ordinary differential equations of the kinetic type. Furthermore, using this approach, we can tune the model, maximizing the fit to the actual clinical and experimental data with little risk of overfitting. In the course of data fitting, different kinds of nonlinearities in the right-hand parts of the differential equations were tested. The main objective was to keep the model physiologically meaningful, simple as possible, and universal in the sense that the same values of the model parameters would be used to simulate the normal and different pathological conditions of the brain tissue. The goodness of-fit was examined and confirmed in numerous independent runs using various laboratory and clinical data published earlier (Mayevsky, 1990; Mayevsky et al., 1984, 1996, 1998b; Meilin et al., 1999).

10.5.2 The Mathematical Model

In the framework of our research, we refer to the tissue as a whole, i.e., as a self-contained system with inputs and outputs. This simplification enables us to utilize the so-called integral- or point-model approach, when the physiological variables are regarded as functions of time only (their interdependence apart), rather than as space-distributed variables. Hence, the "diffusion" terms and the spatial derivatives that appear in the space-distributed models (Revett et al., 1998; Ruppin et al., 1999) are omitted, and we arrive at a classical dynamic system, whose states are described by ordinary differential equations of the kinetic type. Furthermore, using this approach, we can tune the model, maximizing the fit to the actual clinical and experimental data with little risk of overfitting. In the course of data fitting, different kinds of nonlinearities in the right-hand parts of the differential equations were tested. The main objective was to keep the model physiologically meaningful, simple as possible, and universal in the sense that the same values of the model parameters would be used to simulate the normal and different pathological conditions of the brain tissue. The goodness-of-fit was examined and confirmed in numerous independent runs using various laboratory and clinical data published earlier (Mayevsky et al., 1984, 1996, 1998a, b; Mayevsky, 1990; Meilin et al., 1999). The mathematical model manifests many of the essential characteristics of the experimental data with regard to certain observable variables, providing a quantitative fit, which is indicative of the model's reliability. We have tested the ability of the model to correctly reproduce the dynamics of a reduced number of the physiological variables when the dynamics of the remaining variables are inputted in the model. This property of our model may be used in clinical practice—to complement real-time measurements. Our model is based on the assumption that the tissue functional state is fully determined by the following metabolic variables:

1. Concentration of extracellular potassium (K).
2. Rate of potassium reuptake (R).
3. Blood flow (F).
4. Oxygen (O).
5. NADH (N).
6. Partial impairment (P).

Such a formulation of the model was selected because quantitative data on the rates of change of the model variables are scarce. In clinical conditions, only some of the above tissue parameters, such as the blood flow, the concentration of extracellular potassium, and the level of NADH, can be measured. Other, hidden variables (the rate of potassium reuptake, partial impairment, and the level of oxygen) are either nonmeasurable or not measured at the present. Their meaning of the model variables is detailed below.

The model is constructed as a set of nonlinear differential equations (see below (1)–(6)) that govern the temporal evolution of the above metabolic variables. Each of the main tissue variables, being a function of time, depends also on other physiological variables. Its dynamics are governed by an equation that describes the variable's behavior under both normoxic and pathological conditions. All variables of the model (including time) are nondimensional, i.e., they are normalized through some scales. To enable computations, these variables are discretized. Thus, each of the above six variables ranges from 0 to 1 while time runs from 0 to 500 units (with some small step).

In the equations, the variables are denoted by a single capital letter (e.g., K, R), while their specific levels—by the same letter with a subscript (e.g., Krest); multiplicative constant factors are denoted by a capital "C" and subscripted by a number of indices—capital letters usually designating the related variables. The first index identifies the variable whose dynamic equation contains this constant, while the remaining indices (those appearing after the comma) identify the variables that the constant C modulates. The subscripts "rest" and "max," respectively, designate the resting value and the absolute ceiling of a variable. The resting value Frest = Fmax/2 is equal to the normal level of the blood flow. The initial values of the variables in the model reflect normoxic tissue state.

The equations governing the temporal behavior of the metabolic parameters indicated above are constructed on the following grounds.

The rate of change of the extracellular potassium concentration, K, is modeled as a reaction-diffusion process, which is governed by four terms:

$$\mathrm{dK/dt = CK,K(K - Krest)(K - }K\vartheta\mathrm{)(K - Kmax) + (}\delta\mathrm{ + P)(Kmax - K)} \quad (1)$$
$$\mathrm{- CK,RKR + Kinj.}$$

The first "reaction" term in Eq. (1) models the regenerative processes that generate the sharp rise of extracellular potassium concentration occurring when K crosses the thresholds level Kϑ. The second term represents the diffusion of intracellular potassium in undamaged tissue and intracellular potassium release to the extracellular space resulting from various tissue stresses; here constant δ represents the rate of permanent intracellular potassium leakage in normal tissue; for the meaning of P see Eq. (6) and the explanation appearing immediately before it. The third term reflects the reuptake of potassium by Na/K ATPase pumps. The last term represents an external injection of potassium.

The rate of potassium reuptake, R, is modeled by the following equation:

$$\mathrm{dR/dt = CR,NOKR\,NO(K - Krest)(Rmax - R) - CR,RK\,R(Kmax - K)} \quad (2)$$

Two terms in Eq. (2) affect the rate of potassium reuptake, reflecting the functioning of Na/K pumps. The first term in the right-hand part of (2) models the increase of reuptake with increasing levels of NADH and oxygen. The second term gradually brings Na/K ATPase activity back to its resting level when K values are restored.

The blood flow F is regulated by the equation

$$\mathrm{dF/dt = CF,FO(Fmax - F)(Orest - O) + CF,F\,Fax\,2 - F - CF,} \quad (3)$$
$$\mathrm{KF(K - Crest)(Frest - F).}$$

Here the first term represents the dependence of blood supply on the oxygen status and the current flow level. The second term in (3) self-regulates the blood flow toward its base rate. The third term describes variations in the blood flow caused by an irregular extracellular potassium rise (that may result, e.g., from an injection of potassium). When constructing the last term, the previous tissue state has been taken into account. For that purpose, the coefficient CFFK = 0, if the previous value of the blood flow is less than the resting value Frest; otherwise it is taken to equal 0.5.

The energy status of the tissue is determined by oxygen, O. Its level is determined by a competition between the supply and demand terms. The supply term is proportional to the blood flow and current oxygen level. The demand term is an equivalent consumption rate; it involves the oxygen, potassium reuptake rate, and NADH. The resulting equation is:

$$\mathrm{dO/dt = CO,FO\,F(Omax - O) - CO,NRO\,N(RO)0.5.} \quad (4)$$

The rate of NADH change is determined by its level, while the consumption rate depends on the potassium reuptake as well as on oxygen:

$$dN / dt = CN,N\left(Nmax - N\right) - CN,NRON\left(RO\right)0.5. \quad (5)$$

Finally, the level of stresses experienced by the tissue is represented by the partial impairment variable P. Its meaning can be understood from Eq. (1): stresses upset the integrity of tissue cells, which results in increasing diffusion of the intracellular potassium to the extracellular space. In a damaging situation, the stores of oxygen deplete, and O becomes insufficient for the Na/K pump functioning. The concentration of extracellular potassium rises and the partial impairment augments. The corresponding equation is:

$$dP / dt = CP,OK\left(Omin - O\right)\left(K - Krest\right). \quad (6)$$

The model is implemented using "Stella" software (High Performance Systems, Inc., http://www.hps-inc.com/). The differential equations are replaced with a set of their finite-difference analogs and are integrated numerically. Using this model, we investigate how various pathophysiological manipulations interrelate and influence the recovery process. Note that the same constants are used for all modeled situations.

10.5.3 Results

Spreading Depression The pathological state of spreading depression can be induced by washing the cerebral cortex with a diluted solution of KCl (the "potassium injection"). This is modeled by putting Kinj >0 within the time interval t1 < t < t2. When spreading depression is induced, the increase in energy requirement leads to the activation of Na^+-K^+-ATPase in order to restore the normal potassium distribution and causes an increase in blood flow and a decrease in NADH (Fig. 10.6).

Spreading Depression After Partial Ischemia In this case, the blood flow decreases mildly ($F_{max} = 0.5$ within the time interval from t1 to t2), but within the ischemic period, an infusion of potassium is modeled; i.e., Kinj is assumed to be positive within a relatively short time interval contained in the interval t1 < t < t2 during which the main events occur. Accordingly, the blood flow slightly decreases as a result of the infusion, and then increases to the new resting level (Fig. 10.7).

In the cases of global ischemia and CSD, the temporal behavior of the model variables was compared with the corresponding published clinical and experimental data (Mayevsky et al., 1998a; Meilin et al., 1999). In order to compare the simulated and experimental dynamics, the scales were adjusted. The comparison (Figs. 10.8

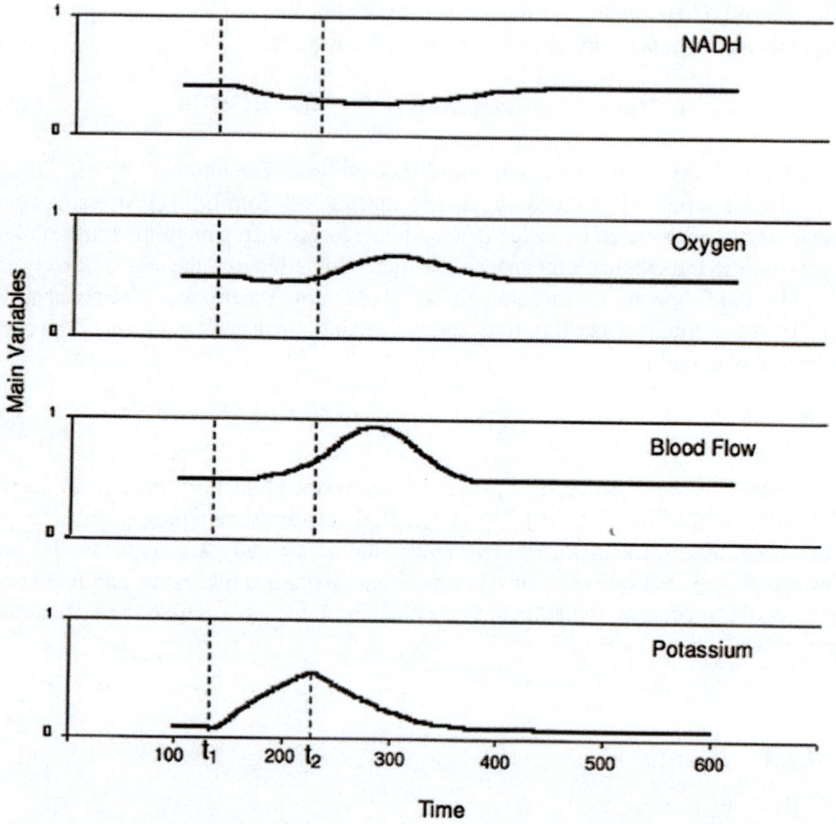

Fig. 10.6 Simulated spreading depression, induced by an injection of potassium when it is at the resting level. The duration of injection is from t1 to t2 (Vatov et al., 2006)

and 10.9) testifies to the qualitative agreement between the computed and experimental values.

10.5.4 Model Discussion

Our results demonstrate the potential of the computational mathematical model of tissue metabolism to describe the pathogenesis of tissue sustained in various clinical and experimental situations. The model simulates five pathophysiological tissue conditions, shown in Figs. 1–5 of our previous publication (Vatov et al., 2006) and Figs. 10.6, 10.7, 10.8, and 10.9 in the current chapter. This includes (1) global ischemia, (2) partial ischemia, (3) partial ischemia without depolarization, (4) spreading depression, and (5) spreading depression induced after partial ischemia. Note that the various states of tissue activity are described using the same values of constants

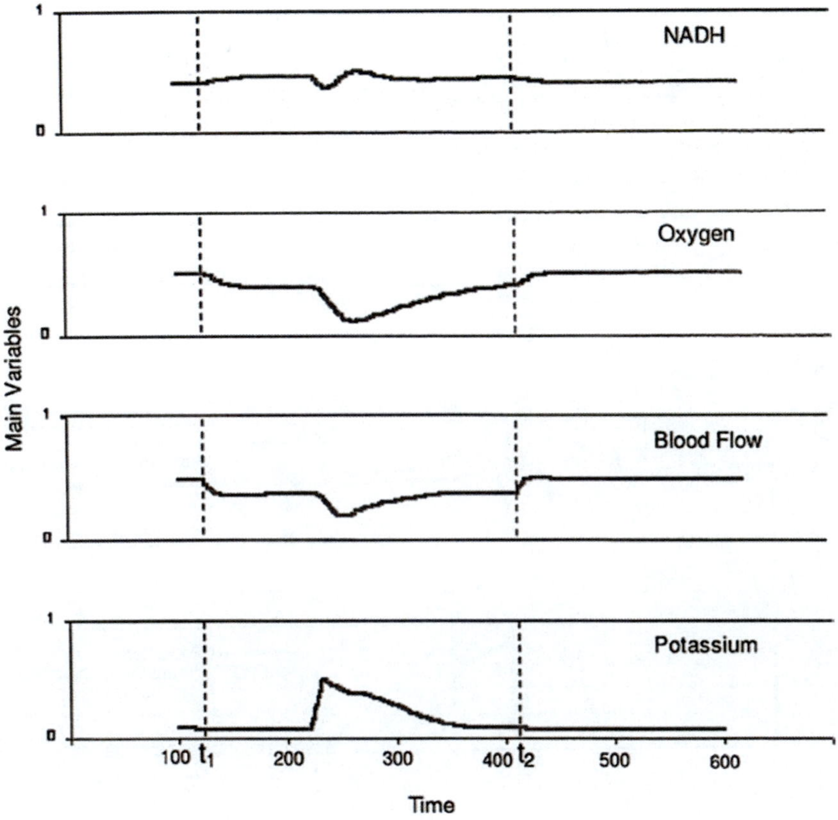

Fig. 10.7 Simulated spreading depression induced after partial ischemia, resulting from a mild decrease in blood flow with a short injection of potassium (Vatov et al., 2006)

(Table 1 in (Vatov et al., 2006)). This indicates the model's adequacy for the processes occurring in the tissue. The considerable qualitative and quantitative agreement with the experimental/clinical data (Figs. 10.8 and 10.9) served as tests to the model's reliability. The number of variables routinely monitored in clinical situations is limited. Using our model in clinical and laboratory practice, the number of measured parameters can be minimized, since the nonmeasured variables can be evaluated by the model in real time. Moreover, in principle, the comparison of the computed and clinical data can be used to test the tissue state and predict the recovery process. When utilized along with real-time clinical measurements, the model may provide the physician with a means for early diagnosis of pathological processes at their early stages.

A wide range of pathological events may ensue due to the limitation of oxygen supply to the tissue by a restriction of blood flow (ischemia) or by a lowering of O_2 level in the inspired air (hypoxia). As a result of oxygen restriction, the mitochondrial activity is inhibited, the ATP production decreases, and the ionic homeostasis

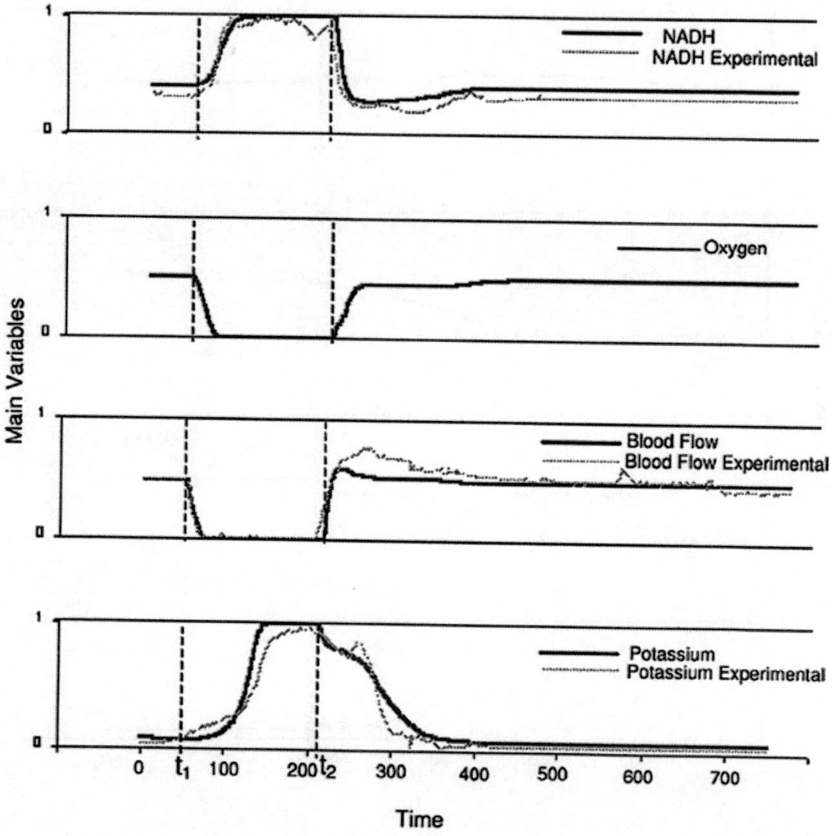

Fig. 10.8 The comparison of the temporal behavior of the model variables to the corresponding experimental data for global ischemia. The simulated dynamics (shown in Fig. 1 of the original publication (Vatov et al., 2006)) overlap with the experimental data from (Mayevsky, 1992)

is disturbed, leading to extracellular K^+ accumulation. A high correlation has been found between changes in the cerebral blood flow and mitochondrial NADH redox state during ischemia (Mayevsky, 1990). On the other hand, high levels of extracellular potassium produced in CSD lead to an increase in the tissue metabolic demands, as indicated by a rise in O_2 consumption (Mayevsky & Weiss, 1991). This process is usually accompanied by an increase in blood flow; as a result, the activation of mitochondrial activity leads to NADH oxidation. In our recent publications (Mayevsky et al., 2002, 2004; Pevzner et al., 2003), we have described the principles behind a new medical device, the TiSpec-Tissue Spectroscope. The TiSpec provides real-time data on three optical variables, namely, the tissue blood flow, mitochondrial NADH redox state, and tissue reflectance. It is expected that by using this device in various clinical situations, the three parameters provided will permit diagnosing the tissue metabolic state, even though extracellular K^+ and O_2 levels are not monitored. Joining these measurements with the data processing by our model

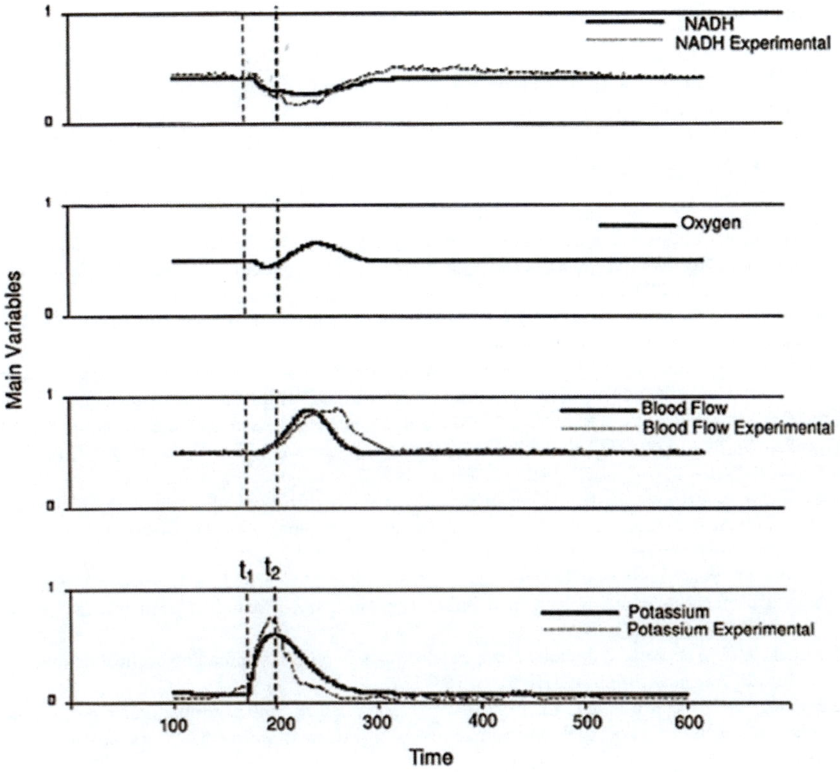

Fig. 10.9 The comparison of the temporal behavior of the model variables to the corresponding clinical data for spreading depression from (Mayevsky et al., 1996; Vatov et al., 2006)

will open up the possibility of expanding the scope of experiments. The aim of such experiments will be the evaluation of the optimal values (or frames) of the variables that determine the required behaviors of NADH and oxygen.

10.6 Future Perspectives

In the previous Sections (10.4 and 10.5), we presented the responses of the brain to CSD in control animal as well as in partial ischemic brains (Fig. 10.5). The use of mitochondrial NADH response to CSD as an indicator to the metabolic state of the brain was proven to be a reliable parameter in vivo. It is clear that under limitation in oxygen supply, due to various factors, the CSD may lead to NADH reduction and not to oxidation as was found by Chance and Williams (1955) under in vitro conditions. Therefore, we are suggesting to repeat those studies in vivo under conditions were oxygen supply to the brain will be elevated by infusing oxygen carrier such as BXT25 developed by BioxyTran Inc.

References

Ayata, C., & Lauritzen, M. (2015). Spreading depression, spreading Depolarizations, and the cerebral vasculature. *Physiological Reviews, 95*, 953–993.

Bures, J., Buresova, O., & Krivanek, J. (1974). *The mechanism and applications of Leão's spreading depression of electroencephalographic activity*. Academic.

Buresova, O. (1957). Changes in cerebral circulationin rats during spreading EEG depression. *Physiologia Bohemoslovenica, 6*, 1–11.

Chance, B., & Baltscheffsky, H. (1958). Respiratory enzymes in oxidative phosphorylation (VII – Binding of intramitochondrial reduced pyridine nucleotide). *The Journal of Biological Chemistry, 233*, 736–739.

Chance, B., & Williams, G. R. (1955). Respiratory enzymes in oxidative phosphorylation (III- The steady state). *The Journal of Biological Chemistry, 217*, 409–427.

Chance, B., & Williams, G. R. (1956). The respiratory chain and oxidative phosphorylation. In F. F. Nord (Ed.), *Advances in enzymology* (pp. 65–134). XVII Interscience Publisher.

Crowe, W., Mayevsky, A., & Mela, L. (1979). The dynamics of K^+ leakage and recovery in cerebral ischemia. In A. Leter, T. Saba, & L. Mela (Eds.), *Advances in shock research* (pp. 221–232). Alan R. Liss.

Dora, E., & Kovach, A. G. B. (1979). Reactivity of the cerebrocortical vasculature and energy metabolism to direct cortical stimulation in haemorrhagic shock. *Acta Physiologica Academiae Scientiarum Hungaricae, 54*, 347–361.

Fabricius, M., Fuhr, S., Bhatia, R., Boutelle, M., Hashemi, P., Strong, A. J., & Lauritzen, M. (2006). Cortical spreading depression and peri-infarct depolarization in acutely injured human cerebral cortex. *Brain, 129*, 778–790.

Krivanek, J. (1961). Some metabolic changes accompanying Leão's spreading cortical depression in the rat. *Journal of Neurochemistry, 6*, 183–189.

LaManna, J. C., & Rosenthal, M. (1975). Effect of ouabain and phenobarbital on oxidative metabolic activity associated with spreading cortical depression in cats. *Brain Research, 88*, 145–149.

Lauritzen, M., & Strong, A. J. (2017). Spreading depression of Leão and its emerging relevance to acute brain injury in humans. *Journal of Cerebral Blood Flow and Metabolism, 37*, 1553–1570.

Leão, A. A. P. (1944a). Pial circulation and spreading depression of activity in the cerebral cortex. *Journal of Neurophysiology, 7*, 391–396.

Leão, A. A. P. (1944b). Spreading depression of activity in cerebral cortex. *Journal of Neurophysiology, 7*, 359–390.

Lewis, D. V., & Schuette, W. H. (1975). NADH fluorescence and $[K^+]_o$ changes during hippocampal electrical stimulation. *Journal of Neurophysiology, 38*, 405–417.

Lim, D. J., Park, J. Y., Park, Y. K., Chung, H. S., Chu, J. H., Lee, K. C., & Lee, H. K. (1995). The changes of cortical spreading depression during hypothermia and rewarming in rat. *Journal of Korean Neurosurgical Association, 24*, 733–742.

Lothman, E., LaManna, J., Cordingley, G., Rosenthal, M., & Somjen, G. (1975). Responses of electrical potential potassium levels, and oxidative metabolic activity of the cerebral neocortex of cats. *Brain Research, 88*, 15–36.

Marshall, W. H. (1959). Spreading cortical depression of Leão. *Physiological Reviews, 39*, 239–279.

Mayevsky, A. (1976). Brain energy metabolism of the conscious rat exposed to various physiological and pathological situations. *Brain Research, 113*, 327–338.

Mayevsky, A. (1978). Ischemia in the brain: The effects of carotid artery ligation and decapitation on the energy state of the awake and anesthetized rat. *Brain Research, 140*, 217–230.

Mayevsky, A. (1984). Brain NADH redox state monitored *in vivo* by fiber optic surface fluorometry. *Brain Research Reviews, 7*, 49–68.

Mayevsky, A. (1990). Level of ischemia and brain functions in the Mongolian gerbil *in vivo*. *Brain Research, 524*, 1–9.

Mayevsky, A. (1992). Cerebral blood flow and brain mitochondrial redox state responses to various perturbations in gerbils. *Advances in Experimental Medicine and Biology, 317*, 707–716.

Mayevsky, A., 2015. Mitochondrial function *in vivo* evaluated by NADH fluorescence. Springer.

Mayevsky, A., & Chance, B. (1973). In H. I. Bicher & D. F. Bruley (Eds.), *A new long-term method for the measurement of NADH fluorescence in intact rat brain with implanted cannula* (pp. 239–244). Plenum Press.

Mayevsky, A., & Chance, B. (1974). Repetitive patterns of metabolic changes during cortical spreading depression of the awake rat. *Brain Research, 65*, 529–533.

Mayevsky, A., & Rogatsky, G. (2007). Mitochondrial function *in vivo* evaluated by NADH fluorescence: From animal models to human studies. *American Journal of Physiology. Cell Physiology, 292*, C615–C640.

Mayevsky, A., & Weiss, H. R. (1991). Cerebral blood flow and oxygen consumption in cortical spreading depression. *Journal of Cerebral Blood Flow and Metabolism, 11*, 829–836.

Mayevsky, A., & Ziv, I. (1991). Oscillations of cortical oxidative metabolism and microcirculation in the ischaemic brain. *Neurological Research, 13*, 39–47.

Mayevsky, A., Zeuthen, T., & Chance, B. (1974). Measurements of extracellular potassium, ECoG and pyridine nucleotide levels during cortical spreading depression in rats. *Brain Research, 76*, 347–349.

Mayevsky, A., Crowe, W., & Mela, L. (1980a). The interrelation between brain oxidative metabolism and extracellular potassium in the unanesthetized gerbil. *Neurological Research, 1*, 213–226.

Mayevsky, A., Lebourdais, S., & Chance, B. (1980b). The interrelation between brain PO_2 and NADH oxidation- reduction state in the gerbil. *Journal of Neuroscience Research, 5*, 173–182.

Mayevsky, A., Zarchin, N., & Tannenbaum, B. (1984). Brain responses to experimental oxygen deficiency in the Mongolian gerbil. *Advances in Experimental Medicine and Biology, 180*, 191–202.

Mayevsky, A., Yoles, E., Zarchin, N., & Kaushansky, D. (1990). Brain vascular ionic and metabolic responses to ischemia in the Mongolian gerbil. *Journal of Basic and Clinical Physiology and Pharmacology, 1*, 207–220.

Mayevsky, A., Flamm, E. S., Pennie, W., & Chance, B. (1991). A fiber optic based multiprobes system for intraoperative monitoring of brain functions. *SPIE, 1431*, 303–313.

Mayevsky, A., Rogatsky, G. G., Zarchin, N., & Thom, S. R. (1993). Interrelation between hyperbaric oxygenation and carbon monoxide intoxication in the rat brain *in vivo*. In P. B. Bennett & R. E. Marquis (Eds.), *Basic and applied high pressure biology* (pp. 409–420). University of Rochester Press.

Mayevsky, A., Doron, A., Manor, T., Meilin, S., Zarchin, N., & Ouaknine, G. E. (1996). Cortical spreading depression recorded from the human brain using a multiparametric monitoring system. *Brain Research, 740*, 268–274.

Mayevsky, A., Manor, T., Meilin, S., Doron, A., & Ouaknine, G. E. (1998a). Real-time multiparametric monitoring of the injured human cerebral cortex – a new approach. In R. Marmarou & R. Bullock (Eds.), *Acta Neurochirurgica Supplement (Wien.)* (pp. 78–81). ICPX.

Mayevsky, A., Meilin, S., Manor, T., Ornstein, E., Zarchin, N., & Sonn, J. (1998b). Multiparametric monitoring of brain oxygen balance under experimental and clinical conditions. *Neurological Research, 20*, S76–S80.

Mayevsky, A., Meilin, S., Manor, T., Zarchin, N., & Sonn, J. (1999). Optical monitoring of NADH redox state and blood flow as indicators of brain energy balance. *Advances in Experimental Medicine and Biology, 471*, 133–140.

Mayevsky, A., Rogatsky, G. G., & Sonn, J. (2000a). New multiparametric monitoring approach for real-time evaluation of drug tissue interaction *in vivo*. *Drug Development Research, 50*, 457–470.

Mayevsky, A., Sonn, J., Manor, T., Razon, N., & Ouaknine, G. E. (2000b). Responses to cortical spreading depression during normoxia and ischemia: Multiparametric monitoring study in ani-

mals and the human brain. In Y. Fukuuchi, M. Tomita, & A. Koto (Eds.), *Ischemic blood flow in the Brain* (pp. 343–350). Keio University Symposium.
Mayevsky, A., Manor, T., Pevzner, E., Deutsch, A., Etziony, R., & Dekel, N. (2002). Real-time optical monitoring of tissue vitality in vivo. *Proceeding of SPIE, 4616*, 30–39.
Mayevsky, A., Manor, T., Pevzner, E., Deutsch, A., Etziony, R., Dekel, N., & Jaronkin, A. (2004). Tissue spectroscope: A novel *in vivo* approach to real time monitoring of tissue vitality. *Journal of Biomedical Optics, 9*, 1028–1045.
Meilin, S., Zarchin, N., & Mayevsky, A. (1999). Inter-relation between hemodynamic, metabolic, ionic and electrical activities during ischemia and reperfusion in the gerbil brain. *Neurological Research, 21*, 699–704.
Penney, D. G., Verma, K., & Hull, J. A. (1989). Cardiovascular, metabolic and neurologic effects of acute carbon monoxide poisoning in the rat. *Toxicology Letters, 45*, 207–213.
Perez-Pinzon, M. A., Mumford, P. L., Rosenthal, M., & Sick, T. J. (1997). Antioxidants, mitochondrial hyperoxidation and elecrical recovery after anoxia in hippocampal slices. *Brain Research, 754*, 163–170.
Pevzner, E., Deutsch, A., Manor, T., Dekel, N., Etziony, R., Derzy, I., Razon, N., & Mayevsky, A. (2003). Real-time multiparametric spectroscopy as a practical tool for evaluation of tissue vitality *in vivo*. In T. Vo-Dinh, W. S. Grundfest, D. A. Benaron, & G. G. Cohn (Eds.), *Proceedings of SPIE* (pp. 171–182).
Revett, K., Ruppin, E., Goodall, S., & Reggia, J. A. (1998). Spreading depression in focal ischemia: A computational study. *Journal of Cerebral Blood Flow and Metabolism, 18*, 998–1007.
Rosenthal, M., & Jobsis, F. F. (1971). Intracellular redox changes in functioning cerebral cortex. II. Effects of direct cortical stimulation. *Journal of Neurophysiology, 34*, 750–762.
Rosenthal, M., & Martel, D. L. (1979). Ischemia-induced alterations in oxidative "recovery" metabolism after spreading cortical depression *in situ*. *Experimental Neurology, 63*, 367–378.
Rosenthal, M., & Somjen, G. (1973). Spreading depression, sustained potential shifts, and metabolic activity of cerebral cortex of cats. *Journal of Neurophysiology, 36*, 739–749.
Rosenthal, M., Martel, D. L., & LaManna, J. C. (1976). Effects of incomplete and complete ischemia on mitochondrial functioning measured in intact cerebral cortex of cats. *Experimental Neurology, 52*, 433–446.
Ruppin, E., Ofer, E., Reggia, J. A., Revett, K., & Goodall, S. (1999). Pathogenic mechanisms in ischemic damage: A computational study. *Computers in Biology and Medicine, 29*, 39–59.
Sharpless, S. K. (1970). Hypnotics and sedatives. In L. S. Goodman & A. Gilman (Eds.), *The pharmacological basis of therapeutics* (pp. 98–105). MacMillan Company, Collier MacMillan Limited.
Somjen, G. G., Rosenthal, M., Cordingley, G., LaManna, J., & Lothman, E. (1976). Potassium, neuroglia, and oxidative metabolism in central gray matter. *Federation Proceedings, 35*, 1266–1271.
Sramka, M., Brozek, G., Bures, J., & Nadvornik, P. (1977). Functional ablation by spreading depression: Possible use in human stereotactic neurosurgery. *Applied Neurophysiology, 40*, 48–61.
Strong, A. J., & Dardis, R. (2005). Depolarisation phenomena in traumatic and ischaemic brain injury. *Advances and Technical Standards in Neurosurgery, 30*, 3–49.
Strong, A. J., Fabricius, M., Boutelle, M. G., Hibbins, S. J., Hopwood, S. E., Jones, R., Parkin, M. C., & Lauritzen, M. (2002). Spreading and synchronous depressions of cortical activity in acutely injured human brain. *Stroke, 33*, 2738–2743.
Sundt, T. M., & Anderson, R. E. (1975). Reduced nicotinamide adenine dinucleotide fluorescence and cortical blood flow in ischemic and nonischemic squirrel monkey cortex. I. Animal preparation, instrumentation, and validity of model. *Stroke, 6*, 270–278.
Van Harreveld, A., & Ochs, S. (1957). Electrical and vascular concomitants of spreading depression. *The American Journal of Physiology, 189*, 159–166.
Van Harreveld, A., & Stamm, J. S. (1952). Vascular concomitants of spreading cortical depression. *Journal of Neurophysiology, 15*, 487–496.

Van Harreveld, A., & Stamm, J. S. (1953). Effect of pentobarbital and ether on the spreading cortical depression. *The American Journal of Physiology, 173*, 164–170.
Vatov, L., Kizner, Z., Ruppin, E., Meilin, S., Manor, T., & Mayevsky, A. (2006). Modeling brain energy metabolism and function: A multiparametric monitoring approach. *Bulletin of Mathematical Biology, 68*, 275–291.
Vern, B. A., Schuette, W. H., Mutsuga, N., & Whitehouse, W. C. (1979). Effects of ischemia on the removal of extracellular potassium in cat cortex during pentylenetetrazol seizures. *Epilepsia, 20*, 711–724.
Warner, D. S. (1997). Cortical spreading depression recorded from the human brain using a multiparametric monitoring system. *Journal of Neurosurgical Anesthesiology, 9*, 292.

Appendix

Introduction

In order to provide the reader of this book with historical scientific background, we decided to reprint three chapters that were published in the main book on CSD published in 1974.

The three chapters (Chapter 1, 5 and part of 6) that are presented in this Appendix (to the current new book) got the permission from Elsevier (Academic Press) to be used.

The cited material was published in the book entitled *The Mechanism and Applications of Leão's Spreading Depression of Electroencephalographic Activity*. The authors of the book published in 1974 were Jan Bures, Olga Buresova, and Jiri Krivanek. The publishers of the book were ACADEMIA, Publishing House of the Czechoslovak Academy of Sciences, Prague, and ACADEMIC PRESS INC. New York and London (Fig. A.1).

The three chapters were reprinted here in the same format as was published in the 1974 book.

Chapter 1: Electrical Properties of Spreading Depression

Spreading depression (SD) affects all basic types of electrical activity of the brain: spike activity of single neurons, synchronized synaptic potentials, and steady potentials generated by neurons or glia. The following paragraphs deal systematically with different electrical manifestations of SD stressing mainly recent findings, controversial problems, and interpretation of experimental evidence.

A. Mayevsky, J. Sonn, *Cortical Spreading Depression of Leao*,
https://doi.org/10.1007/978-3-031-08068-5

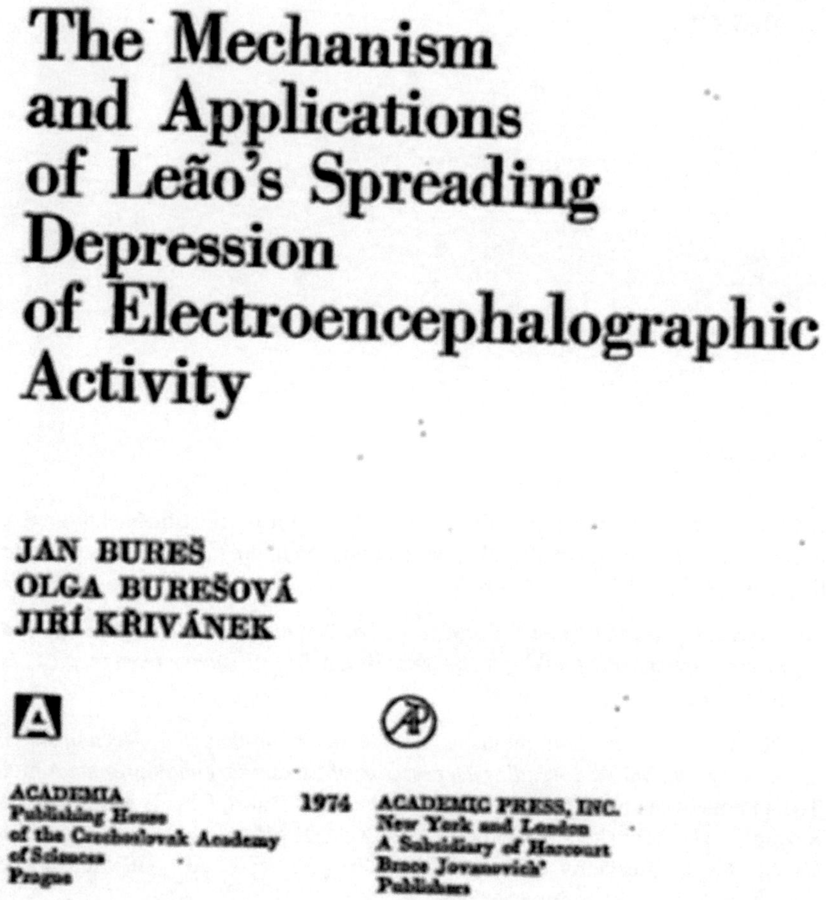

The Mechanism
and Applications
of Leão's Spreading
Depression
of Electroencephalographic
Activity

JAN BUREŠ
OLGA BUREŠOVÁ
JIŘÍ KŘIVÁNEK

ACADEMIA
Publishing House
of the Czechoslovak Academy
of Sciences
Prague

1974

ACADEMIC PRESS, INC.
New York and London
A Subsidiary of Harcourt
Brace Jovanovich'
Publishers

Fig. A.1 The front page of the book on CSD published in 1974 by Bures et al. (1974)

Spontaneous Electroencephalographic Activity

Depression of spontaneous electroencephalographic (EEG) activity, spreading at a rate of 3 mm/min across the cortical surface, accounts for the descriptive name of the phenomenon. At each cortical point, the EEG decreases over approximately 30 s to a level of maximum depression, which maintained for 2–3 min. Recovery proceeds more slowly, the predepression amplitude of the EEG being asymptotically attained after 5–10 min. In Leão's (1944b) classical paper, the EEG was recorded with a row of bipolar electrodes applied to the exposed pia-arachnoid surface of the rabbit brain but essentially similar results were also obtained in rats (Fig. A.2) and in other animals and other neural structures. The EEG depression is most pronounced when recorded with electrode pairs, the interelectrode distance of which

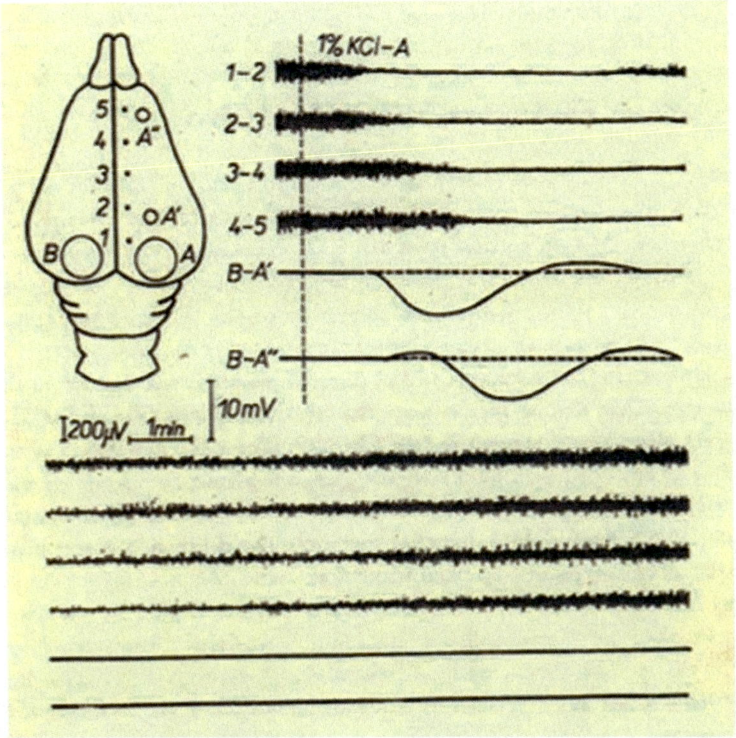

Fig. A.2 Cortical spreading depression in the rat. Position of electrodes indicated on the brain diagram (1, 2, 3, 4, 5). Steel needle electrodes for EEG recording (A', A"). Trephine openings for application of wick calomel cell electrodes for slow potential recording. (A) Large trephine opening for application of KCl solution (1-2, 2-3, 3-4, 4-5). Bipolar EEG recordings with the respective electrode pairs (B-A', B-A"). Slow potential recording from electrodes (A') and (A") against the common reference electrode situated in the contralateral hemisphere (B). The upper and lower recordings are continuous. Note that the onset of the negative SPC coincides with the onset of the EEG depression in the respective areas (Bures et al., 1974)

does not exceed the wavelength of SD (duration times velocity) and which are parallel to the front of the SD wave. In monopolar recording the depression is less pronounced because potentials are picked up from wider areas.

Typical EEG depression is obtained in animals under moderate barbiturate anesthesia, characterized by high-amplitude EEG and occasional "barbiturate spindles." During very deep anesthesia, or in unanesthetized aroused animals, the EEG amplitude is so low that it is not much affected by SD. This phenomenon can be detected by other signs, however.

Depression of the EEG is never complete. Some activity remains even at the height of the process and is probably due to (a) remote potentials from subcortical generators, (b) impulses coming through corticopetal pathways, and (c) in some cases incomplete depression of activity in deep cortical layers.

In addition to amplitude, the shape and frequency of the EEG is also affected by SD. The latter changes may occur in cortical regions not yet invaded by the actual depression, when elimination of a substantial part of the cortex affects the activity of the subcortical pacemaker and energizing centers and thus indirectly influences the remaining intact cortical regions. These mediated remote effects (see Chap. 9), usually characterized in anaesthetized animals by a slowing of the EEG and suppression of spindling, can be easily differentiated from the local direct effects consisting of various signs of hyperactivity in the intervals immediately preceding and following the depression. Under certain conditions these symptoms are so enhanced that EEG depression is fully substituted by paroxysmal activity, which spreads over the cortical surface at the same rate (3 mm/min) as SD. Spreading depression may gradually change into "spreading convulsion" when repeatedly evoked at brief intervals (Whieldon & Van Harreveld, 1950). Convulsion instead of depression appears in cortical regions topically treated with acetylcholine or pilocarpine (Van Harreveld & Stamm, 1953b) when they are invaded by SD. Hypercapnia (7–15% CO_2) has a similar effect. Although the exact mechanism of spreading convulsions is not fully understood, it is obvious that the activity of some neural elements, especially in deep cortical layers, is not blocked. Selective depression of upper cortical layers may evoke transcortical steady potential gradients generating the convulsive activity.

Evoked Potentials

Whereas the variability of the spontaneous EEG prevents exact evaluation of depression, evoked potentials offer a better opportunity for quantitative measurement. Since these potentials are generated in circumscribed cortical areas, they have known afferents pathways and reproducible waveforms corresponding to definite intracortical dipoles.

Severe reduction or complete abolition of sensory evoked potentials and callosal responses during SD were first described by Leão (1944b). Both positive and negative components of the evoked responses are depressed for 2–3 min and recover predepression amplitudes by 5–7 min. The surface negative component of the response, representing the activity of the most superficial cortical layers, is usually the slowest to recover. The less regular late components of the evoked response appearing in unanesthetized animals are influenced for a longer period. When the averaging technique is used for their evaluation, complete recover is often seen only after 20–30 min. In some evoked potentials it is possible to separate from the initial positivity a brief spikelike potential corresponding to the arrival of a highly synchronized volley of afferent impulses into the subcortical white matter and into the deep cortical layers. This potential, particularly prominent in cortical responses to single-shock stimulation of the optic nerve or optic radiation in the cat, is usually resistant to SD which abolishes all later components of the response (Marshall et al., 1950). This component is also reduced by SD, however, when the cat cortex has been exposed for several hours to a Tyrode solution with K^+ concentration increased to 27 mEq/1 and when several SD waves have passed through the area of recording (Freygang Jr. & Landau, 1955; Marshall, 1959).

In a slightly different way SD affects the direct cortical response evoked by electrical stimulation of an adjacent cortical area (Grafstein, 1956b; Ochs, 1958; Ochs & Hunt, 1960). The surface positive component of the response is usually depressed longer than the negative one, full recovery being attained after approximately 5 min in the isolated cortical slab in the cat (Grafstein, 1956b) and after approximately 10 min in the intact rabbit cortex (Ochs, 1958). The early spikelike component of the direct cortical response, which is believed to be generated by activation of afferent fibers, was also briefly depressed at the height of SD (Ochs et al., 1961). The direct cortical response is affected by CSD not only at the recording site but also in the stimulated region. Ichijo and Ochs (1970) succeeded in separating both effects by surgical and pharmacological means. When CSD was confined to the area of stimulation, block of axons directly affected by the electrical stimulus lasted for 2 min. Cortical SD restricted to the responding region caused a more prolonged suppression (3–13 min), probably due to interference with the function of dendrites. Evoked responses in the motor cortex elicited by antidromic stimulation of the pyramidal tract were shown to be depressed by mechanical stimulation of the cortex, which evidently elicited SD (Chang, 1955). The finding was confirmed by Bures et al. (1967) who recorded the antidromically evoked pyramidal responses in the motor cortex of rats.

Slow Potentials

In 1947 Leão described the most characteristic sign of SD, the slow potential change (SPC) accompanying the advancing front of EEG depression. This consists of a surface negative wave attaining 5–15 mV in amplitude, followed after 1–2 min by a less pronounced, but longer lasting positivity (Fig. A.2). Sometimes a small positive wave may precede the negative potential, thus completing a triphasic waveform typical for a moving dipole in a volume conductor. Recording from different depths of the cortex by semi-microelectrodes indicated (Grafstein, 1956a; Ichijo & Ochs, 1970; Leão, 1961; Monakhov et al., 1962a, b) that the negative slow potential invaded all cortical layers, the superficial ones about 20 s earlier than the deep ones. The amplitude of the intracortical SPC depended on the size of the capillary electrodes. With fine tips it may reach 30–40 mV. Positivity in deep cortical layers accompanying the main negative wave at the surface was found only exceptionally in cats (Munez-Martines, 1970). The large positive slow potentials (up to 250 mV) recorded under similar conditions with platinized indium electrodes (Rosenblueth et al., 1967) are due to the electrochemical properties of the metal electrode (Garcia Ramos & Pradac, 1968) and reflect changes in the local concentration of metabolic products (Huizar Sanchez & Garcia Ramos, 1969) rather than field potentials. It appears that the current source corresponding to the focal negativity in the cortex must be sought in distant parts of the corticofugal axons (Collewijn & Harreveld, 1966; Marshall, 1959; Monakhov et al., 1962b) which give rise to a widely distributed electric field that easily escapes detection. Small intracortical neurons with short axons are less important for slow potential generation because they only form short-lasting weak dipoles of random orientation. A possible role of glia in the slow potential generation is discussed elsewhere.

The negative SPC also accompanies SD evoked in other structures. At all sites it has approximately the same spatiotemporal characteristics, **indicating** a remarkable uniformity of the underlying mechanisms.

Unit Activity

Grafstein (1956b) using semi-microelectrodes demonstrated that the onset of EEG depression is accompanied by a discharge of unit activity lasting for a few seconds and followed by prolonged inhibition. More systematic studies using extracellular recording from well-isolated units in the cerebral cortex of rats (Buresova et al., 1963), cats (Morlock et al., 1964), rabbits (Johnson et al., 1965), and in the striatum of pigeons (Buresova et al., 1963) essentially confirmed Grafstein's results: a burst of high-frequency impulse activity lasting for 2–5 s abruptly changed into complete silence lasting for 1–2 min and followed by slow recovery to the predepression firing rate, which was usually attained after 5–7 min (Fig. A.3). The initial discharge is often preceded by short reduction of activity (Morlock et al., 1964) due to elimination of an excitatory influx from the cortical areas already depressed and/or to the decreased effectiveness of presynaptic structures slightly depolarized by the approaching SD wave. The striking silence accompanying SD was also reported by Krnjevic, Randic, and Straughan (Dubner & Brown, 1968; Gartside, 1968; Goldensohn et al., 1967; Krnjevic et al., 1966; Li & Salmoiraghi, 1963; Rosenblueth & Garcia Ramos, 1966) and can be considered a well-established fact. It seems that not a single neuron remains active in the area invaded by SD during the height of the negativity. In deep layers of cat cerebral cortex (Munez-Martines, 1970) which are not invaded by the SPC, depression of unit activity becomes less prominent, whereas the initial acceleration of firing rate is markedly prolonged.

Slightly different results were obtained with evoked unit activity. Whereas responses to an orthodromic input were blocked with a similar time course as the spontaneous activity, antidromically activated Betz cells displayed a steady attenuation of both the initial segment and somatodendritic spikes up to complete blockade persisting for 10 to 20 s. The initial segment spike then started to recover and was followed by intermittent and considerably delayed somatodendritic invasion (Morlock et al., 1964). The changes were completed within 1 min and closely corresponded to the time course of the SPC. Reversible blockade of the antidromically activated Betz cells by CSD was confirmed by Ichijo and Ochs (1967, 1970) who found the response depressed for 1.5 min on the average.

Krnjevic and Randic (1965) and Philis and Ochs (1971) tested the response of nerve cells to microelectrophoretically administered glutamate during different phases of the SD effect. The unit was unresponsive to the amino acid during the silent interval following the initial burst but could be activated in a nearly normal

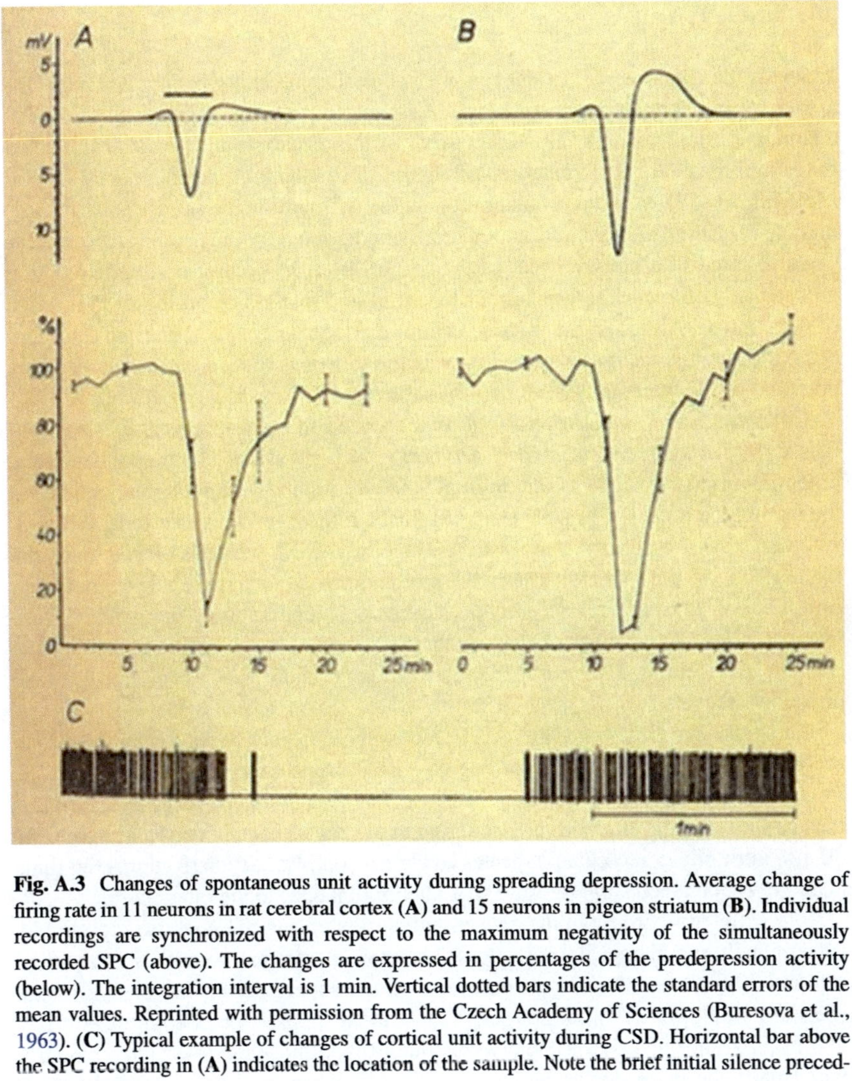

Fig. A.3 Changes of spontaneous unit activity during spreading depression. Average change of firing rate in 11 neurons in rat cerebral cortex (**A**) and 15 neurons in pigeon striatum (**B**). Individual recordings are synchronized with respect to the maximum negativity of the simultaneously recorded SPC (above). The changes are expressed in percentages of the predepression activity (below). The integration interval is 1 min. Vertical dotted bars indicate the standard errors of the mean values. Reprinted with permission from the Czech Academy of Sciences (Buresova et al., 1963). (**C**) Typical example of changes of cortical unit activity during CSD. Horizontal bar above the SPC recording in (**A**) indicates the location of the sample. Note the brief initial silence preceding the burst

manner after 1–2 min, that is, at a time when its spontaneous activity was still markedly depressed. The slow recovery of spontaneous unit activity is not due, therefore, to slow restitution of the postsynaptic threshold, but rather to changes in the presynaptic structures. Krnjevic et al. (1966) suggested that the late phase of unit activity blockade is at least partly due to a hyperpolarizing type of inhibition, perhaps due to conversion of glutamic acid, released during SD, into the depressant gamma-amino-butyric acid (Philis & Ochs, 1971).

Membrane Potentials

Whereas the experimental evidence reviewed in the preceding paragraphs indicates a widespread depolarization of neural elements during the negative SPC of SD, technical difficulties delayed a direct proof of this assumption. Brozek (1966) used the sampling method of membrane-potential measurement, developed by Li and McIlwain (1957), to avoid problems connected with prolonged intracellular recording. He measured all penetration potentials encountered by the capillary microelectrode inserted at a constant rate of 2 mm/min into the central cortex of rats and guinea pigs before, during, and after SD, monitored by SPC recording from the area of penetration. Although the method cannot distinguish between neurons and glial cells, the average membrane potential was found to be reduced during the negative potential of SD to about 20% of control values.

Collewijn and Van Harreveld (1966) succeeded in recording membrane potentials in 14 cortical cells throughout the period of depression. The mean membrane potential dropped to 35% of the original value during maximum depolarization. As the membrane potential was measured against a remote referent electrode, the measurements have to be corrected for the extracellular SPC, which is picked up by the microelectrode even when it enters the cell and causes an apparent hyperpolarization. The corrected membrane potentials were considerably lower, approaching zero in most cases. The change of the membrane potential has a different time course than the extracellular SPC: abrupt depolarization is followed by a prolonged plateau and slow recovery.

Karahashi and Goldring (1966) succeeded to hold 6 "idle" cells (probably glia) in the cat cerebral cortex during the negative SPC. Spreading depression shifted the intracellular potential by 20 to 40 mV positive, but no attempt was made to correct this change for the extracellular SPC. Assuming extracellular negativity of approximately 30 mV, intracellular recording indicates an almost complete depolarization. Higashida et al. (1971) obtained similar results in 11 neuroglial cells in the rabbit cortex.

Intracellular recording in the cat cerebral cortex during SD was also reported by Rosenblueth and Garcia Ramos (1966). As their membrane potentials were small (around 20 mV) and the extracellular SPC in the superficial layers relatively high (negativity up to 25 mV), the statement that "the responses recorded intracellularly had usually the same sign as, and were smaller than those recorded intracellularly had usually the same sign as, and were smaller than those recorded extracellularly" indicates considerable depolarization at least in the upper cortical layers. Depolarization of cortical neurons was demonstrated by Goldensohn and Walsh (1968) who found a rough correlation between intraocularly recorded SPC and the changes of excitability, synaptic potentials, and frequency of firing.

Impedance Changes

Besides affecting active electrogenesis, SD also influences the passive electrical properties of the depressed neural tissue. Increase of cortical impedance during SD in the rabbit was first described by Leão and Martins-Ferreira (1953). Using a bridge arrangement and sinusoidal measuring current in the range of 0.05 to 10 kHz and with a platinum electrode applied to pia and the referent electrode on bone, they found an impedance increase from 2070 to 2940 ohms when the active electrode region was invaded by SD. The impedance changes closely followed the slow potential variation. Van Harreveld and Ochs (1957) measured cortical impedance between two Ag-AgCl discs (1 to 2 mm in diameter) placed on the rabbit cortex 3 to 4 mm apart. At 1000 Hz, SD decreased cortical conductivity by 17%. Weiss et al. (1966) obtained a slightly higher change using smaller electrodes in the rat. In all these studies both the resistive and capacitive components of impedance increased, but the former influenced the overall change in a decisive way. Sometimes SPC was observed without the corresponding impedance shift indicating the possible dissociation of the two phenomena. Although the bridge method gives reproducible relative data, the complex current path involving both cortex and subcortical structures prevents a more accurate interpretation of results.

Changes of specific impendence were studied by Freygang Jr. and Landau (1955) who used constant current pulses to generate a field with parallel lines of flux normal to the cerebral surface and measured potential differences in this field between two capillary electrodes inserted to different depths of the cortex. The specific resistance of the cat cortex (222-ohm cm) increased during SD by 10–20%, the increase being most marked in the superficial cortical layers. Impedance of the sub-cortical white matter decreased at the same time by approximately 5%. Ranck (1964) established the specific impedance of rabbit cerebral cortex by passing sinusoidal currents (5–50,000 Hz) between a platinum electrode on the cortical surface and a remote indifferent electrode. The potential difference of two similar electrodes placed within 1.3 mm of the current electrode was recorded and displayed together with current in the form of Lissajous figures on a cathode-ray oscilloscope. During SD specific impedance increased from 250 to 400-ohm cm and the phase angle from 3.5 to 7.0° at 5000 Hz, whereas at 50 Hz a similar impedance increase (from 300 to 500-ohm cm) was accompanied by a phase angle decrease from 6 to 0°. Asynchronous changes of impedance and phase angle during SD in the cat cerebral cortex were reported by Wong Jr. (1969).

Chapter 5: Chemical Aspects of Spreading Depression

The purpose of the chemical exploration of the brain is to study the interrelationship between functional manifestations of neural activity and the underlying chemical properties, and to describe the mechanism of such correlations. Spreading depression possesses many features that make it particularly suitable for chemical analysis. Spreading depression evoked in one hemisphere does not spread to the contralateral one, which can thus serve as a control. Furthermore, the rate of SD propagation is slow enough to allow chemical analysis under proper experimental conditions.

Several manifestations of SD indicate that chemical changes may play an important role in this remarkable pattern of functional alteration in the cerebral cortex. As has been already mentioned, the temperature coefficient (Q_{10}) for spreading rate and duration of SPC was found to be about 2, which fits fairly well with that for chemical reaction ($Q_{10} = 2$ to 4). No temperature dependence was observed for the SPC amplitude. Since the first SD studies, it has been known that partial oxygen pressure decreases in the depressed region of the cerebral cortex (Lukyanova & Bures, 1967; Marshall, 1959; Van Harreveld, 1959; Van Harreveld & Stamm, 1952). Originally this decline of oxygen tension was ascribed to vasoconstriction of the cortical vessels. This assumption seemed to be supported by the observation that in the rabbit cortex initial vasoconstriction took place during SD (Van Harreveld & Ochs, 1957; Van Harreveld & Stamm, 1952). However, no initial vasoconstriction has been observed in the cat and rat cerebral cortex (Marshall, 1959; Van Harreveld & Ochs, 1957). The predominant, and in the case of rat and cat, exclusive, vasomotor reaction is vasodilation, as demonstrated in studies using either cooled (Buresova, 1957; Seylas & Molnaz, 1966) or heated (Sonnenschein & Walker, 1956) thermocouples, histological technique (Van Harreveld & Ochs, 1957), determination of hemoglobin content, and isotope technique (Buresova, 1957) as well as measurements of cortical volume (Marshall, 1959). Therefore, decrease in pO_2 in the cortical area invaded by SD can be explained only by increased O_2 consumption. Apparently changes taking place in the cerebral cortex during SD trigger biochemical processes controlling oxidative metabolism in the tissue required to provide additional production of energy.

More direct evidence for a possible role of some chemical substance(s) in the mechanism of SD propagation has been postulated by Grafstein (1956b). The spreading rate of SD evoked in isolated slabs of the cat's cerebral cortex could be affected by polarization of the tissue. Propagation of SD toward the cathode was found to be faster than in the opposite direction. It was suggested that SD is a neurohumeral reaction transmitted by some positively charged particle, the diffusion of which is influenced by the electric field. Taking into consideration that the wave of negativity is preceded by a short neuronal discharge, Grafstein came to the conclusion that SD represents a combination of both electrical and chemical mechanisms. The potassium ion has been suggested in the role of the SD "transmitter."

Supporting evidence for this potassium hypothesis has been presented by Bures (1956) who showed that the chemical compounds evoking a typical metabolic reaction in cerebral cortex slices in vitro (increased aerobic glycolysis) elicit SD in vivo. Substances preventing this reaction in vitro also prevented initiation of SD in vivo when added to effective KCl solution in the proper concentration.

Regardless of the indirect nature of all of this evidence, a hypothesis including the following general points seemed to be justified: potassium ions released from the nerve cells by the action of SD-eliciting agents (mechanical, thermal, chemical, or electrical stimulations) accumulate in the extracellular space and affect the metabolism of cortical cells in a manner similar to that demonstrated in in vitro experiments when the K^+ concentration is increased to the appropriate level above the normal extracellular concentration. This is obviously an oversimplification, but perhaps it is sufficiently clear to show the way in which chemical analysis of the problem has been carried out to verify this hypothesis experimentally. Apparently two main problems required solution. First, whether release of potassium ions and their migration actually take place (if so, whether the concentration achieved is sufficient to exert an effect on adjacent cells to ensure the propagation of SD): second, the chemical changes shown in vitro in cerebral cortex slices should be experimentally demonstrated in vivo in the course of SD. Were these elementary processes shown to occur, various additional questions would arise. (a) Are the presumed ion shifts and chemical changes great enough to be analyzed quantitatively? (b) What is the origin of potassium ions supposedly accumulating in fairly high concentration in the exta-neuronal space? (c) Are all these changes reversible and is the rate of recovery sufficient to explain the autoregenerative nature of SD? (d) Are they correlated with the steady potential changes and is such a correlation experimentally demonstrable?

Ion and Water Transport and Distribution During Spreading Depression

The most commonly used and simplest methods of determination of ions in tissue are the atomic absorption flame photometric measurements preceded by digestion of the tissue in a mineral acid. This approach could not be applied in the case of SD elicited by a single KCl application. It was found that no net changes in water, sodium, or potassium contents took place in the whole cortex after 30 or 60 min of repeated SD waves evoked by 25% KCl (Krivanek, unpublished). The possibility remained, however, that shifts in ion distribution between cellular and extracellular compartments might occur. Such changes are more difficult to detect. One possible approach is histochemistry, as used by Van Harreveld and Schade (1959), to demonstrate chloride ion shifts in the course of SD. This method is based on treatment of the frozen tissue with an ethanolic solution of $AgNO_3$, which reacts with chloride ions to form an AgCl precipitate. Subsequent reduction with pyrogallic acid shows a dark silver precipitate in areas corresponding to chloride location. Accumulation of chloride in the apical dendrites during SD has been demonstrated in the rabbit cortex situated laterally to the parasagittal sulcus. The adjacent retrosplenial area, into which SD does not normally propagate (see p. 118), served as a control (Van Harreveld, 1958). Together with the previously demonstrated increase in impedance, the shift of chloride ions obviously reflects transport of NaCl into apical dendrites. By this method chloride transport can be visualized only in structures with a radial arrangement of neurons. It is therefore not applicable to the whole cortex or to other structures with a different structural organization. Another disadvantage is the obvious difficulty of making quantitative measurements.

Some attempts have also been made to study histochemically the distribution of potassium ions during ischemization (Collewijn & Schade, 1962). The method resembles the histochemical technique for chloride detection. It is based on precipitation of K^+ ions by sodium cobalt hexanitrate, which is then transformed into black cobalt sulfide with ammonium sulfide. The known drawbacks of histochemical methods in general, and particularly of those concerned with the determination of soluble substances of small molecular size, seriously limited the use of this method for potassium detection.

A more accurate method, which would allow not only more precise measurements but also a more dynamic approach to electrolyte analysis during SD, was desirable. Two such methods were introduced at the same time independently by Bures and Krivanek (1958, 1960) and Brinley Jr. et al. (1960). In Brinley's procedure the cortical surface was loaded topically with a radioactive isotope of potassium; the output of ^{42}K into Ringer's solution washing the rabbit's cortical surface was followed in both depressed and normal cortices. Significantly elevated release of ^{42}K was found in cortex through which the slow potential wave was passing. Each wave was accompanied by a striking increase of ^{42}K output with a maximum up to 400% above baseline. In the second method (Bures & Krivanek, 1960; Krivanek & Bures, 1960) an isotonic solution (usually 0.25 M sucrose) was perfused under low pressure through a small metal or plastic cannula fixed to the edges of a trephine opening 5 mm in diameter in the rat's skull. The solution was allowed to flow at a constant rate through the cannula and after contact with the cerebral cortex was gathered for analysis (Fig. A.4). Concentration of both sodium and potassium in the outflowing fluid was determined by flame photometry. The results were practically identical with those of Brinley Jr. et al. (1960). When SD-evoking drugs were added to the washing fluid (mainly 2,4-dinitrophenol), an increased outflow of potassium from the cerebral cortex was indicated by an increase in concentration in the outflow solution. As expected, a similar increase (up to 400%) could be found when the cannula was placed in a region not directly affected by the SD-eliciting agent. Efflux of sodium was also increased, but to a much smaller extent. Using a flame photometer with an extremely small jet that allowed measurements of ion concentrations in very small volumes, it was possible to take samples of the outflow at relatively short time intervals (2 min as compared with 15 to 20 min in the previous case). This setup made it possible to correlate K^+ output with different phases of the steady potential change passing through the washed cortical region. As in the experiments of Brinley Jr. et al. (1960), a close correlation between the occurrence of the wave and increased outflow of potassium was found. Moreover a correlation between SPC voltage and the percentage increase of potassium outflow was found in some cases (Krivanek & Bures, 1960) (Fig. A.5). Analogous ion shifts were also observed during cerebellar SD (Fifkova et al., 1961) and in the striatum of chick embryos (Sedlacek, 1969). A similar correlation between recovery of the resting potential (2.5 min) and of potassium content (1 min) in isolated cerebral cortex slices following electrical stimulation (10 min) has been demonstrated in vitro (Cummins & McIlwain, 1961; Hillman et al., 1963).

To employ this method for quantitative estimation of ion shifts in the cerebral cortex, the assumption was made that the potassium concentration in the washing fluid is in equilibrium with that in the compartment from which it is freely diffusible. It may be in contact with the "extraneuronal space." A relative increase in potassium concentration in the washing fluid would then correspond to that in this

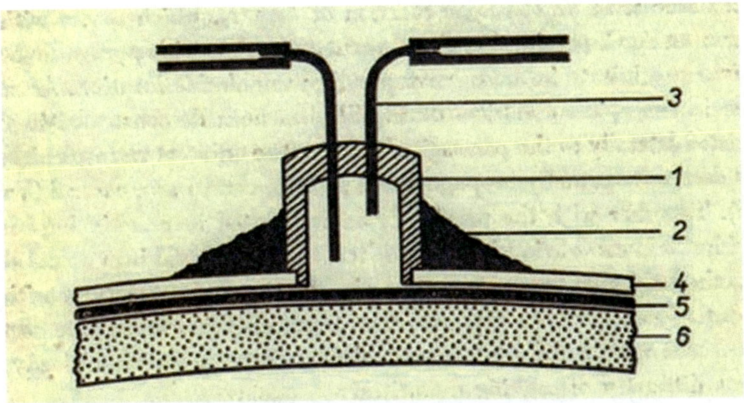

Fig. A.4 Diagram of the epidural cannula. (1) The body of the cannula; (2) phosphate cement of acrylic resin; (3) tubes through which the washing fluid flows; (4) cranial bone; (5) dura; (6) cortex. (Reprinted with permission from the Czech Academy of Sciences (Bures & Krivanek, 1960))

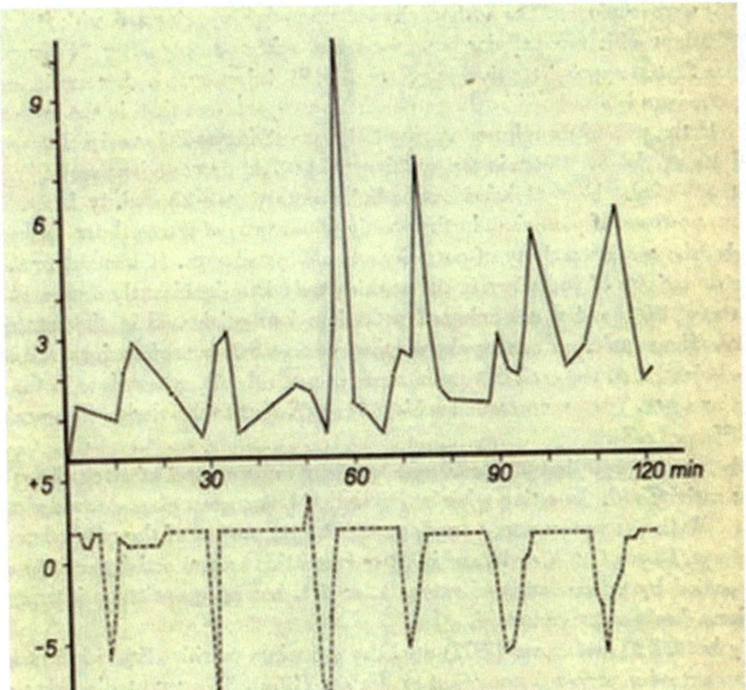

Fig. A.5 Time relation between the course of SPCs (below, ordinate in mV) and K^+ release into 0.9% NaCl (above, ordinate in $\mu M \times 10^2$). (Reprinted with permission from the Czech Academy of Sciences (Krivanek & Bures, 1960))

compartment. An elevation of K^+ concentration by a factor of five or six that was found experimentally (Krivanek & Bures, 1960) was presumed to reflect a similar increase in extracellular concentration, that is, from the usual 3 mEq/1 (estimated from the potassium concentration in the cerebrospinal fluid (Van Harreveld, 1966)) to 15 to 20 mEq/1. This figure is necessarily an underestimation, since the inertia of the superfusion technique and the fact that at any time the negative SPC peak did not occupy more than 50% of the superfused area, tend to smooth the sharp temporal and spatial gradients. The extracellular K^+ levels reached during SD must probably exceed 30 to 40 mEq/1 to account for the observed changes of K^+ efflux. To estimate whether this increase would be sufficient to reach the threshold concentration for eliciting SD, another assumption was made: that the mechanism of potassium action on cortical cells during SD is identical irrespective of whether the K^+ ions were added externally or released endogenously into the extraneuronal space. It would then be possible to compare the estimated threshold concentration of K^+ ions for evoking SD with that reached by extracellular K^+. Correction would have to be made for the dilution of externally applied potassium ions that obviously takes place before the latter reach their site of action by diffusion. The threshold concentration of 80 mM KCl found with epidural application in rats (see p. 33) is then quite compatible with the extracellular K^+ increase inferred from the superfusion experiments. Similar results were reported by Fertziger and Ranck Jr. (1970) who exposed the cat suprasylvian gyrus to ^{42}K and superfused it with 3–120 mM solution of unlabeled KCl. The ^{42}K efflux did not change until the KCl content in the perfusate reached 45–60 mM. These concentrations triggered SD and caused a stereotyped increase in ^{42}K efflux (about 2.3 times) indicating SD initiation.

One crucial question still remained to be solved. What is the origin of the potassium accumulated in the compartment in contact with the washing fluid? The changes found could be a mere reflection of simultaneous changes in the blood vessel dimension that have been shown to take place during SD. An altered permeability of the blood-brain barrier could be another possibility. The proposed mechanism of SD propagation could be borne out only by proof that the released K^+ is of intracellular origin. To solve this problem, the observations reported by Katzman and Leiderman (1953) were employed. The authors showed that after injecting rats with ^{42}K, isotope equilibrium with intracellular potassium was achieved only after 24 hours. This means that at appropriate time intervals after ^{42}K injection, a higher specific activity of potassium is measured in the extracellular compartment than in the intracellular one. If the potassium released during SD were of intracellular origin, the specific activity of the potassium in the washing fluid would decrease because of "contamination" of the highly labeled extracellular potassium with slightly labeled intracellular cations. If potassium in the washing fluid was significantly decreased in the course of SD, so that the released potassium ions originated in the intracellular space. The conclusion can be drawn that during SD potassium ions released by depolarization of the excitable membranes of cortical cells accumulate in the extracellular space. The concentration achieved is sufficient to depolarize the membranes of adjacent cells.

The above conclusion was confirmed by direct measurement of extracellular potassium activity with the cation selective electrodes. Using gross glass electrodes covered with collodion (sensitive area 3 mm^2) applied to pial surface of the cerebral cortex of monkeys, Meyer et al. (1961) found that seizure activity and SD are accompanied by accumulation of extracellular K^+, but no quantitative interpretation of these data was attempted.

Vyskocil et al. (1972) used the potassium sensitive liquid ion exchanger microelectrodes prepared according to Walker (1971). To prevent distortion of the potassium electrode readings by the steady potential fields arising in the area of recording, the potassium microelectrode was glued to another capillary microelectrode filled with 1 M NaCl, the distance between the parallel tips of the two capillaries being less than 10 μm. The K^+ concentration was estimated from the potential difference between the two capillary electrodes. The SP changes in the area of recording were related to a wick calomel cell electrode placed on the occipital muscles. The potassium electrode introduced into the parietal cortex stabilized after a few minutes at a potential corresponding to 3.0 mM K^+. As the SD wave elicited from the occipital cortex approached the potassium electrode inserted 1 mm below the cortical surface, its potential rose at first slowly and then abruptly (10 mV/s) to about 60 mV. This potential change corresponded to a K_e^+ increase to 60 mM and was maintained for 20 s. The potassium potential then started to decline along an S-shaped curve. The slope of the steepest part of the recovery phase was several times lower (3 mV/s) than the slope of the initial K^+ increase. The SP recording reflected the major features of the potassium potential, but was of lower when the K_e^+ was still 6 mM above the resting value. This is in good agreement with Goldman's equation from which it follows that the membrane potential is influenced at K_e^+ levels below 6 mM mainly by the Cl^- and Na^+ gradients. The amplitude of the SD potassium potential did not markedly change over the first millimeter. At 1.5 mm depth the maximal K_e^+ shift was less marked (40 mM), and the potassium rise time was prolonged indicating K^+ diffusion from a remote active region into the passive space around the recording electrode. The SD potassium potential disappeared 2 mm below the cortical surface, where the SP changed to low positivity. The above findings were confirmed in cat cerebral cortex by Prince et al. (1973) who found with microelectrodes of the same type that extracellular potassium increased during SD from the resting level of 3.25 mM up to 80 mM.

The duration of SD evoked by a single application of even 25% KCl is limited to about 2 to 3 hours (see p. 224). Since the concentration of topically applied KCl progressively decreases due to diffusion from the site of application, the frequency of the isolated slow potential waves also decreases. For some purposes, it is desirable to keep the SD process at a maximum level for several hours. This can be done by repeated application of KCl at about 40 min intervals, usually for 4 to 5 hours. Relatively prolonged overloading of central cortical tissue with hypertonic KCl solutions could affect ionic and water metabolism to a serious extent. In contrast to a single application of 25% KCl, a net increase of K^+ content was found after 5 hours of repeated KCl application. A relatively fast recovery occurred after KCl application had been discontinued. On the other hand, water content, which can be measured more accurately, appeared to recover in two phases. The major fraction (about 70%) recovers in 24 hours, whereas a residual portion required much more time to reach normal values. During this interval some residual irregularities in distribution that are undetectable by the analytical method used may also be reversed. Even though a relatively long time is necessary for complete recovery, the electrolyte and water shifts seem to be completely reversible. This concerns only cortical regions outside the area of KCl application that is seriously damaged (Krivanek, 1969a).

Spreading Cortical Depression and Labile Chemical Compounds in the Brain

In the previous section data were summarized that support the view that potassium ions may play a decisive role in the SD phenomenon. The hypothesis was advanced that the chemical events underlying the SD process are analogous to those obtained in in vitro experiments with cerebral cortex slices exposed to an increased concentration of K^+ in the incubation medium. Ashford and Dixon (1935) and Dickens and Greville (1935) found a stimulating action of potassium ions on respiration and aerobic glycolysis of cortical slices. Later this finding was extended to studies of various aspects of brain metabolism as affected by changes in the electrolyte composition of the medium, in particular potassium and calcium. In spite of the great effort invested in the study of the K^+ effect on the metabolism of glucose, glycogen, ion and water transport, transmitters, lipids and proteins in cerebral cortex slices, its mechanism has not yet been fully elucidated.

To verify a possible analogy between this effect of potassium in vitro and the manifestations of SD in vivo, it was necessary to carry out more direct analyses of those chemical systems that respond most clearly to the presence of K^+ in vitro, that is, systems engaged in the metabolism of glucose and compounds involved in the storage and transfer of chemical energy. To measure the concentration of these labile substances in the brain, it is essential to protect them from the extremely fast decomposition that takes place in the immediate post mortem period. The most common approach is to freeze rapidly the dissected brain or severed head, or sometimes even the whole animal, by immersion into a liquid gas. At low temperatures chemical reactions proceed extremely slowly, so that under appropriate conditions, it is possible to recover in the test tube brain constituents in amounts approaching those existing in vivo. To follow changes in glucose metabolism the levels of glycogen, free glucose, and lactic acid were determined in various phases of SD.

The level of glycogen 20–30 min after a single application of 2 or 25% KCl decreased by 30% as compared to that in the opposite hemisphere which had been kept relatively intact. The change was large enough and the rate of recovery sufficiently slow for the decrease to be detected in the whole hemisphere (Krivanek, 1958). On the other hand, changes in the content of free glucose and lactic acid were much smaller and so did not reach the level of statistical significance (Krivanek, 1958, 1961).

Since SD elicited in the cortex is limited to cortical tissue, only cerebral cortices were taken for analysis in further experiments. For this purpose, cortical slabs were dissected from the frozen brain under continuous cooling with liquid nitrogen. Cooling was maintained until enzymes were inactivated by homogenizing frozen cortical samples in protein precipitating agents such as trichloroacetic acid, perchloric acid, or $Ba(OH)_2 + ZnSO_4$. With this technique the SD-elicited decrease of glucose level was fairly high (−30%), especially when a more specific (chromatographic) method was used for determination. Under similar experimental conditions, the lactic acid content in cortical slabs was found to have increased up to 100% above the normal level (Krivanek, 1961; Ruscak & Duda, 1959).

Accumulation of lactate raises the question of possible pH changes during SD. Methodologically this is a rather difficult problem (Marshall, 1965) since the potential corresponding to the pH change must be distinguished from the SPC of SD. Rapoport and Marshall (1964) measured the pH of the rabbit cortex during SD elicited by both electrical (tetanic stimulation) and chemical (2% KCl) stimuli. Potential changes were recorded independently with a glass membrane pH electrode and with a pore Ag-AgCl electrode applied onto the cerebral cortex against a reference electrode. From the difference between the two records (pH electrode recorded both SPC and pH), an acidic shift of -0.027 ± 0.005 pH unit was found in the positive phase of SPC after making corrections for possible errors. This finding shows that only very small (0.05) pH changes occur during SD. The previous assumption by Tschirgi et al. (1957) and Tschirgi and Taylor (1958) that SPC is accompanied by alkalinization and can be entirely accounted for by a jugular-cortical pH difference formed across the blood-brain barrier appears, therefore, to be erroneous. It seems that the great buffer capacity of the cerebral cortex prevents excessive pH shifts in spite of the rather high accumulation of lactate.

Since vasodilation occurs during SD, a decrease in the glucose level is unlikely to be due to a decrease of the glucose supply. Changes in the content of the compounds mentioned above support the view that glucose consumption is increased to such an extent that supply from the blood stream becomes insufficient to cover demands and a net decline in the glucose level takes place. Moreover, mobilization of glycogen reserves takes place as has been shown experimentally by measuring incorporation of ^{14}C from glucose into glycogen during CSD. Specific activity of glycogen in the depressed cortex was higher by 47% as compared with that in the contralateral, intact cortex. It has been concluded that increased rate of glycogen breakdown rather than inhibition of glycogen synthesis is responsible for the fall of glycogen level during SD (Krivanek, 1972a). No matter how small the reserves of glycogen are in the brain (about an order of magnitude lower than in liver or muscle), glycogen can be transformed to glucose-6-phosphate and serve as an additional source of energy (Kerr & Ghantus, 1936). Obviously two possible general mechanisms can be activated by the potassium action. First, energy demand is suddenly increased (enhancement of oxygen consumption), and substances that could serve as an energy source are utilized. These compounds are then catabolized at a higher rate. This enhancement of glucose uptake and catabolism could be considered as a consequence of the potassium effect and corresponds to restoration of the membrane polarity by reestablishment of the asymmetric distribution of ions across the neuronal membrane. The process is obviously an energy-dependent one. The theoretical possibility remains, however, that the increase of glycolysis, in spite of elevated oxygen consumption, is an expression of some kind of reversal of the Pasteur effect induced by potassium action. Under normal conditions an increased consumption of oxygen is followed by a decreased rate of glycolysis (smaller amount of lactic acid formed). Potassium-induced uncoupling of oxidation from the synthesis of macroergic phosphates decreases the efficiency of the energy-producing processes. In this case the potassium ions would be engaged in the chemical changes of SD more directly than in the first one.

In this connection it was important to know whether SD is accompanied by any changes in the level of energy-rich phosphates. An attempt was made to measure levels of creatine phosphate (CrP), adenosine triphosphate (ATP), and inorganic phosphate during SD. Estimation of CrP in the whole hemisphere revealed only a slight decrease during SD elicited by mechanical or chemical stimuli (Zachar & Ruscak, 1958). Thanks to the relatively low rate of SPC propagation through the cortex, it is possible to freeze the particular area of the cortex in which the SPC undergoes a definite phase of development. To demonstrate experimentally a possible correlation of the labile phosphate level with the development of SD, propagation of a single SPC was monitored electrophysiologically. At various phases of de- or repolarization, the corresponding cortical region was frozen in liquid nitrogen. It was shown that in the region in which the amplitude of negativity reached its maximum, the CrP level was decreased by more than 40% (Krivanek, 1961). This change is fairly high if we consider the fact that in the slab of the cortex taken for analysis the degree of depolarization varies and certainly is not maximal at the time of fixation. The maximum of the slow potential wave is limited to a narrow band that is not large enough to allow chemical analysis by the method used. As soon as polarity of the cortex started to recover, CrP was found to be at its normal level. Inorganic phosphate content increased by an amount that is stoichiometrically related to the CrP decline. In spite of a very rapid change in CrP no changes in ATP were seen in the depolarized regions of the cerebral cortex. It is generally accepted that CrP can serve in vivo as an "energy buffer" reversibly offering its phosphate to the ADP and thus maintaining of ATP utilization are covered by CrP, the level of which drops transiently. A similar effect (rapid drop of CrP) can be achieved in cerebral cortex slices in vitro by raising the potassium concentration in the medium (Heald, 1954; Sugawara & Wada, 1957) or by electrical stimulation (Heald, 1954). There is a close correlation between the recovery of polarity in depressed tissue and the CrP level. In the cortical region invaded by the negative slow potential, the CrP level is decreased by about 40%, but as soon as the maximum negativity passes there is a rapid recovery to the normal value. It seems that complete recovery of CrP even precedes that of the slow potential (Krivanek, 1961). A similar observation was made during recovery of cortical polarity from the terminal depolarization evoked by brain ischemia.

A close correlation between recovery rates of both resting potential and CrP was also demonstrated in vitro after electrical stimulation of cerebral cortex slices (Hillman et al., 1963). Once the relatively rapid changes in the levels of the labile compounds had been demonstrated, an attempt was made to answer the question of whether the recovery rate of the slow potential change correlates with the recovery rate of the chemical changes occurring during SD. The aim of the this study was to analyze the chemical basis of the refractory period of SD, that is, the time interval between two successive waves of the slow potential. For this purpose, only one SPC wave was elicited with 1% KCl. The development of SD was checked electrophysiologically. At various time intervals (minutes) after recovery of cortical polarity, the brain was frozen in liquid nitrogen, and glucose, lactic acid, and glycogen were determined in the cortical slabs. The polarity was considered as completely recovered when the curve of the slow negative potential crossed the baseline. From that point it takes approximately 5 min before another wave can be initiated from the same area of cerebral cortex by the same stimulus (see p. 28). Obviously repolarization alone is not a sufficient indicator of recovery. As shown in Fig. A.6, 5 min after restoration of cortical polarity neither glucose nor lactate content have reached normal value. There is fairly good correlation of glucose and lactic acid recovery with the recovery of EEG amplitude. The 10 min period into which recovery of all the above parameters fits well (except the glycogen level which returns after long delay) corresponds to the relative refractory period of SD. Summation of the glucose and lactic acid changes should

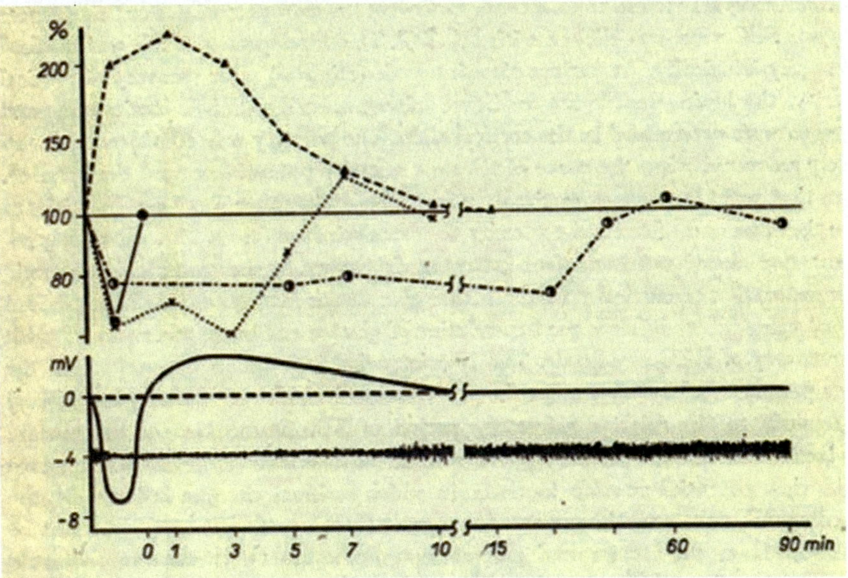

Fig. A.6 Percent change in lactic acid (Δ - - Δ), phosphocreatine (• - •), glycogen (O- . -O), and glucose (x......... x) accompanying changes to doe potential (▬) and EEG activity. Abscissa; time after complete repolarization. (Reprinted with permission from the Czech Academy of Sciences (Krivanek, 1962))

appear with short interwave intervals when waves of the slow potential develop in tissue in which residual changes induced by the preceding SD wave are still present. (An equal efficiency of each SD wave must be assumed.) However, lactate and glucose were shown to be affected to the same degree irrespective of SD duration (Krivanek, 1961). The possibility remains that the origin of the electrophysiological and biochemical parameters is not exactly the same. Steady potential differences are recorded in the uppermost layers of the cortex, whereas the EEG might reflect electrical events located in more remote structures. The location of the chemical compounds analyzed is obviously not only in the upper portion of the cortex. Hence only some shifts in phase of differentially located events might take place.

Lack of cumulative biochemical effects of single slow potential waves has been demonstrated also in studies of the SD effect on levels of both glycogen (see p. 95) and protein synthesis (p. 107). Neither different KCl concentrations nor prolongation of the action for hours by repeated application of 25% KCL made the effects more pronounced. As this is obviously a general phenomenon, it was tempting to speculate that the SD effect is confined only to some compartment(s) or specialized structures of cortical tissue, the rest of it being relatively intact. Apical dendrites might be such structures. They represent a large membrane that could be primarily involved in responses to ionic changes in their environment. This assumption is supported by the finding that chloride ions and water shifts caused by SD are mainly localized in the apical dendrites of the cerebral cortex (Van Harreveld, 1958). Also Ochs and Hunt (1960) came to the conclusion that apical dendrites play a major role in the propagation of SD. Quantitative histochemical studies of the distribution of glycolytic and oxidative enzymes throughout the cortical layers led Lowry et al. (1954) to the assumption that regions rich in dendrites are particularly active and that "brain metabolism is quantitatively the metabolism of dendrites."

Amino Acids and Spreading Depression

Since the metabolism of amino acids is closely bound with that of carbohydrates, and since some amino acids are thought to play an important role in brain function (GABA, glutamic acid), possible involvement of amino acids in the mechanism of SD was studied.

Since amino acids are potent SD-eliciting agents, glutamate, glutamine, and aspartate being particularly active. No changes in their content in the cerebral cortex 30 or 60 min after topical application of 23 or 25% KCl could be found. The same was true for alanine, aspartate, and GABA (Krivanek, 1960; Ruscak, 1964). It was demonstrated in cerebral cortex slices in vitro that incorporation of ^{14}C from uniformly labeled glucose into glutamic acid and GABA increased after concentration of K^+ ions in the medium had been elevated to 100 mM (Kini & Quastel, 1959; Rybova, 1959). Using intraperitoneally ^{14}C-labeled glucose and evoking SD by 25% KCl, specific activities of glutamate, glutamine, aspartate, alanine, and GABA were determined and related to the total activity of the soluble fraction (trichloroacetic acid extracts) to determine the relative specific activity. Only a slight increase of relative specific activity of GABA and alanine was observed (Fig. A.7) (Krivanek, 1972a). No changes in the ammonia level were found (Krivanek, 1960).

While SD alone does not exert any appreciable effect on brain amino acids, a different picture was obtained with SD elicited in the oligemic brain after ligature of both carotids. Electrical and biochemical characteristics are qualitatively the same as in the case of conventional SD, only more pronounced. The slow potential wave recovers more slowly, and the decrease of creatine phosphate level is even detectable in the whole hemisphere (Duda et al., 1961). The decline of free glucose and glycogen levels as well as enhancement of the lactic acid content has been found. Moreover, citrate content was slightly decreased while pyruvate and oxoglutarate remained unchanged (Ruscak & Ruscakova, 1971). Altered levels of some amino acids are the main feature distinguishing this kind of SD. After 30 min SD elicited in the cerebral cortex of rats by 3 M KCl leads to a decrease in amino nitrogen, glutamate, glutamine, and aspartate, whereas alanine as well as gamma-aminobutyric acid levels were found to be increased (Ruscak, 1961, 1962a). Not only KCl but even salts not normally evoking SD (LiCl, NaCl, $CaCl_2$) were shown to cause an increase of alanine and gamma-aminobutyrate when applied topically in a 3 M concentration on the cortical surface of rats with carotid ligature (Ruscak, 1962b).

In rabbits (Van Harreveld, 1959) but not in rats (Bures et al., 1960), glutamate has been reported to evoke SD at a lower threshold concentration than KCl. Using the technique of cortical superfusion (Bures & Krivanek, 1960), Van Harreveld and

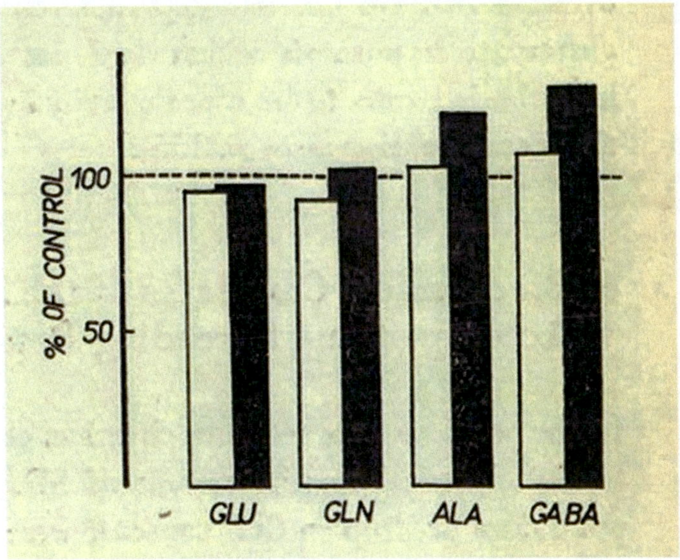

Fig. A.7 Levels of some free amino acids (white columns) and specific activities of their ^{14}C (from ^{14}C glucose) as related to the total radioactivity of the soluble fraction (black columns). Duration of SD 1 h. *GLU* glutamic acid, *GLN* glutamine, *ALA* alanine, *GABA* gamma-aminobutyric acid. (Reprinted with permission from the Czech Academy of Sciences (Krivanek, 1972a))

Koolman (1965) demonstrated an enhanced release of ninhydrin-positive material from the cerebral cortex during SD. When SD was evoked in vitro in the isolated retina that had previously been loaded with radioactive glutamate, increased output of ^{14}C glutamate and to lesser extent ^{14}C aspartate were observed (Van Harreveld & Fifkova, 1969, 1970). On the basis of these experiments, Van Harreveld advanced the hypothesis that glutamate rather than potassium is responsible for the membrane changes occurring during SD. After being released into the extracellular space, glutamate increases the permeability of the cellular membranes to sodium, which is the first step in the action of all agents evoking SD. No definite experimental data have yet been furnished to prove the glutamate hypothesis which does not deny the role of potassium. The potassium hypothesis seems to be more convincingly supported by experiments and quantitative considerations.

Biochemical Correlates of the Functional Changes Accompanying Spreading Depression

In the previous two sections chemical changes were described that are closely related to the electrical manifestations of SD. Some of them, for example, the outflux of potassium or drop in CrP content, were shown to coincide with the development of the negative slow potential. As pointed out in Chaps. 2 and 10, SD is accompanied by a serious impairment of the basic functions of the cerebral cortex. Evoked responses are suppressed and unit activity is inhibited. Conditioned reflexes disappear and learning ability is abolished. It seems therefore that generation and transmission of electric impulses are severely impaired. Possible changes in the enzymatic systems which are supposed to play an important role in chemical synaptic transmission have been studied under conditions of SD.

Of interest was the question of the role of the cortical cholinergic systems in the functional changes accompanying SD. The failure to find a significant correlation between changes in functional activity of the brain and acetylcholinesterase (AChE) activity may often be due to the fact that chemical analysis was performed in large samples of tissue in which modification of functional activity was not uniform and the corresponding effect on the AChE activity was rather small. A method has been developed therefore for colorimetric estimation of AChE activity in very small samples of the brain tissue (1 mg wet weight) in which the change of functional activity and the corresponding response of AChE was expected to be maximally developed (Krivanek, 1972b). The samples for analysis were cut with an injection needle from 1-mm-thick frozen coronal brain sections dissected in a stereotaxic apparatus from the frozen head of the rat. Portions of the motor cortex obtained in this way were homogenized, and 10-µl aliquots were then subjected to the procedure for determination of AChE activity based on Hestrin's method (Hestrin, 1949).

Since it has been shown that barbiturate anesthesia does not interfere with SD and because most of the previous experiments have been carried out in anesthetized rats, the same experimental setup was used in the case of AChE measurements. A single application of 25% KCl on the surface of one hemisphere did not induce statistically significant changes in AChE activity although some tendency to rise above the normal levels as compared with the contralateral cortex and expressed per unit of protein content appeared after 30, 60, and 120 min of SD. However, prolonged treatment of the rat cerebral cortex by repeated application of 25% KCl (PCSD see p. 93) onto the cortical surface resulted in a significant increase of AChE activity. After 4 h of SD, a 16% increase in AChE activity was found. In spite of this relatively high activity increase, the return to normal was rapid. Less than 2 h appeared to be sufficient for recovery from 4 h of KCl action (Fig. A.8).

Since the tissue samples had the same volume, enzymatic activity could be expressed per volume; and thus progressive decrease of AChE with longer intervals after SD initiation was found. The decrease of protein content, to which the AChE activity was related, was much more marked than the decrease of AChE activity; that is, the ratio of AChE to protein increased (Krivanek, 1972b). The experiment performed in unanesthetized rats showed that protein content after 5 h of SD was only 90% of normal (Krivanek, 1969a). The difference in the protein level changes as induced by prolonged KCl action in the two experimental series mentioned is obviously caused by prolonged anesthesia in the former one. It is difficult to maintain the same depth of narcosis for a long time without additional administration of anesthetic (barbiturate in this case). Very often some hypoxic symptoms could be observed. Nevertheless, this experiment demonstrates a relatively high stability of AChE. When total protein content decreases, the decline of AChE activity is much slower. So far it is difficult to determine the basis of this stability. It is possible that only a small portion of cortical synapses is cholinergic in nature. On the other hand, the stability may indicate that the functional changes induced by SD are inadequate to influence the AChE activity.

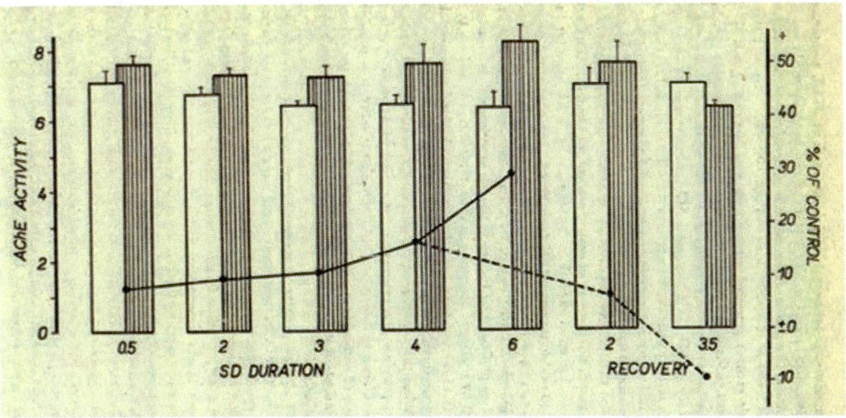

Fig. A.8 Acetylcholinesterase activity (as related to the protein content) in the rat motor cortex in the course of the prolonged CSD (2–6 h), and during recovery (2 and 3.5 h after removal of 25% KCl). White columns are control; shaded columns are PCSD; solid line is percent of controls during PCSD; dashed line is percent of controls during recovery. (Reprinted with permission from the Czech Academy of Sciences (Krivanek, 1972b))

The adrenergic system, especially in the brain stem, has also been subjected to chemical analysis during SD. In 1966 Donoso and Stefano (1966) measured the noradrenaline content in hypothalamus, mesencephalon, cortex, and pons-medulla in rats, while SD was evoked by topical application of 20% KCl on both cortices. Ten minutes of SD caused a 30% decrease in the noradrenaline level in the hypothalamus and 48% increase in the mesencephalon. No changes in cortex and pons were found. In a more recent study a method for the estimation of noradrenaline turnover was used (Schanberg et al., 1967) to study the SD effect on the metabolism of this transmitter. After intracisternal injection, the tritiated noradrenaline (pulse labeling) rapidly penetrated into the surrounding tissue and was metabolized. The rate of disappearance from the brain tissue is an expression of the metabolic rate. Two hours after injecting tritiated noradrenaline, SD was elicited by topical application of 25% KCl onto one hemisphere while the opposite one served as a control. One hour later the animals were killed. In the brain stem (cerebrum devoid of cerebral cortex, subcortical white matter, and cerebellum), the specific activity of noradrenaline as well as that of normetadrenaline was determined. In the brain stem ipsilateral to the depressed cortex the specific activity of both [^{3}H] noradrenaline and [^{3}H] or metadrenaline was decreased by 32 and 17%, respectively. Increased turnover of the transmitter seems to accompany cortical SD. In the opposite brain stem (contralateral to the depressed cortex), the specific activity of noradrenaline was significantly higher (+33%) indicating decreased turnover as compared to controls (both hemispheres treated by a single application of NaCl) (Schanberg et al., 1968). Cortical SD has been recently shown to accelerate the α-methyl-p-tyrosine-induced disappearance of dopamine and to increase homovanillic acid content (dopamine level unchanged) in cerebrum. Therefore, dopamine turnover appears to be enhanced (Keller et al., 1972).

The mechanism by which SD may affect the metabolism of noradrenaline in the areas studied is unclear. It might reflect some functional alterations in brain-stem structures, the electrophysiological, morphological, and behavioral aspects of which are described in Chap. 9. For example, Keller et al. (1972) speculate that disinhibition of the nigro-striatal dopaminergic pathway could be the reason of increased dopamine turnover.

Protein Metabolism in the Course of Spreading Depression

Since Hyden's (1959) formulation of his hypothesis of the chemical basis of memory, a possible role of protein and RNA metabolism in brain function and particularly in memory trace formation and storage has been repeatedly studied in a number of laboratories. A study of possible changes in protein metabolism accompanying the learning process can take advantage of the SD technique which makes it possible to compare in vitro data from the trained and untrained cortex in the same animal. However, it was necessary to determine the possible effect of SD on protein metabolism in the cerebral cortex.

There were no apparent changes in the total protein content of the cerebral cortex after a single epidural application of 25% KCl. This, however, does not mean that there are no changes in conformation of protein structure. According to Martins-Ferreira and de Oliveira-Castro (1966), SD elicited in the isolated retina is accompanied by changes of light scattering, spreading from the focus in all directions over the retinal surface. This optical change may be an expression of structural rearrangement of the retinal macromolecules caused by SD.

The most studied aspect of protein metabolism is the synthesis of protein during various functional states or during learning. According to the theory mentioned above, memory storage is due to the formation of specific protein that ensures the specificity of response to appropriate stimuli. The incorporation rate of radioactive amino acids into proteins of the brain tissue was used as a measure of the rate of protein synthesis.

When SD was elicited by a single application of KCl, protein radioactivity in the depressed cerebral cortex was considerably decreased (Bennett & Edelman, 1969; Krivanek, 1968; Ruscak, 1964). Incorporation of ^{32}P into protein and nucleic acid fractions was also diminished (Krivanek, 1959). The decrease of the incorporation rate is maximal 1 h after KCl application when the specific radioactivity of total protein in the depressed cortex is lower by 40% than that in the control (contralateral) cortex. Recovery started between the first and second hour, and after 2 h the radioactivity of proteins from the depressed cortex reached 80% of the control values. Four to five hours were required for complete recovery.

When maximum SD was maintained for 5 h by repeated application of KCl and radioactive leucine was injected again 60 min before killing the animal (Krivanek, 1969b), incorporation of the labeled amino acid was inhibited in the same way as after a single KCl application. Prolonged action of KCl would obviously not result in accumulation of effects of a single KCl application. Even 8 h of repeated application of KCl gave the same 30–40% inhibition. The absence of a greater effect of the longer-lasting SD may be due to partial recovery of the incorporation rate between the single SD waves. Accordingly, a series of SD waves passing successively over the cortex have no cumulative effects but result in a steady state.

The changes in the protein synthesis found during prolonged SD do not cause any permanent damage, since recovery of incorporation of [^{14}C] leucine can be shown. Control levels of incorporation were reached in the depressed cortex 6 h after KCl had been removed from the cortical surface. The degree of inhibition and recovery as well as the time course of both processes appeared to be identical in the cortical region directly treated with KCl and in the cortex unaffected by the externally applied KCl. Therefore, inhibition of protein synthesis resulting from the action of externally added KCl does not differ from the effect of SD evoked by a single KCl application. However, prolonged inhibition can cause a small but consistent decrease in the total protein content of the cerebral cortex by about 10%. This may be partly due to increased hydration, which could account for at most 40% of the change. Recalculation of protein per dry weight or DNA still showed a statistically significant decrease of about 6%.

After the main facts on the specific activity of the protein had been established, several other problems remained to be solved: (a) What changes are there in the leucine pool? (b) Would it be possible to demonstrate some specificity of this reaction to potassium, that is, are there any subcellular structures or even some individual proteins the synthesis of which would be specifically sensitive to the increased concentration of potassium ions? (c) Is the inhibition of protein synthesis due to the specific effect of K^+ on some critical link in a complex chain of biochemical reactions leading from individual amino acids to the formation of complex polypeptide macro-molecules?

To ascertain whether the lower specific activity of the protein during SD was attributable to decreased specific activity of leucine in the free pool or to a slower rate of incorporation into protein, specific activity of the leucine was determined. After 20 min SD, the specific activity of leucine was not decreased; rather it was higher than that of the control samples by about 45% (Krivanek, 1970). The same trend was seen in the specific activity of the whole TCA extract (Bennett & Edelman, 1969) and in the specific activity of the total free pool amino acids (the radioactivity of the acid-soluble extract related to the total amino acid content as determined by ninhydrin) for an SD of the same duration (Krivanek, 1970). This finding, together with the fact that the levels of the free pool amino acids in the cerebral cortex were not affected by SD (Ruscak, 1964), justified the use of specific activity of the total free pool amino acid as an indication of the changes in specific activity of the free pool leucine at various time intervals after initiation of SD. The data indicated the lower specific activity of leucine in the free pool. The specific activity of the latter was increased during the whole period studied (10–120 min after SD initiation) (Krivanek, 1970).

Distribution of the SD inhibitory effect on the protein labeling among the crude subcellular fractions (mitochondrial, microsomal, cell sap) was also studied. Thirty to 40% inhibition has been found to be evenly distributed (Krivanek, 1970). To see whether proteins of some mitochondrial subfractions, namely, nerve endings particles and pure mitochondria, could reveal a differential susceptibility to the SD action, a double labeling study was undertaken to ensure identical conditions during particles preparation. Practically the same isotope ratios were found, demonstrating again a uniform distribution of the inhibition of the protein labeling through the fractions studied—two nerve endings particles fractions and pure mitochondria (Avaria & Krivanek, 1973).

A possible effect of SD on the composition of the transcallosally transported proteins into synaptosomes has been studied by Bao and Ochs (1972). Proteins had been labeled by topical application of [^{3}H] leucine to the parieto-occipital cortex of anaesthetized rabbits. The decrease of labeling varied, but some implications of the morphological aspects of SD have been considered by these authors.

To see whether the inhibitory effect on protein synthesis in vivo would persist under in vitro conditions and also to test whether the entire protein synthetic activity of the cerebral cortex had been affected, a cell-free system containing mitochondria microsomes, and supernatant, was prepared and augmented with factors essential for its optimum operation.

The cell-free system from the cerebral cortex, which had been treated 60 min with 25% KCl in vivo, did not differ from that prepared from the control cortex in its capability to incorporate ^{14}C leucine added in vitro.

The failure to find any difference between the in vitro incorporation rates of the in vivo depressed and control cortices could be because the tissue was damaged by preparation to such an extent that a “depressed” brain was studied in both cases. Another possibility remained, however, namely, that shortage of some factor(s) engaged in protein synthesis occurs in vivo, but after transfer of the tissue into the medium supplied with an excess of all the essential components, the depressed tissue rapidly becomes “normal.” The assumption was made that sensitivity to the depletion would be much more pronounced in the system prepared from the depressed cortex. After decrease of the Mg^{2+} concentration to 20% of its optimum or after complete removal of Mg^{2+} from the incubation medium, the same inhibition of the incorporation rate was found in both normal and depressed homogenate. Identical percentage inhibition of both systems was also found in the medium completely devoid of guanosine triphosphate (GTP). Different results, however, were obtained after depletion of oxidizable substrate. Inhibition of labeled leucine incorporation was more apparent in the homogenate from the depressed cortex when no substrate was added to the medium. With progressively increasing substrate concentrations in the medium, there was increasing recovery of protein synthetic activity in the homogenates from the potassium-treated cortex. Hence response in the depletion of the oxidizable substrate is the only one in which both types of homogenates differ from each other (Krivanek, 1969b).

The above results indicate that the lower radioactivity of cortical proteins caused by SD is presumably due to a lower incorporation rate of the precursor amino acid into cortical protein. Obviously protein synthetic activity of the brain was inhibited. However, no individual proteins or their groups localized in various subcellular structures were found that would possess an extra sensitivity to SD. The even distribution of the inhibitory effect among all subcellular fractions in adult animals supports the view that there is a rather general nonspecific mechanism by which SD exerts its effect on protein synthesis in the brain. The many lines of evidence mentioned previously lead to the view that certain phases of CSD activate processes involved in generation of an additional chemical energy necessary for the restoration of normal ion distribution across excitable membranes. Periodically repeated switching of the supplied energy to ion transfer might result in a shortage for other energy requiring biochemical processes such as synthesis of protein. The results of the in vitro experiments are in accordance with the view that reactions involving oxidation of substrate in the mitochondria are affected rather than protein synthesis per se.

As already mentioned, APC is preceded by a short-lasting discharge of neuronal activity. It is difficult to establish which of these phenomena is predominantly responsible for the chemical changes hitherto demonstrated. The possibility of separately evoking spreading depression and spreading convulsion by electrical stimulation of the rabbit cerebral cortex has been claimed by Shirane (1961). Decreased level of glucose and elevated content of lactate was found during spreading convulsion whereas SD was accompanied by an increase of glucose and a fall of lactate.

After the experiments hitherto mentioned have furnished supporting evidence for the suggestion that chemical changes caused by SD in the cerebral cortex in vivo are closely related to changes elicited in the brain cortex slices in vitro by increasing the potassium ion concentration in the medium above 20 mM. This assumption makes it possible to include in the discussion data from in vitro experiments concerned with the potassium effect.

A great number of papers have been published dealing with the mechanism of K^+ action on metabolism since the first reports by Ashford and Dixon (1935) and Dickens and Greville (1935). Various aspects of the problem were studied, but no definite solution has yet been found. Increased oxygen consumption and lactate production (enhancement of aerobic glycolysis), augmented glucose utilization, and a decrease in creatine phosphate and very often also in ATP content have been consistently demonstrated and may be considered as firmly established. It is not clear, however, whether this shift reflects stimulation of cerebral metabolism of some specific impairment of glucose metabolism (reversal of the Pasteur effect). Under normal conditions increased respiration should be accompanied by decreased glycolysis. This so-called Pasteur effect seems to be reversed in the potassium-treated brain. Two approaches were used to study changes in energy metabolism as revealed by potassium action. First, those chemical processes were studied that are known to require chemical energy in the form of ATP and CrP (endergonic reactions), followed by more direct processes, in which turnover rates of ATP and CrP were measured. If some kind of general stimulation of metabolism occurred, the rate of endergonic reactions as well as turnover of ATP and CrP would be increased and vice versa. The first approach was to follow the synthesis of glycogen and incorporation of radioactive precursors into the macromolecules of lipids or proteins. Increasing potassium concentration above 50 mM results in inhibition of glycogen synthesis (Kleinzeller & Rybova, 1957) in vitro as well as in vivo (see p. 96). Incorporation of radioactive amino acids ([^{35}S]-methionine (Folbergerova, 1961);] [^{14}C] glycine, (Bassi & Bernelli-Zazzera, 1960; Mase et al., 1962) has been shown to be diminished in vitro as well as in vivo under similar conditions. On the other hand, another endergonic process seems to be stimulated when potassium content is augmented in the medium. In corporation of ^{32}P into phosphoproteins (Tsukada et al., 1958) as well as into phospholipids (Yoshida & Quastel, 1962) was reported to be stimulated by potassium in vitro. The earlier experiments by Findlay et al. (1954) showing an inhibitory effect of potassium on the last two processes are irrelevant since analyses were performed at the end of 4 h of incubation when reversal of the potassium effect had taken place (Hertz & Clausen, 1963; Tsukada et al., 1958).

No definite conclusion has been drawn as far as turnover rates of CrP and adenine nucleotides are concerned. Diminished turnover of "nucleotide labile P," inorganic phosphate, and creatine phosphate was reported by Fonyo et al. (1958), whereas Tsukada, Takagaki and Hirano (Brossard & Quastel, 1963; Tsukada et al., 1958) presented the opposite results, that is, stimulation of ^{32}P incorporation into "acid-soluble P" and "7-min nucleotide phosphate." The SD was demonstrated to be accompanied by an increase of CrP turnover. Although short incubation times were used in all three papers cited, they differ in composition of medium and in tonicity as well. It is therefore difficult to make any proper conclusion from these data in spite of the fact that this problem is a crucial one.

The increased oxygen consumption and glycolysis acceleration as a consequence of increased potassium concentration are usually ascribed to the increased level of phosphate acceptors such as ADP and/or AMP as a result of augmented cleavage of CrP and ATP. These are known to stimulate both oxidative and glycolytic pathways of glucose metabolism. Oxidative utilization of pyruvate seems to be accelerated. Potassium-stimulated respiration could be demonstrated with glucose, fructose, pyruvate, and lactate as substrates mentioned above (Ito, 1960). Accordingly, pyruvate oxidation through tricarboxylic acid cycle would be accelerated. As far as glycolysis is concerned, even less is known about the mechanism of its stimulation in the cerebral cortex by potassium ions. Only recently two papers have been published dealing with changes in glycolytic intermediates in brain slices of the guinea pig (Rolleston & Newsholme, 1967; Takagaki, 1968). Unfortunately, different media were used, and also the absolute values of the levels of the critical intermediates (fructose diphosphate, triosephosphate, pyruvate) achieved by added potassium cannot be compared. However, from the concentration ratios of the single intermediates, it appeared that in the presence of potassium, glycolysis is affected at the sites previously suggested (Lowry & Passonneau, 1964), that is, fructose diphosphate and pyruvate-kinase. To what extent the shifts in levels of these substances are caused exclusively by changes in ADP and/or AMP levels is not clear. A possible additional direct stimulation effect of potassium on pyruvate kinase was suggested (Lipsett & Crescitelli, 1950; Takagaki, 1968) but not experimentally proved. Intracellular concentrations of neither iP nor ADP and AMP were measured in the Takagaki paper after iP addition to the medium. Therefore, an additional effect of high potassium concentration on glycolysis does not seem to prove a specific action. Also confusing is the finding by Rolleston and Newsholme (1967) that a decrease not an increase of ADP and only a slight increase of AMP occurred with high potassium.

The chemical exploration of the phenomenon discussed is hindered by the fact that it is probably confined to only some compartment(s) as suggested earlier (see p. 99). Accurate measurements of the intracellular concentration of the chemical substances are a fairly difficult problem while determination of the compartment most involved in the particular response is almost impossible. This is especially true for brain tissue, the cellular inhomogeneity of which is considerable. No definite proof has been furnished as yet on the extent to which glial cells could participate in the cerebral cortical response to the action of potassium. Van Harreveld and Stamm (1953a) have shown that in vivo glial elements alone are not sufficient to ensure SD propagation. Cerebral cortex damaged by transient ischemia did not react with SD to common eliciting agents when the reduction in the number of neurons was excessive. Also SD would not pass through a narrow band of the cortex that had previously suffered local injury resulting in a disappearance of neurons and an increase in glia (Van Harreveld et al., 1956). On the other hand, respiration of the neurons isolated by manual dissection (Hyden, 1959) was shown to be insensitive to a high potassium medium, whereas that of surrounding isolated glial cells ("oligodendroglia") was increased (Hertz, 1968). However, this kind of preparation suffers from the serious drawback that dissected neurons are completely devoid of their processes. Clear-cut evidence as to the extent to which different kinds of brain cells are affected by potassium ions remains to be presented.

Very little is known about the connection of the chemical events associated with SD with the electrical correlates. Spreading depression in vivo and stimulation of glucose metabolism in vitro can be elicited by a variety of agents like NH_4^+, Rb^+, a decrease in Ca^{2+} (or the use of EDTA), and by some amino acids (p. 36) as well as by electrical stimulation. Depolarization will result in all these cases as demonstrated in vivo by both a shift in the polarity of the cerebral cortex surface and by direct measurements of the resting potential of cortical cells (Brozek, 1966; Collewijn & Harreveld, 1966). In vitro the resting potential of cerebral cortex slices has been shown to decline in the presence of potassium ion concentrations comparable with those that could accumulate in vivo during SD (Hillman & Mc, 1961) and during electrical stimulation (Hillman et al., 1963). Conceivably these agents act by altering the permeability of the excitable membranes in the cerebral cortex. As in peripheral nerves, the presence of sodium ions is essential to evoke the potassium effect in brain preparations. Increased permeability to sodium ions is essential to evoke the potassium effect in brain preparations. Increased permeability to sodium ions is suggested as being a first step in the action of potassium and related agents (Van Harreveld, 1966). Consequently, potassium is released and can elicit depolarization of membranes. Potassium flux rates seem to be enhanced, which might be a result of increased transport rather than increased permeability (Hertz, 1968). Sodium fluxes are claimed to be unaffected by potassium ions in the medium of cerebral cortex slices (Hertz, 1966). Increased potassium fluxes were also shown by Cummins and McIlwain (1961) in isolated cerebral cortex slices during and immediately after electrical stimulation.

The mechanism by which potassium exerts its effect on the membrane is unclear; Na, K-activated ATPase, which is located in the membranes and seems to be responsible for a variety of transport phenomena, has been shown to be stimulated by the agents increasing brain respiration mentioned above, and is inhibited by the factors abolishing stimulated respiration such as ouabain (Wollenberger, 1947). Transport ATPase might therefore be the site of potassium action. Stimulation of its activity would lead to the increased rate of ATP cleavage, which would stimulate glycolysis and respiration, as mentioned above.

Accordingly, transport ATPase can be considered as a crucial point at which the ion transport couple electrical and biochemical phenomena.

It seems, therefore, that the presence of high concentrations of potassium or of related compounds evoking SD in vivo leases to rearrangement of the ions across the excitable membrane by changing the permeability to these ions. This will cause depolarization. Increased uptake of potassium and extrusion of sodium against their concentration gradients require energy, which is generated by stimulation of glucose metabolism by enhancement of ADP or other phosphate acceptors. The metabolism of compounds involved in ion transport is also increased (some phospholipids, phosphoproteins), whereas that of the others not directly related to transport phenomena is inhibited (incorporation of amino acids into proteins) due to a shortage of energy-rich substances. Involvement of cyclic adenosine 3′,5′-monophosphate should also be considered since Sattin and Rall (1967) and Shimizu et al. (1970) have demonstrated that potassium ions and some other depolarizing agents cause an accumulation of cyclic AMP in incubated slices of cerebral cortex.

Chapter 7: Mechanism of Spreading Depression

At different stages of the history of SD research, hypothetical mechanisms have been suggested to account for the available experimental evidence.

Vascular Mechanisms

The critical role of the vascular system in the mechanism of propagation was postulated by Leão (1944a) and by Leão and Morison (1945). According to the first vague hypothesis, cortical vessels release a transmitter substance acting on pial nerve fibers. Marshall's observations (Marshall & Essig, 1951; Marshall et al., 1951; Marshall et al., 1950) that pathological conditions of the pia-arachnoid membrane facilitate SD in cats and monkeys also provided indirect support for the above assumption. The pia-arachnoid explanation became untenable, however, when it was shown that subpial transcortical cuts, leaving the pial circulation intact, block SD movement (Grafstein, 1956a; Ochs, 1958; Van Harreveld et al., 1956).

Although marked dilation of pial vessels was seen during SD (Leão, 1944a), decreased O_2 concentration was found in blood samples taken from the sagittal sinus (Leão & Morison, 1945). Vasoconstriction of intracortical vessels was also indicated by displacement of the cortical surface and by decreased pO_2 in the depressed region (Van Harreveld & Stamm, 1952). It was proposed that the electrophysiological signs of SD are due to a wave of vasoconstriction mediated by vasomotor fibers (Roibak, 1955). This explanation contrasted with the finding, however, that high-amplitude SPCs of SD type appear only after more than 2 min of anoxia (Bures & Buresova, 1957; Leão, 1947; Ochs & Van Harreveld, 1956). Vasodilation rather than vasoconstriction was demonstrated in other studies using heated (Freygang Jr. et al., 1954; Sonnenschein & Walker, 1956) or cooled (Buresova, 1957) thermocouples, histological technique (Van Harreveld & Ochs, 1957), estimation of hemoglobin content (Buresova, 1957), and measurement of cortical volume (Marshall, 1959). Under these conditions the decrease of pO_2 confirmed by

Marshall (1959) and Lukyanova and Bures (1967) cannot be due to a decreased supply but to an increased demand of oxygen. The changes of blood flow must therefore be considered a concomitant rather than the cause of SD.

Electric Fields

The electric-field hypothesis, advanced by Sloan and Jasper (1950), was based on the finding that SD can propagate across transcortical cuts. According to this hypothesis the negative slow potential generates current flow causing depolarization of adjacent regions. Most later attempts to demonstrate SD transmission across transcortical cuts failed, however (Grafstein, 1956a; Ochs, 1958; Van Harreveld et al., 1956). The electric-field hypothesis is incompatible with the fact that the extracellular current flow induced by the voltage drop between depressed and undepressed regions is considerably lower than the current density required to trigger an SD wave.

Humoral Transmission

Whereas the above hypotheses are of historical interest only, the neurohumoral hypothesis proposed by Burns and Grafstein (1953) has considerable experimental support, and its several slightly different versions became widely accepted. According to this hypothesis SD is a chain reaction consisting of a series of causally linked events. Local stimulation evokes depolarization of neurons and release of potassium ions (or of some other transmitter substance) into the extracellular space. When the extracellular potassium concentration reaches a level sufficient to depolarize adjacent neurons, additional potassium is released and the process becomes self-maintained.

Potassium Hypothesis

Most electrophysiological signs of SD pointed directly or indirectly towards depolarization of neural elements. Evidence for the accumulation of the depolarizing transmitter substance in the extracellular space during SD was obtained by Bures and Krivanek (1958); Krivanek and Bures (1960); Bures et al. (1960); and Brinley Jr. et al. (1960) with the superfusion technique. The efflux of K^+ from the depressed region increased four to five times above the control level, evidently because the extracellular potassium concentration increased in a similar way, that is, to at least 15–20 mEq/L. This figure probably is an underestimation, however, since the inertia and limited spatial resolution of the method tend to smooth out sharp peaks. If the membrane potential during SD corresponds to the equilibrium potential for potassium, the extracellular K^+ can be calculated from the membrane potential values reached during the peak depolarization of SD. Assuming an intracellular K^+ level of 150 mM (Van Harreveld, 1966), an average depolarization of 75% (Brozek, 1966; Collewijn & Harreveld, 1966) corresponds to an increase of the extracellular K^+ level to 60–70 mEq/L. Similar figures can be derived from the finding that bundles of myelinated fibers passing through the gray matter invaded by SD develop a conduction block (Bures & Fifkova, 1968; Bures et al., 1967), which starts only with extracellular concentrations exceeding 30 mEq/L (Orkand et al., 1966).

The increase in extracellular K^+ level during SD must be compared with the K^+ concentration required to elicit SD. For KCl solution applied onto the exposed cortical surface covered by dura, the threshold was found to be around 80 mEq/L (Bures & Buresova, 1956). This seems to be a considerable overestimation, however, since the gradient across the meninges and the escape of KCl into pial capillaries reduce the KCl concentration in the upper cortical layers to a fraction of the surface concentration. With intracortical administration the threshold amount of injected KCl was 2 to $4\text{–}10^{-8}$ gEq K^+, that is, 20–40 times less than the threshold amount required with surface application. Assuming equal distribution around the point of the injection, the overall potassium concentration would be raised to 20–40 mEq/1 in a 1 µl volume of tissue. It follows from such considerations that during SD extracellular K^+ reaches or exceeds a level sufficient to elicit the response.

Glutamic Acid Hypothesis

As another candidate for SD transmitter, Van Harreveld (1959) suggested glutamic and which is present in relatively high concentration (10 mM/kg tissue) in the brain and can elicit SD in slightly lower concentration than KCl. There was no evidence, however, that glutamic acid is released into the extracellular space during SD. Superfusion of the cerebral surface in rats (Bures & Krivanek, 1962) did not reveal the expected increase of glutamate outflux. With the same technique Van Harreveld and Koolman (1965) found in the rabbit that the excessed ninhydrin-positive material released into the washing fluid during SD mainly stems from blood in pial vessels.

More recently Van Harreveld and Fifkova (1970) reported, however, that release of labeled glutamate from the isolated chicken retina preloaded with ^{14}C glutamate is markedly enhanced when diffuse SD is evoked by superfusion with unlabeled glutamate, KCl, or by electrical stimulation. It was not clear, however, whether the glutamate concentration (0.2 mM) attained during retinal SD was high enough to cause SD. Subsequent experiments showed that the correlation between glutamate release and SD presence (indicated by the change of retinal transparency, see p. 137) was not perfect (Van Harreveld & Fifkova, 1971, 1972). Thus $MgCl_2$ and GABA suppressed the retinal SD without decreasing the efflux of glutamate. On the other hand, retinal SD elicited by ouabain, EDTA, glutamine, and DL-homocysteic acid was not accompanied by increased liberation glutamate. Tetrodotoxin blocked the transparency change elicited by glutamate and ouabain, but not that elicited by KCl. These dissociations indicate that glutamate release may be an important, but not indispensable link of the SD mechanism. The tetrodotoxin experiments suggest that potassium induced electrolytic shifts are independent on glutamate release and can proceed in spite of the blockade of the glutamate sensitive sodium transport. Unlike potassium, glutamate does not depolarize the myelinated axons and cannot account, therefore, for the conduction block caused by SD in the fibers penetrating through the depressed gray matter (Bures & Fifkova, 1968). Whereas the experimental evidence completes the case for the transmitter role of potassium, the postulated glutamate role has yet to be demonstrated. Although the potassium and glutamate hypotheses are not mutually exclusive, an experiment by Grafstein (1956b) indicates that the transmitter is a positively rather than a negatively charged particle. When a tangential polarization gradient was established in the isolated cortical slab, SD moved faster from positive to negative pole than vice versa. For all these reasons, we shall use the potassium hypothesis in the further considerations. It must be pointed out, however, that the general mathematical analysis of the mechanism of SD is independent of the nature of the transmitter and therefore also applies to glutamate.

Concomitant Phenomena

Although the increment of K^+ concentration in the extracellular fluid is crucial for the mechanism of spreading, it is only a link of a complex chain of metabolic and physiological events. At first the increasing potassium level causes a net inward K^+ flux. When depolarization reaches the critical level, the membrane permeability to sodium suddenly increases and action potentials are generated. Net inward movement of sodium is followed by net outward movement of potassium. As more sodium enters than potassium leaves the cells, Na^+ is accompanied by Cl^-, the accumulation of which in cortical dendrites has been histochemically demonstrated by Van Harreveld and Schade (1959). The net movement of NaCl into cells induces shifts of water, resulting in an increase of intracellular volume (Van Harreveld, 1958) and in a corresponding reduction of extracellular space, thus causing a marked increase in impedance (Van Harreveld & Ochs, 1957). Hydrolysis of intracellular components, for example, of high-energy phosphates (Krivanek, 1958), may also increase the intracellular osmolarity and thus contribute to swelling of neurons.

All of these changes continue until the increasing extracellular K^+ concentration depolarizes the membrane to a level at which action potentials cease to be generated. At this point the potassium efflux starts to decrease and influx to increase until the net potassium movement is reversed. As active transport removes sodium ions from the intracellular compartment, potassium returns to the cells until the normal ionic equilibrium is again obtained. Excess extracellular potassium is also removed by blood flow.

References

Ashford, C. A., & Dixon, K. C. (1935). The effect of potassium on the glucolysis of brain tissue with reference to the Pasteur effect. *The Biochemical Journal, 29*, 157–168.

Avaria, L. G., & Krivanek, J. (1973). Double labelling study of the spreading depression effect on the incorporation of labelled leucine into the proteins of the brain mitochondrial subfractions. *Journal of Neurochemistry, 20*, 905–908.

Bao, D. C., & Ochs, S. (1972). Incorporation of ^{3}H-keucine in rabbit cortex and transport into synaptosomes. *Federation Proceedings, 31*, 304.

Bassi, M., & Bernelli-Zazzera, A. (1960). Effects of potassium ions on brain respiration and aminoacid incorporation into brain proteins *in vitro*. *Experientia, 16*, 430.

Bennett, G. S., & Edelman, G. M. (1969). Amino acid incorporation into rat brain proteins during spreading cortical depression. *Science, 163*, 393–395.

Brinley, F. J., Jr., Kandel, E. R., & Marshall, W. H. (1960). Potassium outflux from rabbit cortex during spreading depression. *J. Neurophysiology, 23*, 246–256.

Brossard, M., & Quastel, J. H. (1963). Studies of the cationic, and acetylcholine, stimulation of phosphate incorporation into phospholipids in rat brain cortex *in vitro*. *Canadian Journal of Biochemistry and Physiology, 41*, 1243–1256.

Brozek, G. (1966). Changes in the membrane potential of cortical cells during spreading depression. *Physiologia Bohemoslovaca, 15*, 98–103.

Bures, J. (1956). Some metabolic aspects of Leão's spreading depression. *Journal of Neurochemistry, 1*, 153–158.

Bures, J., & Buresova, O. (1956). The question of ionic antagonism in spreading depression. *Physiologia Bohemoslovaca, 5*, 195–205.

Bures, J., & Buresova, O. (1957). Anoxic terminal depolarization as an indicator of cerebral cortex vulnerability in anoxia & ischemia. *Pflügers Archiv für die Gesamte Physiologie des Menschen und der Tiere, 264*, 325–334.

Bures, J., & Fifkova, E. (1968). The effect of potassium ions liberated from nerve structures invaded by spreading depression on conduction in fibres of the adjacent white matter. *Physiologia Bohemoslovaca, 17*, 405–410.

Bures, J., & Krivanek, J. (1958). *Washing of brain surface by epidural fistula – A method for the study of brain electrolytes in vivo*. Uber den Mechansmus der Erregung. VEB Deutscher Verlag der Wissdenschafter, Berlin, pp. 1–3.

Bures, J., & Krivanek, J. (1960). Ionic movements in the brain as studied with the aid of washing the cortical surface with an epidural cannula. *Physiologia Bohemoslovenica, 9*, 488–493.

Bures, J., & Krivanek, J. (1962). *Uber die rolle einiger aminosauren im Mechanismus der sich verbreitenden EEG-depression, Struktur and Biologische Fungtion der Eiweisse*. 111Humboldt Symposium uber Grandfragen der Biologie, Gustav Fischer Verlag, Berlin, pp. 181–184.

Bures, J., Buresova, O., & Krivanek, J. (1960). Some metabolic aspects of Leão's spreading cortical depression. In D. B. Tower & J. P. Schade (Eds.), *Structure and function of the cerebral cortex* (pp. 257–265). Elsevier.

Bures, J., Hartmann, G., & Lukyanova, L. D. (1967). Blockade of thalamocortical and pyramidal pathways by striatal spreading depression in rats. *Experimental Neurology, 18*, 404–415.

Bures, J., Buresova, O., & Krivanek, J. (1974). *The mechanism and applications of Leão's spreading depression of electroencephalographic activity*. Academic.

Buresova, O. (1957). Changes in cerebral circulationin rats during spreading EEG depression. *Physiologia Bohemoslovenica, 6*, 1–11.

Buresova, O., Shima, I., Bures, J., & Fifkova, E. (1963). Unit activity in regions affected by the spreading depression. *Physiologia Bohemoslovaca, 12*, 488–494.

Burns, B. D., & Grafstein, B. (1953). *Spreading depression in isolated cerebral cortex*. In XIXth International Physiological Congress, Montreal, Abstract of Communication, pp. 251–252.

Chang, H. T. (1955). Cortical response to stimulation of medullary pyramid in rabbit. *Journal of Neurophysiology, 18*, 332–352.

Collewijn, H., & Harreveld, A. V. (1966). Membrane potential of cerebral cortical cells during reading depression and asyxia. *Experimental Neurology, 15*, 425–436.

Collewijn, H., & Schade, J. P. (1962). Potassium movements in the cerebral cortex during asphyxia. *Acta Morphologica Neerlando-Scandinavica, 5*, 11–20.

Cummins, J. T., & McIlwain, H. (1961). Electrical pulses and the potassium and other ions of isolated cerebral tissues. *The Biochemical Journal, 79*, 330–341.

Dickens, F., & Greville, G. D. (1935). The metabolism of normal and tumour tissue: Neutral salt effects. *The Biochemical Journal, 29*, 1468–1483.

Donoso, A. O., & Stefano, F. J. (1966). Cerebral changes in the noradrenaline content during cortical spreading depression in the rat. *Acta Physiologica Latino Americana, 16*, 22–25.

Dubner, R., & Brown, F. J. (1968). Response of cells to restricted visual stimuli in an association area of cat cerebral cortex. *Experimental Neurology, 20*, 70–86.

Duda, P., Ruscak, M., & Zachar, J. (1961). Spreading cortical depression and the polarization gradient of the cerebral cortex. *Physiologia Bohemoslovaca, 10*, 438–447.

Fertziger, A. P., & Ranck, J. B., Jr. (1970). Potassium accumulation in interstitial space during epileptiform seizures. *Experimental Neurology, 26*, 571–585.

Fifkova, E., Bures, J., Koshtoyants, O. K., Krivanek, J., & Weiss, T. (1961). Leão's spreading depression in the cerebellum of rat. *Experientia, 17*, 572–573.

Findlay, M., Magee, W. L., & Rossiter, R. J. (1954). Incorporation of radioactive phosphate into lipids and pentosenucleic acid of cat brain slices; the effect of inorganic ions. *The Biochemical Journal, 58*, 236–243.

Folbergerova, J. (1961). The effect of potassium and some other factors upon the incorporation of ^{35}S-methionine into protein of brain corte slices. *Physiologia Bohemoslovaca, 10*, 130–138.

Fonyo, A., Kovach, A. G., Maklary, E., Leszkovszky, G., & Meszaros, J. (1958). The effect of potassium ions and glutamate on the incorporation of ^{32}P into nucleotides and phosphocreatine in brain slices. *Acta Physiologica Academiae Scientiarum Hungaricae, 14*, 305–307.

Freygang, W. H., Jr., & Landau, W. M. (1955). Some relations between resistivity and electrical activity in the cerebral cortex of the cat. *Journal of Cellular and Comparative Physiology, 45*, 377–392.

Freygang, W. H., Jr., Landau, W. M., & Marshall, W. H. (1954). Changes in resistance and electric potetial in the cortex. *Federation Proceedings, 13*, 48.

Garcia Ramos, J., & Pradac, J. (1968). Electrochemistry of the fine indium-platinized electrodes used in biology. *Currents in Modern Biology, 2*, 165–170.

Gartside, I. B. (1968). Mechanisms of sustained increases of firing rate of neurons in the rat cerebral cortex after polarization: Reverberating circuits or modification of synaptic conductance? *Nature, 220*, 382–383.

Goldensohn, E. S., & Walsh, G. (1968). Sequential functional changes in single cortical neurons during spreading cortical depression. *EEG Clinical Neurophysiology, 24*, 290–291.

Goldensohn, E., Escueta, A., & Runk, L. (1967). Unit membrane shifts and firing patterns during spreading depression. *EEG Clinical Neurophysiology, 23*, 77–97.

Grafstein, B. (1956a). Locus of propagation of spreading cortical depression. *Journal of Neurophysiology, 19*, 308–316.

Grafstein, B. (1956b). Mechanism of spreading cortical depression. *Journal of Neurophysiology, 19*, 154–171.

Heald, P. J. (1954). Rapid changes in creatine phosphate level in cerebral cortex slices. *The Biochemical Journal, 57*, 673–679.

Hertz, L. (1966). Neuroglial localization of potassium and sodium effects on respiration in brain. *Journal of Neurochemistry, 13*, 1373–1387.

Hertz, L. (1968). Potassium effects on ion transport in brain slices. *Journal of Neurochemistry, 15*, 1–16.

Hertz, L., & Clausen, T. (1963). Effects of potassium and sodium on respiration: Their specificity to slices from certain brain regions. *The Biochemical Journal, 89*, 526–533.

Hestrin, S. (1949). The reaction of acetylcholine and other carboxylic acid derivatives with hydroxylamine, and its analytical application. *The Journal of Biological Chemistry, 180*, 249–261.

Higashida, H., Miyake, A., Tarao, M., & Watanabe, S. (1971). Membrane potential changes of neuroglial cells during spreading depression in the rabbit. *Brain Research, 32*, 207–211.

Hillman, H. H., & Mc, I. H. (1961). Membrane potentials in mammalian cerebral tissues *in vitro*: Dependence on ionic environment. *The Journal of Physiology, 157*, 263–278.

Hillman, H. H., Campbell, W. J., & Mc, I. H. (1963). Membrane potentials in isolated and electrically stimulated mammalian cerebral cortex. Effects of chloropromazine, cocaine, phenobarbitone and protamine on the tissue's electrical and chemical responses to stimulation. *Journal of Neurochemistry, 10*, 325–339.

Huizar Sanchez, P., & Garcia Ramos, J. (1969). Spreading redox potentials in the cat's cortex. *Acta Physiologica Latino Americana, 19*, 80–42.

Hyden, H. (1959). Biochemical changes in glial cells and nerve cells of varying activity. In F. Brucke (Ed.), *Biochemistry of the central nervous system* (p. 64). Pargamon Press.

Ichijo, M., & Ochs, S. (1967). Effect of cortical spreading depression on the N-wave DCR and Betz cells. *Physiologist, 10*, 209.

Ichijo, M., & Ochs, S. (1970). Spreading depression of negative wave of direct cortical response and pyramidal tract responses. *Brain Research, 23*, 41–56.

Ito, K. (1960). Metabolic relationship of succinate and pyruvate in brain tissue with special reference to the potassium effect. *The Japanese Journal of Experimental Medicine, 30*, 261–277.

Johnson, W. L., Goldring, S., & O'Leary, J. L. (1965). Behavioral, unit and slow potential changes in methionine sulfoximine seizures. *EEG Clinical Neurophysiology, 18*, 229–238.

Karahashi, Y., & Goldring, S. (1966). Intracellular potentials from "idle" cells in cerebral cortex of cat. *EEG Clinical Neurophysiology, 20*, 600–607.

Katzman, R., & Leiderman, P. (1953). Brain potassium exchange in normal adult and immature rats. *The American Journal of Physiology, 175*, 263–270.

Keller, H. H., Bartholini, G., Pieri, L., & Pletscher, A. (1972). Effects of spreading depression on the turnover of cerebral dopamine. *European Journal of Pharmacology, 20*, 287–290.

Kerr, S. E., & Ghantus, M. (1936). The carbohydrate metabolism of brain. II. The effect of varying the carbohydrate and insulin supply on the glycogen, free sugar, and lactic acid in mammalian brain. *Journal of Biological Chemistry, 116*, 9–20.

Kini, M. M., & Quastel, J. H. (1959). Carbohydrate-amino-acid inter-relations in brain cortex *in vitro*. *Nature, 184*, 252–256.

Kleinzeller, A., & Rybova, R. (1957). Glycogen synthesis in brain cortex slices and some factors affecting it. *Journal of Neurochemistry, 2*, 45–57.

Krivanek, J. (1958). Changes of brain glycogen in the spreading EEG-depression of Leão. *Journal of Neurochemistry, 2*, 337–343.

Krivanek, J. (1959). Incorporation of ^{32}P into phosphorus fractions of rat brain during spreading EEG depression. *Ceskoslovenská Fysiologie, 8*, 417–418.

Krivanek, J. (1960). *On the metabolic background of the spreading cortical depression*. Institute of Physiology, Czechoslovak Academy of Sciences, Prague.

Krivanek, J. (1961). Some metabolic changes accompanying Leao's spreading cortical depression in the rat. *Journal of Neurochemistry, 6*, 183–189.

Krivanek, J. (1962). Concerning the dynamics of the metabolic changes accompanying cortical spreading depression. *Physiologia Bohemoslovaca, 11*, 383–391.

Krivanek, J. (1968). Changes in the metabolism of protein, water and potasium in the course of repeated spreading cortical depression (in Czech). *Ceskoslovenská Fysiologie, 17*, 221.

Krivanek, J. (1969a). Changes in protein, electrolyte, and water metabolism caused by prolonged spreading cortical depression in rats. *Journal of Neurobiology, 1*, 147–154.

Krivanek, J. (1969b). *Concerning the mechanism of potassium ions effect onto incorporation of ^{14}C-leucine into protein of the brain cortex in vivo* (in Czech). Summaries of Vth National Biochemistry Congress, Brno 1969, pp. 6–17.

Krivanek, J. (1970). Effects of spreading cortical depression on the incorporation of [^{14}C] leucine into proteins of rat brain. *Journal of Neurochemistry, 17*, 531–538.

Krivanek, J. (1972a). Incorporation of ^{14}C from glucose into glycogen and some amino acids of cerebral cortex influenced by potassium ions *in vivo*. *Physiologia Bohemoslovaca, 21*, 615–622.

Krivanek, J. (1972b). Spreading cortical depression and acetylcholinesterase activity in rat cerebral cortex. *Physiologia Bohemoslovaca, 21*, 163–169.

Krivanek, J., & Bures, J. (1960). Ion shifts during Leão spreading cortical depression. *Physiologia Bohemoslovenica, 9*, 494–503.

Krnjevic, K., & Randic, M. (1965). *Cortical inhibition and spreading depression*. Abstract of XXIII International Congress of Physiology. Science, Tokyo, 395.

Krnjevic, K., Randic, M., & Straughan, D. W. (1966). An inhibitory process in the cerebral cortex. *The Journal of Physiology, 184*, 16–48.

Leão, A. A. P. (1944a). Pial circulation and spreading depression of activity in the cerebral cortex. *Journal of Neurophysiology, 7*, 391–396.

Leão, A. A. P. (1944b). Spreading depression of activity in cerebral cortex. *Journal of Neurophysiology, 7*, 359–390.

Leão, A. A. P. (1947). Further observations on the spreading depression of activity in the cerebral cortex. *Journal of Neurophysiology, 10*, 409–414.

Leão, A. A. P. (1961). Composicao ionica extracellular a depressao alstrante. *Anais da Academia Brasileira de Ciências, 33*, XXXVIII–XXXIX.

Leão, A. A. P., & Martins-Ferreira, H. M. (1953). Alteracao da impedancia eletrica no decurso da depressio alastrante da ativadade do cortex cerebral. *Anais da Academia Brasileira de Ciências, 25*, 259–266.

Leão, A. A. P., & Morison, R. (1945). Propagation of spreading cortical depression. *Journal of Neurophysiology, 8*, 33–45.

Li, C. L., & McIlwain, I. H. (1957). Maintenance of resting membrane potentials in slices of mammalian cerebral cortex and other tissues *in vitro*. *The Journal of Physiology, 139*, 178–190.

Li, C. L., & Salmoiraghi, G. C. (1963). Cortical steady potential changes: Extracellular microelectrode investigations. *Nature, 198*, 858–859.

Lipsett, M. N., & Crescitelli, F. (1950). The effects of increased potassium concentration on the metabolism of rat cerebral cortical slices. *Archives of Biochemistry, 28*, 329–337.

Lowry, O. H., & Passonneau, J. V. (1964). The relationships between substrates and enzymes of glycolysis in brain. *The Journal of Biological Chemistry, 239*, 31–42.

Lowry, O. H., Roberts, N. R., Leiner, K. Y., Wu, M. L., Farr, A. L., & Albers, R. W. (1954). The quantitative histochemistry of brain. III. Ammon's horn. *The Journal of Biological Chemistry, 207*, 39–49.

Lukyanova, L. D., & Bures, J. (1967). Changes in pO_2 due to spreading depression in cortex and nucleus caudatus of rat. *Physiologia Bohemoslovaca, 16*, 449–455.

Marshall, W. H. (1959). Spreading cortical depression of Leão. *Physiological Reviews, 39*, 239–279.

Marshall, W. H. (1965). Some problems of general neurophysiology of the brain. *Actual Neurophysiology (Paris), 6*, 43–54.

Marshall, W. H., & Essig, C. F. (1951). Relation of air exposure of cortex to spreading depression of Leão. *Journal of Neurophysiology, 14*, 265–273.

Marshall, W. H., Hanna, C., & Barnard, G. (1950). The relation of dehydration of the brain to the spreading depression of Leão. *EEG Clinical Neurophysiology, 2*, 177–185.

Marshall, W. H., Essig, C. F., & Dubroff, S. J. (1951). Relation of temperature of cerebral cortex to spreading depression of Leão. *Journal of Neurophysiology, 14*, 153–166.

Martins-Ferreira, H., & de Oliveira-Castro, G. O. (1966). Light-scattering changes accompanying spreading depression in isolated retina. *Journal of Neurophysiology, 29*, 715–726.

Mase, K., Takahashi, U., & Ogata, K. (1962). The incorporation of (^{14}C) glycine into protein of Guinea-pig brain cortex slices. *Journal of Neurochemistry, 9*, 281–288.

Meyer, J. S., Gotch, F., & Tazaki, Y. (1961). Inhibitory action of carbone dioxide and acetazolamide in siezure activity. *EEG Clinical Neurophysiology, 13*, 762–775.
Monakhov, K. K., Fifkova, E., & Bures, J. (1962a). Steady potential field of hippocampal spreading depression. *Journal of Cellular and Comparative Physiology, 59*, 155–161.
Monakhov, K. K., Fifkova, E., & Bures, J. (1962b). Vertical distribution of the slow potential change of spreading depression in the cerebral cortex of the rat. *Physiologia Bohemoslovaca, 11*, 269–276.
Morlock, N. L., Mori, K., & Ward, A. A., Jr. (1964). A study of single cortical neurons during spreading depression. *Journal of Neurophysiology, 27*, 1192–1198.
Munez-Martines, E. J. (1970). Facilitation of cortical cell activity during spreading depression. *Journal of Neurobiology, 2*, 47–60.
Ochs, S. (1958). Effects of spreading depression on direct cortical response studied with an island technique. *Journal of Neurophysiology, 21*, 159–170.
Ochs, S., & Hunt, K. (1960). Apical dendrites and propagation of spreading depression in cerebral cortex. *Journal of Neurophysiology, 23*, 432–444.
Ochs, S., & Van Harreveld, A. (1956). Cerebral impedance changes after circulatory arrest. *The American Journal of Physiology, 187*, 180–192.
Ochs, S., Hunt, K., & Booker, H. (1961). Spreading depression using chronically implanted electrodes. *The American Journal of Physiology, 200*, 1211–1214.
Orkand, R. K., Nicholls, J. G., & Kuffler, S. W. (1966). Effect of nerve impulses on the membrane potential of glial cells in the central nervous system of amphibia. *Journal of Neurophysiology, 29*, 788–806.
Philis, J. W., & Ochs, S. (1971). Excitation and depression of cortical neurones during spreading depression. *Experimental Brain Research, 12*, 132–149.
Prince, D. A., Lux, H. D., & Neher, E. (1973). Measurement of extracellular potassium activity in cat cortex. *Brain Research, 50*, 489–495.
Ranck, J. B. (1964). Specific impedance of cerebral cortex during spreading depression and an analysis of neuronal, neuroglial, and interstitial contributions. *Experimental Neurology, 9*, 1–16.
Rapoport, S. I., & Marshall, W. H. (1964). Measurement of cortical Ph in spreading cortical depression. *The American Journal of Physiology, 206*, 1177–1180.
Roibak, A. I. (1955). *Bioelectrical phenomena in the Cerebrak Cirtex (in Russian)*. AN GSSR.
Rolleston, F. S., & Newsholme, E. A. (1967). Control of glycolysis in cerebral cortex slices. *The Biochemical Journal, 104*, 524–533.
Rosenblueth, A., & Garcia Ramos, J. (1966). Some phenomena usually associated with spreading depression. *Acta Physiologica Latino Americana, 15*, 142–179.
Rosenblueth, A., Garcia Ramos, J., & Nima, L. F. (1967). The relation between the impedance and e.m.f. changes in the cerebral cortex. *Acta Physiologica Latino Americana, 17*, 76–87.
Ruscak, M. (1961). Changes in amino nitrogen during EEG depression. *Journal of Neurochemistry, 7*, 305–307.
Ruscak, M. (1962a). Changes in the level of gamma-aminobutyric acid (GABA) in the ischaemic brain of rats following application of some stimuli evoking spreading EEG depression. *Physiologia Bohemoslovaca, 11*, 192–198.
Ruscak, M. (1962b). Influence of various stimuli on lactic acid, alanine and gamma-aminobutyric acid levels in rat brains. *Nature, 195*, 290–292.
Ruscak, M. (1964). Incorporation of 35s-methionine into proteins of the cerebral cortex *in situ* in rats during spreading EEG depression. *Physiologia Bohemoslovaca, 13*, 16–20.
Ruscak, M., & Duda, P. (1959). On cerebral metabolism in EEG depression and in cerebral anemia. *Ceskoslovenská Fysiologie, 8*, 429–430.
Ruscak, M., & Ruscakova, D. (1971). *Metabolism of the nerve tissue in relation to ion movements in vitro and in situ*. University Park Press.
Rybova, R. (1959). Effect of cations on g-aminobutyric acid level in slices of brain cortex. *Nature, 185*, 542–543.

Sattin, A., & Rall, T. W. (1967). The effect of brain extracts on the accumulation of cyclic 3′,5' AMP in slice of Guinea pig cerebral cortex. *Federation Proceedings, 26*, 707.

Schanberg, S. M., Schildkraut, J. J., & Kopin, I. J. (1967). The effects of pentobarbital on the fate of intracisternally administered norepinephrine ^{3}H. *The Journal of Pharmacology and Experimental Therapeutics, 157*, 311–318.

Schanberg, S. M., Schildkraut, J. J., Krivanek, J., & Kopin, I. J. (1968). Effect of cortical spreading depression on norepinephrine ^{3}H metabolism in brain stem. *Experientia, 24*, 909–910.

Sedlacek, J. (1969). Kinetics of substances in developing cerebral tissue of chick embryos [in Czech]. *Sbornik Lekarsky, 71*, 37–43.

Seylas, J., & Molnaz, L. (1966). Observations recentes sur les correlations entre le debit cerebral et la depresssion en vahlesante. C.R. *Social Biology, 160*, 294–298.

Shimizu, H., Creveling, C. R., & Daly, J. W. (1970). Cyclic adenosine 3′,5′-monophosphateformation in brain slices: Stimulation by batrachotoxin, ouabaine, verstridine, and potassium ions. *Molecular Pharmacology, 6*, 184–186.

Shirane, Y. (1961). Biochemical studies on Leão's phenonmenon (in Japanese). *Fukuoka Acta Medica, 52*, 135–147.

Sloan, N., & Jasper, H. (1950). The identity of spreading depression and "suppression". *Electroencephalography and Clinical Neurophysiology, 2*, 59–78.

Sonnenschein, R. R., & Walker, R. M. (1956). Local cerebral blood flow in spreading cortical depression. *Federation Proceedings, 15*, 175.

Sugawara, H., & Wada, E. (1957). Effect of potassium and malonate on the content of labile phosphate in the brain tissue. *Scientific Papers of the College of General Education, University of Tokyo, 7*, 121–130.

Takagaki, G. (1968). Control of aerobic glycolysis and pyruvate kinase activity in cerebral cortex slices. *Journal of Neurochemistry, 15*, 903–916.

Tschirgi, R. D., & Taylor, J. L. (1958). Slowly changing bioelectric potentials associated with the blood-brain barrier. *The American Journal of Physiology, 195*, 7–22.

Tschirgi, R. D., Inanaga, K., Taylor, J. L., Walker, R. M., & Sonnenschein, R. R. (1957). Changes in cortical pH and blood flow accompanying spreading cortical depression and convulsion. *The American Journal of Physiology, 190*, 557–562.

Tsukada, Y., Takagaki, G., & Hirano, S. (1958). Incorporation of readioactive phosphate into protein-bound phosphorus fractions of brain slices in reference to its relation the metabolic activity. *Journal of Biochemistry (Tokyo), 45*, 489–501.

Van Harreveld, A. (1958). Changes in the diameter of apical dendrites during spreading depression. *The American Journal of Physiology, 192*, 457–463.

Van Harreveld, A. (1959). Compounds in brain extracts causing spreading depression of cerebral cortical activity and contraction of crustacean muscle. *Journal of Neurochemistry, 3*, 300–315.

Van Harreveld, A. (1966). *Brain tissue electrolytes*. Butterworth.

Van Harreveld, A., & Fifkova, E. (1969). Glutamate release from the retina during spreading depression. *International Society with Neurochemistry. Communication Abstracts, 206*.

Van Harreveld, A., & Fifkova, E. (1970). Glutamate release from the retina during spreading depression. *Journal of Neurobiology, 2*, 13–29.

Van Harreveld, A., & Fifkova, E. (1971). Effects of glutamate and other amino acids on the retina. *Journal of Neurochemistry, 18*, 2145–2154.

Van Harreveld, A., & Fifkova, E. (1972). Effects of metabolic inhibitors on the release of glutamate from the retina. *Journal of Neurochemistry, 19*, 1439–1450.

Van Harreveld, A., & Koolman, M. (1965). Amino acid release from the cerebral cortex during spreading depression and asphyxation. *Journal of Neurochemistry, 12*, 431–439.

Van Harreveld, A., & Ochs, S. (1957). Electrical and vascular concomitants of spreading depression. *The American Journal of Physiology, 189*, 159–166.

Van Harreveld, A., & Schade, J. P. (1959). Chloride movements in cerebral cortex after circulatory arrest and during spreading depression. *Journal of Cellular and Comparative Physiology, 54*, 65–84.

Van Harreveld, A., & Stamm, J. S. (1952). Vascular concomitants of spreading cortical depression. *Journal of Neurophysiology, 15*, 487–496.
Van Harreveld, A., & Stamm, J. S. (1953a). Cerebral asphyxiation and spreading cortical depression. *The American Journal of Physiology, 173*, 171–175.
Van Harreveld, A., & Stamm, J. S. (1953b). Spreading cortical convulsions and depressions. *Journal of Neurophysiology, 16*, 352–366.
Van Harreveld, A., Terres, G., & Dernburg, E. A. (1956). Cortical discontinuity and propagation of spreading depression. *The American Journal of Physiology, 184*, 233–238.
Vyskocil, F., Kritz, N., & Bures, J. (1972). Potassium-selective microelectrodes used for measuring the extracellular brain potassium during spreading depression and anoxic depolarization in rats. *Brain Research, 39*, 255–259.
Walker, J. A. (1971). Ion specific liquid ion exchanger microelectrodes. *Analytical Chemistry, 43*, 89A–93A.
Weiss, H. R., Kado, R. T., & Adey, W. R. (1966). Impedance and d-c- potential shifts during cortical spreading depression. *Activitas Nervosa Superior, 8*, 194–195.
Whieldon, J. A., & Van Harreveld, A. (1950). Comulative effects of minimal cortical stimulation. *EEG Clinical Neurophysiology, 2*, 49–57.
Wollenberger, A. (1947). Metabolic action of the cardiac glycosides; influence on respiration of heart muscle and brain cortex. *The Journal of Pharmacology and Experimental Therapeutics, 91*, 39–51.
Wong, E. K., Jr. (1969). Changes in impedance and phase angle in the cerebral cortex during spreading depression. *Acta Physiologica Latino Americana, 19*, 63–66.
Yoshida, H., & Quastel, J. H. (1962). Effects of metabolic inhibitors on potassium- and acetylcholine-stimulated incorporation of phosphate into phospholipids of rat-brain cortex slices. *Biochimica et Biophysica Acta, 57*, 67–72.
Zachar, J., & Ruscak, M. (1958). Cerebral metabolism in spreading depression. *Ceskoslovenská Fysiologie, 7*, 191–192.

Index

A. Mayevsky, J. Sonn, *Cortical Spreading Depression of Leao*,
https://doi.org/10.1007/978-3-031-08068-5